HEALTH PROFESSIONALS' EDUCATION IN THE AGE OF CLINICAL INFORMATION SYSTEMS, MOBILE COMPUTING AND SOCIAL NETWORKS

HEALTH PROFESSIONALS' EDUCATION IN THE AGE OF CLINICAL INFORMATION SYSTEMS, MOBILE COMPUTING AND SOCIAL NETWORKS

Edited by

AVIV SHACHAK
University of Toronto, Toronto, ON, Canada

ELIZABETH M. BORYCKI
University of Victoria, Victoria, BC, Canada

SHMUEL P. REIS
The Hebrew University of Jerusalem, Jerusalem, Israel

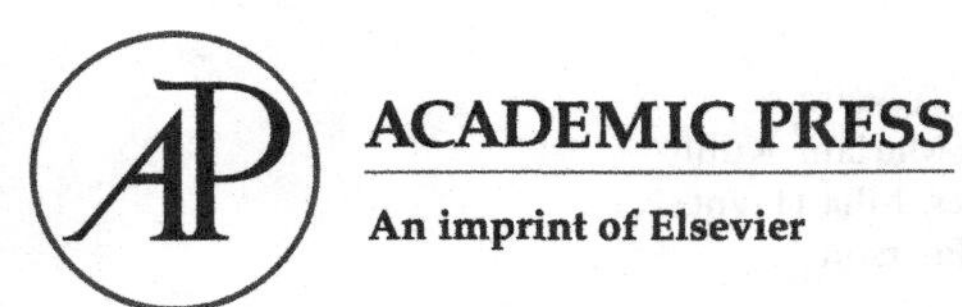

Academic Press is an imprint of Elsevier
125 London Wall, London EC2Y 5AS, United Kingdom
525 B Street, Suite 1800, San Diego, CA 92101-4495, United States
50 Hampshire Street, 5th Floor, Cambridge, MA 02139, United States
The Boulevard, Langford Lane, Kidlington, Oxford OX5 1GB, United Kingdom

British Library Cataloguing-in-Publication Data
A catalogue record for this book is available from the British Library

Library of Congress Cataloging-in-Publication Data
A catalog record for this book is available from the Library of Congress

ISBN: 978-0-12-805362-1

For Information on all Academic Press publications
visit our website at https://www.elsevier.com/books-and-journals

Publisher: Mica Haley
Acquisition Editor: Rafael Teixeira
Editorial Project Manager: Mariana Kuhl
Production Project Manager: Julia Haynes
Cover Designer: Victoria Pearson

Typeset by MPS Limited, Chennai, India

Contents

Section IV

Health Information Literacy and Credibility Assessment

PART II

EXPERIENCES FROM THE FIELD

Section V

Training Clinicians in Informatics and Practicing in IT-Enabled Settings

Subsection V.A

Curriculum Design and Implementation Strategies

Section VI

Local and Regional Interventions

List of Contributors

Maria L. Alkureishi University of Chicago, Chicago, IL, United States
Kathryn E. Anthony University of Southern Mississippi, Hattiesburg, MS, United States
Antonia Arnaert McGill University, Montreal, QC, Canada
Hanan Asiri Armed Forces Hospitals Southern Region, Khamis Mushait, Saudi Arabia
Timothy D. Aungst MCPHS University, Worcester, MA, United States
Tamara J. Bahr University of Toronto, Toronto, ON, Canada
Josef Bartels Family Medicine Residency of Idaho, Boise, ID, United States
Nicolet H. Bell University of Southern Mississippi, Hattiesburg, MS, United States
Rashaad Bhyat Canada Health Infoway, Toronto, ON, Canada
Fran Biagioli Oregon Health & Science University, Portland, OR, United States
Richard G. Booth Western University, London, ON, Canada
Elizabeth M. Borycki University of Victoria, Victoria, BC, Canada
Elizabeth A. Breeden Lipscomb University College of Pharmacy and Health Sciences, Nashville, TN, United States
Kerryn Butler-Henderson University of Tasmania, Hobart, TAS, Australia
Ryan Chan Western University, London, ON, Canada
Dawn Choo University of Melbourne, Melbourne, VIC, Australia
Kevin A. Clauson Lipscomb University College of Pharmacy and Health Sciences, Nashville, TN, United States
Miguel Tavares Coimbra Instituto de Telecomunicações (IT), Lisboa, Portugal; University of Porto, Porto, Portugal
Amanda Condon Winnipeg Regional Health Authority, Winnipeg, MB, Canada
Ovídio Costa University of Porto, Porto, Portugal
Noah H. Crampton University of Toronto, Toronto, ON, Canada
Ricardo Cruz-Correia Center for Health Technology and Services Research (CINTESIS), Porto, Portugal; University of Porto, Porto, Portugal
Elizabeth Cummings University of Tasmania, Hobart, TAS, Australia
Zoumanan Debe Consultant, Ontario, QC, Canada
Sharon Domb University of Toronto, Toronto, ON, Canada
Prerna Dua Louisiana Tech University, Ruston, LA, United States
Rachel H. Ellaway University of Calgary, Calgary, AB, Canada
Gerard Farrell Memorial University, St. John's, NL, Canada
Margarida Figueiredo-Braga University of Porto, Porto, Portugal
Brent I. Fox Auburn University, Harrison School of Pharmacy, Auburn, AL, United States
Richard M. Frankel Indiana University School of Medicine, Indianapolis, IN, United States; Richard L. Roudebush Veterans Administration Medical Center, Indianapolis, IN, United States; Cleveland Clinic, Cleveland, OH, United States
Candace Gibson Western University, London, ON, Canada

Jeffery Gold Oregon Health & Science University, Portland, OR, United States
Pedro Gomes Instituto de Telecomunicações (IT), Lisboa, Portugal; University of Porto, Porto, Portugal
Paul Gorman Oregon Health & Science University, Portland, OR, United States
Kathleen Gray University of Melbourne, Melbourne, VIC, Australia
Jodi Hall Fanshawe College, London, ON, Canada
Seana-Lee Hamilton Fraser Health Authority, Surrey, BC, Canada
Robert Hayward University of Alberta, Edmonton, AB, Canada
William Hersh Oregon Health & Science University, Portland, OR, United States
Kendall Ho University of British Columbia, Vancouver, BC, Canada
Mowafa Househ King Saud Bin Abdulaziz University for Health Sciences, Riyadh, Saudi Arabia
Steven Kassakian Oregon Health & Science University, Portland, OR, United States
Stephanie Kerns Oregon Health & Science University, Portland, OR, United States
Paulette Lacroix PC Lacroix Consulting, North Vancouver, British Columbia
Julie A. Lasslo Eastern Kentucky University, Richmond, KY, United States
Wei Wei Lee University of Chicago, Chicago, IL, United States
Sharon Levy University of Edinburgh, Edinburgh, United Kingdom
John Liebert Private Practice of Psychiatry, Scottsdale, AZ, United States
Brittany Loggie Western University, London, ON, Canada
Anthony Maeder Flinders University, Bedford Park, SA, Australia
Carey Mather University of Tasmania, Hobart, TAS, Australia
Sandra Mattos Heart Institute of Pernambuco, Pernambuco, Brazil
Vishnu Mohan Oregon Health & Science University, Portland, OR, United States
Zilma Silveira Nogueira Reis Federal University of Minas Gerais, Minas Gerais, Brazil
Christopher Pearce University of Melbourne, Parkville, VIC, Australia
Daniel Pereira Instituto de Telecomunicações (IT), Lisboa, Portugal; Center for Health Technology and Services Research (CINTESIS), Porto, Portugal; University of Porto, Porto, Portugal
Norma Ponzoni McGill University, Montreal, QC, Canada
Kalyani Premkumar University of Saskatchewan, Saskatoon, SK, Canada
Anne Redmond Trinity Western University, Langley, BC, Canada
Carrie E. Reif University of Southern Mississippi, Hattiesburg, MS, United States
Shmuel Reis The Hebrew University of Jerusalem, Jerusalem, Israel
Bev Rhodes Alberta Health Services, Medicine Hat, AB, Canada
Marcy Rosenbaum University of Iowa Carver College of Medicine, Iowa City, IA, United States
Carla Sá University of Porto, Porto, Portugal
Gretchen Scholl Oregon Health & Science University, Portland, OR, United States
Tracy Shaben Alberta Health Services, Edmonton, AB, Canada
Aviv Shachak University of Toronto, Toronto, ON, Canada
Anne Short Alberta Health Services, Brooks, AB, Canada
Roger Simard Uniprix, Inc., Montreal, QC, Canada
Barbara Sinclair Western University, London, ON, Canada
Dilermando Sobral University of Porto, Porto, Portugal

Gillian Strudwick Centre for Addiction and Mental Health (CAMH), Toronto, ON, Canada

Anupam Thakur University of Toronto, Toronto, ON, Canada; Centre for Addiction and Mental Health, Toronto, ON, Canada

Maggie Theron Trinity Western University, Langley, BC, Canada

James Tong Western University, London, ON, Canada

David Topps University of Calgary, Calgary, AB, Canada

Maureen Topps University of Calgary, Calgary, AB, Canada

Dongwen Wang Arizona State University, Scottsdale, AZ, United States

Victoria Wangia-Anderson University of Cincinnati, Cincinnati, OH, United States

Sue Whetton Sue Whetton Consulting, Launceston, TAS, Australia; University of Tasmania, Hobart, TAS, Australia

David Wiljer University Health Network, Toronto, ON, Canada; University of Toronto, Toronto, ON, Canada

Foreword

The book *Health Professionals' Education in the Age of Clinical Information Systems, Mobile Computing and Social Networks* edited by Shachak, Borycki, and Reis focuses on how technology is changing health professions education. The text looks at this important issue through three distinct and important lenses. The first lens looks at the new challenges that health information and communication technologies are bringing to health professions education. The second lens provides examples from the field as academic educators prepare future health professionals for practicing in the digital era and the lessons they have learned. The third of these lenses examines how student learning outcomes and educational programs could be evaluated.

Internationally, there are considerable changes and progress taking place with the introduction of new health information and communication technologies. As health information and communication technologies are being developed and adapted for use in health care around the world so there has emerged a need for a technology literate health professional workforce that can effectively use the technology implemented in health care organizations (e.g., hospitals) to care for patients. There has also emerged a need to integrate health information technology competencies into health professions education internationally. Such work is important and critical to the advancement of health care systems.

In my work as the President of the International Medical Informatics Association (IMIA), I have promoted the importance of health information and communication technology use to improve health care quality and to achieve health equity. In keeping with this work, IMIA has promoted the development of biomedical and health informatics as a discipline and the development of health professional informatics competencies.

Today, the work of IMIA members includes not only developing health information and communication technologies for health care, and educating the future academics who teach biomedical and health informatics professionals at the undergraduate and graduate levels, but also educating academics who teach Medicine, Nursing, Pharmacy, and other health professions about the technologies that are part of today's health care system and developing what it is to be a health professional in an ever changing health care system.

This book is very important to read as it not only celebrates rapid development of health information and communication technologies in health care, but describes how educators of health professions are teaching students about how to effectively use and integrate these technologies into patient care. Lastly, the book closes the loop by describing evaluation approaches involving information technology in health professions education.

As an international organization, IMIA promotes the advancement of health informatics around the globe. This book provides perspectives from several regions of the world including North America, Europe, Australia, and the Middle East. These contributions are brought by academics who provide us with knowledge about how to improve and educate health professionals about health information technologies in health care and turn our readings and thought toward the lessons learned so that we can improve technology and education over time.

We have all seen the impact of health information and communication technologies. Today, we are learning more about how we can educate health professionals about using these technologies to improve patient care!

Hyeoun-Ae Park
International Medical Informatics Association (IMIA)

Preface

BACKGROUND

Information and communication technology (ICT) relentlessly transforms all aspects of modern life globally and health care is no exception. Electronic Health Records, mobile applications, internet use for health queries, big data collection and analysis, personalized medicine, and bioinformatics are all turning ubiquitous. This is certainly the case for high income countries where even in countries such as the United States and Canada that previously lagged behind about 80% of primary care physicians now use electronic medical records [1,2]. In medium- and low-income countries, too, there has been a steady growth in the use of ICT for health care, and some noteworthy initiatives, especially in the use of mobile technology [3–5].

A lot has been written in the health informatics literature about the implementation of clinical information systems. Through this research we learned a great deal about the importance of socio-technical issues and particularly the role of engaging stakeholders, support from administration, training, and "special people" such as champions and super users [6,7]. We also learned about the unintended consequences of health information technology that are not always positive including impact on clinicians' workflow and workload, overdependence on the technology, health information technology safety issues, and privacy breaches [8–10]. As clinical information systems are widely adopted, it is now time to switch our attention from implementation issues toward effective use of these systems, realizing their potential to improve care, and minimizing their negative unintended consequences.

Beyond clinical information systems, ICT transforms traditional power relationships within the health care system. Patients now have access to vast amounts of health information literally at their fingertips and are increasingly interested in viewing and contributing to their own medical records. Patient engagement, where patients are informed, involved, empowered, partnered with, and form communities to exchange knowledge and support each other, is now viewed as an essential component of high-quality health care; especially for people living with chronic diseases. Technology can be a great enabler of this, but it also brings with it multiple challenges including design of the technology to match patient needs and characteristics, reliability and credibility of health information online, health information literacy of patients and health care practitioners alike, and the tension between the level of service expected of health ICT and the need to protect people's privacy [11–13]. For example, there have been cases where patients pressured providers and policy makers to authorize unproven treatments based on information distributed on social media; of people self-(mis)diagnosing their conditions; and of privacy and security breaches [10,14,15]. And there have also been the opposite examples of expert patients partnering with their health care providers to achieve better outcomes and patients becoming more active in their own care [16].

While ICT applications in health care make headways, complementary ripples ensuring their effective, efficient, and moral use are called for. Central to these is the need to educate healthcare professionals about their brave, new technology-soaked field of practicing. From high to low income countries, ICT opens new options for affecting the health of individuals and communities, and forms new roles based on novel competencies that these options mandate. In addition, these ripples make ample use of ICT in order to educate, monitor, and evaluate this drive. The purpose of this book is to provide a comprehensive resource for health professions education in, and for, the digital era. It summarizes contemporary knowledge of the challenges and lessons learned from educators around the globe and several disciplines (Medicine, Pharmacy, Nursing, and Health Informatics) who are researching these issues and have developed educational interventions to deal with them.

ORGANIZATION OF THE BOOK

This book is organized around three intersecting themes: challenges for health professions education that are brought forth by the use of ICT in health care; experience from the field of educators who have been addressing these challenges and the lessons they have learned; and, finally, evaluation of students and educational programs intended to enhance health professionals' competence for the digital era.

In **Part 1**, our contributors have identified several challenges related to the use of health information technology. The first is the impact of using ICT on the patient–clinician relationships. In Chapter 1, Computers, Patients, and Doctors—Theoretical and Practical Perspectives, Christopher Pearce provides an overview of the impact of electronic medical records on the patient–clinician relationship. Josef Bartels in Chapter 2 specifically addresses the topic of conversational silence in the computer-assisted patient–clinician encounter. They highlight the new triadic patient–clinician–computer relationship that replaces the traditional dyadic patient–clinician interaction and the positive and negative impacts that this may have on the consultation. Chapter 3, Overcoming Health Disparities: The Need for Communication and Cultural Competency Training for Healthcare Providers Practicing Virtually in Rural Areas, by Lasslo, Anthony, Reif, and Bell, takes us beyond the traditional face-to-face interactions and discusses the challenges of virtual encounters via telemedicine. While common challenges with telemedicine such as health practitioners' compensation and licensing issues have been discussed quite extensively [17], these authors highlight a different challenge: the need for communication and cultural competency training for those who practice virtually via telemedicine in remote rural areas. While the chapter mostly focuses on rural areas in the United States, it also has multiple implications for low-resource countries were telemedicine has become a viable option to providing health services.

The second set of challenges has to do with the professional and ethical use of ICT in health care, and particularly with regards to communication forms such as email, text messaging, and social media. In Chapter 4, The Facets of Digital Health Professionalism:

Defining a Framework for Discourse and Change, Bahr, Crampton, and Domb provide a framework for defining and discussing digital professionalism. Among other issues, their chapter explores the definitions of digital professionalism, privacy, and reputation issues, as well as teaching and modeling professional digital behavior. Following this broad discussion of digital professionalism, Lacroix and Hamilton specifically address privacy challenges in Chapter 5, Privacy and the Hi-Tech Healthcare Professional. Their chapter highlights the need to find balance between the often conflicting demands of utilizing health ICT effectively, and compliance with privacy legislation. Chapter 6, Ethics, Obligations, and Health Informatics for Clinicians, by Wiljer and Thakur completes the discussion of ethical and professional aspects by focusing on the learning environment of future health care professionals with the unique challenges and learning opportunities it provides, including the digital responsibilities of the learners in that context.

Finally, the last set of challenges deals with effective use of ICT in health care. Arnaert, Ponzoni, Liebert, and Debe in Chapter 7, Transformative Technology: What Accounts for the Limited Use of Clinical Decision Support Systems in Nursing Practice when Compared to Medicine? ask "What accounts for the limited use of clinical decision support systems in Nursing practice when compared to Medicine?" and call the profession to address the barriers to use of these systems in Nursing, including via education. In Chapter 8, Developing Digital Literacies in Undergraduate Nursing Studies: From Research to the Classroom, Theron, Borycki, and Redmond discuss the growing need for students in health professions in general, and nursing in particular, to be able to search for, retrieve, and critically appraise the quality of online health information. They highlight the need for health information literacy education so that future professionals can make better use of various forms of online health information.

Part 2 of the book features curricular approaches and designs, and educational interventions to address the challenges of the digital era as well as use of ICT to deliver health professions education. In this part, educators and researchers describe their initiatives, experiences, and the lessons they have learned from them. In Chapter 9, Lessons Learned and Looking Forward With Pharmacy Education: Informatics and Digital Health, Clauson, Aungst, Simard, Fox, and Breeden describe the opportunities that ICT opens for transforming the practice of pharmacy. They provide direction for pharmacy education to adapt to the changing practice and examples of how pharmacy informatics topics can be introduced to students.

The next two chapters take us back to one of the challenges introduced in the first chapters of the book: integrating the electronic medical record into the patient–clinician interaction. In face of a growing frustration and dissatisfaction with this aspect of EMR use, especially in the United States [18], Alkureishi, Lee, and Frankel (Chapter 10: Patient-Centered Technology Use: Best Practices and Curricular Strategies) outline the best practices identified through research and describe curricular approaches to teach these skills. Figueiredo-Braga, Sorbal, and Rosenbaum complement this picture in Chapter 11, Incorporating Patient's Perspectives in Educational Interventions: A Path to Enhance Family Medicine Communication in the age of Clinical Information Systems, by bringing in the patient's perspective and discussing their experience with specific educational interventions to enhance patient–clinician–computer communication in Portugal.

To complete this subsection on curricular approaches and designs, Chapter 12, Strategies Through Clinical Simulation to Support Nursing Students and Their Learning of Barcode Medication Administration (BCMA) and Electronic Medication Administration Record (eMAR) Technologies, by Booth et al. focuses on the specific topic of barcode medication administration and electronic medication administration record use in nursing practice. They propose a simulation-based approach to teaching these processes in Nursing education, and highlight key considerations and best practices in developing such educational interventions.

Chapters 13–16 take us through local and regional education and training programs in the United States, the United Kingdom, and Canada. First, Hersh et al. (Chapter 13: From Competencies to Competence: Model, Approach, and Lessons Learned from Implementing a Clinical Informatics Curriculum for Medical Students) take us through the conceptualization and implementation of the exemplary program in clinical informatics at Oregon Health and Science University (OHSU) medical school. Their insights on the obstacles and success factors would provide multiple valuable insights for similar initiatives. Next, Levy in Chapter 14, Nurse Education in the Digital Age—A Perspective From the United Kingdom, provides an overview of the Nursing Informatics landscape in the United Kingdom, followed by a case study on embedding informatics competencies into postgraduate Nursing education at the University of Edinburgh in Scotland. In Chapter 15, Effectiveness of Training Strategies That Support Informatics Competency Development in Healthcare Professionals, Rhodes, Shaben, and Short reflect on their experience at Alberta Health Services (Canada) with the training provided in the implementation of health information technology to build "healthcare professional competency in using ICT to provide safe and quality patient care." Finally, Cummings, Whetton, and Mather (Chapter 16: Integrating Health Informatics Into Australian Higher Education Health Profession Curricula) demonstrate the strategies employed by one Australian university for introducing nursing informatics contents into programs at the undergraduate, postgraduate, and continuing professional development levels as well as how they developed mobile learning opportunities for clinical preceptors.

Scaling up these local and regional experiences is a challenging task. Two chapters in this book provide insight into how this can be done. In Chapter 17, Implementing Informatics Competencies in Undergraduate Medical Education: A National-Level "Train the Trainer" Initiative, Bhyat et al. describe how two Canadian organizations with a national focus—one representing medical schools and another that is responsible for the national eHealth strategy—partnered to develop and deliver health informatics contents to medical educators, taking a "train the trainer" approach. Wang (Chapter 18: Development and Evaluation of a Statewide HIV-HCV-STD Online Clinical Education Program for Primary Care Providers) provides another example of a large-scale initiative: a statewide continuing medical education program on Human Immunodeficiency Virus (HIV), Hepatitis C (HCV), and other sexually transmitted diseases (STDs) in New York State. The program created a large repository of multimedia educational items and disseminated them using multiple channels. It demonstrates innovative use of health informatics tools such as classification systems and clinical guideline representation model in an educational context.

As highlighted in Chapter 18, Development and Evaluation of a Statewide HIV-HCV-STD Online Clinical Education Program for Primary Care Providers, described above, ICT not only introduces new challenges, but it also provides opportunities and can be used as a vehicle for delivering educational interventions. While this may, in and of itself, be the topic of a whole different book, we included a number of chapters in this book that provide insight into how ICT may be used in health professions education. First, Pereira (Chapter 19: IS4Learning—A Multiplatform Simulation Technology to Teach and Evaluate Auscultation Skills) describes a novel simulation technology for teaching heart auscultation skills. While some argue that auscultation is largely replaced by ultrasonic imaging and Doppler techniques these days, Pereira submits that it remains a simple cost-effective method and thus, teaching the "almost lost art" of auscultation is still valuable. This may be especially important for low resource settings.

In Chapter 20, The Use of Mobile Technologies in Nursing Education and Practice, Asiri and Househ provide an overview of the use of mobile technology in nursing educations and discuss opportunities, challenges, and future directions. Wangia-Anderson and Dua (Chapter 21: Leveraging Social Media for Clinician Training and Practice) complete this section by discussing the use of social media in health professions education highlighting opportunities and calling on educators to take advantage of them.

Part 3, the last part of the book, presents the topic of evaluating the above-mentioned efforts. While evaluation has been included within many of the interventions discussed in previous chapters, we have identified a gap in systematically embedding it within program designs. This gap is addressed at two levels. First, in Chapter 22, Using Activity Data and Analytics to Address Medical Education's Social Contract, Topps, Ellaway, and Topps propose an intriguing and potentially controversial approach to evaluating individual students' performance using ambient surveillance tools. They argue that smarter use of data could overcome the biases and inconsistencies of preceptors when it comes to students' performance assessment. Second, Shachak and Reis (Chapter 23: Evaluating Educational Interventions for Health Professions in the Digital Age) discuss the evaluation of educational modules and programs. They highlight the need for evaluation at multiple levels and the specific challenges involved in evaluating interventions aimed at preparing clinicians for the challenges of the digital era including choice of study design, small number of participants, and lack of specific measure instruments. Case studies within the chapter, including a contribution from Gray, Choo, Butler-Anderson, Whetton, and Maeder, illustrate these challenges and approaches to handle them.

We believe that this smorgasbord in three sections gives a comprehensive overview of the current state of the field, and will supply insights and innovative examples to the readers. We hope that the continued professional conversation will be fostered by this volume and look forward to hear from our readership.

References

[1] The Commonwealth Fund. Commonwealth Fund International Survey of Primary Care Physicians in 10 Nations; 2015. Available from: <http://www.commonwealthfund.org/interactives-and-data/surveys/international-health-policy-surveys/2015/2015-international-survey> [cited 04.03.17].

[2] National Physician Survey. National Physician Survey; 2014. Available from: <http://www.nationalphysiciansurvey.ca/> [cited 28.01.15].
[3] Lewis T, Synowiec C, Lagomarsino G, Schweitzer J. E-health in low- and middle-income countries: findings from the Center for Health Market Innovations. Bull World Health Organ 2012;90(5):332–40.
[4] Bastawrous A, Armstrong MJ. Mobile health use in low- and high-income countries: an overview of the peer-reviewed literature. J R Soc Med 2013;106(4):130–42.
[5] Hall CS, Fottrell E, Wilkinson S, Byass P. Assessing the impact of mHealth interventions in low- and middle-income countries---what has been shown to work? Glob Health Action 2014;7:25606.
[6] Ash JS, Stavri PZ, Dykstra R, Fournier L. Implementing computerized physician order entry: the importance of special people. Int J Med Inform 2003;69(2–3):235–50.
[7] Shachak A, Montgomery C, Dow R, Barnsley J, Tu K, Jadad AR, et al. End-user support for primary care electronic medical records: a qualitative case study of users' needs, expectations, and realities. Health Syst 2013;2(3):198–212.
[8] Campbell EM, Sittig DF, Ash JS, Guappone KP, Dykstra RH. Types of unintended consequences related to computerized provider order entry. J Am Med Inform Assoc 2006;13(5):547–56.
[9] Kushniruk AW, Triola MM, Borycki EM, Stein B, Kannry JL. Technology induced error and usability: the relationship between usability problems and prescription errors when using a handheld application. Int J Med Inform 2005;74(7–8):519–26.
[10] Mlinek EJ, Pierce J. Confidentiality and privacy breaches in a university hospital emergency department. Acad Emerg Med 1997;4(12):1142–6.
[11] Irizarry T, DeVito Dabbs A, Curran CR. Patient portals and patient engagement: a state of the science review. J Med Internet Res 2015;17(6):e148.
[12] O'Grady L. Future directions for depicting credibility in health care web sites. Int J Med Inform 2006;75 (1):58–65.
[13] Järvinen OP. Privacy management of patient-centered e-health. In: Wilson EV, editor. Patient-centered e-health. New York: Hershey; 2009.
[14] Chafe R, Born KB, Slutsky AS, Laupacis A. The rise of people power. Nature 2011;472(7344):410–11.
[15] White RW, Horvitz E. Cyberchondria: studies of the escalation of medical concerns in Web search. ACM Trans Inf Syst 2009;27(4):1–37.
[16] deBronkart D. How the e-patient community helped save my life: an essay by Dave deBronkart. BMJ 2013;346:f1990.
[17] Weinstein RS, Lopez AM, Joseph BA, Erps KA, Holcomb M, Barker GP, et al. Telemedicine, telehealth, and mobile health applications that work: opportunities and barriers. Am J Med 2014;127(3):183–7.
[18] Rosenbaum L. Transitional chaos or enduring harm? The EHR and the disruption of medicine. N Engl J Med 2015;373(17):1585–8.

List of Reviewers

Maria (Lolita) Alkureishi University of Chicago, USA
Tamara J. Bahr University of Toronto, Canada
Elizabeth Cummings University of Tasmania, Australia
Sharon Domb Sunnybrook Health Sciences Centre and the University of Toronto, Canada
Raya Gal York University, Canada
Candace Gibson Western University, Canada
Drew McArthur The Office of the Information and Privacy Commissioner of British Columbia, Canada
Janessa Griffith University of Toronto, Canada
Mowafa Househ King Saud bin Abdulaziz University for Health Sciences, Kingdom of Saudi Arabia
Avi Hyman University of Toronto, Canada
Christopher M. Pearce Australasian College of Health Informatics, Monash University and University of Melbourne, Australia
Peter Pennefather University of Toronto, Canada
Gurprit K. Randhawa Vancouver Island Health Authority and University of Victoria, Canada
Esther Sangster-Gormley University of Victoria, Canada
Laurel M. Schwartz Eastern Kentucky University, USA
Gillian Strudwick Centre for Addiction and Mental Health, Canada
Danica Tuden University of Victoria, Canada
Dongwen Wang Arizona State University, USA
David Wiljer University Health Network and University of Toronto, Canada

PART I

THE CHANGING LANDSCAPE OF INFORMATION AND COMMUNICATION TECHNOLOGY (ICT) IN HEALTH CARE: IMPLICATIONS FOR HEALTH PROFESSIONALS' EDUCATION

SECTION I

THE CHANGING NATURE OF THE PATIENT-CLINICIAN RELATIONSHIPS

CHAPTER

1

Computers, Patients, and Doctors—Theoretical and Practical Perspectives

Christopher Pearce

University of Melbourne, Parkville, VIC, Australia

INTRODUCTION

The core activity of medicine remains the interaction between humans, although this is increasingly being challenged, or at least changed, by the revolution brought by the digitization of society. Whilst other aspects of societal transactions (such as banking and law) can now be conducted almost entirely in an online environment, the practice of medicine still requires personal contact, both because it is done "to" the person and increasingly "with" the person. In parallel with this development in health has come academic research to understand these changes at a deeper level. Such research has included both empirical analysis and theory development.

Theory development is essential to academic understanding yet in this particular area, theory development has lagged behind empirical analysis [1]. The object of this chapter is to outline the conceptual development of this interaction, both as a dyadic (patient and doctor), and now a triadic environment (including the computer), and to relate these concepts to the implications for educating health professionals into the future. It shall include history, models, and theory, to better understand where the social practice of medicine has come from, and were therefore it might go into the future.

The rapid computerization of the world in general has changed many of the fundamentals of society. Computers—as characterized as the box that sits on a desktop, are now taking a conceptual back seat as the internet (itself nothing more than networked computers) and devices such as phones and tablets come to mediate our relationship with the digital world. Increasingly, it is about information, not technology. Human society has experienced such changes in modernization before—the age of enlightenment, the industrial revolution, and each has been associated with upheaval as society adjusts to new ways of

Health Professionals' Education.
DOI: http://dx.doi.org/10.1016/B978-0-12-805362-1.00001-2

doing things and new ways of thinking. In each case, the practice of healthcare has made adjustments to integrate these new ways of thinking, but in each case, it has maintained the human interaction at its core.

In many ways, the largest most recent challenge to this human interaction has been the introduction of the scientific method in the latter part of the 19th century. The concept of the "body as a machine" has permeated much of medical science in the last century. The understanding of the "body as a machine" and the promise of truly effective cures, led to a belief that medicine no longer needed to worry about the "nonmechanical" parts of the human, that a pill would fix all things. Because of this, over the last century science has in many ways attempted to remove the doctor–patient relationship from our political and social structures [2]. As a result, modern medicine is an extraordinary work of reason: an elaborate system of specialized knowledge, technical procedures, and rules of behavior. However, by no means is medicine purely a rational process; our conceptions of disease and responses to it unquestionably show the imprint of our culture.

Whilst any interaction between humans can be characterized as a social interaction, what is unique about this relationship? The relationship even has a specific name, "the consultation," a name based on one of its key features being that it is (usually) initiated by the patient, who "consults" the doctor as to how their problem can be addressed [3]. The common thread throughout the doctor–patient relationship is the presence of, or more correctly a concern about the presence of, disease. And it is this concern, this oft unspoken presence, which underpins the whole relationship. The concern about disease has different meanings for the participants of the relationship. Diseases can have different natures according to the perspective of the participants [4]. The patient can see it as a threat or a burden to their life, or a source of fear, while the doctor may approach it more as a riddle, to be organized and classified in order to be understood; an intellectual puzzle. So disease may not necessarily imply a physical problem, but has been more correctly described as disease, or the presence of a disturbance in the perception of the patient [5]. We will see later how computers are now involved in this process of translation. Indeed, it is the computer's role in information flow that is central to its involvement.

HISTORY AND MODELS

The relationship between physicians and their patient has long been recognized as being of central importance to the practice of medicine [6]. In 1991, the Toronto Consensus Statement on Doctor-Patient Communication [7] was created in response to the recognition that roughly 50% of patient complaints and concerns are often not identified in the consultation [8], and that trainees [9] and practicing physicians [10] often have significant deficiencies in effective communication skills toward their patients. Seven principles have been described that underpin good doctor–patient communication [11]; that it should:

- serve the patient's need to tell the story of his or her illness, and the doctor's need to hear it;
- reflect the special expertise and insight that the patient has into his or her physical state and well-being;

- reflect and respect the relationship between a patient's mental state and his or her physical experience of illness;
- maximize the usefulness of physicians' expertise;
- acknowledge and attend to its emotional content;
- openly reflect the principle of reciprocity of the standing of those involved—doctors and patients; and
- help participants overcome stereotyped roles and expectations so that both participants gain a sense of power and freedom to change.

Michael Balint, an English psychiatrist in the 1950s, was one of the first to examine the patient–doctor relationship in detail. He worked with large numbers of inner London general practitioners to identify some of the details of the relationship in a groundbreaking study. He presented the results of this work in his book *The Doctor. His Patient and the Illness* [12]. In his introduction, he says:

> The first topic chosen for discussion at one of these seminars happened to be the drugs usually prescribed by practitioners. The discussion quickly revealed...that by far the most frequently used drug in general practice was the doctor himself, i.e. that it was not only the bottle of medicine or the box of pills that mattered, but the way the doctor gave them to his patient – in fact the whole atmosphere in which the drug was given and taken. (12: 1)

Balint's description above was the first time that anyone had specifically identified the doctor as an important, independent part of the relationship. Nevertheless, his view took a very doctor-centric approach, and in many ways followed the doctor-dominant paradigm of the time in which doctors *did* things to patients; whether they used themselves, or drugs, or other procedures. Whilst Balint was emphasizing the role of the doctor, others were attempting to emphasis the relational aspects of the interaction, by describing it as "an abstraction embodying the activities of two interacting people" [13], thereby suggesting that the consultation was in some way different to just a simple interaction, that it was in some way unique.

Stott and Davis [14] highlighted the potential positive effect that could be generated by each primary care consultation. They outlined how each interaction had the potential to manage both the presenting and potential problems, provided the opportunity for health promotion activities and the modification of help-seeking behavior. This was modified [15] to identify how in each consultation there exists a patient and a doctor agenda, which must then merge into a negotiated plan or outcome from the consultation.

George Engels, an American psychiatrist, described the failings of the biomedical model, and suggested a broader approach that included both psychological and social issues. Not surprisingly, this method was called the "biopsychosocial" model [16]. The outcome of this concept was to shift the focus from diseases *doing something* to a person, to illnesses as something being *experienced* by a person. Development of this model of patient-centered method continued, with many interpretations of this term. Moira Stewart and colleagues [17] have provided the most comprehensive definition of Patient Centered Clinical Method (PCCM), with six key components:

1. Exploring both the disease and the illness experience.
2. Understanding the whole person.
3. Finding common ground.

4. Incorporating prevention and health promotion.
5. Enhancing the doctor–patient relationship.
6. Being realistic.

Whilst PCCM is a practical approach, others have been continuing the theory development, describing different models of the relationship [13]. Emanuel and Emanuel [18] followed the notion of *ideal types* [19], and constructed a typology for the doctor–patient relationship. An ideal type is formed from characteristics and elements of the given phenomena, but it is not meant to correspond to all of the characteristics of any one particular case. It is designed to stress certain elements common to most cases of the given phenomenon. From this work, we can generate four ideal types of the relationship:

- Paternalistic: also called the parental or priestly model. In this model, the doctor acts as the patient's guardian, articulating and implementing what they feel is in the patient's best interest.
- Informative: also called the scientific, engineering or consumer model. The obligation here is for the doctor to provide all the available facts, and then the patient, with their own value set, determines the outcome.
- Interpretive: here the aim is for the doctor to elucidate the patients' value system, and thereby help select the best intervention.
- Deliberative: In this final model, the doctor acts as teacher or friend.

These models can then be expanded: including the further elements of patient values, doctor's obligations, patient autonomy, and physician role (given in Table 1.1). The subsequent matrix represents a comprehensive framework to understand the patient–doctor relationship.

THE DOCTOR–PATIENT RELATIONSHIP AS A SOCIAL INTERACTION

In approaching a physician for help, a patient brings not only a physical problem but also a social context. [20]

Beyond the creation of models of the consultation, social theorists have analyzed the interaction from a social perspective. A patient's experience of their physical problems is inseparable from the wider social context in which these problems occur [21] and it is impossible to separate the patients from the context in which they exist. Their context includes relationships at work, in the family, and in the wider community. This environment has best been described by Jurgen Habermas as the *lifeworld*. Lifeworld is a term first used in a phenomenological description of human society [22]. Lifeworld is the stock of skills, competencies and knowledge that ordinary members of society use in order to negotiate their way through everyday life, to interact with other people and ultimately to create and maintain social relationships [23]. Lifeworld contrasts with the system, which is a rules-governed element, usually representing either the economy or the state [24]. It is in the consultation that the system interacts with the patient's lifeworld.

TABLE 1.1 Aspects of the Doctor–Patient Relationship

Types of doctor-patient relationship

Elements of doctor-patient interaction

	Paternalistic	Informative	Interpretive	Deliberative
Patient values	Objective and shared by doctor and patient	Defined, fixed, and known to the patient	Inchoate and conflicting, requiring elucidation	Open to development and revision through moral discussion
Doctor's obligation	Promoting the patients well-being independent of the current circumstances	Providing relevant factual information and implementing patients selected intervention	Elucidating and interpreting patients values as well as informing and implementing	Articulating and persuading the patient of the most admirable values as well as informing and implementing
Concept of patient autonomy	Assenting to objective values	Choice of, and control over, medical care	Self-understanding relevant to medical care	Moral self-development relevant to medical care
Concept of physicians role	Guardian	Competent technical expert	Counselor or advisor	Friend or teacher

Dramaturgy

Alongside Habermas, the theories of Dramaturgy by Goffman offer a useful lens to examine the consultation relationship. Goffman held that the entire structure of society is made of rituals, and thus the "self" is in fact a socially enacted ritual. When an individual plays a part, he or she implicitly request their observers to take seriously the impression that is presented to them. It allows that the performer can be taken in fully by their own act, or may not be taken in at all [25], and it is the effect of the act on the audience that has significance.

Within the medical consultation therefore, we can see that the actions can stem from several motivations according to this schema. Effectual action for instance, is based on emotional states, seen when doctors become angry with patients for not following their treatment recommendations. Traditional action is seen in many of the ritualized encounters, perhaps for worker's compensation certificates. Actions can be based on values, where the patient believes the doctor is seeking "important" information from the computer. Finally, we can find means-end rationality in the PCCM approach of finding common ground. Multiple actions require the participants (now including the computer) to adopt multiple roles, often at different times throughout the consultation.

Goffman describes how the interactions between physician and patient can be elucidated in detail, and describes units of social interaction called *frames, which are* "the schemata of interpretation which individuals use to organize their everyday perceptions" [26]:

> When an individual. . .recognizes a particular event, he tends. . .to imply in this response (and in effect employ) one or more frameworks or schemata of interpretation of a kind that can be called primary. Frame analysis is based on his view that it is how humans interact with each other that is important, that an individual can be defined by his interactions with others. [27]

Therefore, a frame is a description of a socially derived reality, and is usually discussed in terms of the physical world, social ecology, and institutional setting; all of which have relevance in the highly ritualized interactions involved in the consultation.

Power

Important in the consultation is the role of power and authority. If we maintain that the computer is changing the power structure in the consultation [28,29], then how power is manifest is important to understand. Power manifests itself by proxy in the consultation, and this discussion is an analysis of those manifestations. From Marx, through to recent discussions of information technology and culture [30], power has been ascribed to institutions and can be sought by them. Power, at the most rudimentary and personal level, originates in dependence, which has also been used to describe the patient's place in the consultation [31]. The medical profession has had an especially persuasive claim to authority. The dominance of the medical profession in its interactions with patients goes considerably beyond simply the rational foundations provided by science. Its authority spills over its clinical boundaries into an arena of moral and political action for which medical judgment is only partially relevant and often incompletely equipped [20]. The profession derives its power from two sources, personal authority by dint of "character" and intimate knowledge of patients, and institutional authority conferred by the standards of the profession. The first role of the doctor has been described as a political one—"the struggle against disease must begin with a war against bad government" [2]. Thus, power can have many manifestations.

Power in social theory begins with the work of Marx. In brief, Marx introduced the concepts of control by use of language, and the power inherent in social structures such as class. Weber discussed power in the context of "the probability an actor in a social relationship will be able to exert his will." Balint [12] terms this the *apostolic function* of the doctor, where the doctor exhibits an almost religious zeal to convert the patient to the doctors way of thinking. Foucault refocused the discussion away from the Marxist view of power related to economic concepts, to a study of power relations. For instance, Foucault believed that, in the modern world, the methods of power have assumed responsibility for life processes, [32] and he discussed how the power in medicine derives as much from social and cultural issues as it does from science [2]. Giddens examines the influence of structure on social interactions, and vice versa, in what is termed *structuration theory*. Structuration theory holds that social structures are both constituted by human agency, and yet at the same time are the very medium through which structures are created. Thus structure is the medium and outcome of the conduct it recursively organizes, and the actors in the structure are knowledgeable and competent agents who reflexively monitor their action, and adjust their action accordingly [33].

One way of thinking about power in the doctor–patient relationship is to consider the extent to which the relationship revolves around four elements or types of power [34]. *Structural* power arises from the speaker's affiliation with the social institution of medicine; *Charismatic power* is based on personal characteristics; the influence of *Social* power is based on social prestige; whilst *Aesculapian* power is the ability to heal based on medical knowledge. Doctors need specialized knowledge and the power to be their patients' advocates. Patients are unlikely to choose a doctor whom they perceive or know to be powerless. Therefore, doctors need the ability (1) to share information with patients, respond to patients' cues, and obtain a full understanding of patients' wants (accountability); (2) to help patients tell their stories, formulate and express preferences, and make informed decisions on treatments (autonomy); (3) to act in a trustworthy manner in healthcare matters on behalf of, and for, patients (fidelity); and (4) to interact with patients with sensitivity and compassion, bearing in mind the increased emotional vulnerability that illness and fear of death can produce (humanity).

We begin to see how the human actors in the consultation will be able to ascribe power to the computer for a variety of reasons, and through a variety of frames. It is not just about the knowledge it contains, the position it occupies on the desk, or its very role as a participant in the consultation. To do so the computer acquires cultural and social standing. The summation of this section is this: in the past, the doctor embodied power in the consultation for several reasons. The doctor had knowledge and training, a revered position in society, and this power manifest itself in the consultation, in creating the paternalistic model of the interaction. They represented a Habermasian system, which interacted with the patient's lifeworld. However, this balance is changing, with other factors coming into play [35]. Patient centeredness has shifted power closer to the patient, with a greater recognition of the importance of both the patient's perspective and the information that the patient brings [36]. Into this dyad, we have introduced the computer, which can and does manifest power as well. If "knowledge is power," then the computer also represents a significant source of power by dint of its knowledge, limited only by the human's ability to find and interpret that knowledge. Doctors have become much more managers of information rather than repositories of same, as we will see later in this chapter.

Communication

Good communication is clearly essential to an effective doctor–patient relationship. This section will discuss communication from two perspectives that apply both to the doctor–patient relationship and the increasing role of the computer-language and nonverbal behavior. The doctor–patient relationship is a subset of the many and varied interactions between humans that can occur. It particularly falls into the category of one on one communication, differentiating it from one to many (group communication) and the align = "center" communication that occurs within one's own mind [37].

People communicate for many reasons [38]:

- As a means of reducing uncertainty
- To achieve social influence
- As part of our membership of certain groups (work groups, etc.)
- To achieve a certain identity or self-concept
- To identify with one another
- As an expression of culture
- To improve our relationships
- As entertainment
- To know each other better

Many of these can be applied in differing circumstances and with different emphases in the consultation.

Language

The importance of the spoken word should not be forgotten when considering power in the consultation. Conversation as a form of communication forms part of a greater process of the narrative or discourse. So to contribute to the relationship, the computer must also contribute to these processes. The computer does not use human language, and cannot communicate in the same way. Nor should it interfere with the expression of the narrative; it must not turn the interaction into a technical one. The computer needs to find ways to communicate effectively and not interfere. It can do so in a variety of ways. The humans can act as transcribers and interpreters ("the computer says you have had your test"). We can use means such as paper (written information), or audible warnings to notify one of the results of a test.

Language can be an expression of knowledge and power. This is influenced by the social and cultural background (and rules) that the participants bring to any conversation [39], and by our understanding of discourse is influenced by our perception of the agent of delivery [40]. Thus on legal matters we give more weight to the statements of a lawyer than a plumber. Discourse is different to communicative action: the latter occurs when there is meaningful interaction between persons, and the former when there is imbalance in communication, when one party challenges the assumptions of the other [23,24].

Language is characterized by rules that may change according to the background and cultural perspectives of the participants [41]. It is also a key expression of the concept of narrative. Narrative is underpinned by the use of language. Narrative theory gives us the concept of the "fabula" [42], which is the basic story material that we try to interpret, and to which we give structure. In the modern medical context, that fabula can be organized into an oral form, then a written form, and now may undergo a further processing, as the doctor places information about the patient into the computer [43]. The information changes in form at each level of processing. The conversational process aims for doctors and patients to come to a common understanding [44], and it is by listening to their stories that doctors can come to a deeper understanding of their patient's lifeworld [45]. The narrative provides much more information than the bare bones of the disease.

The dichotomy created here is in the different lifeworlds—where humans think in terms of this fabula—which in turn is reduced to data by the doctor placing information in the record [46]. A complex human process is therefore transformed into a series of data points that the computer can understand—and must therefore be similarly transformed. This constant transformation and interpretation is at the heart of the new relationship.

Nonverbal Behavior

There has been considerable research on the significance of nonverbal cues in medical practice. We know that there is much in the consultation that is unsaid [47]. We know, for instance, that humans are good at assessing deception in speech, purely by body language [48] and can even distinguish between different types of deception. Whilst facial cues alone can be unreliable, the combination of body cues, facial cues, and tone of voice, are all better than chance at assessing deception. Body language and gaze are a key factor in generating empathy in the clinical context [49,50]

Gaze plays a major role in regulating the exchange of speaking turns in conversation, thus influencing the turn-by-turn structure [51]. The more one looks at the other, the more regard they have for that subject [52], and the more powerful look less, and the less powerful look more. Thus, gaze and body language, as with conversation, can be an expression of power or authority in the consultation.

Body movement does not necessarily work on the turn-by-turn structure of conversation [38], and instead we must look at three levels of the body—face, trunk, and legs—each of which can function differently. This connection with the patient may be maintained by keeping the legs pointed toward them, even when the gaze is pointed away. This variation in direction between legs and head has been described as "torque" [53]. Gaze can be used to direct the emphasis, indicate beginning, continued involvement, and the end of consultation.

When a computer is introduced to the consultation, the doctor faces a dilemma [54]. The doctor must interact with two different representations of the patient: *the patient embodied* in the chair, and *the patient inscribed* in the medical record before them. Shifting the gaze from patient to medical record introduces a dysfluency in the patient's narrative [53]. Indeed, the amount of patient-directed gaze has been used as a measure of how well GPs deal with psychosocial problems [55].

THE TRIADIC RELATIONSHIP

The previous discussion introduced the work that had been done largely looking at the previous dyadic relationship. However, we have already introduced the concept that this no longer applies, and it is time to broaden the definition. Scott and Purves [56] first used the term "triadic" to describe the relationship between the humans and computers in the consulting room. Their work was the first time in which the concept that the computer may be a partner, rather than just another tool, was introduced into the literature. Something more than a stethoscope or a pen:

> In the GP consultation scenario, it is inadequate to talk in terms of HCI as simply a dyadic interaction; either between doctor and computer, or as a problem of attitudes of the patient towards the machine. Nor is it sufficient any longer to analyse the consultation, whether by describing it in terms of conversation analysis or by way of traditional doctor-training appraisal schedules, without considering the third ubiquitous component. We must see all components collectively, all sides simultaneously.

The Coming of the Computer

Computerization of medical practices is part of a global trend in western societies, or at least those with centralized health systems. Whilst computers have been involved in medicine almost since they were first invented, significant impacts into the interaction did not start until the advent of desktops in the late 1980s—computers capable of sitting in the doctor's room and assisting them in providing care. Programs written to assist doctors began to appear at this time as well [57]. Usually, the first action of a computer was to handle the billing and accounting side of a practice, including patient registration and demographic details. This was an action well suited to the early computers, and could easily be adapted from existing programs. Once in place, other functions could be used that benefited from the computer's ability to automate repetitive tasks. Actions such as: maintaining patient registers and recall lists, and sending automated reminders for reviews, for instance. These were usually back office functions paving the way for the next step, programs to assist the doctor in the consulting room. The exact purpose then depended on the context in which they were introduced [57], be it complex prescribing regulations or complex care needs. Doctors spend anywhere from 20% [58] to 40% [59] of their time interacting with the computer, either with the patient involvement, or without. It is this variety of tasks that is the reason we use the term "computer" rather than electronic patient record, as the computer presence is more than just the record.

Computerization across the world remains patchy, but is increasing [60]. In western countries, electronic health records have become almost ubiquitous, although developing countries still lag behind [61]. However, the metric is no longer the presence of a system or a computer, but what functions they perform, within the healthcare facility [62,63].

The Concept of a Triadic Relationship

From Scott and Purves, there have been several studies outlining the impacts of the computer. Greatbatch [64] was the first to identify that patients could respond to computers as well as doctors. Als [65] was the first to attempt to categorize behaviors, including ascribing the computer to actions beyond simply information gathering. From these early beginnings now come an ever-increasing number of studies exploring the relationship from several perspectives. Most have been descriptive in nature, using ethnographic [66] or conversation analysis [67] methodologies, and video observation is the most common method [68]. Grounded theory has been the most used theory, but more recently studies have used Goffman's dramaturgy [68,69].

Early work tended to create simple observations of behavior, often divided into categories and usually from the perspective of the doctor. Thus, we had described "black box" behaviors, where doctors would refer to the computer as a source of magic [65]. Following on from this, an English study [70] identified three "characteristic behaviors" and three "strategies" used by the doctors. The behaviors were controlling, responsive/opportunistic, and ignoring, while the strategies are sign posting, "blather" or pointless conversation, and responding every time to the patient interaction.

A US based study observing how doctors changed their behaviors after the adoption of a computer system, divided the changes into four domains: visit organization, verbal and nonverbal behavior, computer navigation, and spatial organization of the exam room [71]. Also in the United States, Ventres and colleagues [66,72] completed a video study of US doctors in a primary care clinic, again describing specific styles of interaction behavior by the doctors:

Interpersonal: where the doctors remained primarily focused on the patient
Informational: where the doctor seemed primarily driven by the needs of the computer
Managerial: a state alternating between the two

A further Australian study offered a categorization of behaviors from the perspective of all three actors, the doctor, patient, and computer. In this, the computer could be seen to shape the consultation by its own actions—calling for information, providing information, or simply distracting from the interaction [73]. All of these classifications have been combined into a single structure, in an attempt to provide clarity [74] (Fig. 1.1).

Since then studies have multiplied, with increasingly smaller foci on different elements. Not surprisingly, different computer programs have different effects [75], and this has been ascribed to both the "face" that each program presents to the human actors [76], and to how computers structure the data within the record [77]. Simple decisions such as the

Actor	Preferred term from classification		Similar or related concepts
	Orientation	Interaction style	
Doctor	Unipolar	Engaging	Explaining [18] Responsive [20] Minimal IT user [25] Magic box [18] Time out [18] Ignoring [20] Controlling [20] Managerial [10]
	Bipolar	Disengaging	Block or conversational IT user [25] Informational [10]
		Cogitating	Lookers [18]
Patient	Dyadic	Screen controlling Screen watching Screen ignoring	
	Triadic	Screen sharing	
Computer	Excluded Controlled Shared	Informative Prompting Distracting	Divided attention [19], Cognitive dissonance [30]

FIGURE 1.1 Classifications.

use of "interruptive" prompts versus flags can change the flow of the consultation. The program us just one of the factors that affects the cognitive loads on both the doctor [78]. Another is the new sources of information, and consequent changes to authority, introduced by the computer [79].

How the type of consultation affects computer use [80], with less computer use during consultations with a significant psychological component, although patients themselves can encourage the use of the computer, in a way that shares the responsibility for information provision [28]. Having to accommodate the needs of the computer has even changed how the room needs to be set up, and may reflect the doctors' style of computing [81]. With authority as an element in the consultation as a dyad, this authority is changing in the world where information is only a click away [79]. Two papers have attempted to summarize the observational literature [74,82]. It is this knowledge that we must pass onto the health professionals of the future.

The overall interpretation of the literature is that the presence of the computer is fundamentally changing the patient–doctor interaction as we know it. Not only is the physical setting changing, but also those areas traditionally the preserve of the profession, knowledge, and authority. No longer is the doctor the repository of all information, and the computer shapes the consultation form the very beginning [83,84]. In Habermasian terms, the computer acts as an ever-increasing bridge between the system and the lifeworld, affecting both communicative and strategic actions that play out in the consultation.

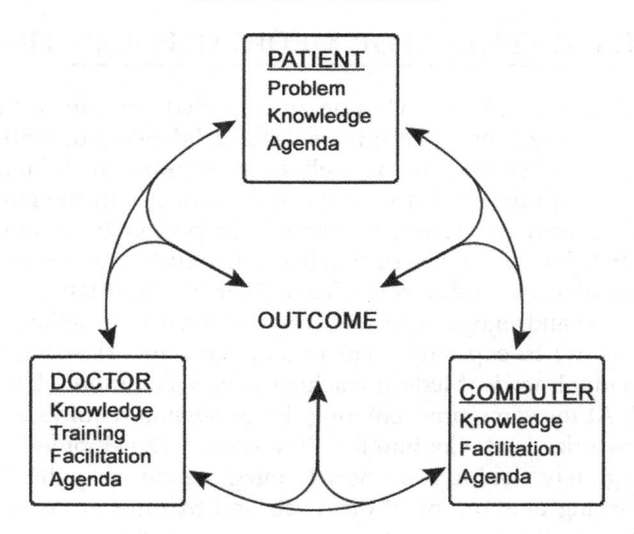

FIGURE 1.2 The Triadic consultation.

Building on the Middleton model, we can now describe the relationship as a true triad [28], increasing the range from simply patient and doctor agendas to include the areas that each actor brings to the consultation, including their knowledge, training and influence (Fig. 1.2).

By democratizing data flow in the consultation, the computer is shifting the balance of power. Government regulatory (and social) agendas can be manifest through the computer—in some cases preventing actions (such as prescribing some drugs) as well as facilitating actions [85,86]. By expressing information needs in the consultation, it can shape the agenda [66,84], another form of power. All of these changes require significant adjustments to the nature of the consultation, how this is expressed as an interaction, and the meanings that the humans interpret [69].

It should be emphasized at this point that the above discussion takes place at an early stage of evolution of the information revolution. Whilst the computer (as screen and keyboard) is likely to remain the predominant model in the office settings, hospitals and other environments will require other solutions [87]. The need for face to face care may also be reduced [88,89]. The computer is providing ever more sources of information into the consultation [90]. As the boundaries between sources break down, patients gain greater access to their own information [91,92]. mHealth (healthcare mediated by mobile devices) is the new buzzword [93] as is "big data." The demands on doctors and patients will change, and the education process must evolve to meet these new demands. If power and authority is shifting within the interaction, then the shifts will have as yet unrealized effects on healthcare delivery. Humans will never be excluded, but they will have to negotiate new ways to maintain the focus on patient-centered care. In turn, this will increase the need to learn new ways of doing a wide variety of tasks and activities.

IMPLICATIONS FOR MEDICAL EDUCATION

Medical education has long been styled on an extended apprenticeship model. Students were attached usually to teaching institutions—either hospitals or universities, but learnt their craft by watching at the bedside as well as in lectures. In the latter half of the 20th century, medical teaching identified a number of deficiencies in the current methods, and began slowly to move from conventional teaching, or pedagogy, to adult learning techniques (Androgogy) [94] based on the principles that adults are motivated and that their learning is both motivated by, and derived from, their life experiences.

The summary of this androgogic approach was to move from telling students what they need to know (pedagogy) to exposing them to situations and allowing them to learn from them, per their individual needs. Medical teaching is now commonly based around Problem Based Learning [95]. At the same time, entering the profession requires a commitment to lifelong learning, characterized as Continuing Professional Development [96]. Doctors who graduated 20 years ago have had to learn how to integrate the computer into their practice, at the same time as learning new means of diagnosis and treatments. As well they have learnt ways on how to integrate it into the conversation, how to balance the demands of electronic medical record (EMR) use and documentation with patient education and discussion, so this requires a slightly different but specific set of communication skills to be developed.

Despite this, there has been little emphasis on the implications of this "new consultation" and medical education, although calls for its inclusion are increasing [84,97]. For example the curriculum of the Royal Australian College of General Practitioners makes no mention of computer-mediated communication in its section under communication (see <https://www.racgp.org.au/Education/Curriculum/CS16/Domain-1/gp-under-supervision/communicate-effectively-and-appropriately-to-provide-quality-care>.

Given we know that doctors with good baseline communication skills are more successful at integrating the computer [98] it behooves us to ensure that these skill are ingrained not only into practicing physicians, but also our trainee doctors as well. Yet we know that models of training are few [99]. Often they represent "tips and hints" [100], or more comprehensive guides [101] and are limited in that they are not longitudinal in nature (i.e., providing training from the student trainee through to the practicing physician). While some institutions are beginning to provide this type of training [99], more widespread implementation is needed in order to promote a culture of positive technology use within this triad.

Furthermore, none of these guides yet fully emphasize the impact of this transformational change on all aspects—communication, power, even the basic role of the doctor. Doctors traditionally assessed, diagnosed, treated, and prognosticated. Now to those skills they must add information management [102]. Doctors have always been information managers [103], but have had to evolve from keeping information in their head, to paper, and now computer. The computer inserts, as well as manages, information. To do so requires undergraduate teaching to include the basic principles of information management, in the same way anatomy and pathology are basic skills. An undifferentiated graduate must be able to understand these basic principles. From there postgraduate training is required to add the extended skills of marrying information management to the social constructions of the consultation (as well as understanding the roles of all three actors in the consultation) with practical advice.

The academic community therefore has a significant role in advancing the field of research and informing the development of better models of care—for it's not just the doctors that need information and training, but patients, and the third element; software and information providers.

It is not just about the computer.

References

[1] Scambler G, Britten N. System, lifeworld and doctor-patient interaction. In: Scambler G, editor. Habermas, critical theory and health. London; New York: Routledge; 2001. p. viii, 212.
[2] Foucault M. The birth of the clinic: an archaeology of medical perception. New York: Vintage Books; 1994. xix, 215 p.
[3] Gask L, Usherwood T. ABC of psychological medicine. The consultation. BMJ 2002;324(7353):1567–9.
[4] Mol A. The body multiple: ontology in medical practiceIn: Herrnstein Smith B, Weintraub R, editors. 1 ed Durham, NC: Duke University Press; 2002. 196 p.
[5] Caplan AL, Engelhardt HT, McCartney JJ. Concepts of health and disease: interdisciplinary perspectives. Reading, Mass.: Addison-Wesley Advanced Book Program/World Science Division; 1981. xxxi, 756 p.
[6] Bliss M. William Osler: a life in medicine. Oxford; New York: Oxford University Press; 1999. xiv, 581 p.
[7] Simpson M, Buckman R, Stewart M, Maguire P, Lipkin M, Novack D, et al. Doctor-Patient Communication—the Toronto Consensus Statement. BMJ 1991;303(6814):1385–7.
[8] Stewart MA, McWhinney IR, Buck CW. The doctor/patient relationship and its effect upon outcome. J R Coll Gen Pract 1979;29(199):77–81.
[9] Platt FW, McMath JC. Clinical hypocompetence: the interview. Ann Intern Med 1979;91(6):898–902.
[10] Byrne PS, Long BEL. Doctors talking to patients: a study of the verbal behaviour of general practitioners consulting in their surgeries. London: H. M. Stationery Off; 1976. 194 p.
[11] Roter D, Hall JA. Doctors talking with patients/patients talking with doctors: improving communication in medical visits. Westport, Conn.; London: Auburn House; 1993. xii, 203 p.
[12] Balint M. The doctor, his patient and the illness. 2nd ed Edinburgh: Churchill Livingstone; 1964. xii, 395 p.
[13] Szasz T, Hollander M. A contribution to the philosophy of medicine: the basic models of the doctor-patient relationship. Arch Intern Med 1956;97(5).
[14] Stott NC, Davis RH. The exceptional potential in each primary care consultation. J R Coll Gen Pract 1979;29 (201):201–5.
[15] Middleton JF. The exceptional potential of the consultation revisited. J R Coll Gen Pract 1989;39(326):383–6.
[16] Engel GL. The need for a new medical model: a challenge for biomedicine. Science. 1977;196(4286):129–36.
[17] Stewart M, Brown J, Weston W, McWhinney L, McWilliam C, Freeman T. Patient-centred medicine: transforming the clinical method. 2nd ed. Oxford: Radcliffe Medical Press; 2003.
[18] Emanuel EJ, Emanuel LL. Four models of the physician-patient relationship. JAMA 1992;267(16):2221–6.
[19] Weber M. Economy and society; an outline of interpretive sociology. New York: Bedminster Press; 1968. 3 v. (cviii, 1469, lxiv) p.
[20] Starr P. The social transformation of American medicine. New York: Basic Books; 1982. xiv, 514 p.
[21] World Health Organization (WHO). The Alma Ata Declaration on Primary Health Care 1978 [Available from: www.who.int/hpr/NPH/docs/declaration_almaata.pdf.
[22] Husserl E. The crisis of European sciences and transcendental phenomenology: an introduction to phenomenological philosophy. Evanston: Northwestern University Press; 1970. xliii, 405 pp.
[23] Habermas J. The theory of communicative action, v 1: Reason and the rationalisation of society. Boston: Beacon Press; 1984. 2 v. p.
[24] Habermas J. The theory of communicative action v2; lifeworld and system: a critique of functionalist reason. Cambridge: Polity Press; 1987.
[25] Goffman E. The presentation of self in everyday life. Harmondsworth: Penguin; 1971. 251 p.
[26] Goffman E. Frame analysis: an essay on the organization of experience. York, Penns.: Northeastern University Press; 1974ix, 586 p
[27] Drew P, Wootton AJ. Erving Goffman: exploring the interaction order. Cambridge: Polity; 1988. iv, 298 p.

[28] Pearce C, Arnold M, Phillips C, Trumble S, Dwan K. The patient and the computer in the primary care consultation. J Am Med Inform Assoc 2011;18(2):138–42.
[29] Swinglehurst D, Roberts C, Li S, Weber O, Singy P. Beyond the 'dyad': a qualitative re-evaluation of the changing clinical consultation. BMJ Open 2014;4(9):e006017.
[30] Gallivan M, Srite M. Information technology and culture: identifying fragmentary and holistic perspectives of culture. Inf Organ 2005;15(4):295–338.
[31] Brody H. The healer's power. New Haven: Yale University Press; 1992. xiii, 311 p.
[32] Foucault M. The history of sexuality. London: Penguin; 1990. 3 v. p.
[33] Ritzer G. The Blackwell companion to major classical social theorists. Malden, MA: Blackwell; 2003. xii, 436 p.
[34] Brody M. The clinician as ethnographer: a psychoanalytic perspective on the epistemology of fieldwork. Cult Med Psychiatry 1981;5:273–301.
[35] Fochsen G, Deshpande K, Thorson A. Power imbalance and consumerism in the doctor-patient relationship: health care providers' experiences of patient encounters in a rural district in India. Qual Health Res 2006;16 (9):1236–51.
[36] Goodyear-Smith F, Buetow S. Power issues in the doctor-patient relationship. Health Care Anal 2001;9 (4):449–62.
[37] Littlejohn S. Theories of Human Communication. first ed. Belmont, California: Wadsworth; 1983. 309 p.
[38] Heath RL, Bryant J. Human communication theory and research: concepts, contexts, and challenges. 2nd ed. Mahwah, New Jersey: L. Erlbaum; 2000. 454 p.
[39] Foucault M. The archaeology of knowledge. London; New York: Routledge; 1989. 218 p.
[40] Bourdieu P, Thompson JB. Language and symbolic power. Cambridge: Polity Press in association with Basil Blackwell; 1991. ix, 302 p.
[41] McLaughlin ML. Conversation: how talk is organized. Beverly Hills: Sage Publications; 1984. 296 p.
[42] Onega Jaén S, García Landa JÁ. Narratology: an introduction. London: New York: Longman; 1996. xii, 324 p.
[43] Kay S, Purves I. The electronic medical record and "the story stuff": a narrativisitic model. In: Greenhalgh T HB, editor. Narrative Based Medicine. 1 ed. London: BMJ Books; 1998. p. 185–201.
[44] Edwards D. Discourse and cognition. London: Sage Publications; 1997. vii, 356 p.
[45] Cameron I. The importance of stories. Canadian Family Physician 1991;37:2617.
[46] Purves IN. Facing future challenges in general practice: a clinical method with computer support. Family Pract 1996;13(6):536–43.
[47] Andre M, Borgquist L, Foldevi M, Molstad S. Asking for 'rules of thumb': a way to discover tacit knowledge in general practice. Family Pract 2002;19(6):617–22.
[48] Zuckerman MDB, Rosenthal R. Humans as deceivers and lie detectors. In: Blanck PD, Buck R, Rosenthal R, editors. Nonverbal communication in the clinical context. University Park: Pennsylvania State University Press; 1986. p. viii, 320
[49] Blanck PD, Buck R, Rosenthal R. Nonverbal communication in the clinical context. University Park: Pennsylvania State University Press; 1986. viii, 320 p.
[50] Frankel R. The laying on of hands: aspects of the organisation of gaze, touch and talk in a medical encounter. In: Todd AD, Fisher S, editors. The Social organization of doctor-patient communication. Norwood, New Jersey: Ablex Pub. Corp.; 1993. p. xi, 306.
[51] Kendon A. Some Functions of Gaze Direction in Social Interaction 1967.
[52] Argyle M, Dean J. Eye-contact, distance and affiliation. Sociometry 1965;28(3):289–304.
[53] Ruusuvuori J. Looking means listening: coordinating displays of engagement in doctor-patient interaction. Soc Sci Med 2001;52(7):1093–108.
[54] Robinson JD. Getting down to business—talk, gaze, and body orientation during openings of doctor-patient consultations. Human Commun Res 1998;25(1):97–123.
[55] Bensing JM, Kerssens JJ, Vanderpasch M. Patient-directed gaze as a tool for discovering and handling psychosocial problems in general-practice. J Nonverbal Behav 1995;19(4):223–42.
[56] Scott D, Purves I. Triadic relationship between doctor, computer and patient. Interacting Comput 1996;8 (4):347–63.
[57] Pearce C. The adoption of computers by Australian general practice—a complex adaptive systems approach. OMICS J. Gen Pract 2013;1(3):1–3.

[58] Bui D, Pearce C, Deveny E, Liaw T. Computer use in general practice consultations. Aust Fam Physician 2005;34(5):400.

[59] Kumarapeli P, de Lusignan S. Using the computer in the clinical consultation; setting the stage, reviewing, recording, and taking actions: multi-channel video study. J Am Med Inform Assoc 2013;20(e1):e67–75.

[60] Jha AK, Doolan D, Grandt D, Scott T, Bates DW. The use of health information technology in seven nations. Int J Med Inform 2008;77(12):848–54.

[61] Shu T, Liu H, Goss FR, Yang W, Zhou L, Bates DW, et al. EHR adoption across China's tertiary hospitals: a cross-sectional observational study. Int J Med Inform 2014;83(2):113–21.

[62] McCoy A, Wright A, Eysenbach G, Malin B, Patterson E, Xu H, et al. State of the art in clinical informatics: evidence and examples. Yearb Med Inform 2013;8(1):13–19.

[63] McInnes DK, Saltman DC, Kidd MR. General practitioners' use of computers for prescribing and electronic health records: results from a national survey. Med J Aust 2006;185(2):88–91.

[64] Greatbatch D, Luff P, Heath C, Campion P. Interpersonal-communication and human-computer interaction—an examination of the use of computers in medical consultations. Interact Comput 1993;5(2):193–216.

[65] Als A. The desk-top computer as a magic box: patterns of behaviour connected with the desk-top computer; GPs' and patients' perceptions. Family Pract 1997;14(1):17–23.

[66] Ventres W, Kooienga S, Vuckovic N, Marlin R, Nygren P, Stewart V. Physicians, patients, and the electronic health record: an ethnographic analysis. Ann Fam Med 2006;4(2):124–31.

[67] Greatbatch D, Heath C, Luff P, Campion P. Conversation analysis: human-computer interaction and the general practice consultation. In: Monk A, Gilbert GN, editors. Perspectives on HCI: diverse approaches. London: Academic Press; 1995. p. 198–222.

[68] Pearce C, Arnold M, Phillips C, Dwan K. Methodological considerations of digital video observation: beyond conversation analysis. Int J Multiple Res Approach 2010;4(2):90–9.

[69] Swinglehurst D, Roberts C, Greenhalgh T. Opening up the 'black box' of the electronic patient record: a linguistic ethnographic study in general practice. Commun Med. 2011;8(1):3–15.

[70] Booth N, Kohanned J, Robinson P. Information in the consulting room. Newcastle upon Tyne: University of Newcastle upon Tyne; 2002 July.

[71] Frankel R, Altschuler A, George S, Kinsman J, Jimison H, Robertson NR, et al. Effects of exam-room computing on clinician-patient communication: a longitudinal qualitative study. J Gen Intern Med 2005;20 (8):677–82.

[72] Ventres W, Kooienga S, Marlin R, Vuckovic N, Stewart V. Clinician style and examination room computers: a video ethnography. Family Med. 2005;37(4):276–81.

[73] Pearce C, Dwan K, Arnold M, Phillips C, Trumble S. Doctor, patient and computer---a framework for the new consultation. Int J Med Inform 2009;78(1):32–8.

[74] de Lusignan S, Pearce C, Kumarapeli P, Stavropoulou C, Kushniruk A, Sheikh A, et al. Reporting observational studies of the use of information technology in the clinical consultation. A position statement from the IMIA Primary Health Care Informatics Working Group (IMIA PCI WG). Yearb Med Inform 2011;6(1):39–47.

[75] Refsum C, Kumarapeli P, Gunaratne A, Dodds R, Hasan A, de Lusignan S. Measuring the impact of different brands of computer systems on the clinical consultation: a pilot study. Inform Primary Care 2008;16 (2):119–27.

[76] Pearce C, Arnold M, Phillips CB, Trumble S, Dwan K. The many faces of the computer: an analysis of clinical software in the primary care consultation. Int J Med Inform 2012;81(7):475–84.

[77] Lown BA, Rodriguez D. Commentary: lost in translation? How electronic health records structure communication, relationships, and meaning. Acad Med 2012;87(4):392–4.

[78] Shachak A, Hadas-Dayagi M, Ziv A, Reis S. Primary care physicians' use of an electronic medical record system: a cognitive task analysis. J Gen Intern Med 2009;24(3):341–8.

[79] Swinglehurst D. Displays of authority in the clinical consultation: a linguistic ethnographic study of the electronic patient record. Soc Sci Med 2014;118:17–26.

[80] Chan WS, Stevenson M, McGlade K. Do general practitioners change how they use the computer during consultations with a significant psychological component? Int J Med Inform 2008;77(8):534–8.

[81] Pearce C, Walker H, O'Shea C. A visual study of computers on doctors' desks. Inform Primary Care 2008;16 (2):111–17.

[82] Crampton NH, Reis S, Shachak A. Computers in the clinical encounter: a scoping review and thematic analysis. J Am Med Inform Assoc 2016;23(3):654–65.
[83] Pearce C, Trumble S, Arnold M, Dwan K, Phillips C. Computers in the new consultation: within the first minute. Family Practice 2008;25(3):202–8.
[84] Pearce C, Kumarpeli P, de Lusignan S. Getting seamless care right from the beginning—integrating computers into the human interaction. Stud Health Technol Inform 2010;155:196–202.
[85] Pearce CM, de Lusignan S, Phillips C, Hall S, Travaglia J. The computerized medical record as a tool for clinical governance in Australian primary care. Interact J Med Res 2013;2(2):e26.
[86] de Lusignan S. Clinical Governance. In: Bevir M, editor. Encyclopaedia of governance. Thousand Oaks: Sage; 2007. p. 99–101.
[87] Alsos OA, Das A, Svanæs D. Mobile health IT: The effect of user interface and form factor on doctor–patient communication. Int J Med Inform 2012;81(1):12–28.
[88] Adamson SC, Bachman JW. Pilot study of providing online care in a primary care setting. Mayo Clin Proc 2010;85(8):704–10.
[89] Mansfield SJ, Morrison SG, Stephens HO, Bonning MA, Wang S-H, Withers AHJ, et al. Social media and the medical profession. Med J Aust 2011;194(12):642–4.
[90] Pearce C, Bainbridge M. A personally controlled electronic health record for Australia. J Am Med Inform Assoc 2014;21(4):707–13.
[91] Archer N, Fevrier-Thomas U, Lokker C, McKibbon KA, Straus SE. Personal health records: a scoping review. J Am Med Inform Assoc 2011;18(4):515–22.
[92] Falcão-Reis F, Correia ME. Patient empowerment by the means of citizen-managed Electronic Health Records: web 2.0 health digital identity scenarios. Stud Health Technol Inform 2010;156:214–28.
[93] PwC WIUa. Emerging mHealth: paths for growth. 2012:1–44.
[94] Lindeman EC. The meaning of adult education. New York: New Republic; 1926.
[95] Watling CJ, Lingard L. Grounded theory in medical education research: AMEE Guide No. 70. Medical Teacher. 2012;34(10):850–61.
[96] Mazmanian PE, Davis DA. Continuing medical education and the physician as a learner: guide to the evidence. JAMA. 2002;288(9):1057–60.
[97] Hammoud MM, Dalymple JL, Christner JG, Stewart RA, Fisher J, Margo K, et al. Medical student documentation in electronic health records: a collaborative statement from the Alliance for Clinical Education. Teach Learn Med 2012;24(3):257–66.
[98] Ventres WB, Frankel RM. Patient-centered care and electronic health records: it's still about the relationship. Family Med 2010;42(5):364–6.
[99] Duke P, Frankel RM, Reis S. How to integrate the electronic health record and patient-centered communication into the medical visit: a skills-based approach. Teach Learn Med 2013;25(4):358–65.
[100] Ventres W, Kooienga S, Marlin R. EHRs in the exam room: tips on patient-centered care. Fam Pract Manag 2006;13(3):45–7.
[101] Simpson L. E-communication skills: a guide for primary care. Radcliffe Publishing; 2005.
[102] Pearce C, Veil K, Williams P, Cording A, Liaw ST, Grain H. Driving the Profession of Health Informatics: The Australasian College of Health Informatics. Stud Health Technol Inform 2015;216:458–61.
[103] Wright A, Sittig DF, McGowan J, Ash JS, Weed LL. Bringing science to medicine: an interview with Larry Weed, inventor of the problem-oriented medical record. J Am Med Inform Assoc 2014;21(6):964–8.

CHAPTER

2

What's All This Silence? Computer-Centered Communication in Patient-Doctor-Computer Communication

Josef Bartels

Family Medicine Residency of Idaho, Boise, ID, United States

INTRODUCTION

Your patient is describing some new urinary symptoms that he's been having at night, and you're narrowing down your differential. You ask some clarifying questions and as the patient is answering you turn toward your screen to check whether a Prostate-Specific Antigen (PSA) level was ever checked. During your hunt through the results, your attention wanders from your patient's words, and even before you find the PSA level, a white blood cell count turns your thoughts toward lymphoma. Your next question has nothing to do with urination.

Like medicine, aviation has experienced an exponential increase in complexity as pilots adapted from simple wood and canvas crafts to subsonic-computerized super machines. A number of tragic accidents deemed "pilot error" led to a demand for a clear division of labor and a focus on human factors [1]. Pilots are trained to minimize multitasking interference either by careful planning and delegation or by slowing the situation down intentionally and performing the necessary tasks sequentially. The discussion within aviation is ongoing and each new aircraft in each new mission is evaluated for the ideal distribution of cognitive tasks across the crew [2]. Though the errors caused by insufficient division of cognitive resources across tasks may not be immediately fatal in many medical contexts, they should still motivate a root cause analysis of distracted and disengaged clinicians.

Clinicians are faced with increasingly complex data management responsibilities while simultaneously attending to verbal and nonverbal cues to maintain some moments of

Health Professionals' Education.
DOI: http://dx.doi.org/10.1016/B978-0-12-805362-1.00002-4

patient-centered care (PCC) within triadic patient-doctor-computer communication (PDCC) [3,4]. The improved quantity and quality of the data provided by electronic health records (EHRs) has led to decreased errors, improved coordination of care, and greater patient satisfaction concerning accuracy of information retrieval [5–8]. Increasing demands on clinicians to contribute to and access data in real time during patient visits suggest that our performance metrics should consider workload, cognition, interruption, and multitasking as decision makers have in aviation, combat, and policing [9,10]. However, the existing studies and our own data presented later in this chapter suggest that the interaction between clinician and computer detracts from the communication behaviors and outcomes that comprise PCC [11–21]. These competing tasks challenge clinicians to speak to patients while recording data; health systems to balance tradeoffs between communication quality and data improvement; EHR designers to minimize clinician workload while maximizing data entry; investigators to elucidate mechanisms through which computer-centered care (CCC) correlates with PCC and health outcomes; and educators to align specific competencies with best practice [21,22]. This chapter will focus specifically on the challenges the researchers face in elucidating the effects of CCC on PCC and healthcare quality.

Two decades of effort on PCC has advanced our knowledge tremendously while unveiling the complexity that underlies the patient-doctor relationship and all interactions encompassed by it. Though PCC is well defined, and many specific elements have been identified, it's effects can be blunted or augmented unpredictably by parts of PDCC that we do not yet understand. A recent study by Epstein et al. randomized clinicians to receive PCC training, and though the intervention group was more patient centered after the intervention, those clinicians did not achieve better health status for their patients [23]. Since so little communication during each encounter is truly patient centered [20], it's the other types of communication that comprise communication's contribution to healthcare quality. Most of us who have seen a clinician recently know that CCC has a tremendous effect on the content and style of communication in the clinic or hospital [24]. Computers are becoming ubiquitous across healthcare encounters and so it falls upon researchers to elucidate CCC with the same precision and persistence that we have applied to PCC.

BACKGROUND

Previous investigations at the communication-computing nexus have used video, mouse/keyboard tracking, and direct observation to qualify and quantify the EHR's effect on communication [7,14,15,17–20,25–28]. Time-intensive analysis has limited the amount of data used, and collection methods may have limited participant enrollment due to potential intrusions on privacy. Even in those studies where a clear link between computing and specific communication outcomes has been established, no direct link to health outcomes can be pursued through such time-intensive analyses. A large-scale analysis of computing/linguistic interaction during clinician–patient communication will require automated data collection and processing to get enough data to detect potentially small effects on health outcomes.

I piloted a new approach using audio data alone that together with mobile phone technology may allow this type of longitudinal data collection. Based on an observation made by Gibbings-Isaac et al. regarding the increased amount of time where neither human is speaking yet the clinician is busy [29], I hypothesized this silence to be one type of computer-associated multitasking interference that may adversely impact the quality of both PCC and data utilization [29–31]. You can think of this interference as a spectrum from minor distraction to gross interruption. It would be audacious to suggest that one encounter with your clinician where he or she was completely distracted by the computer would result in measurable change to any health outcome, so we need to quantify this distraction over multiple encounters. As prelude to this chapter, let me underline the existence of computer-associated verbal silences through a small pilot study that may give us the ability to measure multitasking interference using an audio-only collection technique.

It is my sincere hope that this chapter helps to illuminate the importance of CCC in communication research, training, health systems, and EHR design. Improving communication and healthcare will not be as simple as ensuring an appropriate ratio between a computer-centered and patient-centered approach. Both orientations have strengths and weaknesses in the age of data-driven healthcare. Finally, I will suggest that audio data alone will allow us to identify CCC as well as interruption and displacement of PCC within a large variety of activities that occur during a healthcare visit.

MULTITASKING INTERFERENCE AND INTERRUPTION

Though Pearce et al. proposed a triadic relationship between doctor, computer, and patient [17], communication in the exam room cannot be truly triadic if patients neither enter nor extract significant amounts of information from the computer. This leaves the clinician balancing their attention between two sources of information, both of which demand attention in different ways. This creates tension between data gathering and data documentation. One of the largest questions that remains to be answered is whether simultaneous completion of computer tasks and communication tasks is beneficial for either patient care or accurate data entry.

Simultaneous completion sounds like multitasking, but before I get there, I'll review the foundations. Brixey defines interruption by the steps of a task: perception, interpretation, evaluation, goal, intention, and task execution [22]. Interruption is defined as anything that causes discontinuous task performance. Even more nuanced, an interruption consists of a notification either internal or external, then an interruption lag, the interruption itself, and finally a resumption lag before the original task is resumed. Multitasking attempts to explain how multiple tasks are performed by humans and how the performance of specific tasks differs when combined. Early research assumed a single channel bottleneck that limited the speed of task processing [32], and proposed that performance was limited by the time available and the speed of a central processor in the brain. It followed that the more processing required by a second task, the more it would interfere with the original task. The task characteristics were captured by a variety of ratings such as bandwidth, working memory requirement, and skill level, to predict the performance changes of adding one task on top of the original task. Experiments showed that additional time had to be added for task switching and that parallel performance of tasks resulted in supra-additive time to completion with increased error rates [33].

More recently, functional MRI during classic multitasking challenges led to the understanding that there are multiple co-processer areas in the brain [32], each of which specialized in performing certain types of tasks. Based on these limited data, it is no longer sufficient to suggest that PCC and CCC simply compete for the same brain space therefore should not be completed simultaneously. Yet just because we may have separate processing centers for these tasks does not mean that they will not interfere with each other. It may be unrealistic to suggests that maintaining eye contact and orienting the computer screen toward the patient will somehow keep the typing part of the brain from interfering with the listening part [4,34].

If we borrow from extensive research in crew resource management in other fields, we might conclude that communication interruption is a primary cause of many errors in healthcare [35]. However others have documented the number of errors prevented by timely interruptions [36]. An awareness of the competing tasks piled upon clinicians is already leading to a more nuanced understanding of when clinicians need a "sterile cockpit" to facilitate PCC versus CCC and how to designate when each will contribute most to improving the health of the patient [37]. Any attempt to correlate cognitive patterns with patient care will require some method of differentiating and quantifying these patterns.

Could Silence Be the Key?

Silence has many different forms and functions; some types are associated with presence and attunement while other types may mark a wandering mind or computer usage. One unique type of silence may reliably mark the cognitive pattern that clinicians use when focused on the computer. Because existing literature purports that tasks can be interrupted or performed in parallel, measuring silence would isolate those tasks that are performed without simultaneous verbal communication. This specific phenomenon within PDCC, this increased amount of time where neither human is speaking, is defined broadly here as silence despite the presence of ambient sounds such as computer clicking or typing [19,21,29]. Within the triadic relationship of PDCC, a distinction between computer-associated silences and other more patient-centered silences would provide a marker and quantification of CCC.

Other work on silence has described therapeutic, productive, compassionate, and eloquent silences occurring in a medical setting [30,38–40]. This pilot study describes quantitatively a very different type of silence that is not associated with patient-doctor communication, but instead with doctor-computer communication. The existence of a non-communicative type of silence and its association with non-communicative tasks is unequivocal [40–49]. My literature review of the structure and etiology of conversational silence revealed a qualitative split between those silences that were fundamentally communicative and those that were a result of other tasks such as typing or reading from a computer. This small study confirms computer silence's presence and ubiquity and suggests that audio data alone is sufficient to study the relation between the distortion in patient-doctor communication and quality of care/communication/health outcomes.

Methods: I used 124 audio recordings from phase one of the Values and Options in Cancer Care (VOICE) Study, a National Cancer Institute funded randomized controlled trial of communication in cancer care conducted in Rochester/Buffalo, NY and Sacramento, CA regions [50]. First I visualized audio intensity over time using *Atlas.ti* software and screened for low amplitude of >2 seconds duration. Two seconds were chosen to specifically capture abnormally long conversational pauses by exceeding two standard deviations above normal turn-taking and speech hesitations [51]. These low intensity moments were then screened by trained listeners for presence of low amplitude verbal activity and then consensus coded as silent or not silent, computing sounds or no computing sounds. All silences present during physical examinations were excluded. All remaining silences >2 seconds were classified as "computer-associated silence" or "conversational silence" based on the presence or absence of typing and clicking sounds anytime 5 seconds before, during, or up to 5 seconds after the silence. I used *Atlas.ti* to code and measure the silence lengths. I calculated location as proportion through visit and checked for correlation using a mixed effects model to account for physician identity and consultation length.

RESULTS: 47% of verbal silences >2 seconds were associated with sounds of physician computing. These silences were longer than other silences and occurred in all parts of the consultations. Physician identity was not an independent predictor of computer-associated silence quantity. Tables 2.1 and 2.2 list the characteristics of the physicians and patients included.

Identification of computer-associated silences was reliable between three separate coders (kappa = 0.82). Tables 2.3 and 2.4 list descriptive and univariate statistics for the

TABLE 2.1 Patient Demographics

	N	%
All	124	100
RACE		
White	112	90
Other	12	10
SITE		
URMC	78	65
UCD	42	35
PATIENT EDUCATION		
Some college or more	85	68
HS or less	39	33
AGGRESSIVE CANCER		
Nonaggressive	66	53
Aggressive	58	47
PATIENT GENDER		
Female	70	56
Male	54	44

TABLE 2.2 Physician Demographics

	N	%
All	41	100
PHYSICIAN GENDER		
Male	29	71
Female	12	29
PHYSICIAN RACE		
Asian	16	39
Black/AA	1	2
White	18	44
Other	1	2
Missing	5	12
BREAST CANCER PHYSICIAN		
No	33	80
Yes	8	20
PHYSICIAN AGE		
Mean	Median	std
44.7	44	9.8

TABLE 2.3 Descriptive Statistics for Visits

Silence Counts by Visit	Mean	Median	SD	Min	Max
Consultation length (minutes)	19.8	19.2	10.2	2.0	56.4
Conversational silence count	5.12	3.00	6.13	0	33
Computer-associated silence count	4.59	2.00	6.44	0.0	28.0

visits, respectively. Both types of silence were distributed evenly throughout the visit and did not display significant clustering (Fig. 2.1). The mixed effects model showed no significant coefficient for physician identity or visit length.

Conclusion: Computer-associated silence may be a proxy for interference between computer tasks and patient communication tasks. These results suggest that automated analysis may be able to identify when computing impacts fluent communication and provide real-time feedback to the clinician. I was limited in studying the effect of these silences by the cross-sectional nature of my data, as having only one visit for each patient recorded did not allow measuring the effect of these silences over time on health outcomes. In addition, misclassification bias may exist due to a lack of video corroboration. A closer analysis

TABLE 2.4 Univariate Statistics

Computer-associated Silence Length (sec)	$n = 573$	*Conversational Silence Length (sec)*	$n = 635$
Mean	7.75	Mean	4.019450394
Standard error	0.43	Standard error	0.1512286297
Median	4.80	Median	2.91
Mode	4.30	Mode	2.2
Standard deviation	10.41	Standard deviation	3.810841443
Sample variance	108.42	Sample variance	14.5225125
Kurtosis	147.02	Kurtosis	70.60600444
Skewness	9.58	Skewness	6.730209431
Range	182.37	Range	55.32
Minimum	2.00	Minimum	2
Maximum	184.37	Maximum	57.32
Sum	4,441.41	Sum	2552.351
Count	573.00	Count	635
Largest(1)	184.37	Largest(1)	57.32
Smallest(1)	2.00	Smallest(1)	2
Confidence level (95%)	0.85	Confidence level (95%)	0.2969695774

of these silences may identify specific computer tasks that are interfering with clinician–patient communication to facilitate workflow redesigns to minimize multitasking interference.

COMPUTER-CENTERED CARE

PCC has been linked to improved health outcomes in numerous studies over the past two decades. Over this same time, technology has become central to the physician workflow, augmenting everything from clinical decision-making to chronic disease management and population surveillance. These additional data have made it possible to spend hours managing a patient's medical conditions without collaborating directly with the patient to learn values, set goals, and plan care. We cannot underestimate the positive influence of many decision-support tools and clinical reminder systems that have substantially improved preventive healthcare and medication accuracy [52], but perhaps the benefits of increased CCC are dulled by the corresponding decrease in PCC. Interruptions in patient communication from computer tasks have the potential for both positive and negative influence on patient care, depending on the context. For example, an automatic pop-

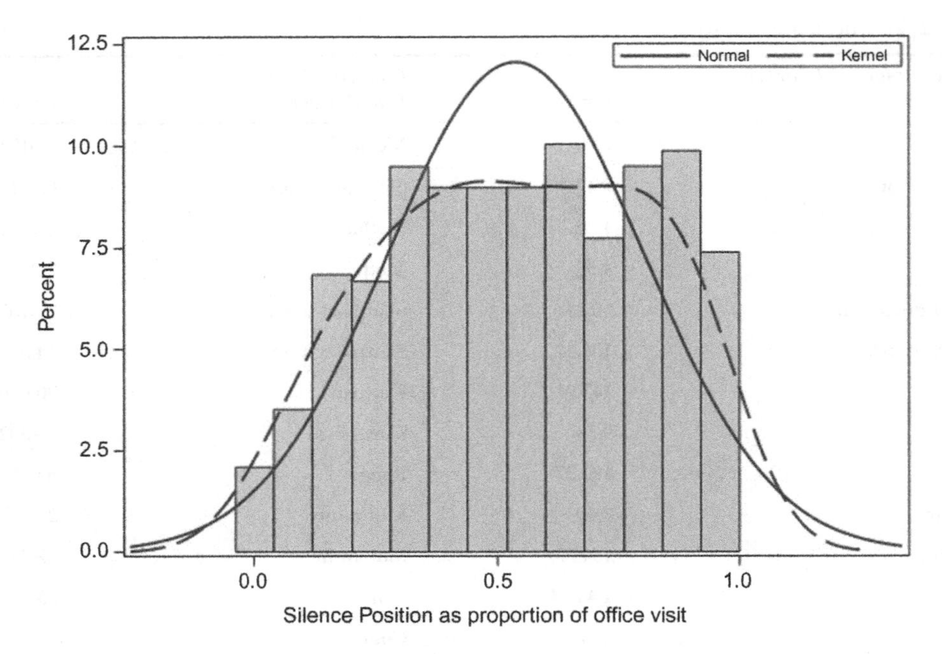

FIGURE 2.1 Computer-associated silence distribution within outpatient oncology office visits.

up alerting the physician to an overlooked drug–drug interaction when prescribing a new medication has the potential to improve patient care. In contrast, repeated difficulties in accessing a CT image while navigating the EHR may solely serve as a distraction and thus negatively impacts patient care. To understand the current threat to PCC, we've got to be able to measure and understand its converse; CCC.

CCC may be balanced with PCC, but an increase in either means a necessary decrease in the other given the time pressures of modern medicine. Because neither can ever occur in isolation, we may want to start measuring their relative contributions by tracking various communication and health outcomes. There may be no golden ratio between data management and relationship management, but instead a unique mixture appropriate for each particular context and specific clinical question. Some encounters may benefit from an orientation toward the data in the computer, for example, an unconscious ICU patient, while others need more focus on the person (assessment for depression for example).

IMPLICATIONS

Why do we need to hold PCC in balance with CCC? Why can't we maximize both PCC and CCC to provide maximal benefit to our patients? Health practitioners already have many cognitive tasks to complete during each visit, and tasks associated with operating and exchanging information with a computer may interfere with verbal communication

[53]. Emerging evidence suggests that even dissimilar tasks can have profound interference effects depending on the particular cognitive resources required to complete them [54]. For example, a procedure requiring considerable dexterity may be compromised by listening to a colleague simultaneously. Our healthcare systems demand that clinicians both contribute to and access data in real time while attempting to interact meaningfully with our patients. This conflict should motivate us to focus more energy on workload, cognition, interruption and multitasking as decision-makers have in aviation, combat, and policing [32]. Computer-associated silence could represent interference to communication. Such a relationship and any resulting health effects could be tested more easily if identification and measurement of computer-associated silences during clinical visits was automated.

Implications for researchers: While conducting this study, I relied partially on my own ears to verify silences after *Atlas.ti* identified them. It is not hard to imagine applications such as MATLAB that could be easily programmed with algorithms to sift large amounts of audio data. Besides identifying silences, these algorithms can reliably detect events such as mouse clicking and typing to further classify silences that occur in the exam room. It is quite feasible to design a mobile application that could be distributed directly to clinicians' cell phones across an entire health system to explore how PCC and CCC are balanced and correlate with health outcomes. Future research could focus on exploring the relationship between computer-associated silences and specific communication and health outcomes as well as clinician–patient relationship. This will require automated silence identification and classification. The answers may allow for tailored communication training for clinicians, cost effectiveness calculations for interventions such as scribes, and rapid quality improvement loops regarding EHR modifications.

Implications for clinicians and clinician training: Data management and CCC could be separated from PCC, and training could focus on the deliberate combination or separation of PCC from CCC as indicated depending on the clinical scenario. If alternations between patient-centered to CCC result in lower errors in both domains, training should focus on separating these tasks effectively. This is not to discourage attaining higher skills in both arenas; from taking advantage of keyboard shortcuts to learning how to add compassion onto empathy, but these findings reinforce others that suggest we should not assume that multitasking is either necessary nor sufficient for the highest clinical performance and efficiency.

Health systems: As we discover the cognitive flows that work best to care for patients at various levels of acuity, health systems may adjust their crew resource management allocations accordingly. It seems likely that in primary care, where relationship is so central to preventive medicine adherence [55], a clinician may be most successful with a high ratio of PCC to CCC while the patient is in the room. Therefore, the clinic may want to incentivize this attention on the patient by eliminating the pressure to complete documentation simultaneously. Alternatively, in more data-driven situations, hospitals may want to protect and incentivize time devoted purely to data extraction and synthesis.

Electronic Health Record: A lofty goal for clinical reminders and decision-support tools would be a context-aware system that anticipates the complexity of the data required in each situation. Sometimes interruptions and reminders are critical to patient safety while at other times they lead to errors in communication. Real time measurement of cognitive load required by specific work flows will enable programmers to modify these flows to augment the type of clinician attention that is desired in each situation.

DISCUSSION AND CONCLUSIONS

Ultimately the widespread and blind belief in multitasking may be hurting our patients as well as our medical workforce. Physician satisfaction and patient safety demand a careful reflection about the divide between PCC and CCC. Burnout is largely blamed on the EHR and documentation burden while highly effective practices have successfully separated some relationship tasks from the data-driven tasks [56]. Existing evidence does not support the use of multitasking in medicine because the quality of each task is decreased, and when those two tasks are PCC versus accurate data extraction and recording, our patients cannot afford to have either one compromised.

Clinicians are asked to multitask, but as we have seen in aviation, it may not be healthy for clinicians or their patients. What's actually happening in a lot of modern healthcare is CCC whether or not the patient is present. We know a lot about PCC but as of yet have few ways to measure cognitive demands on clinicians that span both emotional and technological task domains. Measuring and defining CCC is essential to healthcare quality because both PCC and CCC are valuable in their own contexts. One particular type of silence may provide a window into multitasking interference and allow us to compare and contrast PCC from CCC. If using audio to measure CCC can be reliably reproduced, a new awareness of the balance between these two major flavors of clinician activities may change clinical workflow, research methods, medical education, EHRs design, and human resource allocation throughout our healthcare systems.

Acknowledgments

The authors gratefully acknowledge the contributions of Beau Abar, Katherine Ciesinski, James Dolan, Mary Dombeck, Ronald Epstein, Kevin Fiscella, Marie Flannery, Robert Gramling, and Rachel Rodenbach to the research that was the basis of this chapter as well as support from the National Cancer Institute (Grant number: R01-140409-05, Ronald Epstein and Richard Kravitz, coprincipal investigators). The authors thank for the editing contributions of Liam Bartels, Nancy Gillard Bartels, Silas Bartels, Betsy Lombardi, and Sarah Porter.

References

[1] Loukopoulos LD, Dismukes RK, Barshi I. Cockpit interruptions and distractions: a line observation study. Proceedings of the 11th international symposium on aviation psychology. Columbus: Ohio State University Press; 2001.

[2] Stanton NA, Harris D, Starr A. The future flight deck: modelling dual, single and distributed crewing options. Appl Ergon 2016;53:331–42.

[3] Pearce C, Trumble S, Arnold M, Dwan K, Phillips C. Computers in the new consultation: within the first minute. Fam Pract June 2008;25(3):202–8.

[4] Assis-Hassid S, Heart T, Reychav I, Pliskin JS, Reis S. Enhancing patient-doctor-computer communication in primary care: towards measurement construction. Israel J Health Policy Res 2015;4(1):4.

[5] Alkureishi MA, Lee WW, Lyons M, Press VG, Imam S, Nkansah-Amankra A, et al. Impact of electronic medical record use on the patient–doctor relationship and communication: a systematic review. J Gen Intern Med 2016;31(5):548–60.
[6] Arar N, Wen L, McGrath J, Steinbach R, Pugh J. Communicating about medications during primary care outpatient visits: the role of electronic medical records. J Innov Health Inform 2005;13(1):13–21.
[7] Kumarapeli P, de Lusignan S. Using the computer in the clinical consultation; setting the stage, reviewing, recording, and taking actions: multi-channel video study. J Am Med Inform Assoc June 2013;20(e1):e67–75.
[8] Furukawa MF, King J, Patel V, Hsiao CJ, Adler-Milstein J, Jha AK. Despite substantial progress in EHR adoption, health information exchange and patient engagement remain low in office settings. Health Aff (Millwood) September 2014;33(9):1672–9.
[9] McFarlane DC, Latorella KA. The scope and importance of human interruption in human-computer interaction design. Hum Comput Interact 2002;17(1):1–61.
[10] Mancero G, Wong W, Loomes M. Change blindness and situation awareness in a police C2 environment. European conference on cognitive ergonomics: designing beyond the product—understanding activity and user experience in ubiquitous environments. VTT Technical Research Centre of Finland; 2009.
[11] Greatbatch D, Heath C, Campion P, Luff P. How do desk-top computers affect the doctor-patient interaction. Fam Pract 1995;12(1):32–6.
[12] Makoul G, Curry RH, Tang PC. The use of electronic medical records: communication patterns in outpatient encounters. J Am Med Inform Assoc November–December 2001;8(6):610–15.
[13] Booth N, Robinson P, Kohannejad J. Identification of high-quality consultation practice in primary care: the effects of computer use on doctor–patient rapport. Inform Prim Care 2004;12(2):75–83.
[14] Frankel R, Altschuler A, George S, Kinsman J, Jimison H, Robertson NR, et al. Effects of exam-room computing on clinician–patient communication. J Gen Intern Med 2005;20(8):677–82.
[15] Margalit RS, Roter D, Dunevant MA, Larson S, Reis S. Electronic medical record use and physician–patient communication: an observational study of Israeli primary care encounters. Patient Educ Couns 2006;61(1):134–41.
[16] Chan W, Stevenson M, McGlade K. Do general practitioners change how they use the computer during consultations with a significant psychological component? Int J Med Inf 2008;77(8):534–8.
[17] Pearce C, Arnold M, Phillips C, Trumble S, Dwan K. The patient and the computer in the primary care consultation. J Am Med Inform Assoc March–April 2011;18(2):138–42.
[18] Shachak A, Reis S. The impact of electronic medical records on patient–doctor communication during consultation: a narrative literature review. J Eval Clin Pract 2009;15(4):641–9.
[19] Noordman J, Verhaak P, van Beljouw I, van Dulmen S. Consulting room computers and their effect on general practitioner–patient communication. Fam Pract 2010 Dec;27(6):644–51.
[20] Dowell A, Stubbe M, Scott-Dowell K, Macdonald L, Dew K. Talking with the alien: interaction with computers in the GP consultation. Austr J Prim Health 2013;19(4):275–82.
[21] Street RL, Liu L, Farber NJ, Chen Y, Calvitti A, Zuest D, et al. Provider interaction with the electronic health record: the effects on patient-centered communication in medical encounters. Patient Educ Couns 2014;96(3):315–19.
[22] Brixey JJ, Robinson DJ, Johnson CW, Johnson TR, Turley JP, Zhang J. A concept analysis of the phenomenon interruption. Adv Nurs Sci 2007;30(1):E26–42.
[23] Epstein RM, Duberstein PR, Fenton JJ, Fiscella K, Hoerger M, Tancredi DJ, et al. Effect of a patient-centered communication intervention on oncologist-patient communication, quality of life, and health care utilization in advanced cancer: The VOICE randomized clinical trial. JAMA Oncol 2017;3(1):92–100.
[24] Frankel RM. EHR and physician–patient communication. Safety of health IT. Switzerland: Springer; 2016. p. 129–41.
[25] Booth N, Robinson P, Kohannejad J. Identification of high-quality consultation practice in primary care: the effects of computer use on doctor–patient rapport. J Innov Health Inform 2004;12(2):75–83.
[26] Ventres W, Kooienga S, Vuckovic N, Marlin R, Nygren P, Stewart V. Physicians, patients, and the electronic health record: an ethnographic analysis. Ann Fam Med March–April 2006;4(2):124–31.
[27] Rouf E, Whittle J, Lu N, Schwartz MD. Computers in the exam room: differences in physician–patient interaction may be due to physician experience. J Gen Intern Med 2007;22(1):43–8.
[28] Weibel N, Rick S, Emmenegger C, Ashfaq S, Calvitti A, Agha Z. Lab-in-a-box: semi-automatic tracking of activity in the medical office. Pers Ubiq Comput 2015;19(2):317–34.

[29] Gibbings-Isaac D, Iqbal M, Tahir MA, Kumarapeli P, de Lusignan S. The pattern of silent time in the clinical consultation: an observational multichannel video study. Fam Pract October 2012;29(5):616–21.
[30] Bartels J, Rodenbach R, Ciesinski K, Gramling R, Fiscella K, Epstein R. Eloquent silences: a musical and lexical analysis of conversation between oncologists and their patients. Patient Educ Couns 2016;99(10):1584–94.
[31] Salvucci DD, Taatgen NA. Threaded cognition: an integrated theory of concurrent multitasking. Psychol Rev 2008;115(1):101.
[32] Wickens CD. Multiple resources and performance prediction. Theoret Issue Ergon Sci 2002;3(2):159–77.
[33] Takeuchi N, Mori T, Suzukamo Y, Tanaka N, Izumi S. Parallel processing of cognitive and physical demands in left and right prefrontal cortices during smartphone use while walking. BMC Neurosci 2016;17(1):9.
[34] Reis S, Sagi D, Eisenberg O, Kuchnir Y, Azuri J, Shalev V, et al. The impact of residents' training in Electronic Medical Record (EMR) use on their competence: report of a pragmatic trial. Patient Educ Couns 2013;93(3):515–21.
[35] Dunn EJ, Mills PD, Neily J, Crittenden MD, Carmack AL, Bagian JP. Medical team training: applying crew resource management in the Veterans Health Administration. Jt Comm J Qual Patient Saf 2007;33(6):317–25.
[36] Grundgeiger T, Dekker S, Sanderson P, Brecknell B, Liu D, Aitken LM. Obstacles to research on the effects of interruptions in healthcare. BMJ Qual Saf June 2016;25(6):392–5.
[37] Green LA, Nease Jr D, Klinkman MS. Clinical reminders designed and implemented using cognitive and organizational science principles decrease reminder fatigue. J Am Board Fam Med May–June 2015;28 (3):351–9.
[38] Martyres G. On silence: a language for emotional experience. Aust N Z J Psychiatry March 1995;29(1):118–23.
[39] Frankel Z, Levitt HM, Murray DM, Greenberg LS, Angus L. Assessing silent processes in psychotherapy: an empirically derived categorization system and sampling strategy. Psychother Res 2006;16(5):627–38.
[40] Back AL, Bauer-Wu SM, Rushton CH, Halifax J. Compassionate silence in the patient–clinician encounter: a contemplative approach. J Palliat Med 2009;12(12):1113–17.
[41] Ephratt M. The functions of silence. J Pragmat 2008;40(11):1909–38.
[42] Ephratt M. Linguistic, paralinguistic and extralinguistic speech and silence. J Pragmat 2011;43(9):2286–307.
[43] Levitt HM. Sounds of silence in psychotherapy: the categorization of clients' pauses. Psychother Res 2001;11 (3):295–309.
[44] Sharpley CF, Munro DM, Elly MJ. Silence and rapport during initial interviews. Couns Psychol Q 2005;18 (2):149–59.
[45] Lane RC, Koetting MG, Bishop J. Silence as communication in psychodynamic psychotherapy. Clin Psychol Rev 2002;22(7):1091–104.
[46] Kurzon D. Discourse of silence. John Benjamins Publishing; 1998.
[47] Rowland-Morin PA, Carroll JG. Verbal communication skills and patient satisfaction: a study of doctor-patient interviews. Eval Health Prof 1990;13(2):168–85.
[48] Bruneau TJ. Communicative silences: forms and functions. J Commun 1973;23(1):17–46.
[49] Jensen JV. Communicative functions of silence. ETC: Rev Gen Semant 1973;30(3):249–57.
[50] Hoerger M, Epstein RM, Winters PC, Fiscella K, Duberstein PR, Gramling R, et al. Values and options in cancer care (VOICE): study design and rationale for a patient-centered communication and decision-making intervention for physicians, patients with advanced cancer, and their caregivers. BMC Cancer 2013;13(1):188.
[51] Ten Bosch L, Oostdijk N, Boves L. On temporal aspects of turn taking in conversational dialogues. Speech Commun 2005;47(1):80–6.
[52] Murphy EV. Clinical decision support: effectiveness in improving quality processes and clinical outcomes and factors that may influence success. Yale J Biol Med 2014;87(2):187–97.
[53] Beasley JW, Wetterneck TB, Temte J, Lapin JA, Smith P, Rivera-Rodriguez AJ, et al. Information chaos in primary care: implications for physician performance and patient safety. J Am Board Fam Med November–December 2011;24(6):745–51.
[54] Scott SK, McGettigan C, Eisner F. A little more conversation, a little less action—candidate roles for the motor cortex in speech perception. Nat Rev Neurosci 2009;10(4):295–302.
[55] Street Jr RL, Makoul G, Arora NK, Epstein RM. How does communication heal? Pathways linking clinician–patient communication to health outcomes. Patient Educ Couns 2009;74(3):295–301.
[56] Sinsky CA, Willard-Grace R, Schutzbank AM, Sinsky TA, Margolius D, Bodenheimer T. In search of joy in practice: a report of 23 high-functioning primary care practices. Ann Fam Med May–June 2013;11(3):272–8.

CHAPTER

3

Overcoming Health Disparities: The Need for Communication and Cultural Competency Training for Healthcare Providers Practicing Virtually in Rural Areas

Julie A. Lasslo[1], *Kathryn E. Anthony*[2], *Carrie E. Reif*[2] *and Nicolet H. Bell*[2]

[1]**Eastern Kentucky University, Richmond, KY, United States** [2]**University of Southern Mississippi, Hattiesburg, MS, United States**

TELEMEDICINE: AN ACCESS TO CARE INTERVENTION STRATEGY

The days of doctors making house calls are over, but well into the 20th century, it was common for health practitioners to traverse great distances to provide treatment to patients. However, healthcare has changed dramatically since the days of traveling physicians, and great disparities associated with access to medical care, particularly in rural areas, have become more prominent in recent decades. The number of hospitals and other health professionals in rural areas has dwindled significantly over the last 25 years. In the United States, one quarter of the population lives in rural areas where less than 10% of physicians practice [1].

Also, persons living in rural communities are more likely to experience greater levels of poverty and higher rates of chronic disease than those residing in urban centers. The dual burden of disease further complicates efforts to meet the healthcare needs of rural

Health Professionals' Education.
DOI: http://dx.doi.org/10.1016/B978-0-12-805362-1.00003-6

residents. Phillips and McLeroy clearly articulate the public health challenges inherent in rural healthcare:

> The delivery of public health services in rural areas faces daunting challenges, including low population density, transportation issues, lack of access to grant funding, lower public funding levels for rural services and programs, difficulties in recruiting staff, and potential fragmentation of scarce resources. This suggests that we cannot simply rescale public health programs and services from urban areas and expect them to be successful in rural areas. Rather, we need to consider alternative models for program delivery. [2, p. 1663]

As we will discuss below, one viable, and so far proven, alternative model for program delivery to rural communities is telemedicine.

Today's technological advances can now enable health practitioners to not only give much needed care to underserved populations remotely, but to provide that care with a substantial degree of interpersonal exchange. In the following sections, we will describe the ways in which telemedicine is a successful alternative model for providing care to rural communities.

Telemedicine is defined by the American Telemedicine Association as, "the use of medical information exchanged from one site to another via electronic communications to improve a patient's clinical health status" [3]. Telemedicine facilitates access to the medical community through text messaging, video chat, virtual house calls, monitoring of vital signs, and the transfer of medical information via the web [4–6]. Engaging with patients via smart phone, tablet, or computer effectively eliminates the geographic and transportation barriers that directly affect so many residing in rural areas. The field of telemedicine encompasses a myriad of disciplines including psychology, pharmacology, radiology, physician consultation, and education. No longer must a patient travel to a major city several hours away to see a specialist or obtain a second opinion. Through telemedicine, local doctors and their patients can connect with specialists anywhere in the world [7]. This not only decreases time and money spent traveling, but telemedicine necessitates the inclusion of the patients' local doctors in the decision-making process, promoting more patient advocacy and ultimately bettering patient outcomes [8].

Telemedicine can be both synchronous and asynchronous. Asynchronous telemedicine could be explained by sending an email or text to a physician or healthcare provider and waiting for his or her response. Text-messaging reminders from providers to patients could also be considered asynchronous telemedicine. Synchronous communication would more likely employ the use of teleconferencing technologies to facilitate more immediate communication between patients and their healthcare providers.

The practice of telemedicine is on the rise, particularly for treating individuals in rural communities both in low-income and high-income countries. It offers a wide array of advantages and applications that can increase healthcare efficiency, effectiveness, and reach. In terms of patient perceptions of telemedicine, a growing number of patients have expressed positive perceptions, including an increased perceived accessibility to care, the potential for efficiency provided by telemedicine, and the perceived convenience it ascribes to patients [9]. Additionally, patients who have received care via telemedicine have indicated high levels of satisfaction with their experiences. Agha et al. [10]

discovered that 98.3% of patients were satisfied with their telemedicine experiences and many actually preferred it to in-person consultations. The high satisfaction rates may be attributed to growing frustration with traditional healthcare services, including appointment scheduling, travel time, and patient involvement.

Telemedicine virtual house calls offer healthcare workers unprecedented access to patients while providing those in rural populations the necessary and more frequent contact with healthcare providers, which is inextricably linked to improved patient outcomes [11]. Communication channels in the form of text messages, phone calls, and email equip practitioners to efficiently care for patients dealing with chronic disease(s) on a consistent basis, often resulting in improved patient care, improved quality of life, and money saved by eliminating frequent emergency room visits [12,13]. Given the shrinking global technology gap and subsequent increase in the use of telemedicine, the opportunities for treating rural patients encumbered by geographical barriers, transportation challenges, and health literacy constraints continue to improve [14].

Traditionally, healthcare providers have interacted with patients in a face-to-face medical encounter to build relationships with patients. Establishing a productive patient-provider interaction entails the consideration of several factors, including the physical environment of the encounter, the communication styles and cultural backgrounds of both of the provider and patient, and the health literacy of the patient [14]. However, the introduction of telemedicine is dramatically reshaping the ways in which patients and their providers interact. For example, medical consultations are no longer confined to in-person office visits, but they can now occur synchronously and asynchronously through video conferencing, Web Chat, text messaging, email, or other applications.

Telemedicine is rapidly gaining acceptance in the medical community as a secure and easily accessible platform to provide healthcare. Consumers are also turning to telemedicine to address their health needs, especially those living in rural and remote areas [15]. It is estimated that the telemedicine market will grow to $1.9 billion in 2018 from $240 million in 2013, indicating an annual growth rate of 56% [16]. As telemedicine continues to become an integral part of healthcare services, physicians must adapt communication practices to meet the needs of their patients.

PERCEIVED COST ADVANTAGES OF TELEMEDICINE

Telemedicine offers an array of cost benefits to healthcare providers and patients. Telemedicine has been found to lower medical and pharmaceutical costs and reduce hospital admission [17]. Patients who use telemedicine have also experienced an increase in their healthcare benefits by $40 per visit, and reduced their loss from physician rationing by 20% [18]. Horn et al. [19] disclosed that treating rural Native American populations with telemedicine was the most cost-effective means of providing care to those communities. The scholars found that the implementation of telemedicine resulted in an average savings of 138.34 dollars per patient.

Smith [20] revealed that telemedicine consultations were able to reduce average travel time for rural patients in Australia by 50 minutes or more, and 96% of families in the study reported meaningful cost savings attributed to fuel, meal purchases, parking expenses, and time away from work. In some areas of Kenya, a trip to the physician can cost a day's wage or more in travel costs alone. However, telemedicine is relieving that burden [21]. The use of mental health synchronous teleconsultations has greatly reduced the need for patient travel among patients of U.S. Veterans Hospitals, producing a savings in travel reimbursements by as much as 63,000 annually in some cities [22]. Additionally Callahan et al. [23] attribute record hospital savings as the result of the implementation of telemedicine. The scholars noted that these savings occurred through the reduction of the number of hospital admissions and a decrease in the length of hospital stays.

TELEMEDICINE IN RURAL AREAS

Rural and remote areas, though encumbered by challenges that potentially inhibit the use of telemedicine, are uniquely positioned to benefit the most from its use. Mars [24] explains the dire need for telemedicine throughout Africa citing that 60% of its population lives in rural areas with limited access to healthcare. Consequently, living in a rural or remote area greatly reduces access to health care. For example, in some areas of the country of Ghana there are fewer than 1 doctor per every 93,000 people which undoubtedly effects health outcomes [25]. Larger, more urbanized cities, offer a plethora of well- paying employment opportunities for medical personnel. In rural areas, medical practitioners face lower pay and greater demands on their time [26]. Less-appealing employment opportunities means there are fewer medical professionals vying for employment in rural areas and because of this, the population that needs medical attention suffers.

The World Health Organization [27] (WHO) describes that for people living in rural and remote regions, ensuring them access to trained healthcare workers is one of the most complex challenges in achieving health for all. The shortage of healthcare workers in rural areas affects nearly every country, regardless of income level [27]. In the global policy briefing "*Increasing Access to Health Workers in Remote and Rural Areas Through Improved Retention,*" the WHO reveals the challenge of providing care to rural populations:

> Approximately one half of the global population lives in rural areas, but these are served by only 38% of the total nursing network force and by less than a quarter of the total physician workforce. The situation is especially dire in 57 countries where a critical shortage of trained health workers means an estimated one billion people have no access to essential health-care services. [27, p. 7]

The brief goes on to explain that the challenge in providing adequate care to individuals living in rural regions is present in both low-income and high-income nations. For instance, in Canada where much of the country is rural, only 9.3% of practitioners provide care to rural areas in which 24% of the population resides. And in some countries, like South Africa, 46% of the population lives in areas deemed to be rural. In that country, a mere 12% of physicians provide care to rural areas [27]. And while countries in Africa overall experience 24% of the global burden of disease, they only have access to roughly 3% of the world's healthcare providers [24]. Clearly, there is an international epidemic concerning the shortage of healthcare providers who practice in rural areas.

The negative health consequence resulting from a shortage of healthcare providers are further accentuated by the burden of disease in low-income nations. Not only do low-income nations experience a much greater rate of morbidity and mortality related to communicable disease, but they now face the double burden of rapidly increasing rates of noncommunicable disease [28]. Currently, 79% of deaths worldwide are caused by noncommunicable diseases, once a trend common in high-income nations now occurs with increasing frequency in low-income countries [29]. This double burden of communicable and noncommunicable diseases will only continue to tax the many fragile and inadequate healthcare systems globally. Despite this potentially dismal outlook for global health, telemedicine offers strategies that can help address these issues through increased efficiency, effectiveness, and reach of healthcare in diverse settings and among diverse cultures specifically in already underserved rural areas [30].

APPLICATION OF TELEMEDICINE IN LOW-INCOME AND MEDIUM-INCOME NATIONS

Although telemedicine is currently more widely used in nations with higher income, its implementation in low-income countries is on the rise and is already offering a wealth of evidence to support its proliferation. Recently, Kenya launched a national telemedicine initiative aimed at connecting rural dwelling patients with resources and expertise from the nation's leading hospital in Nairobi [31]. Sponsored by the pharmaceutical company, Merck, this program is designed to address issues related to lack of healthcare providers and to facilitate faster diagnoses of noncommunicable diseases like cancers and diabetes. Previous studies in Kenya have demonstrated success in telemedicine interventions aimed at connecting poor rural patients with nurses for medical consultation [21]. Scholars found that this telemedicine intervention was able to provide cost-effective quality care similar to what the patients experienced with in-person consultation [21].

In Minas Gerais, Brazil, a medium-income country, telemedicine services have become an integral piece of the healthcare system. In partnership with government and university entities, the telemedicine program uses simple low-cost technologies to reach rural communities with medical services including tele-echocardiography. In 2012, the telemedicine program reported high patient satisfaction scores, a significant economic return on investment, and in that year, a record milestone of 1 million tele-echocardiograms were completed [32].

In India, the use of telemedicine is exploding. The government sponsored Development of Telemedicine Technology (DTT) project funds a myriad of telemedicine initiatives. One of the DTT initiatives aims to equip the nation with a telemedicine network through several phases [33,34]. This government initiative involves local physicians and medical staff in the design process to ensure rural community needs are addressed. The programs equip rural and remote medical facilities with simple streamlined technology that allows for synchronous and asynchronous communication with specialists in the major city hospitals specifically regarding radiology, pathology, and cardiology. As India's population continues to expand and its physician density decreases, government telemedicine initiative in the private and public sector will necessarily proliferate. Government telemedicine initiative in the private and public sector will necessarily proliferate.

APPLICATION OF TELEMEDICINE IN HIGH-INCOME NATIONS

There have also been several exemplars of the application of telemedicine within high-income nations. Telemedicine has been employed in an effort to level the playing field between the medical care available in urban centers versus medical care accessibility in rural communities. For example, many Native Americans in the United States face major health disparities due to the geographic barriers of their reservations [35]. Native Americans living in rural areas face higher rates of chronic diseases, including diabetes and cardiovascular disease, communicable diseases, including tuberculosis, and substance abuse.[35] However, employing telemedicine has enabled Native Americans to get more consistent access to care, as well as more affordable care. For example, the implementation of telemedicine has resulted in the decline of both the tuberculosis death rate and the infant mortality rate among Native Americans [36]. Additionally, Sequist et al. [36] argue that the implementation of telemedicine has improved the overall life expectancy of those communities.

The implementation of telemedicine in Native American communities has also revealed economic benefits. For instance, for every one-dollar spent on telemedicine, an individual patient's expenses are reduced by an estimated $11.50. Additionally, the general health care expenses for the Native American population could be decreased by $36 billion per year with the employment of telemedicine initiatives [37]. Telemedicine not only increases access to providers by decreasing patient travel and cost, but it also reduces the demand for extensive travel and costs for providers [38].

Other high-income countries have also implemented successful telemedicine initiatives [39,40]. For example, in a pilot study conducted with multiple sclerosis (MS) patients in Haifa, Israel, at the Carmel Medical Center for MS, patients showed a decrease in symptom severity, resulting in increased positive clinical outcomes [41]. Additionally, 67% of the patients experienced at least a 35% decrease in healthcare costs for their treatment [41]. Spinal cord injury patients in Belgium, Italy, and England have also benefited from the inclusion of teletherapy as part of their care [39]. In fact, the group initially exposed to the telemedicine intervention showed increased fine motor scores over the control group [39].

In Japan, a telemedicine program connects 14 regional cancer centers to the main cancer center at the national hospital, which allows teleconferences, high-resolution image sharing, remote consultation, and regular medical conferences between the main clinic and the regional clinics throughout their country [40]. This provides a nationwide streamlined approach to cancer education and treatment. Moreover, this network also allows for international teleconference projects where the network in Japan has connected with Duke University, Georgetown University, and Singapore University for high definition telemedicine conferences [40].

Beyond the physical and economic benefits of these programs, several studies report that the quality of healthcare remained constant, or even increased, during the implementation of telemedicine [41,42]. For example, when counseling Native American cancer survivors in Alaska via telecommunication, researchers reported effective results and satisfaction ratings as high as 4.59 out of a possible 5 points [38]. Patients also appreciated the convenience accrued to them because of the care they received via telemedicine. Asynchronous telecommunication, like email or health messaging systems, allowed flexibility and little time commitment on the part of the Native American patient and provider yielding patient satisfaction with the quality of the interaction [36,43]. Similarly, when spinal cord injury patients in Belgium, Italy, and England received telemedicine as part of their follow-up care, researchers found an overall higher satisfaction with telemedicine as opposed to traditional care across all three countries [39]. Finally, MS patients in Israel reported high satisfaction with telemedicine treatment, and the patients reported that they would recommend the services to others [41]. Asynchronous telecommunication, like email or health messaging systems, allowed flexibility and little time commitment on the part of the Native American patient and provider yielding patient satisfaction with the quality of the interaction [36,43]. Thus, telemedicine provides economic, geographic, and overall health benefits to various populations within higher income nations across the globe.

CHALLENGES IN THE DIFFUSION OF TELEMEDICINE IN RURAL POPULATIONS

Despite the clear advantages of employing telemedicine in both low and high-income nations for improving access to healthcare for rural communities, there are some challenges that exist in the diffusion of telemedicine to these rural areas. Thankfully, Bagchi [44] contends over 90% of patients needing treatment do not need to be touched to receive adequate care. In other words, if physical contact is not necessary, telemedicine can adequately facilitate the patient-provider medical encounter. However, low technical literacy among patients and a lack of technical expertise among providers as well as concerns of privacy and liability can significantly jeopardize the ability to adequately employ telemedicine strategies [45]. In addition to these human and technological interaction concerns, research suggest that many patients and providers are resistant to using these technologies because of perceived costs and also because of the potential for the legal concerns associated with the transmission of sensitive information. Elderly patients, in particular, have expressed concern for the overall security of their information when using telemedicine [46].

Technological infrastructure. Currently, telemedicine is practiced to a much greater degree in high-income countries. There are many difficult challenges to the diffusion of telemedicine in a large section of the world [47]. Unreliable and slow bandwidth connection can impact the utility of telemedicine in a variety of ways including poor imaging for radiological consults and an inability to video conference [48,49]. Obtaining clinical opinions locally and globally can be greatly hindered if image resolution is compromised by low bandwidth [50,51]. Furthermore, even when basic technology like a computer, Internet connection, adequate bandwidth, and electricity are available, the software maintenance and other technological glitches can stall the continued use of telemedicine in rural areas [44]. These significant differences in technology infrastructure between high- and low-income countries continue to impede the adoption of telemedicine practices globally [47].

Patient technological literacy. Even in areas where technological infrastructure is available, technical literacy among patients proves to be an issue. If a person has low self-efficacy in regard to technology they are far less likely to utilize it. Training in the use of technology is essential to the success of telemedicine and is one of the barriers to its use in rural and poverty-stricken areas. Masucci et al. [52] revealed that basic computer and Internet skills were the best predictors of success for telemedicine use among study participants. Encouragingly, this same study demonstrated that despite having little or no technological skills participants were able to effectively learn to use a cardio monitoring telemedicine system after attending only one two hour workshop [52].

Beyond technical literacy, another barrier to telemedicine has been described by Bagchi [44] as "technophobia," or a fear of the unknown when handling computers and other equipment. Bhatta et al. [53] articulated that both Nepalese healthcare providers and patients experienced some technophobia when trying to use telemedicine because they were uncomfortable with the technology. This research demonstrates that the proliferation of telemedicine requires that patients must be equipped with technological skills to build self-efficacy and eliminate technophobic tendencies. Further, it is paramount that easy to use and affordable technologies be provided to target populations to ensure adoption.

Perception of cost. Perception of cost is another barrier to the adoption of telemedicine practices for providers and patients alike. However, research suggest that possible setup costs are tempered by savings attributed to reduced length of hospital stays, decreases in the amount of travel, and overall improved patient outcomes [5]. For example, Marcin et al. [54] revealed that in an intervention with pediatric patients, telemedicine was cost saving for hospitals who relied on the technology rather than transporting patients who did not necessitate transport to another care facility. Utidjian and Abramson [55, p. 372] second this notion by stating, "a direct financial benefit of telemedicine's ability to overcome the distance between patients and providers is the avoidance of the costs of transporting patients between health care facilities."

Additionally, as the technology gap shrinks and the adoption of handheld technologies like smart phones increases, so does the ease and cost effectiveness of telemedicine interventions. The use of smart phones and other handheld devices creates a medium for telemedicine medicine known as Mhealth, or mobile health. Patients who are already familiar with these technological devices need not purchase a new device or attend any trainings. Telemedicine has been shown to be advantageous in cost-benefit analysis and may prove to be more cost effective than traditional medical care [56,57]. For instance, Ifejika et al. [58] discovered that weight loss interventions for stroke patients implemented by telemedicine via smart phones not only resulted in significant weight loss for participants, but the use of smart phones proved to be a cost effective approach to care.

Legal concerns. As with most technological advancements, telemedicine is not without its legal concerns and challenges. Telemedicine faces several pragmatic concerns, including a "lack of reimbursement, tedious licensing and credentialing requirements, and concerns about security." [7, p.1092] Transmitting confidential patient information virtually around the globe carries with it the inherent risk of violating privacy laws. Dart et al. [59] describe the legal concerns of privacy and confidentiality associated with telemedicine. The scholars describe that not only must providers be aware of the state and federal laws related to Health Insurance Portability and Accountability Act (HIPPA), but they must also be familiar with Family Educational Rights and Privacy Act (FERPA) laws. For instance, it is nearly impossible for providers to guarantee the security of a text message while it is equally difficult to ensure the confidentiality of an email [59].

Additionally, because telemedicine affords cross-state and cross-country consultation, it begs the question of who has jurisdiction over the patient and who answers to litigious inquiries. Furthermore, the utilization of telemedicine initiates a host of complex issues such as whether or not medical professionals require extra licensing to cover multiple states and countries. At current, these concerns are insufficiently addressed [60,61]. Additionally, Dart et al. [59] describe how very few laws and legal precedents concerning the storage of private medical content on electronic servers exist. These various sources of uncertainty may impact the decision to adopt telemedicine practices.

Resistance to a care model different from the traditional care model. Employing telemedicine often involves a consult with a physician from outside of the patient's state or country. For some, this consultation with someone outside of their culture or country is perceived as losing control or even as antinationalism [5]. Furthermore, resistance to telemedicine involving practitioners outside of a patient's community is exacerbated by sociocultural difference that may not be well understood by providers, specifically providers in higher income nations [33,49,62]. Alverson [62] suggests that a telemedicine system should seek to blend into each community's existing systems while taking into account the various cultural perspectives and needs of each area. Additionally, because patients fear that patient-provider encounters will be far more impersonal than face-to-face interactions, providers must attempt to put patients at ease despite the potential geographic distance between them. Encounters will be far more impersonal than face-to-face interactions, providers must attempt to put patients at ease despite the potential geographic distance between them.

Clearly, some perceptions and barriers have inhibited the diffusion of telemedicine in certain rural areas. Because of concerns associated with telemedicine, including securing data, provider capability with technology, fear of impersonal interactions, and Western providers treating individuals in rural communities without sufficient knowledge of the culture in those regions, we advocate for extensive training programs for telemedicine providers in both cultural competency and interpersonal communication skills for engaging more effectively in patient-provider interactions. In the following sections, we will discuss the specific need for greater training among telemedicine providers particularly concerning cultural competency and patient-provider communication skills.

THE NECESSITY OF TRAINING FOR TELEMEDICINE PROVIDERS IN RURAL COMMUNITIES

Training is necessary for telemedicine providers for a myriad of reasons. First, from a technology perspective, some telemedicine providers have reported the frustrations that can accompany the addition or increased use of technology in their medical practices [63]. Additionally, healthcare providers need to have a sufficient understanding of the cultures of the rural communities they serve. Finally, providers need to be more well skilled in effective patient-provider communication. The Healthy People 2020 objectives emphasize the need to "improve the health literacy of the population" and "increase the proportion of persons who report that their healthcare providers have satisfactory communication skills." [64] For healthcare practitioners practicing with rural communities via telemedicine, effective training must necessarily include (1) technological training with providers; (2) cultural competency training for providers working with patients in rural, underserved areas; and (3) communication skills for effectively communicating with patients. The remaining focus of this chapter revolves around cultural competency and patient-provider communication training for practitioners. The following sections will explore these topics in-depth.

CULTURAL COMPETENCY IN THE VIRTUAL HEALTHCARE SETTING

The benefits of technologies like telemedicine and text message outreach are well documented [13,65]. However, despite the sophistication of many of these technologic innovations, they will be of little help to providers who do not appreciate or understand certain cultures, including the beliefs, values, and rituals of the communities they serve. Agha et al. [10] suggest that patients' perceived satisfaction with the care received via telemedicine may directly affect adherence to provider recommendations and future use of telemedicine. Healthcare provider education must focus on training providers to be culturally competent in order to be effective among underserved populations.

According to the Office of Minority Health [66], cultural competency is defined as "a set of congruent behaviors, attitudes, and policies that come together in a system, agency, or among professionals that enables work in cross-cultural situations." This demand makes cultural competency training for providers who practice medicine virtually of particular importance for communities in rural regions that have a history of distrust of the medical community [67] and lower health literacy levels than persons residing in urban areas [68,69].

For example, in the rural Appalachian region of Kentucky the sense of "place" is paramount in understanding the relationship between the population and the medical community [70]. This area is plagued with staggeringly high levels of health-related disparities including poor oral health [71], high levels of cardiovascular disease [72], high levels of poverty [73] low levels of education, low levels of health literacy [74], and few primary care providers [1]. Furthermore, over the last few decades, this area has experienced a demographic shift as young adults move away from the region looking for better opportunities, leaving behind culturally proud but chronically ill older patients [73]. Poor communication from providers, a general distrust of the medical community, and fear of exploitation all negatively affect the region's engagement with the healthcare system historically and in present day [67]. Practitioners who are to be successful in this area must understand the multifaceted aspects of the culture and the conceptualization of health within the region.

Despite a low population density, isolated rural communities like those in Appalachia manage to "pull together" in hard times [75]. This "pulling together" is an important sociocultural component that aids in caring for individuals and families but one that has also been shown to negatively affect health-seeking behavior as residents report they get most of their health knowledge from neighbors and friends rather than health professionals. Corresponding with this idea of information sharing and "pulling together" is the understanding that isolated rural communities "take care of their own" and are often reluctant to take charity and outside aid [70]. This greatly affects attitudes toward healthcare professionals and necessitates that providers recognize and take care to build trust and respect [70].

Howitt et al. [76] remind us that even when technologies and human resources are made available the difficult task of persuading people to adopt a new way of doing things remains a major barrier to change. Incorporating telemedicine technologies in rural communities must begin with a firm understanding of the local culture and proceed by educating the users of its benefits while engaging and giving answers to their concerns [76]. Moreover, the WHO identifies misunderstandings of sociocultural differences as a core challenge to the effectiveness of care given by practitioners from outside of the service community [29]. Though expert advice is often necessarily provided by practitioners outside of the service area, it is of paramount importance that those with local knowledge be consulted as well. When cultural subsystems between the service and provider sites do not align, meaningful patient-provider interaction is hindered. A provider must be attuned to the local cultural context to successfully reach the patient with care especially in the virtual setting.

Several examples exist concerning the training of providers in cultural competency. For instance, Patel et al. [77] trained healthcare providers in the United Kingdom who were working with diabetes patients of South Asian descent. The patients had misconceptions surrounding insulin use, including great pain associated with insulin use and a stigma associated with needles and drug use. Part of the training included educating providers on the perceived barriers within the community and helping them understand how to better overcome those barriers with their patients through the use of DVDs. Schouten et al. [78] also acknowledged a greater need for attention to the cultural competency of providers toward their patients. The scholars implemented a training program in the Netherlands among providers working closely with minority populations. In the program, practitioners were asked to consider their own personal cultural norms and biases and consider their sensitivity to the norms of other cultures. Following the training, the medical encounters typically lasted longer, and physicians typically engaged in more utterances with their patients.

Patient Health Literacy The world health organization defines health literacy as "the cognitive and social skills which determine the motivation and ability of individuals to gain access to, understand, and use information in ways which promote and maintain good health." [79] Having a proficient level of health literacy is an important component to maintaining one's overall health. Unfortunately, low health literacy is commonplace in both high-income and low-income countries [80]. It is closely linked with low health outcomes, poor disease management, increased hospitalizations, and overall diminished quality in patient health [81]. In fact, the American Medical Association (AMA) revealed that health literacy level is a stronger indicator of mortality risk than education level, socioeconomic status, race, or employment status [82]. According to the Institutes of Medicine (IOM), older adults, minorities, people with low education levels, and the poor are more likely to have low-literacy levels and subsequent health management struggles [83].

Health literacy is not only an issue for those receiving health information but also for those disseminating health information. Rudd [84] describes poor health literacy as a systemic issue. Health literacy is a shared burden between patients and providers, and much of the burden sides with the provider to adequately ensure patient understanding [85]. Improving health literacy must involve building trust between the patient and health professional. Additionally, providers must encourage patients to have ownership over their health and empower patients to take control over factors and barriers that influence their health [86]. Practitioners must be mindful of the health literacy challenges of their patients and adapt their approach accordingly during the medical interaction [87]. A necessary approach to treating patients with low levels of health literacy involves tailoring messages to fit the patient's cultural context. Cultural context shapes the way individuals interact with health information, including the beliefs and values of their community, and the ways in which these beliefs affect how people communicate, understand, and respond to health information [88]. This is especially important for providers treating patients from diverse cultural backgrounds through telemedicine.

The National Patient Safety Counsel outlines simple steps to take to improve patient-provider interactions. The Ask Me 3 [89] tool is "intended to help patients become more active members of their health care team, and provide a critical platform to improve communications between patients, families, and health care professionals." Ask Me 3 encourages providers to help their patients understand the answer to three key questions: (1) What is my main health problem? (2) What do I need to do? (3) Why is it important for me to do this? [89]. Along with these three key questions, practitioners should use plain language and avoid unnecessary details to avoid confusion. Additionally, practitioners are encouraged to avoid technical jargon and uncommon acronyms. The use of pictures and other media is encouraged to aid in patient understanding as these tools often convey instructions better than words [90]. This approach will assist providers and patients in improving health literacy skills.

Culture-centered approach to training. Although cultural competency training is essential for the success of telemedicine providers, training programs should be created by enlisting the help of actual members of the community. Jana et al. [91] argue for the importance of participatory communication when interacting with groups, like many living in rural communities, that face specific health disparities and marginalization [92]. Participatory communication is defined as "an approach based on dialogue, which allows the sharing of information, perceptions, and opinions among the various stakeholders and thereby facilitates their empowerment." [93, p. 3] According to the culture-centered approach, scholars should engage in extensive dialogue with members of a population to deepen their understandings of the health disparities of the community. Additionally, researchers should engage participant voices to more adequately understand the community's cultural understandings of health. In other words, researchers must involve participants in generating ideas for promoting health in their community, for engendering behavior change, implementing initiatives for behavior change, and observing the resulting benefits of change within the community [91]. Thus, participants are encouraged to define their health problems and assist in generating solutions, thereby improving behavioral change and improving health conditions [94].

The culture-centered approach to health communication provides members of communities facing health disparities the opportunity to act as active participants in social change instead of as passive recipients to health communication interventions [95]. Dutta and Basnyat [95] claim that a culture-centered approach to health communication must include the voice of the community, or unheard narratives from the population, and the cultural context of the community. Cultural context includes religion, structure, interpersonal communication customs, and cultural communicative practices. Failing to include the voices of the community, particularly in low-income regions of the worlds, results in "erasures of the voices from the margins are tied to the continuing disenfranchisement of the margins through top-down programs that are often out of touch with the lived experiences of the marginalized." [96, p.160]

Thankfully, there exist a few exemplar cases focusing on the education of telemedicine providers in both technological skills and cultural competency for the rural communities they serve. For instance, Gifford et al. [97] conducted a telemedicine training for practitioners working with rural communities in Alaska. In their program, the scholars trained providers to use technology more proficiently while also employing a participatory communication approach to train the providers on the culture of the rural Alaskan communities they treat. Following the training, the healthcare providers perceived the program to be extremely effective in assisting them with technology demands of telemedicine, responding more effectively to patients, and overwhelmingly, they exhibited a greater cultural competency for the rural Alaskan communities. Similarly, Nelson et al. [98] also conducted telemedicine training with providers practicing with rural Alaskan residents. Their program also addressed technological concerns and a cultural competency component. Like the training program initiated by Gifford et al. [97] the program included a community elder to educate telemedicine practitioners on the health practices and cultural understandings of health throughout the region. Practitioners reported that the presence of the elder was the most helpful part of the training seminar (Table 3.1).

It is obvious that community members should be included in order to voice their understandings of health and then participate in ways that help healthcare practitioners to better

TABLE 3.1 Cultural Training for Healthcare Providers Using Telemedicine

Approaches to Participatory Communication	Citations
Rather than assuming the beliefs and understanding of a culture regarding health and illness, community members must help inform outsiders of the community's health-related beliefs and understandings.	Niles et al. [97]
Involve community members, particularly respected community leaders or elders, to also assist researchers and practitioners in developing educational tools for health care providers.	Nelson et al. [98]
Culture-centered approach: Employing practices like photovoice, which encourages community members to use photography to reveal pressing health concerns within the community.	Dutta and Lahiri [99]

grasp the understandings of health within the community. Dutta and Lahiri [99] recommends the implementation photovoice, or an approach that "empowers participants to take photographs that reflect complex community health and disability issues, and then the photos are used as a source of critical dialogue on the research topic." [100, p. 296] Only through listening to the voices within the community will cultural barriers and understandings of health be adequately conveyed to healthcare providers.

COMMUNICATION TRAINING SPECIFIC TO VIRTUAL HEALTHCARE PROVIDERS

Beyond understanding the cultural beliefs surrounding health practices within rural communities, health care providers must also possess the capacity for effective patient-provider communication. Particularly with telemedicine, in which providers are unable to connect with their patients through physical touch, other aspects of the provider's communication skills should be sharpened to ensure adequate care. Ultimately, the better the patient-provider communication, the more likely that shared meaning will be obtained within the interaction.

PATIENT-CENTERED CARE AND PATIENT-PROVIDER COMMUNICATION

Patient-provider communication plays an enormous role in a patient's path to healing and overall health. Effective patient-provider communication enable physicians to better "facilitate accurate diagnosis, counsel appropriately, give therapeutic instructions, and establish caring relationships with patients." [101] Street [102] articulates that improved patient-provider relationships often result in immediate improvements in health (i.e., reduce anxiety over medical concerns) and ultimately a greater psychological and physiological health for patients, even if mediated through other variables.

Barry and Edgman-Levitan [103] found that a major shift within the field of medicine toward patient-centered care has occurred. According to the IOM [104], patient-centered care is defined as "care that is respectful of and responsive to individual patient preferences, needs, and values" and care in which patients perceive their values to be of the utmost importance in clinical decision-making. A patient-centered approach is an individualized care delivery process that privileges a strong interpersonal physician–patient relationship. Harter and Bochner [105] articulate that providers must both elicit and understand the perspectives of their patients and be understanding and accepting of the psychosocial and cultural contexts from which their patients come. Physicians should also strive to understand patients' illness narratives and experiences.

The necessity of patient-centered care cannot be overstated. Providers striving to practice patient-centered medicine who exhibit effective communication skills are perceived to deliver a higher quality of care than providers who do not [69]· Further, patients' perceptions of a strong interpersonal relationship with their provider may increase their ability to cope with health issues, reduce their illness symptoms, and lower overall referral rates to specialized healthcare providers [106]. Patients have also reported feeling more engaged in the decision-making process, more satisfied with their care [107], and more likely to adhere to recommended medical treatment when receiving patient-centered care [108]. A patient-centered approach has been linked to a higher psychological quality of life and reduced symptoms of depression and anxiety [17]. Finally, a strong physician–patient relationship positively shapes the patients' evaluations of their treatment and reduces the number of medical malpractice lawsuits [109].

However, care that is not patient-centered may severely impede care. Conversely, poor patient-provider communication practices may strain physician–patient relationships and impede patient care. Ineffective communication has been linked to patients feeling misunderstood, insignificant, and rejected by their physician [110]. Patients have also reported increased uncertainty and anxiety surrounding the diagnosis and nature of their condition when they perceived communication from their providers to be ineffective [10]. Poor patient-physician communication has been found to reduce up to 40% of patients' adherence to recommended treatment regimens [110]. Additionally, primary care doctors who do not engage in patient-centered care are more likely to increase patients' utilization of unnecessary health care services and medical expenses [111].

Although the medical community acknowledges the importance of physician–patient communication, medical educators spend a majority of classroom time teaching biomedical information and little time on communication skills [112]. A lack of formal communication training may further complicate the use of telemedicine among rural populations. While physicians who practice the traditional model of care may be able to invoke interpersonal touch to put patients more at ease, providers in telemedicine are unable to do so. Further, providers practicing remotely may live a great distance from the patient, and regional and cultural differences may make the interaction much more difficult than normal (as discussed above). Thus, communication training for providers who practice virtually is necessary to ensure that they are providing patients with the best care possible.

COMMUNICATION TRAINING FOR VIRTUAL PROVIDERS

Communication skills training has been found to improve physicians' ability to interact with patients during face-to-face consultations. Roa et al. [113] describe how physicians trained in effective communicative practices tend to exhibit more behaviors considered to be patient-centered, including asking more open-ended questions, eliciting patients' concerns, expressing empathy and reassurance, and providing a better explanation of diagnosis and treatment options. Following communication training, physicians are also more likely to integrate these skills within their clinical practices up to 15 months following training interventions [114]. Communication training has also been linked to improved patient satisfaction and increased adherence to medical treatment. Because a strong physician–patient relationship is vital for patients to receive quality care, physicians need to receive additional communication training for not only face-to-face interactions, but also for telemedicine consultations.

In training providers, Brown and Bylund [115] state that both communication skills and process skills should be taught. For instance, the scholars provide the example of delivering bad news to patients. They claim that the communication skills needed might include acknowledging that the scenario is difficult and would likely be difficult for anyone, and they also recommend encouraging patients to express their emotions. Additionally, some of the process tasks might include maintaining eye contact with patients during the delivery of bad news, asking them what questions they have, offering expressions of concern, and allowing patients time to process the news.

In the following sections, we describe several areas in which telemedicine providers greatly improve their communication with patients through training. We will discuss the necessary communication for providers, including an emphasis on empathetic communication, motivational interviewing, shared decision-making, and ensuring patient understanding through teach-back methods.

Empathetic communication. Empathy is a powerful force within the medical encounter as it can greatly enhance the patient-provider relationship. Neumman et al. [116, p. 63] define physician empathy as the "socio-emotional competence of a physician to be able to understand the patient's situation, perspective and feelings, to communicate that understanding and check its accuracy, and to act on that understanding with the patient in a helpful (therapeutic) way." Some of the behaviors considered to be empathetic by patients include perceiving that their provider listens to their concerns, allows them to share their concerns, makes them feel comfortable, attempts to get to know them, and remains positive throughout the encounter [117].

Flickinger et al. [118] revealed that providers who exhibited more empathetic communication with their patients were more likely to experience increased patient interaction and observe improved self-management among patients living with HIV. Further, patients who feel their physicians display empathy perceive their overall care to be better [119]. Bayne et al. [120] revealed that when physicians engaged in more empathetic communication with their patients, patients were more likely to participate in their own care and were far more likely to seek clarification from providers if something was unclear. Additionally, patients tend to adhere to the treatment recommendations of the provider if they perceive the provider to be empathetic. Finally, patients are less likely to file malpractice suits against practitioners they perceive to be empathetic. Parkin et al. [121] found that patients are more likely to agree with their clinicians' recommendations for care if patients perceive providers to be empathetic.

Teaching providers to be more capable in trying to understand the background and needs of their patients, and training them to be more effective in conveying empathy in their nonverbal and verbal communication could enhance the patients' overall perception of the encounter. For instance, Brugel et al. [122] revealed that when providers looked directly at their patients when speaking to them and when their body orientation directly faced their patients, patients perceived their providers to be more empathetic to their needs. Bonvicini et al. [123] found that after engaging in an empathy-training program with providers, trained providers increased in their expressions of empathy. Additionally, those trained practitioners were more likely to acknowledge and appropriately respond when patients expressed emotion or challenged the physician by encouraging patients to express their feelings further.

Specifically, Pehrson et al. [124] developed and implemented an empathy-training program that involved four strategies for engaging in empathetic communication with patients. First, practitioners should acknowledge the emotion their patients are attempting to express. For this strategy, providers should pay attention to both the verbal and nonverbal behavior of their patients. Second, providers should attempt to understand patients' emotions by asking open-ended questions and asking for clarification when necessary. During this phase, physicians should refrain from offering any "premature reassurance" to patients. Third, providers should attempt to empathetically respond to the emotions and feelings of patients by acknowledging their situations, validating their concerns or fears, and praising them for the ways in which they have positively managed their situations, and encouraging patients to express their emotions and feelings. Pehrson et al. [124] claim that this strategy is at the heart of the training curriculum. Fourth, the provider should attempt to assist patients in coping with situations and attempt to facilitate social support for patients. Providers can do this by exhibiting a willingness to help their patients and by referring them to other healthcare providers.

Additionally, telemedicine providers could enhance their effectiveness by paying particular attention to their nonverbal communication. Mast [125] states that providers should be better trained in nonverbal communication, and in particular, they should become more aware of their own nonverbal behaviors in interactions with patients. Hannawa [126] revealed a major difference in patient satisfaction among patients treated by providers who displayed appropriate and inappropriate levels of nonverbal engagement when disclosing medical errors to patients. Patients treated by physicians not displaying appropriate levels of nonverbal communication were far less satisfied and more willing to find a new doctor. Providers modeling appropriate nonverbal involvement leaned forward when speaking to and listening to patients, exhibited facial and vocal expressiveness, displayed altercentrism (or affirming head nods and consistent eye contact), and competently maintained the conversation (few interruptions, taking conversational turns). These providers were more likely to be perceived as truly remorseful and trustworthy than providers not adequately engaged nonverbally in the conversation.

Motivational interviewing. In taking a patient-centered approach to obtaining information in the medical encounter, healthcare providers have begun to employ motivational interviewing (MI). MI has been implemented in the treatment of a wide array of conditions, including counseling HIV positive patients [127], treating patients with diabetes [128], smoking cessation programs [129], and medication adherence programs for children with asthma [130]. Rather than a physician or practitioner asking impersonal, closed-ended questions or being "preachy" with patients, motivational interviewing encourages the providers to engage in "open-ended questions, reflective listening, and support for patient autonomy and self-efficacy." [131 p945] Spencer and Wheeler [132, p. 1100] claim that "motivational interviewing encourages reflective listening to help the subject explore their own goals and motivations for change." The scholars claim that this strategy also encourages patients to participate more actively in their healthcare by helping develop solutions to changing their health behaviors and reaching their desired health goals.

While the behaviors implicit in motivational interviewing may not come naturally to all individuals, these communication skills can be taught to clinicians. Pollak et al. [133] provided communication coaching to family medicine practitioners and pediatricians to train them to become more effective in their MI skills. The scholars presented the clinicians with a 1-hour training session, and then provided feedback to the providers based on the sessions. The providers' MI skills were evaluated based on 10 actual sessions with patients. Following the training, the healthcare providers reported feeling more competent when interviewing patients. The providers reported they were less likely to feel burnout and were more likely to feel a greater sense of cohesion within their healthcare team. Additionally, patients reported higher levels of patient satisfaction after meeting with practitioners who participated in the training. Similarly, following their motivational interviewing intervention with providers, Edwards et al. [134] revealed they were able to improve practitioners' knowledge of MI, their skills, and also their confidence to effectively engage in MI. However, Fu et al. [135] warn that one training session is insufficient for training providers, and providers need the opportunity to learn such skills through multiple training sessions, input from MI experts, and practice and feedback with patients. Therefore, with sufficient exposure to training, practitioners' MI skills can undoubtedly be improved.

Teaching MI skills to telemedicine providers is extremely important. Beyond teaching the nonverbal skills to engage in active listening with patients and to not interrupt patients when they are trying to answer, providers also need to be taught the importance of promoting patient self-efficacy. For instance, Jerant et al. [136] developed a training seminar to teach physicians to promote more self-efficacy within their patients during medical interviews. The scholars recommend that once physicians have listened thoroughly to the concerns of their patients and have negotiated the first steps to take toward behavior change, providers should apply "confidence-boosting techniques" including discussing the patients' past successes with behavior change, encouraging patients to reframe previous behavior change failures as learning opportunities, and to discuss successful behavior change outcomes of others. Further, the scholars recommend that providers should convey this information to patients in an upbeat tone.

Additionally, language is an important consideration in MI. Hesson et al. [137] revealed that when practitioners asked more open-ended questions, patients were more likely to have increased participation in the interaction. Additionally, the scholars advocate that in patient-centered interviewing, once practitioners welcome patients to the interaction and solicit from the patient what the agenda will be for the encounter, providers should engage in open-ended questioning of patients. While thoroughly listening to patients, providers should also be closely monitoring patients' body language as they listen to their concerns.

Finally, Oh and Lee [138] acknowledge that while MI has had great success with patients in more western healthcare settings, far less research has been conducted to assess its effectiveness in other cultures. The scholars describe that while successful modifications for MI have been made for behavioral interventions with Hispanic and Native American populations, providers must understand the culture intimately of the group.

Shared decision-making. Shared decision-making is a critical component in providing patient-centered care. Charles et al. [139] define shared decision-making as an interactive process between physicians and patients, where both parties exchange health information, discuss treatment options, and reach an agreement on a treatment plan. In this decision-making model, physicians relinquish their "paternalistic authority" over patients [103]. Instead, patients are encouraged to take an active role in their care, and they are given greater autonomy over their treatment decisions. Physicians provide patients with the risks and benefits of available treatment options and assist patients in making their decisions [140]. To achieve desired health outcomes, patients and providers work together to make health care decisions that incorporate patients' needs, values, and preferences [141].

According to Chewning et al. [142] patients prefer using the shared decision-making model when interacting with their providers. Patients report gaining more knowledge of their illnesses, understanding more fully the risks and benefits of treatment options, and feeling empowered to make more accurate treatment decisions [143]. Incorporating patients in the decision-making process has been shown to reduce patients' decisional conflict and increase perceived satisfaction with their healthcare experiences [144]. Additionally, patients that participate in their own care are far more likely adhere to a treatment regimen they co-constructed with their providers [145]. Alternatively, patients who assume a passive role in their healthcare have indicated reduced health outcomes and significantly lowered quality of life [146].

While shared decision-making is considered the standard in patient-centered care, many providers struggle involving patients in the decision-making process. For instance, Braddock et al. [147] revealed out of a total of 1057 consultations between physicians and patients in which over 3500 clinical decisions were made, only 9.0% of met the standard for informed decision-making. However, training has been shown to improve physicians' ability to effectively use shared decision-making skills when interacting with patients. Bieber et al. [148] conducted two 4-hour training sessions to teach physicians' shared decision-making skills. After completing the training modules, providers reported greater confidence in their ability to interact with patients, recognize their attitudes, and explore their illness beliefs. Training was especially effective for physicians who indicated lacking interpersonal skills, like being overly domineering or hostile to challenges from patients. In addition, patients have reported increased satisfaction with physicians trained in shared decision-making techniques. Patients are more likely to feel engaged in the treatment process, supported, and accepted by their physicians [149,150]. Additionally, shared decision-making training has also been shown to improve physician–patient relationships. According to Bieber et al. [149] physicians who receive training are more likely to display empathy toward their patients and are more willing to involve them in their own care.

Particularly in the realm of telemedicine, a greater emphasis on shared decision-making may promote a greater sense of trust among patients from low-income countries who may be receiving care from providers in high-income nations. A greater focus on including patients in their medical decision-making, if they want to participate, should hopefully make health providers appear more engaged, caring, and concerned for their patients, and ultimately may engender improved results in telemedicine relationships. Just as shared decision-making can enhance trust and adherence to provider recommendations in the traditional medical encounter, shared decision-making should also enhance the relationship between patients and their providers via telemedicine. However, Hawley and Morris [151] found that shared decision-making does not come easily for all patients, particularly minority patients. The scholars claim that a long-standing system of discrimination in healthcare tends to favor white patients over their minority counterparts to feel compelled and able to participate in their own care. The scholars state, "to achieve shared decision-making, patients first must be informed to have accurate understanding of the pros and cons of their options, and then must be engaged to seek knowledge they do not have and to voice values, preferences, and opinions." [152, p. 3] Additionally, the scholars argue that providers should acknowledge culturally specific understandings of health (i.e., distrust of physicians and healthcare systems), and should encourage participation from their patients.

Rowan [152] suggests that brief 15-minute communication training sessions should occur monthly among practitioners in an effort to keep them well versed on current research. These brief training sessions would serve as opportunities for telemedicine providers to reflect on their current communication behaviors with patients, identify and understand the areas in which they could improve, and then integrate both well-established and newly acquired communication techniques into their daily patient practices.

CONCLUSION

As the world population continues to grow, and the rate of chronic diseases continues to rise, the desperate need for access to healthcare for rural communities across the globe is undeniable. Given the prevalence and positive outcomes of telemedicine being practiced around the world, greater research attention should be directed toward overcoming barriers to the implementation of such programs in both high-income and low-income nations. Telemedicine has not only evolved as an important step in improving access to healthcare for rural residents of the global community, it is now a necessity.

The call to expand telemedicine internationally would be incomplete without properly training healthcare providers who are practicing virtually with rural communities. In this chapter, we have demonstrated the need for a greater emphasis on cultural competency training with providers working with rural communities. We have also discussed the importance of the necessary inclusion of voices and participation of community members who are treated by providers. Telemedicine can indeed be an effective method of meeting the health needs of underserved communities and can also be an avenue to promote understanding of health and wellness within the community. Finally, we have advocated for communication training among virtual providers to assist them in their interpersonal interactions with patients. Providers must be able to adequately engage in shared decision-making, convey empathy to patients, effectively obtain and share information with patients, and ensure, through feedback, their patients have understood all interactions.

When healthcare providers learn to develop a deeper understanding of the culture of the populations they serve and also convey appropriate emotions and actions within patient communities, they will surely begin to foster a strong relationship with their patients. Hopefully, it is that relationship which will ultimately lessen any trust gaps between providers and patients and begin to negate the global inequalities in access to healthcare.

References

[1] National Rural Health Association. NRHA—What's Different about Rural Health Care? [Internet]. What's different about rural health care?. Available from: <http://www.ruralhealthweb.org/go/left/about-rural-health/what-s-different-about-rural-health-care> [cited 12.08.16].

[2] Phillips CD, McLeroy KR. Health in rural America: remembering the importance of place. Am J Public Health October 2004;94(10):1661–3. Available from: <http://www.ncbi.nlm.nih.gov/pmc/articles/PMC1448509/>.

[3] American Telemedicine Association. What is Telemedicine [Internet]. Available from: <http://www.americantelemed.org/about-telemedicine/what-is-telemedicine#.V63bc5MrJ0s> [cited 12.08.16].

[4] Craig J, Patterson V. Introduction to the practice of telemedicine. J Telemed Telecare 2005;11(1):3–9. Available from: http://dx.doi.org/10.1258/1357633053430494.

[5] Wootton R. Telemedicine support for the developing world. J Telemed Telecare 2008;14(3):109–14. Available from: http://dx.doi.org/10.1258/jtt.2008.003001.

[6] Zanaboni P, Wootton R. Adoption of telemedicine: from pilot stage to routine delivery. BMC Med Inform Decis Mak 2012;12:1. Available from: http://dx.doi.org/10.1186/1472-6947-12-1.

[7] Parmar P, Mackie D, Varghese S, Cooper C. Use of telemedicine technologies in the management of infectious diseases: a review. Clin Infect Dis December 16, 2014; ciu1143. doi:10.1093/cid/ciu1143.

[8] Kvedar J, Coye MJ, Everett W. Connected health: a review of technologies and strategies to improve patient care with telemedicine and telehealth. Health Aff Proj Hope February 2014;33(2):194–9. Available from: http://dx.doi.org/10.1377/hlthaff.2013.0992.

[9] Bull TP, Dewar AR, Malvey D, Szalma J. Considerations for telehealth systems of tomorrow: An analysis of student perceptions of telehealth technologies. JMIR Med Educ 2016;2(2). Available from: <http://mededu.jmir.org/2016/2/e11/>.

[10] Agha Z, Schapira RM, Laud PW, McNutt G, Roter DL. Patient satisfaction with physician-patient communication during telemedicine. Telemed J E-Health Off J Am Telemed Assoc November 2009;15(9):830–9. Available from: http://dx.doi.org/10.1089/tmj.2009.0030.

[11] McConnochie KM, Wood NE, Herendeen NE, Ng PK, Noyes K, Wang H, et al. Acute illness care patterns change with use of telemedicine. Pediatrics June 2009;123(6):e989–95. Available from: http://dx.doi.org/10.1542/peds.2008-2698.

[12] Singh R, Lichter MI, Danzo A, Taylor J, Rosenthal T. The adoption and use of health information technology in rural areas: results of a national survey. J Rural Health Off J Am Rural Health Assoc Natl Rural Health Care Assoc January 2012;28(1):16–27. Available from: http://dx.doi.org/10.1111/j.1748-0361.2011.00370.

[13] Head KJ, Noar SM, Iannarino NT, Grant Harrington N. Efficacy of text messaging-based interventions for health promotion: a meta-analysis. Soc Sci Med November 2013;97:41–8. Available from: http://dx.doi.org/10.1016/j.socscimed.2013.08.003.

[14] Onor ML, Misan S. The clinical interview and the doctor-patient relationship in telemedicine. Telemed J E-Health Off J Am Telemed Assoc February 2005;11(1):102–5. Available from: http://dx.doi.org/10.1089/tmj.2005.11.102.

[15] Roine R, Ohinmaa A, Hailey D. Assessing telemedicine: a systematic review of the literature. CMAJ Can Med Assoc J J Assoc Medicale Can September 18, 2001;165(6):765–71. Available from: <http://www.cmaj.ca/content/165/6/765.long>.

[16] Graham J. Top health trend for 2014: telehealth to grow over 50%. What role for regulation?—Forbes [Internet]. 2013. Available from: <http://www.forbes.com/sites/theapothecary/2013/12/28/top-health-trend-for-2014-telehealth-to-grow-over-50-what-role-for-regulation/#7cd90d471e7b> [cited 04.08.16].

[17] Rosenberg CN, Peele P, Keyser D, McAnallen S, Holder D. Results from a patient-centered medical home pilot at UPMC Health Plan hold lessons for broader adoption of the model. Health Aff Proj Hope November 2012;31(11):2423–31. Available from: http://dx.doi.org/10.1377/hithaff.2011.1002.

[18] Berman M, Fenaughty A. Technology and managed care: patient benefits of telemedicine in a rural health care network. Health Econ June 1, 2005;14(6):559–73. Available from: http://dx.doi.org/10.1002/hec.952.

[19] Horn BP, Barragan GN, Fore C, Bonham CA. A cost comparison of travel models and behavioural telemedicine for rural, Native American populations in New Mexico. J Telemed Telecare January 2016;22(1):47–55. Available from: http://dx.doi.org/10.1177/1357633X15587171.

[20] Smith AC, Youngberry K, Christie F, Isles A, McCrossin R, Williams M, et al. The family costs of attending hospital outpatient appointments via videoconference and in person. J Telemed Telecare 2003;9(Suppl 2): S58–61. Available from: http://dx.doi.org/10.1258/135763303322596282.

[21] Qin R, Dzombak R, Amin R, Mehta K. Reliability of a telemedicine system designed for rural Kenya. J Prim Care Community Health October 2012;4(3):177–81. Available from: http://dx.doi.org/10.1177/2150131912461797.

[22] Russo JE, McCool RR, Davies L VA. Telemedicine: an analysis of cost and time savings. Telemed J E-Health Off J Am Telemed Assoc March 2016;22(3):209–15. Available from: http://dx.doi.org/10.1089/tmj.2015.0055.

[23] Callahan CW, Malone F, Estroff D, Person DA. Effectiveness of an internet-based store-and-forward telemedicine system for pediatric subspecialty consultation. Arch Pediatr Adolesc Med April 1, 2005;159(4):389–93. Available from: http://dx.doi.org/10.1001/archpedi.159.4.389.

[24] Mars M. Telemedicine and advances in urban and rural healthcare delivery in Africa. Prog Cardiovasc Dis December 2013;56(3):326–35. Available from: http://dx.doi.org/10.1016/j.pcad.2013.10.006.

[25] Central Intelligence Angency. The World Factbook [Internet]. Available from: <https://www.cia.gov/library/publications/the-world-factbook/rankorder/2226rank.html> [cited 19.08.16]

[26] Berk M, Feldman J, Schur C, Gupta J. Satisfaction with practice and decision to relocate: an examination of rural physicians [Internet]. Rural Health Research & Policy Centers 2009;1–15. Available from: <http://www.norc.org/PDFs/Walsh%20Center/Main%20Page/SatisfactionwithPracticeandDecisiontoRelocateAnExaminationofRuralPhysicians.pdf>.

[27] World Health Organization. WHO | Increasing access to health workers in remote and rural areas through improved retention [Internet]. WHO. Available from: <http://www.who.int/hrh/retention/guidelines/en/>; 2010 [cited 19.08.16].

[28] Bygbjerg IC. Double burden of noncommunicable and infectious diseases in developing countries. Science. September 21, 2012;337(6101):1499–501. Available from: http://dx.doi.org/10.1126/science.1223466.

[29] World Health Organization. WHO | Global status report on noncommunicable diseases 2014 [Internet]. WHO. Available from: <http://www.who.int/nmh/publications/ncd-status-report-2014/en/>; 2010 [cited 19.08.16]

[30] Schweitzer J, Synowiec C. The economics of eHealth and mHealth. J Health Commun 2012;17(Suppl 1)):73–81. Available from: http://dx.doi.org/10.1080/10810730.2011.649158.

[31] HIT Consultant. Improving Patient Access to Specialized Health Care in Remote Brazilian Regions [Internet]. INSTICC PORTAL. Available from: <http://www.insticc.org/portal/NewsDetails/TabId/246/ArtMID/1130/ArticleID/423/Improving-Patient-Access-to-Specialized-Health-Care-in-Remote-Brazilian-Regions.aspx>; 2015 [cited 13.08.16].

[32] Alkmim MB, Marcolino M, Figueria R, Maia J, Cardoso C, Abreu M, et al. 1,000,000 Electrocardiograms by distance: an outstanding milestone for telehealth in Minas Gerais, Brazil. Glob Telemed EHealth Update Knowl Resour 2013;6:459–64. Available from: <http://www.crics9.org/es/files/2012/12/Id_13_BR_1000000.pdf>.

[33] Sood SP, Negash S, Mbarika VWA, Kifle M, Prakash N. Differences in public and private sector adoption of telemedicine: Indian case study for sectoral adoption. Stud Health Technol Inform 2007;130:257–68. Available from: <https://www.researchgate.net/profile/Solomon_Negash/publication/5928211_Differences_in_public_and_private_sector_adoption_of_telemedicine_Indian_case_study_for_sectoral_adoption/links/5599d4f508ae99aa62cc6f39.pdf>.

[34] Natarajan M. Impact of information technology to E-health strategies with special reference to India. Asian J Libr Inf Sci 2010;2(1-4):26–35. Available from: <http://www.escienceworld.org/index.php/ajlis/article/view/90/90>.

[35] Kruse CS, Bouffard S, Dougherty M, Parro JS. Telemedicine use in rural native American communities in the era of the ACA: a systematic literature review. J Med Syst April 27, 2016 27;40(6):145. Available from: http://dx.doi.org/10.1007/s10916-016-0503-8.

[36] Sequist TD, Cullen T, Acton KJ. Indian health service innovations have helped reduce health disparities affecting American Indian and Alaska native people. Health Aff (Millwood) October 1, 2011;30(10):1965–73. Available from: http://dx.doi.org/10.1377/hlthaff.2011.0630.

[37] Roh C-Y. Telemedicine: what it is, where it came from, and where it will go. Comp Technol Transf Soc 2008;6(1):35–55. Available from: http://dx.doi.org/10.1353/ctt.0.0002.

[38] Doorenbos AZ, Eaton LH, Haozous E, Towle C, Revels L, Buchwald D. Satisfaction with telehealth for cancer support groups in rural American Indian and Alaska native communities. Clin J Oncol Nurs December 1, 2010;14(6):765–70. Available from: http://dx.doi.org/10.1188/10.CJON.765-770.

[39] Dallolio L, Menarini M, China S, Ventura M, Stainthorpe A, Soopramanien A, et al. Functional and clinical outcomes of telemedicine in patients with spinal cord injury. Arch Phys Med Rehabil December 2008;89(12):2332–41. Available from: http://dx.doi.org/10.1016/j.apmr.2008.06.012.

[40] Mizushima H, Uchiyama E, Nagata H, Matsuno Y, Sekiguchi R, Ohmatsu H, et al. Japanese experience of telemedicine in oncology. Int J Med Inf May 2001;61(2–3):207–15. doi:10.1016/S1386-5056(01)00142-3.

[41] Zissman K, Lejbkowicz I, Miller A. Telemedicine for multiple sclerosis patients: assessment using Health Value Compass. Mult Scler Houndmills Basingstoke Engl April 2012;18(4):472–80. Available from: http://dx.doi.org/10.1177/1352458511421918.
[42] Carroll AE, DiMeglio LA, Stein S, Marrero DG. Using a cell phone–based glucose monitoring system for adolescent diabetes management. Diabetes Educ January 1, 2011;37(1):59–66. Available from: http://dx.doi.org/10.1177/0145721710387163.
[43] Robinson JD, Turner JW, Levine B, Tian Y. Expanding the walls of the health care encounter: support and outcomes for patients online. Health Commun March 2011;26(2):125–34. Available from: http://dx.doi.org/10.1080/10410236.2010.54.1990.
[44] Bagchi S. Telemedicine in rural India. PLoS Med March 2006;3(3):e82. Available from: http://dx.doi.org/10.1371/journal.pmed.0030082.
[45] Brooks E, Turvey C, Augusterfer EF. Provider barriers to telemental health: obstacles overcome, obstacles remaining. Telemed J E-Health Off J Am Telemed Assoc June 2013;19(6):433–7. Available from: http://dx.doi.org/10.1089/tmj.2013.0068.
[46] Cimperman M, Brenčič MM, Trkman P, Stanonik M, de L. Older adults' perceptions of home telehealth services. Telemed J E-Health Off J Am Telemed Assoc October 2013;19(10):786–90. Available from: http://dx.doi.org/10.1089/tmj.2012.0272.
[47] World Health Organization. WHO | Telemedicine opportunities and developments in member states [Internet]. WHO. 2009. Available from: <http://www.who.int/goe/publications/goe_telemedicine_2010.pdf> [cited 13.8.16].
[48] Jang-Jaccard J, Nepal S, Alem L, Li J. Barriers for delivering telehealth in rural australia: a review based on Australian trials and studies. Telemed J E-Health Off J Am Telemed Assoc May 2014;20(5):496–504. Available from: http://dx.doi.org/10.1089/tmj.2013.0189.
[49] Geissbuhler A, Ly O, Lovis C, L'Haire J-F. Telemedicine in Western Africa: lessons learned from a pilot project in Mali, perspectives and recommendations. AMIA Annu Symp Proc 2003;2003:249–53. Available from: <http://www.ncbi.nlm.nih.gov/pmc/articles/PMC1479936/>.
[50] Steele R, Lo A. Telehealth and ubiquitous computing for bandwidth-constrained rural and remote areas. Pers Ubiquitous Comput January 21, 2012;17(3):533–43. Available from: http://dx.doi.org/10.111/j.1445-2197.2007.04299.
[51] Stutchfield BM, Jagilly R, Tulloh BR. Second opinions in remote surgical practice using email and digital photography. ANZ J Surg November 2007;77(11):1009–12.
[52] Masucci MM, Homko C, Santamore WP, Berger P, McConnell TR, Shirk G, et al. Cardiovascular disease prevention for underserved patients using the Internet: Bridging the digital divide. Telemed J E-Health Off J Am Telemed Assoc February 2006;12(1):58–65. Available from: http://dx.doi.org/10.1089/tmj.2006.12.58.
[53] Bhatta R, Aryal K, Ellingsen G. Opportunities and challenges of a rural-telemedicine program in Nepal. J Nepal Health Res Counc August 2015;13(30):149–53. Available from: <http://www.jnhrc.com.np/index.php/jnhrc/article/view/640/499>.
[54] Marcin JP, Nesbitt TS, Struve S, Traugott C, Dimand RJ. Financial benefits of a pediatric intensive care unit-based telemedicine program to a rural adult intensive care unit: impact of keeping acutely ill and injured children in their local community. Telemed J E-Health Off J Am Telemed Assoc. 2004;10(Suppl 2)):S-1–5. Available from: <http://www.amdtelemedicine.com/telemedicine-resources/documents/FinancialImpactofaPediatricTelemedProgramina competitivehospitalregion_2013.pdf>.
[55] Utidjian L, Abramson E. Pediatric telehealth: opportunities and challenges. Pediatr Clin North Am April 2016;63(2):367–78. Available from: http://dx.doi.org/10.1016/j.pcl.2015.11.006.
[56] Pradhan MR. ICTs application for better health in Nepal. Kathmandu Univ Med J KUMJ June 2004;2(2):157–63. Available from: <http://imsear.hellis.org/bitstream/123456789/46434/2/kumj2004v2n2p.157.pdf>.
[57] Qaddoumi I, Mansour A, Musharbash A, Drake J, Swaidan M, Tihan T, et al. Impact of telemedicine on pediatric neuro-oncology in a developing country: the Jordanian-Canadian experience. Pediatr Blood Cancer January 2007;48(1):39–43. Available from: http://dx.doi.org/10.1002/pbc.21085.
[58] Ifejika NL, Noser EA, Grotta JC, Savitz SI. Swipe out stroke: feasibility and efficacy of using a smart-phone based mobile application to improve compliance with weight loss in obese minority stroke patients and their carers. Int J Stroke Off J Int Stroke Soc July 2016;11(5):593–603. Available from: http://dx.doi.org/10.1177/1747493016631557.
[59] Dart E, Whipple H, Pasqua JL, Furlow CM. Legal, regulatory, and ethical issues in telehealth technology. In: Luiselli JK, Fischer AJ, editors. Computer-assisted and web-based innovations in psychology, special

education, and health. Elsevier inc: Academic Press; 2016. p. 339–63. Available from: <https://www.researchgate.net/profile/Fabrizio_Stasolla/publication/303415035_Assistive_Technologies_for_Persons_with_Severe-Profound_Intellectual_and_Developmental_Disabilities/links/5745705408ae9ace8421b72c.pdf#page = 358>.

[60] Pattynama PM. Legal aspects of cross-border teleradiology. Eur J Radiol 2010;73(1):26–30. Available from: http://dx.doi.org/10.1016/j.erad.2009.10.017.

[61] Vassallo DJ, Hoque F, Roberts MF, Patterson V, Swinfen P, Swinfen R. An evaluation of the first year's experience with a low-cost telemedicine link in Bangladesh. J Telemed Telecare June 1, 2001;7(3):125–38. Available from: http://dx.doi.org/10.1258/1357633011936273.

[62] Alverson DC, Swinfen LR, Swinfen LP, Rheuban K, Sable C, Smith AC, et al. Transforming systems of care for children in the global community. Pediatr Ann. 2009;38(10):579–85. Available from: http://dx.doi.org/10.3928/00904481-20090918-11.

[63] Jarvis-Selinger S, Chan E, Payne R, Plohman K, Ho K. Clinical telehealth across the disciplines: lessons learned. Telemed J E-Health Off J Am Telemed Assoc September 2008;14(7):720–5. Available from: http://dx.doi.org/10.1089/tmj.2007.0108.

[64] Office of Disease Prevention and Health Promotion. Health Communication and Health Information Technology | Healthy People 2020 [Internet]. Available from: <https://www.healthypeople.gov/2020/topics-objectives/topic/health-communication-and-health-information-technology/objectives>; 2011 [cited 20.08.16]

[65] Mahmud N, Rodriguez J, Nesbit J. A text message-based intervention to bridge the healthcare communication gap in the rural developing world. Technol Health Care Off J Eur Soc Eng Med 2010;18(2):137–44. Available from: http://dx.doi.org/10.3233/THC-2010-0576.

[66] Office of Minority Health. What is Cultural Competency? [Internet]. Available from: <https://www.cdc.gov/nchhstp/socialdeterminants/docs/what_is_cultural_competency.pdf>; 2011 [cited 20.8.16]

[67] Lemacks JL, Huye H, Rupp R, Connell C. The relationship between interviewer–respondent race match and reporting of energy intake using food frequency questionnaires in the rural South United States. Prev Med Rep June 10, 2015;2:533–7. Available from: http://dx.doi.org/10.1016/j.pmedr.2015.06.002.

[68] McAlearney AS, Robbins J, Kowalczyk N, Chisolm DJ, Song PH. The role of cognitive and learning theories in supporting successful EHR system implementation training a qualitative study. Med Care Res Rev June 1, 2012;69(3):294–315. Available from: http://dx.doi.org/10.1177/1077558711436348.

[69] Street RL, Makoul G, Arora NK, Epstein RM. How does communication heal? Pathways linking clinician-patient communication to health outcomes. Patient Educ Couns March 2009;74(3):295–301. Available from: http://dx.doi.org/10.1016/j.pec.2008.11.015.

[70] Behringer B, Friedell GH. Appalachia: where place matters in health. Prev Chronic Dis [Internet] September 15, 2006;3(4) . Available from: <http://www.ncbi.nlm.nih.gov/pmc/articles/PMC1779277/> [cited 20.8.16]

[71] Gorsuch MM, Sanders SG, Wu B. Tooth loss in Appalachia and the Mississippi delta relative to other regions in the United States, 1999–2010. Am J Public Health February 13, 2014;104(5):e85–91. Available from: http://dx.doi.org/10.2105/AJPH.2013.301641.

[72] Center for Disease Control and Prevention. Kentucky Behavioral Risk Factor Survey [Internet]. Available from: <http://chfs.ky.gov/NR/rdonlyres/3595B54F-2A97-4D2D-95A9-B4430DF68C12/0/2013KyBRFSAnnualReportFinal_05132016.pdf>; 2013 [cited 20.08.16].

[73] Appalachian Regional Commission. The Appalachian region [Internet]. Available from: <http://www.arc.gov/appalachian_region/TheAppalachianRegion.asp> [cited 20.08.16]

[74] Zahnd WE, Scaife SL, Francis ML. Health literacy skills in rural and urban populations. Am J Health Behav October 2009;33(5):550–7. Available from: < http://www.ncbi.nlm.nih.gov/pubmed/19296745>.

[75] Kluhsman BC, Bencivenga M, Ward AJ, Lehman E, Lengerich EJ. Initiatives of 11 rural Appalachian cancer coalitions in Pennsylvania and New York. Prev Chronic Dis October 2006;3(4):A122. Available from: < http://www.ncbi.nlm.nih.gov/pmc/articles/PMC1779286/>.

[76] Howitt P, Darzi A, Yang GZ, Ashrafian H, Atun R, Barlow J, et al. Technologies for global health. The Lancet 2012;38(9840):507–35 doi:10.1016/S0140-6736(12)61127-1.

[77] Patel N, Stone MA, Hadjiconstantinou M, Hiles S, Troughton J, Martin-Stacey L, et al. Using an interactive DVD about type 2 diabetes and insulin therapy in a UK South Asian community and in patient education and healthcare provider training. Patient Educ Couns September 2015;98(9):1123–30. Available from: http://dx.doi.org/10.1016/j.pec.2015.04.018.

[78] Schouten BC, Meeuwesen L. Cultural differences in medical communication: a review of the literature. Patient Educ Couns December 2006;64(1–3):21–34. Available from: http://dx.doi.org/10.1016/j.pec.2005.11.014.

[79] World Health Organization. WHO | Track 2: Health literacy and health behaviour [Internet] Available from: <http://www.who.int/healthpromotion/conferences/7gchp/track2/en/> WHO. 1998 [cited 20.08.16].

[80] Nutbeam D. The evolving concept of health literacy. Soc Sci Med December 2008;67(12):2072–8. Available from: http://dx.doi.org/10.1016/j.socscimed.2008.09.050.

[81] Dewalt DA, Berkman ND, Sheridan S, Lohr KN, Pignone MP. Literacy and health outcomes: a systematic review of the literature. J Gen Intern Med December 2004;19(12):1228–39. Available from: http://dx.doi.org/10.111/j.1525-1497.2004.40153.

[82] Weiss BD. Removing barriers to better, safer care health literacy and patient safety: help patients understand manual for clinicians [Internet]. 2nd ed Chicago, IL: American Medical Association; 2007. Available from: <http://med.fsu.edu/userFiles/file/ahec_health_clinicians_manual.pdf> [cited 20.08.16].

[83] Nielsen-Bohlman L, Panzer AM, Kindig DA, editors. Health literacy: a prescription to end confusion. Washington, DC: National Academies Press; 2004 367 p. Available from: <http://www.ncbi.nlm.nih.gov/books/NBK216035/>.

[84] Rudd RE. Improving Americans' health literacy. N Engl J Med December 9, 2010;363(24):2283–5. Available from: http://dx.doi.org/10.1056/NEJMp1008755.

[85] VanGeest JB. Identifying and improving care for patients with limited health literacy. In: APHA. Available from: <https://apha.confex.com/apha/143am/webprogram/Paper339492.html>; 2015 [cited 20.08.16]

[86] Mefalopulos P. Communication for sustainable development: applications and challenges. In: Hemer O, Tufte O, editors. Media and global change: rethinking communication for development. Nordicom and CLACSO; 2005. p. 241–60. Available from: <http://biblioteca.clacso.edu.ar/clacso/coediciones/20100824072510/20Chapter14.pdf>.

[87] Kreps GL, Neuhauser L. New directions in eHealth communication: opportunities and challenges. Patient Educ Couns March 2010;78(3):329–36. Available from: http://dx.doi.org/10.1016/j.pec.2010.01.013.

[88] US Department of Health and Human Services. Quick guide to health literacy. Washington, DC: US Department of Health and Human Services; 2006. Available from: <https://health.gov/communication/literacy/quickguide/>.

[89] National Patient Safety Foundation. Ask Me 3: Good Questions for Your Good Health [Internet]. Available from: <http://www.npsf.org/?page = askme3> [cited 20.08.16]

[90] Indian Health Service. Healthcare Communications—Patient-Provider Communication Toolkit [Internet]. Available from: <https://www.ihs.gov/healthcommunications/index.cfm?module = dsp_hc_toolkit>; 2016 [cited 20.8.16]

[91] Jana S, Basu I, Rotheram-Borus MJ, Newman PA. The Sonagachi project: a sustainable community intervention program. AIDS Educ Prev October 1, 2004;16(5):405–14. Available from: http://dx.doi.org/10.1521/aeap.16.5.405.48734.

[92] Dutta-Bergman MJ. Primary sources of health information: comparisons in the domain of health attitudes, health cognitions, and health behaviors. Health Commun 2004;16(3):273–88. Available from: http://dx.doi.org/10.1207/S15327027HC1603.

[93] Tufte T, Mefalopulos P. Participatory communication: a practical guide. Washington, DC: World Bank Publications 2009; 62 p. doi:10.1596/978-0-8213-8008-6.

[94] Basu A, Dutta MJ. The relationship between health information seeking and community participation: the roles of health information orientation and efficacy. Health Commun. 2008;23(1):70–9. Available from: http://dx.doi.org/10.1080/10410230701807121.

[95] Dutta MJ, Basnyat I. The Radio Communication Project in Nepal: a culture-centered approach to participation. Health Educ Behav Off Publ Soc Public Health Educ August 2008;35(4):442–54. Available from: http://dx.doi.org/10.1177/1090198106287450.

[96] Dutta MJ, Anaele A, Jones C. Voices of hunger: addressing health disparities through the culture-centered approach. J Commun February 1, 2013;63(1):159–80. Available from: http://dx.doi.org/10.1111/jcom/12009.

[97] Gifford V, Niles B, Rivkin I, Koverola C, Polaha J. Continuing education training focused on the development of behavioral telehealth competencies in behavioral healthcare providers. Rural Remote Health 2012;12:2108. Available from: pmid:232340871.

[98] Nelson D, Hewell V, Roberts L, Kersey E, Avey J. Telebehavioral health delivery of clinical supervision trainings in rural Alaska: an emerging best practices model for rural practitioners. J Rural Ment Health 2012;36(2):10–15. Available from: http://dx.doi.org/10.1037/h0085810.

[99] Dutta S, Lahiri K. . Report No.: ID 2621473. Available from: <http://papers.ssrn.com/abstract = 2621473>; Is provision of healthcare sufficient to ensure better access? An exploration of the scope for public-private partnership in India [Internet]. Rochester, NY: Social Science Research Network; June 2015 [cited 20.08.16]
[100] Villagran M. Methodological diversity to reach patients along the margins, in the shadows, and on the cutting edge. Patient Educ Couns March 2011;82(3):292–7. Available from: http://dx.doi.org/10.1016/j.pec.2010.12.020.
[101] Ha JF, Longnecker N. Doctor-patient communication: a review. Ochsner J 2010;10(1):38–43. Available from: http://dx.doi.org/10.1043/TOJ-09-0040.1.
[102] Street Jr. RL. Mediated consumer–provider communication in cancer care: the empowering potential of new technologies. Patient Educ Couns May 2003;50(1):99–104 doi:10.1016/S0738-3991(03)00089-2.
[103] Barry MJ, Edgman-Levitan S. Shared decision making—the pinnacle of patient-centered care. N Engl J Med March 1, 2012;366(9):780–1. Available from: http://dx.doi.org/10.1056/NEJMp1109283.
[104] Institute of Medicine. Crossing the quality chasm: a new health system for the 21st century. Washington, DC: National Academic Press; 2001. Available from: <https://www.nationalacademies.org/hmd/~/media/Files/Report%20Files/2001/Crossing-the-Quality-Chasm/Quality%20Chasm%202001%20%20report%20brief.pdf>.
[105] Harter LM, Bochner AP. Healing through stories: a special issue on narrative medicine. J Appl Commun Res May 1, 2009;37(2):113–17.
[106] Little M, Jordens CFC, Paul K, Sayers E, Sriskandarajah D. Face, honor and dignity in the context of colon cancer. J Med Humanit December 2000;21(4):229–43. Available from: http://dx.doi.org/10.1023/A:1009077209274.
[107] Fallowfield L. Communication with patients after errors. J Health Serv Res Policy January 2010;15 (Suppl 1)):56–9. Available from: http://dx.doi.org/10.1254/jhsrp.2009.09s107.
[108] Haskard Zolnierek KB, DiMatteo MR. Physician communication and patient adherence to treatment: a meta-analysis. Med Care August 2009;47(8):826–34. Available from: http://dx.doi.org/10.1097/MLR.0b013e31819a5acc.
[109] Levinson W, Roter DL, Mullooly JP, Dull VT, Frankel RM. Physician-patient communication: the relationship with malpractice claims among primary care physicians and surgeons. JAMA February 19, 1997;277 (7):553–9. Available from: http://dx.doi.org/10.1001/jama.1997.03540310051034.
[110] Martin LR, Williams SL, Haskard KB, DiMatteo MR. The challenge of patient adherence. Ther Clin Risk Manag September 2005;1(3):189–99. Available from: <http://www.ncbi.nlm.nih.gov/pmc/articles/PMC1661624/>.
[111] Bertakis KD, Azari R. Patient-centered care is associated with decreased health care utilization. J Am Board Fam Med JABFM June 2011;24(3):229–39. Available from: http://dx.doi.org/10.3122/jabfm.2011.03.100170.
[112] Levinson W, Pizzo PA. Patient-physican communication: it's about time. JAMA May 4, 2011;305 (17):1802–3. Available from: http://dx.doi.org/10.1001/jama.2011.556.
[113] Rao JK, Anderson LA, Inui TS, Frankel RM. Communication interventions make a difference in conversations between physicians and patients: a systematic review of the evidence. Med Care April 2007;45 (4):340–9. Available from: http://dx.doi.org/10.1097/01.mlr.0000254516.04951.d5.
[114] Fallowfield L, Jenkins V, Farewell V, Solis-Trapala I. Enduring impact of communication skills training: results of a 12-month follow-up. Br J Cancer 2003;89(8):1445–9. Available from: http://dx.doi.org/10.1038/sj.bjc.6601309.
[115] Brown RF, Bylund CL. Communication skills training: describing a new conceptual model. Acad Med J Assoc Am Med Coll January 2008;83(1):37–44. Available from: http://dx.doi.org/10.1097/ACM.0b013e31815c631e.
[116] Neumann M, Wirtz M, Bollschweiler E, Mercer SW, Warm M, Wolf J, et al. Determinants and patient-reported long-term outcomes of physician empathy in oncology: a structural equation modelling approach. Patient Educ Couns December 2007;69(1-3):63–75. Available from: http://dx.doi.org/10.1016/j.pec.2007.07.003.
[117] Mercer SW, Maxwell M, Heaney D, Watt GC. The consultation and relational empathy (CARE) measure: development and preliminary validation and reliability of an empathy-based consultation process measure. Fam Pract December 2004;21(6):699–705. Available from: http://dx.doi.org/10.1093/famepra/cmh621.
[118] Flickinger TE, Saha S, Roter D, Korthuis PT, Sharp V, Cohn J, et al. Clinician empathy is associated with differences in patient-clinician communication behaviors and higher medication self-efficacy in HIV care. Patient Educ Couns February 2016;99(2):220–6. Available from: http://dx.doi.org/10.1016/j.pec.2015.09.001.

[119] Steinhausen S, Ommen O, Thüm S, Lefering R, Koehler T, Neugebauer E, et al. Physician empathy and subjective evaluation of medical treatment outcome in trauma surgery patients. Patient Educ Couns April 2014;95(1):53–60. Available from: http://dx.doi.org/10.1016/j.pec.2013.12.007.

[120] Bayne H, Neukrug E, Hays D, Britton B. A comprehensive model for optimizing empathy in person-centered care. Patient Educ Couns November 2013;93(2):209–15. Available from: http://dx.doi.org/10.1016/j.pec.2013.05.016.

[121] Parkin T, de Looy A, Farrand P. Greater professional empathy leads to higher agreement about decisions made in the consultation. Patient Educ Couns August 2014;96(2):144–50. Available from: http://dx.doi.org/10.1016/j.pec.2014.04.019.

[122] Brugel S, Postma-Nilsenová M, Tates K. The link between perception of clinical empathy and nonverbal behavior: the effect of a doctor's gaze and body orientation. Patient Educ Couns October 2015;98(10):1260–5. Available from: http://dx.doi.org/10.1016/j.pec.2015.08.007.

[123] Bonvicini KA, Perlin MJ, Bylund CL, Carroll G, Rouse RA, Goldstein MG. Impact of communication training on physician expression of empathy in patient encounters. Patient Educ Couns April 2009;75(1):3–10. Available from: http://dx.doi.org/10.1016/j.pec.2015.11.021.

[124] Pehrson C, Banerjee SC, Manna R, Shen MJ, Hammonds S, Coyle N, et al. Responding empathically to patients: development, implementation, and evaluation of a communication skills training module for oncology nurses. Patient Educ Couns April 2016;99(4):610–16.

[125] Mast MS. On the importance of nonverbal communication in the physician-patient interaction. Patient Educ Couns August 2007;67(3):315–18. Available from: http://dx.doi.org/10.1016/j.pec.2007.03.005.

[126] Hannawa AF. Disclosing medical errors to patients: effects of nonverbal involvement. Patient Educ Couns March 2014;94(3):310–13. Available from: http://dx.doi.org/10.1016/j.pec.2013.11.007.

[127] Chariyeva Z, Golin CE, Earp JA, Suchindran C. Does motivational interviewing counseling time influence HIV-positive persons' self-efficacy to practice safer sex?. Patient Educ Couns April 2012;87(1):101–7. Available from: http://dx.doi.org/10.1016/j.pec.2011.07.021.

[128] van Eijk-Hustings YJL, Daemen L, Schaper NC, Vrijhoef HJM. Implementation of motivational interviewing in a diabetes care management initiative in The Netherlands. Patient Educ Couns July 2011;84(1):10–15. Available from: http://dx.doi.org/10.1016/j.pec.2010.06.016.

[129] Davis MF, Shapiro D, Windsor R, Whalen P, Rhode R, Miller HS, et al. Motivational interviewing versus prescriptive advice for smokers who are not ready to quit. Patient Educ Couns April 2011;83(1):129–33. Available from: http://dx.doi.org/10.1016/j.pec.2010.04.024.

[130] Riekert KA, Borrelli B, Bilderback A, Rand CS. The development of a motivational interviewing intervention to promote medication adherence among inner-city, African-American adolescents with asthma. Patient Educ Couns January 2011;82(1):117–22. Available from: http://dx.doi.org/10.1016/j.pec.2010.03.005.

[131] Ekong G, Kavookjian J. Motivational interviewing and outcomes in adults with type 2 diabetes: a systematic review. Patient Educ Couns June 2016;99(6):944–52. Available from: http://dx.doi.org/10.1016/j.pec.2015.11.022.

[132] Spencer JC, Wheeler SB. A systematic review of motivational interviewing interventions in cancer patients and survivors. Patient Educ Couns July 2016;99(7):1099–105. Available from: http://dx.doi.org/10.1016/j.pec.2016.02.003.

[133] Pollak KI, Coffman CJ, Tulsky JA, Alexander SC, Østbye T, Farrell D, et al. Teaching physicians motivational interviewing for discussing weight with overweight adolescents. J Adolesc Health July 2016;59(1):96–103. Available from: http://dx.doi.org/10.1016/j.jadohealth.2016.03.026.

[134] Edwards EJ, Stapleton P, Williams K, Ball L. Building skills, knowledge and confidence in eating and exercise behavior change: brief motivational interviewing training for healthcare providers. Patient Educ Couns May 2015;98(5):674–6. Available from: http://dx.doi.org/10.1016/j.pec.2015.02.006.

[135] Fu SS, Roth C, Battaglia CT, Nelson DB, Farmer MM, Do T, et al. Training primary care clinicians in motivational interviewing: a comparison of two models. Patient Educ Couns January 2015;98(1):61–8. Available from: http://dx.doi.org/10.1016/j.pec.2014.10.007.

[136] Jerant A, Kravitz RL, Tancredi D, Paterniti DA, White L, Baker-Nauman L, et al. Training primary care physicians to employ self-efficacy-enhancing interviewing techniques: randomized controlled trial of a standardized patient intervention. J Gen Intern Med March 8, 2016;31(7):716–22. Available from: http://dx.doi.org/10.1007/s11606-016-3644-z.

[137] Hesson AM, Sarinopoulos I, Frankel RM, Smith RC. A linguistic study of patient-centered interviewing: emergent interactional effects. Patient Educ Couns September 2012;88(3):373–80. Available from: http://dx.doi.org/10.1016/j.pec.2012.06.005.

[138] Oh H, Lee C. Culture and motivational interviewing. Patient Educ Couns June 2016;17(3):317–24. Available from: http://dx.doi.org/10.1037/a0024035.

[139] Charles C, Gafni A, Whelan T. Shared decision-making in the medical encounter: what does it mean? (or it takes at least two to tango). Soc Sci Med March 1, 1997;44(5):681–92. Available from: http://dx.doi.org/10.1177/0272878X07306779.

[140] Lin GA, Fagerlin A. Shared decision making state of the science. Circ Cardiovasc Qual Outcomes March 1, 2014;7(2):328–34. Available from: http://dx.doi.org/10.1161/CIRCOUTCOMES.113.000322.

[141] Makoul G, Clayman ML. An integrative model of shared decision making in medical encounters. Patient Educ Couns March 2006;60(3):301–12. Available from: http://dx.doi.org/10.1016/j.pec.2005.06.010.

[142] Chewning B, Bylund C, Shah B, Arora NK, Gueguen JA, Makoul G. Patient preferences for shared decisions: a systematic review. Patient Educ Couns January 2012;86(1):9–18. Available from: http://dx.doi.org/10.1016/j.pec.2011.02.004.

[143] Stacey D, Bennett CL, Barry MJ, Col NF, Eden KB, Holmes-Rovner M, et al. Decision aids for people facing health treatment or screening decisions. Cochrane Database Syst Rev 2011;10:CD001431. Available from: http://dx.doi.org/10.1002/14641858.CD001431.pub3.

[144] Edwards A, Elwyn G, Hood K, Atwell C, Robling M, Houston H, et al. Patient-based outcome results from a cluster randomized trial of shared decision making skill development and use of risk communication aids in general practice. Fam Pract August 2004;21(4):347–54. Available from: http://dx.doi.org/10.1093/fampra/cmh402.

[145] Joosten EA, DeFuentes-Merills L, de Weert GH, Sensky T, van der Staak CP, de Jong CA. Systematic review of the effects of shared decision-making on patient satisfaction, treatment adherence and health status. Psychother Psychosom 2008;77(4):216–26. Available from: http://dx.doi.org/10.1159/000126073.

[146] Hack TF, Degner LF, Watson P, Sinha L. Do patients benefit from participating in medical decision making? Longitudinal follow-up of women with breast cancer. Psychooncology January 2006;15(1):9–19. Available from: http://dx.doi.org/10.1002/pon.907.

[147] Braddock III CH, Edwards KA, Hasenberg NM, Laidley TL, Levinson W. Informed decision making in outpatient practice: time to get back to basics. JAMA December 22, 1999;282(24):2313–20. Available from: http://dx.doi.org/10.1001/jama.282.24.2312.

[148] Bieber C, Nicolai J, Hartmann M, Blumenstiel K, Ringel N, Schneider A, et al. Training physicians in shared decision-making—who can be reached and what is achieved? Patient Educ Couns October 2009;77(1):48–54. Available from: http://dx.doi.org/10.1016/j.pec.2009.03.019.

[149] Bieber C, Müller KG, Blumenstiel K, Schneider A, Richter A, Wilke S, et al. Long-term effects of a shared decision-making intervention on physician-patient interaction and outcome in fibromyalgia. A qualitative and quantitative 1 year follow-up of a randomized controlled trial. Patient Educ Couns November 2006;63(3):357–66. Available from: http://dx.doi.org/10.1016/j.pec.2006.05.00310.1016/j.pec.2006.05.003.

[150] Elwyn G, Edwards A, Wensing M, Hood K, Atwell C, Grol R. Shared decision making: developing the OPTION scale for measuring patient involvement. Qual Saf Health Care April 2003;12(2):93–9. Available from: <http://www.ncbi.nlm.nih.gov/pubmed/12679504>.

[151] Hawley ST, Morris AM. Cultural challenges to engaging patients in shared decision making. Patient Educ Couns July 4, 2016;1–7. Available from: <http://dx.doi.org.lynx.lib.usm.edu/10.1016/j.pec.2016.07.008>.

[152] Rowan KE. Monthly communication skill coaching for healthcare staff. Patient Educ Couns June 2008;71(3):402–4. Available from: http://dx.doi.org/10.1016/j.pec.2008.02.016.

Further Reading

Katzenstein J, Chrispin BR. Designing a telemedicine system in Tanzania: a sociotechnical systems approach. In: Bangert DC, Doktor RH, Valdez M, editors. Human and organizational dynamics in e-health. Radcliffe Publishing; 2005. p. 35–54.

SECTION II

ETHICAL AND PROFESSIONAL CONDUCT IN THE DIGITAL AGE

CHAPTER

4

The Facets of Digital Health Professionalism: Defining a Framework for Discourse and Change

Tamara J. Bahr, Noah H. Crampton and Sharon Domb

University of Toronto, Toronto, ON, Canada

It is a well-known fact that to become a healthcare professional is highly desirable. Year in and year out, with limited yet coveted spots in medical school, aspiring doctors would have needed to excel in both college and their chosen extracurricular activities to succeed in gaining admission. The application process can be grueling and often yields much fewer acceptances than rejections. Yet despite this, medicine as a chosen profession has never been more popular.

Once accepted into an MD program, you will spend four or more years learning everything from anatomy, physiology, diagnostic and procedural skills, to empathy, nonjudgment and altruism. Essentially, the learner is being trained to be an expert scholar, communicator, collaborator, advocate, and leader [1]. But these skills can be misappropriated if they are not used in a "professional" manner. So over the course of history, the raw humanism behind clinical healthcare has necessitated that its practice become one of the most ethically guided professions ever. Healthcare practitioners are held to the highest ethical and professional standards. The training involved to become competent in those standards is rigorous and forces profound character and professional development. Indeed, such positive societal perceptions of physicians and other health professionals can be gleaned from historical texts dealing with the origins of the medical profession as well as in popular culture. The modern medical drama illuminates the complexity of the profession of medicine while helping to maintain societal ideas and acceptance of health professionals as highly educated, competent, compassionate, and ethical beings.

Historically, it has been held that ethical and professional behavior is intrinsic to a medical professional's core set of values and that doctors come about ethical compassionate practice "by osmosis" [2]. But we know that perception is born from a time past where paternalism in the medical profession was commonplace: "The patient was treated like a

Health Professionals' Education.
DOI: http://dx.doi.org/10.1016/B978-0-12-805362-1.00004-8

child; innocent, unschooled, and too simple to know how to take care of himself or herself. This wise father-simple child relationship led to an inherently paternalistic model of the physician-patient relationship." [3] What we considered acceptable in the early to mid-twentieth century does not hold to be societally acceptable today. Just as societal views on gender parity, identity, and racial segregation have evolved as we advance our understanding and acceptance; changes come about in a deliberative way and not just "by osmosis." The notion of change is an important one. Throughout history, over and again, new technologies have forced us to change, to rethink the ways in which we work, learn, and communicate with one another. The Internet and its related technologies has turned many a profession completely upside down. It is a technological paradigm shift that has forced change in so many ways. This has never been so true for the health professions. As we evolve, so too do the tenets by which we practice healthcare. So, it only stands to reason that professionalism, from the digital frame of reference, is constantly evolving as well.

DEFINING DIGITAL PROFESSIONALISM

The descriptions of what it means to be a professional in the health professions provide an overarching set of moralistic principles guiding the role of professionalism. Identifying what those principles look like in practice is much more difficult to peg down. Add to this discourse the idea of *professional digital behavior*, and the waters become even further muddied. The question of what professional behavior looks like in digital environments is paramount to understanding how the health professions address digital media participation. We know what *unprofessional* digital behavior is when we see it but how to directly and specifically address the topic in a cohesive manner, as a preemptive measure remains unclear. Scholars in this area agree on many points, but there currently is a lack of common and cohesive characterization of digital professionalism as it pertains to the health professions. This may, in part, be due to the broad all-encompassing nature of the term *professional* and what it semantically means as a whole. Common definitions characterize professionalism with terms such as *public good, ethical standards, values, integrity, honesty, altruism, humility, respect* for diversity, and *transparency* [1].

Definitions are important to understanding effective ways in which we might assess and address topics of professionalism. Birden's 2014 systematic review of the literature on professionalism in medicine concluded the following:

> "The semantics of professionalism obfuscate more than they clarify, and the continually shifting nature of the medical profession and in the organizational and social milieu in which it operates creates a dynamic situation where no definition has yet taken hold as the definitive one." [4]

Without common language, after all, how can we share a common understanding with regards to what needs to be addressed? It has become increasingly clear through the literature that while professionalism as a concept is broadly understood, there lacks a common understanding or set of definitions to reflect the many concepts and ideas that emerge from this topic, specifically in healthcare; ethics, altruism, duty, privacy, reputation, compassion, communication, patient safety, and relationships. The very same holds true for the newest cousin on the professionalism block, digital professionalism.

While digital professionalism is a very hot topic in medical professions circles, literature that directly and specifically addresses the topic in a cohesive manner is still relatively novel. Scholars in this area agree on many points, which in time will help to build a common and cohesive characterization of digital professionalism as it pertains to the health professions. In preparation for this chapter, we conducted a broad search for literature addressing medical professionalism and online of digital media in the practice setting. The results uncovered the following core themes, which give some shape to the emerging discourse of how to address digital professionalism in health professions settings:

- Privacy and reputation concepts—sustaining public trust in the profession; policy and guidelines regarding the use of electronic communication
- Educational interventions to optimize digital professional behavior
- Positive use cases for clinician-led electronic communication (if digital professionalism is properly adhered to)

We will look at these areas in brief and offer our thoughts on this 21st century dilemma where digital media has begun to blur the boundaries between personal and professional life. Specifically, we wish not only to offer a lens to bring into focus the problems and challenges but also to offer ideas for forming discourse(s) on finding solutions.

Based on the themes uncovered from current literature, we propose digital professional behavior involves a conscious awareness of privacy, policy, fluency through literacy and education, and a common understanding of generational perceptions regarding uses of Information and Communications Technologies (ICTs) and social media platforms professionally and privately. Questions to pose when and before engaging in use of digital media should follow these tenets to ensure professional standards can be met: what are the impacts, if any, on Personal Health Information or patients' privacy when I share this information in online and digital environments? Have I influenced or educated my patient and/or the patient's family of the impact of sharing personal health information via electronic communication? Is there consent by all parties communicating online? Are my colleagues in agreement with the type and nature of communication regarding patient care and treatment planning? What impact does this action potentially hold for my professional reputation and the reputation of the institutions I work for? Is the behavior presented on this social media platform reflective of professional standards? Can this representation online impact the broader public's perception of my practice and or professional integrity? We hope by the end of this chapter that these questions can begin to be thoughtfully answered.

DIGITAL PRIVACY AND SAFETY IN THE HEALTH PROFESSIONAL CONTEXT

We begin this section by restating the ethical high bar of which healthcare must attain. Patients must be open to sharing their deepest and sometimes darkest thoughts, behaviors, and habits with who they presume is an honorable healthcare provider. To do so without hesitation, they must be reassured that whatever is shared between them remains private and confidential. Without this reassurance, gleaned both from the interaction with an individual clinician as well as from the knowledge that the clinician has undergone rigorous professional training to uphold confidentiality, the practice of medicine becomes eminently more difficult. While clinicians are well aware of their legal and moral obligations to protect patient health information in general, lack of awareness of the specific privacy risks inherent in the use of electronic communication tools has become a significant problem.

Indeed, the rapid integration of electronic communication, applications, and tools into the practice of medicine since the turn of the century has had a profound impact on the way health care personnel function. For example, primary care physicians across the Western world have had substantial adoption of use of email communication with patients, from a low of 15% in Canada up to 80% in Switzerland [5]. With these dramatic changes the learning curve for clinicians has certainly been steep. Not only have they had to learn to chart on a computer rather than on paper, but also to manage the myriad complexities involved in adhering to the principles of confidentiality in a computerized world. On the surface, this might seem simple; however, that is far from reality. Clinicians who used to only be concerned about ensuring their paper charts were stored in a locked cabinet, now need to think about far more complex issues such as email security, location of servers, hacking, and social media profiles. Taken together with the widely varying incentive structures to adopt electronic communication among the different nations and the reasons for the still relatively low uptake become clear [6].

In fact, current social media practices provide a clear picture of some of these issues. When Facebook was initially launched in 2004, it was aimed at Harvard university students, to enable them to create social networks. It gradually expanded beyond Harvard, and in 2006 it opened up to the public. This was the dawning of the age of social media. Over the subsequent ten years, use of social media has gone from the purely personal realm, to a point now where it is firmly entrenched in the professional sphere. In medicine, this has presented countless additional challenges as people learn to meld these two worlds and establish appropriate boundaries. Some do so better than others. While younger physicians are, for the most part, quite comfortable with the use of electronic communication and social media, this comfort level does not always translate well to appropriate professional use. These same clinicians, given their familiarity with these electronic tools, may also have less reservations about using them with, or about, patients. In 2015, the Medical Journal of Australia published results of an online survey of 17,000 Australian medical students in all 20 of the country's medical schools. Of the respondents, 99.4% reported using Facebook, of whom 35% reported posting unprofessional content, such as evidence of intoxication (34.2%) and patient information (1.6%) [7].

Various professional bodies have published guidelines and recommendations on how physicians should be using various electronic platforms to communicate with patients; however, as will be described in a later section, many healthcare providers do not intuitively adhere to digital health professionalism, and have asserted that there has been insufficient instruction of it in medical training. As a result, many clinicians conduct their day-to-day activities unaware they may be breaching expected professional standards. Certain extremely controversial cases have even become media fodder. In December 2015 a group of male dentistry students at Dalhousie University in Halifax, Canada formed a "closed" Facebook group called the Dalhousie Dentistry (DDS 2015) gentleman's club. The group discussed, among other things, using chloroform on women and voted on which women from their class they would like to have "hate sex" with [8]. While the students thought their supposedly "closed" Facebook group was completely private, they quickly found out this was not the case. Thirteen students were suspended from clinical practice for two months because the group, with its sexist and derogatory remarks, was discovered and made public. This wildly inappropriate use of a social media platform by professionals in training was nothing short of shocking, and could someday be similarly used in discussing patients. What sort of best practices currently exist to prevent such a failure in professionalism? In this section we will explore the important aspects of digital health professionalism regarding email and messaging, as well as social media use.

EMAIL, TEXT, AND INSTANT MESSAGING

Email, Short Message Service (SMS) text messaging, and popular instant messaging over platforms such as iMessage, BBM, WhatsApp, and Skype are all very convenient, which is why clinicians and patients like them. But imagine for a moment using standard email software to send personal medical information to a patient—and getting the email address wrong. Worse—the email does not bounce back, but rather appears in the mailbox of an unintended recipient. The risks of interception or errors in sending email, texts, or instant messages can be significant. As a result, some understanding of the technical aspects and limitations listed in Text Box 4.1 of sending a message or email securely becomes crucial. Without this knowledge, clinicians would not be able to actively consider important ethical issues that may derive from the use of unsecured electronic communications. Summarized in Text Box 4.2 are some common ethical considerations with which

TEXT BOX 4.1

TECHNICAL CONSIDERATIONS OF ELECTRONIC COMMUNICATIONS BETWEEN CLINICIANS AND PATIENTS

- Are the parties using validated secure email accounts or a secured third party messaging service? For example, while some of the aforementioned popular instant messaging platforms have built-in encryption, they were not programmed to secure sensitive patient health information, which is why specifically designed third-party messaging platforms for interactions between clinicians and patients exist.
- Is the clinician's server secured according to known specifications?
- Mobile devices can be lost or stolen, and may be accessed by others. Is the device, mobile or otherwise, being used to

TEXT BOX 4.1 *(cont'd)*

generate messages adequately protected (password-protected, built-in encryption, multifactor authentication enabled, etc.)?

- Electronic communication with patients should be part of the patient's permanent medical record. If a clinician's EMR system does not have built-in e-mail functionality, additional work such as use of the copy and paste function will be needed to include emails into patients' medical records. Even more laborious is recording mobile-only SMS or instant messaging communications into patients' charts (seemingly, taking a screen shot of the communication and storing it as an image is the only known method). Is that feasible?
- SMS messages specifically pose exceptional technical challenges: they are transmitted unencrypted over unsecured cellphone tower networks. Should SMS ever be used to communicate with patients?
- Have requirements of all relevant governing bodies for use of electronic communication, which typically take all the above into account, been met?

TEXT BOX 4.2

ETHICAL CONSIDERATIONS OF ELECTRONIC COMMUNICATIONS BETWEEN CLINICIANS AND PATIENTS

- Messages can inadvertently go to the wrong recipient—due to a typo in the email address or phone number, change in someone's email address or number, email or instant message account being hacked, etc.
- Even with the correct email address or phone number, the sender has no idea who is viewing the recipient's messages. For example, many people have their email, text or instant messages set up to automatically pop up on their computers or phones when it arrives; however, the sender has no idea who is using the computer or phone at the receiver's end. It could be the patient's spouse, child, or any other user.
- Although clinicians, or the institutions they work for, often add disclaimers at the bottom of an email, such a disclaimer does not negate a clinician's responsibility to protect sensitive patient health information.
- If patients are using an employer's or a third party's email or messaging system, the employer/third party may have the right to access the communications. Regulations in this area are constantly in flux and vary by jurisdiction.

clinicians must become familiar. Use of email and messaging for clinical care clearly requires much more consideration than use of these tools in the personal domain.

For all of the above reasons, as well as concerns about workload issues [9], many clinicians today have chosen not to communicate with patients electronically; however, for clinicians who do decide to do so, there are some basic tenets that should be clearly established. In terms of the clarity of email or message communications, the onus remains on clinicians not to use acronyms and medical jargon, while being mindful that "it is difficult to communicate humor, wit, sensitivity", in electronic written language [10]. In contrast to this, with regard to the timeliness of these communications with patients, the onus is actually on patients to follow up on emails and messages that have not yet been responded to. Electronic messages can arrive hours or days after they are sent, and the receiving party may not review incoming messages in a timely manner. Given this, clinicians should make known to all clinic stakeholders that it is the patients' responsibility for following up. But these same stakeholders generally agree that a target response time is required to maintain a high quality of care. So if a patient has not received an electronic response from a clinician within 48 hours for example, clinic policy should be that she should simply call the clinic at that time and inform that she has yet to receive a response. Furthermore, patients must be advised not to use email for urgent time-sensitive issues. Taken together, patients should be aware of the potential risks of electronic communication and agree to assume those risks. A signed consent form that outlines the risks of email communication, as well as the obligations placed on patients who wish to correspond via email, provides a permanent record of the consent given. See Fig. 4.1 for an example of a Patient Consent for Email and Other Electronic Communications that adheres to these digital health professional themes, issued by the Canadian Medical Protection Association (CMPA) [11].

SOCIAL MEDIA

The key professional issues described above are similar yet even stricter when it comes to social media, as outlined in Text Box 4.3. Social media can be defined as a set of web-based and mobile technologies that allow people to monitor, create, share or manipulate text, audio, photos or video, with others. This information can be shared unidirectionally, such as posting text to one's own blog, or multi-directionally, such as contributing to a discussion on an online forum. Social media places particular emphasis on interactive, user-driven communication. Undeniably, websites such as Facebook, Twitter or LinkedIn have changed the way people build relationships, communicate, interact and gather and disseminate information. As such, consensus among legal experts, biomedical ethicists and professional bodies is that social media must be considered a virtual public space, used by millions and potentially accessible by anyone. Postings are never truly anonymous or private. Furthermore, anything posted on social media is subject to the same laws as any other written media, including "copyright, libel, defamation and plagiarism." [12] Indeed for a professional, postings on social media platforms, even in one-to-one chats, should be perceived as the equivalent of posting on the front page of a newspaper.

CONSENT TO USE ELECTRONIC COMMUNICATIONS

This template is intended as a *basis for an informed discussion*. If used, physicians should adapt it to meet the particular circumstances in which electronic communications are expected to be used with a patient. Consideration of jurisdictional legislation and regulation is strongly encouraged.

PHYSICIAN INFORMATION:

Name:

Address:

Email (if applicable):

Phone (as required for Service(s)):

Website (if applicable):

The Physician has offered to communicate using the following means of electronic communication ("the Services") [check all that apply]:

☐ Email	☐ Videoconferencing (including Skype®, FaceTime®)
☐ Text messaging (including instant messaging)	☐ Website/Portal
☐ Social media (specify):	
☐ Other (specify):	

PATIENT ACKNOWLEDGMENT AND AGREEMENT:

I acknowledge that I have read and fully understand the risks, limitations, conditions of use, and instructions for use of the selected electronic communication Services more fully described in the Appendix to this consent form. I understand and accept the risks outlined in the Appendix to this consent form, associated with the use of the Services in communications with the Physician and the Physician's staff. I consent to the conditions and will follow the instructions outlined in the Appendix, as well as any other conditions that the Physician may impose on communications with patients using the Services.

I acknowledge and understand that despite recommendations that encryption software be used as a security mechanism for electronic communications, it is possible that communications with the Physician or the Physician's staff using the Services may not be encrypted. Despite this, I agree to communicate with the Physician or the Physician's staff using these Services with a full understanding of the risk.

I acknowledge that either I or the Physician may, at any time, withdraw the option of communicating electronically through the Services upon providing written notice. Any questions I had have been answered.

Patient name:	
Patient address:	
Patient home phone:	
Patient mobile phone:	
Patient email (if applicable):	
Other account information required to communicate via the Services (if applicable):	
Patient signature:	Date:
Witness signature:	Date:

FIGURE 4.1 Example of a patient consent for email and other electronic communications. *Reproduced with permission from Canadian Medical Protective Association (CMPA) Canadian Medical Protective Association (CMPA), Consent to Use Electronic Communications [Internet]. CMPA. 2017 [cited 24 February 2017]. Available from: https://www.cmpa-acpm.ca/documents/10179/301287261/com_16_consent_to_use_electronic_communication_form-e.pdf.*

APPENDIX

Risks of using electronic communication

The Physician will use reasonable means to protect the security and confidentiality of information sent and received using the Services ("Services" is defined in the attached Consent to use electronic communications). However, because of the risks outlined below, the Physician cannot guarantee the security and confidentiality of electronic communications:

- Use of electronic communications to discuss sensitive information can increase the risk of such information being disclosed to third parties.
- Despite reasonable efforts to protect the privacy and security of electronic communication, it is not possible to completely secure the information.
- Employers and online services may have a legal right to inspect and keep electronic communications that pass through their system.
- Electronic communications can introduce malware into a computer system, and potentially damage or disrupt the computer, networks, and security settings.
- Electronic communications can be forwarded, intercepted, circulated, stored, or even changed without the knowledge or permission of the Physician or the patient.
- Even after the sender and recipient have deleted copies of electronic communications, back-up copies may exist on a computer system.
- Electronic communications may be disclosed in accordance with a duty to report or a court order.
- Videoconferencing using services such as Skype or FaceTime may be more open to interception than other forms of videoconferencing.

If the email or text is used as an e-communication tool, the following are additional risks:

- Email, text messages, and instant messages can more easily be misdirected, resulting in increased risk of being received by unintended and unknown recipients.
- Email, text messages, and instant messages can be easier to falsify than handwritten or signed hard copies. It is not feasible to verify the true identity of the sender, or to ensure that only the recipient can read the message once it has been sent.

Conditions of using the Services

- While the Physician will attempt to review and respond in a timely fashion to your electronic communication, **the Physician cannot guarantee that all electronic communications will be reviewed and responded to within any specific period of time. The Services will not be used for medical emergencies or other time-sensitive matters.**
- If your electronic communication requires or invites a response from the Physician and you have not received a response within a reasonable time period, it is your responsibility to follow up to determine whether the intended recipient received the electronic communication and when the recipient will respond.
- Electronic communication is not an appropriate substitute for in-person or over-the-telephone communication or clinical examinations, where appropriate, or for attending the Emergency Department when needed. You are responsible for following up on the Physician's electronic communication and for scheduling appointments where warranted.
- Electronic communications concerning diagnosis or treatment may be printed or transcribed in full and made part of your medical record. Other individuals authorized to access the medical record, such as staff and billing personnel, may have access to those communications.
- The Physician may forward electronic communications to staff and those involved in the delivery and administration of your care. The Physician might use one or more of the Services to communicate with those involved in your care. The Physician will not forward electronic communications to third parties, including family members, without your prior written consent, except as authorized or required by law.
- You and the Physician will not use the Services to communicate sensitive medical information about matters specified below [check all that apply]:
  - ☐ Sexually transmitted disease
  - ☐ AIDS/HIV
  - ☐ Mental health
  - ☐ Developmental disability
  - ☐ Substance abuse
  - ☐ Other (specify):
- You agree to inform the Physician of any types of information you do not want sent via the Services, in addition to those set out above. You can add to or modify the above list at any time by notifying the Physician in writing.
- Some Services might not be used for therapeutic purposes or to communicate clinical information. Where applicable, the use of these Services will be limited to education, information, and administrative purposes.
- The Physician is not responsible for information loss due to technical failures associated with your software or internet service provider.

Patient initials______

FIGURE 4.1 (Continued)

APPENDIX CONTINUED

Instructions for communication using the Services
To communicate using the Services, you must:

- Reasonably limit or avoid using an employer's or other third party's computer.
- Inform the Physician of any changes in the patient's email address, mobile phone number, or other account information necessary to communicate via the Services.

If the Services include email, instant messaging and/or text messaging, the following applies:

- Include in the message's subject line an appropriate description of the nature of the communication (e.g. "prescription renewal"), and your full name in the body of the message.
- Review all electronic communications to ensure they are clear and that all relevant information is provided before sending to the physician.
- Ensure the Physician is aware when you receive an electronic communication from the Physician, such as by a reply message or allowing "read receipts" to be sent.
- Take precautions to preserve the confidentiality of electronic communications, such as using screen savers and safeguarding computer passwords.
- Withdraw consent only by email or written communication to the Physician.
- **If you require immediate assistance, or if your condition appears serious or rapidly worsens, you should not rely on the Services.** Rather, you should call the Physician's office or take other measures as appropriate, such as going to the nearest Emergency Department or urgent care clinic.
- Other conditions of use in addition to those set out above: *(patient to initial)*

I have reviewed and understand all of the risks, conditions, and instructions described in this Appendix.

Patient signature ______________________

Date ______________________

Patient initials______

FIGURE 4.1 (Continued)

Given this, what principles should underlie a clinician's rules of engagement on social media? Online boundaries are often difficult to negotiate, even without putting patients in the mix. For example, many try to use Facebook for personal matters, and LinkedIn for professional matters, but even this boundary can be blurred. What does a clinician do when a patient sends them a contact request on either or both of these platforms? Should a clinician even look at a patient's Facebook profile when it comes up as a suggested contact? Other platforms like RateMyMD offer an "anonymous" platform for patients to

TEXT BOX 4.3

PARTICULAR CONSIDERATIONS FOR SOCIAL MEDIA USE BY HEALTHCARE PROFESSIONALS

Patient confidentiality

- The privacy and security of individual patient information is paramount and should never be shared beyond the circle of care. As with email and text, communicating with an individual patient in other than a face-to-face environment requires the use of a secure electronic communication platform. Identifiable patient information, including images, should never be posted or shared online.
- When using social media, physicians should endeavor to use the most stringent security and privacy settings available for the particular platform.
- Clinicians and clinic managers should make all clinic staff aware of issues concerning patient confidentiality in their own use of social media. Consideration should be given to instituting a social media policy for the office or practice.

Professionalism

- Having an online profile or identifiable presence on social media can have the same degree of positive or negative impact on a physician's social reputation as being active in any other public venue. In fact, having access to a global audience can magnify this reputation.
- The most effective use of social media often involves communicating information that is both personal and professional. However, physicians must retain the appropriate boundaries of the patient-physician relationship when dealing with individual patients. The same standards of professionalism that would apply in face-to-face physician-patient interactions also apply in electronic interactions.
- If a physician is an employee of a health care institution or organization that has social media guidelines in place, he or she should review these and act accordingly.

Online communication issues

- Social networking sites cannot guarantee confidentiality. Anything written on a social networking site can theoretically be accessed and made public. For example, the Patriot Act in the United States makes it possible for the U.S. government under certain conditions to access any information posted on a social networking site or website hosted by a U.S. service provider, even if this information is located within the "private or direct message" area of the site. Electronic communications are not anonymous and are always stored in some form. As such, it is possible to trace the author of a comment even if posted anonymously.
- Once their material is published online, authors of comments on social media sites no longer control how and where the information is disseminated, and these comments can sometimes lose context.
- Postings to social media sites are subject to the same laws of copyright, libel, and defamation as written or verbal communications.

comment on their doctor's practice. What should a clinician do when she sees negative comments about her and the care she provides on this public platform? The problems here are myriad. Social Media platforms often change their terms of service without notice so even the most vigilant user might unwittingly open their contacts to a third party or be subjected to unsolicited advertising based on keyword searches. Ratings platforms provide a false sense of security when they offer features such as "anonymous" posting of comments. The truth is that it is quite easy for a physician to be able to identify a patient based on just the comments themselves. Whether we are aware of it or not our common notions of anonymity, privacy, and safety are directly challenged by algorithms we have no control over. Reflecting and addressing these sorts of questions are even more challenging for younger clinicians, who have grown up with and use electronic communication almost reflexively the way they have always used them in their private life. Another scenario to consider is whether it is a clinician's ethical duty to correct misinformation when it is identified on a social media platform or website.

First of all, it is crucial to understand the various social media platforms and their possible audiences. These platforms function in different ways and often have different goals. Even broad types of social media such as social networking sites have different terms and conditions under which they operate. Some social networking sites, for example, Facebook, are intended for use by everyone, with individual users selecting a network of contacts they know personally. Others, like Twitter, are designed for interacting with people the user might not know at all. And others still, such as Asklepios, are intended for peer-to-peer interaction between clinicians only. In order to use social media effectively, it is necessary to have a good grasp of how they function and who the intended audience will be before using them.

Second, transparency when using social media as a health professional should be the modus operandi. A clinician user must clearly identify who they are and any potential conflicts of interest they may have in association with the information they are providing. Being transparent encourages more honest interaction with others and a more productive outcome. If the clinician user is discussing relevant medical issues, it is probably beneficial to themselves as clinicians. If an institution or organization employs them, they should state either that they are reflecting corporate policies or that the views expressed are theirs alone and not those of their employer.

Third, when participating on a social networking site such as Facebook that may include patients from the users' practice, they should avoid communicating personal or private information. In fact, biomedical ethicists continue to debate whether even "friending" on social media is ever truly appropriate between clinician and their patients. Even if the clinician restricted some viewable personal information to the patient, what are the expectations of the clinician if she suddenly has open access to the patient's social media presence and daily personal life? If she sees something there that she would qualify as damaging to that patient's health, is she ethically required to engage with the patient professionally on that issue? To avoid these ethical pitfalls, it is generally recommended that no "friending" take place and that a professional page be established instead, separate from a clinician's personal social media page. On this page, patients can only follow information about the practice and general health information and links, thereby minimizing the risk of any ethical transgressions by the clinician.

Finally, it is strongly recommended that clinicians' posts on social media remain focused on their areas of expertise. As a clinician, one can often bring most value to a forum or conversation by discussing issues on which he or she has a particular expertise. Sharing this information—as long as it does not contravene individual patient confidentiality—raises the level of discourse on social media sites and is likely to be viewed favorably by other participants (as will be discussed later). Clinicians should anticipate that the information they provide on social media may be challenged by both other clinicians and non-clinicians. If this happens, it is imperative for the clinician to keep the debate online to a civilized level. Similarly, clinicians should not be unnecessarily offended if their viewpoint is rejected, even if they do feel it is based on best available evidence. If clinicians choose to use their own website to communicate to a nonmedical audience about medical issues, they should include a terms of use agreement to advise users that the information is intended for citizens of their jurisdiction and that individual health queries will not be addressed.

IMPLEMENTING ELECTRONIC COMMUNICATIONS RECOMMENDATIONS IN THE REAL WORLD

In healthcare, digital professionalism does not exist in a vacuum. Clinicians in their private lives use convenient twenty first century modes of electronic communication just like everyone else. However, the way these tools are used professionally requires more consideration than the way they are used personally. For example, while clinicians might not hesitate to post pictures of their family's vacation on Facebook, posting pictures of what they do at work on the same channel is strictly regulated. As such, there is tremendous interplay between many factors in determining the actual implications on clinical practice—what is professionally acceptable and what is not. Generally speaking, clinicians need to navigate a complex set of relatively new, and sometimes conflicting, rules issued by bodies that make recommendations or guidelines about how to use electronic communication professionally. Separate policies may be issued by a medical professional insurance body, the national regulatory body, the state or provincial privacy commissioner and professional college, as well as the clinician's local hospital or clinic administration. Overarching themes have informed the policies of these regulatory bodies, including defining the legal boundaries and ethics of privacy, confidentiality and security of the electronic communication, clarifying the expected timeliness of responses, and ensuring the clarity of communication. It is the clinicians' responsibility to distill all these themes to a coherent set of policies for their unique set-up in their jurisdiction. For a busy clinician, it is a laborious process to even navigate the plethora of recommendations and guidelines devoted to electronic communications.

So how does implementing these recommendations on electronic communication impact actual clinical workflow? In fact, it makes using email, text, instant messaging, and social media much more challenging, and in many cases, negates the benefits, such as convenience, in the first place. Third party applications do exist to enable communication to meet the technical requirements; however, most require the added step of clinicians and patients signing on to another application in order to communicate. So, while ensuring that all of the technical and ethical aspects are adequately considered will help to make electronic communication with patients safer, the inevitable downside is a decrease in the main benefit most clinicians see in this mode of communication—ease and speed. Currently, many clinicians are using email to communicate with patients without sufficient consideration of existing recommendations—some due to lack of knowledge about the existence of such recommendations, and others due to the perceived negative implications of following them. Furthermore, in academic centers, faculty often use text messages to communicate with residents about patients, sometimes even including patient identifying information. For those who are using email or social media to communicate with patients with appropriate safeguards in place, there still remains the challenge of ensuring any documentation of these communications get inserted into the patient's electronic medical record file. All this adds up to a difficult assessment by clinicians and clinic managers whether to invest the time and resources needed to ensure all electronic communication recommendations are being properly followed. With the increasing prevalence of use of electronic communication in healthcare, regulatory bodies across the world are becoming more proactive in ensuring medical professionals are abiding by their recommendations. As an example, in early January 2016, a nurse in Saskatchewan, Canada was charged with professional misconduct by the Saskatchewan Registered Nurses Association (SRNA) after commenting about the medical care her grandfather received in hospital [13]. She posted both positive and negative comments about the care her grandfather received on her Facebook site, and identified herself as a registered nurse. The SRNA charged her with violation of confidentiality, failure to follow proper channels, impact on reputation of facility and staff, failure to first obtain all the facts, and using status of registered nurse for personal purposes under the Code of Ethics for Registered Nurses. To avoid these sorts of unfortunate stories, it behooves all clinicians to start taking professionalism in digital health seriously.

DIGITAL REPUTATION IN THE HEALTH PROFESSIONS CONTEXT

While texting and email involving patient identifying information remain a challenge to practice workflow, communications activity that happens outside of the clinical context poses added challenges to the digital professionalism fore. As noted in the case of the Saskatchewan Nurse, the SRNA pointed to detrimental reputational issues related to her use of social media. The Internet and social media have fostered a ripe environment for anonymous communication and also novel and frequently excessive public communication. The consequences of this new ability are playing out before us. Law enforcement officers, transit workers, and shop clerks are regularly called out by the public for unprofessional behavior. The medical profession is not immune to this public scrutiny.

Numerous examples have played out in the media and depending on where you are located you will most likely be able to recall an example of how social media shone a spotlight on some of the inherent challenges to institutional and individual reputation; An innocent video of medical learners blowing off steam in parodying life in an anatomy lab [14], the aforementioned dentistry "gentleman's club" on Facebook [8]—fill in the blank, the public can search and view professional misconduct at a moment's notice. It is here, in the public domain, where all context is lost and where observers are free to judge for themselves the appropriateness of a viral video performance, comment, or photo opening the door for all kinds of reputational damage not only for the learners themselves, but also for medical and educational institutions.

In *e-Professionalism: A New Frontier in Medical Education* Kaczmarczyk discusses how physicians might mitigate the risks to personal and institutional reputation:

because e-Professionalism includes an online persona, e-Professionalism is an essential and increasingly important element of professional identity formation and medical education. Similarly e-Professionalism encompasses behaviors involving social media and, therefore, should be included in the development of professional value actions and aspirations in medical education [15].

Kaczmarczyk and Ellaway [15,25] characterize digital reputation as the effects of digitally mediated environments on individual, professional, and or institutional reputation. Distinctions across disciplines, and in different circles of how newer digital technologies are perceived produce what James refers to as *digital tensions* [16]. In his article *Out of the box: The perils of professionalism in the digital age,* James quotes Adams who posed the following set of questions as frame for understanding general perceptions of technology use in professional settings suggesting:

1. everything that's already in the world when you're born is just normal;
2. anything that gets invented between then and before you turn thirty is incredibly exciting and creative and with any luck you can make a career out of it; and
3. anything that gets invented after you're thirty is against the natural order of things and the beginning of the end of civilization as we know it until it's been around for about ten years when it gradually turns out to be alright really [16].

This idea of digital tensions corroborates the *digital native* trope so common today, which has permeated discourses on learners in the last 10 or so years since Mark Prensky coined the term in his article *Digital natives digital immigrants* [16]. In the article, he posits students of today are no longer the people our education system was designed to teach, that this generation of learners were born and have spent their entire life surrounded by the tools of the digital age and essentially think and process information fundamentally differently from their predecessors. The differences go for further and deeper than most educators expect or realize [17].

Adams' quote in the James' summary of understanding perceptions of technology above [16] reflects some of the inherent challenges in medicine when trying to address the teaching of digital professionalism. That is to say those in Adams' group 3 (faculty) are the ones "teaching" those in group 2 (medical students). It's challenging for faculty who are in this headspace about technology to effectively use it, let alone teach it.

TEACHING DIGITAL PROFESSIONALISM IN THE HEALTH PROFESSIONAL CONTEXT

Can Digital Professionalism be taught? Historically professional behavior was thought to be intrinsic to a medical practitioner. Professional behaviors were learned by osmosis and not taught [2]. Today professionalism is a skill set and a cornerstone competency for the modern medical teaching curriculum [18]. One of the most common themes emergent from the professionalism literature is curriculum as formal, informal, and hidden [2,18,22,23]. Now at a time where much change is already taking place within medical education, at a time when curricula are being overhauled and renewed for competency-based models, there is room for institutions and educators to champion professional digital behavior both explicitly within the formal curriculum and implicitly within the informal curriculum and to a degree within the hidden curriculum by attempting to mitigate the risk of unprofessional behavior on social media platforms. While digital professionalism is relatively new on the curriculum renewal landscape there are plenty of scholarly models for teaching medical professionalism to help guide the teaching of professional digital behavior in health sciences curricula. Some potential new models are also embedded within the recent literature specifically dedicated to digital professionalism.

In *Professionalism in Medicine* Thistlethwaite suggests how the informal or hidden curriculum can be addressed by using an appreciative narrative-based approach to bring about organizational change and align the informal curriculum with the formal one through promoting mindfulness on the part of every faculty member, residents, and staff members about the values we exhibit and thereby teach in our everyday interactions [2]. In *Teaching Medical Professionalism*, Cruess and Cruess put forth some guiding principles for supporting professional identity formation [18], and in *Understanding Medical Professionalism*, Levinson et al. explore the teaching professionalism first characterizing formal vs. informal vs. hidden curriculum and then by looking at educational theories to help guide the teaching of professionalism in these domains [19].

Ellaway and Bates address digital professionalism in two papers written as part of an environmental scan on the Future of Medical Education in Canada [20]. The literature produced from that scan posited that faculty development is a vital component to integrating digital technology within medical education settings. The medical education literature includes many topics that touch on the area of digital and or online professionalism but there is a gap in the literature in terms of or approaches to explicitly and implicitly addressing such content within the curriculum.

IMPLEMENTING EDUCATIONAL INTERVENTIONS FOR DIGITAL PROFESSIONALISM IN THE HEALTH PROFESSIONS

There are opportunities to explore current models for teaching professionalism and identify which methods best suit the teaching of professionalism in the digital context. To further this idea, the following models can act as a catalyst for looking at ways in which to address the teaching of digital professionalism within the health professions: exploring threads of digital behavior in current Competency-Based Education frameworks (CBE), exploring the concepts of focused faculty development, formal curriculum, informal curriculum, and the hidden curriculum in the professionalism literature. Table 4.1 illustrates teaching points and examples for each area of focused faculty development strategies.

TABLE 4.1 Guide for Addressing Digital Professionalism from a Curricular Standpoint

	Faculty Development	Formal Curriculum	Informal Curriculum	Hidden Curriculum
Teaching Points	Faculty need to know foundations of and gain fluency in technology usage in order to role model in their teaching.	Learners though often adept at new technologies often don't use them in a way that upholds the tenets of professionalism	When institutions don't acknowledge digital media as a part of the teaching milieu it is difficult to role model professional and positive digital behavior	Many unintended lessons are learned
Approach	Sessions on the following: • Modes and uses of electronic communication/social media for teaching • How to separate personal and professional identities and boundaries • How to set up their privacy settings on social media? • Guidelines Risks and Benefits (e.g., What bodies provide guidance, and have legal jurisdiction, on how electronic communication and social media are used in the professional context?)	Approach curriculum reviews with a goal to incorporate digital professionalism when possible Adapt existing lessons on professionalism to include professional behavior in digital environments Set an expectation for fluency in professional digital behavior	Identify which media are acceptable to use and make it known Provide faculty workshops on how to incorporate digital media into their teaching practice Identify faculty who are fluent in digital media to guide colleagues	Review policies and guidelines within institution Are there any involving digital professionalism? If yes how know are they? If no develop some Elicit feedback from learners and faculty around how digital professionalism can become a cultural norm Aim for cultural change that fosters positive lessons learned
Methods	Focused faculty development sessions	Any method that is currently used in the formal curriculum—here it is the content that matters most	Role modeling	Surveys Focus Groups

HOW IS THE DIGITAL ADDRESSED IN CURRENT COMPETENCY-BASED EDUCATION FRAMEWORKS?

Since the early 2000s, health professions education in North America and Europe have moved from time-based teaching models to competency-based medical education (CBE, CBME). CBME as a model places an emphasis on two key areas: competency on a continuum and competence across multiple domains of practice. Though frameworks vary slightly across jurisdictions and health professions, generally speaking, competency models emphasize the intrinsic competencies as much as the domain of (Medical) Expert. The intrinsic competencies include professionalism, communication, health advocacy, leadership, collaboration, and scholarship. This model is an increasingly familiar one in the health professions and most cogently reflected in the CanMEDS 2015 Physician Competency Framework [1].

Within the CanMEDS framework many aspects of digital professionalism can be identified as embedded within a broader "cross-cutting concept" referred to as eHealth which has been defined as follows: "the appropriate use of information and communication technologies for health service delivery, education, and research" [21].

The inclusion of eHealth as an element that streams across competencies is a very good start to the discourse around defining digital professionalism within the health professions. eHealth as it appears in the CanMEDS framework introduces subjects such as data handling, adoption of ICTs, information sharing, information technology standards, and use of health informatics to enhance quality of patient care, to name a few. There can be no argument that all practitioners should be proficient in the use of electronic tools and that is made quite clear with the inclusion of eHealth to the competency framework. Digital professionalism, however, as a well-defined concept should be reflective of how the above areas are understood, addressed, and carried out in practice. Take for example the following eHealth enabling competency for the role of Medical Expert: "*Adopt a variety of information and communication technologies to deliver patient-centered care and provide expert consultation to diverse populations in a variety of settings.*" (21) The adoption of such technologies may demonstrate the utility of a selected tool in terms of achieving the end goal of expert consultation and patient-centered care. From a digital professionalism standpoint, however, there should also be present a demonstrated rationale for technology choice such as privacy, security, and patient consent to electronic communications.

A defined digital professionalism framework will bring a set of minimum guidelines to affect behavior change around ICT usage across the competence continuum. Examples of such guidelines might include instilling in practitioners an intrinsic understanding of the most effective uses of ICTs, email, EHRs, telephone communications, and even some social media platforms, that will lend themselves better to patient care within different contexts. That is to say, that each vehicle differs in very nuanced ways at times and so the efficacy of each is largely dependent on a clinician understanding which fits the right context. Some areas to address within a potential framework might include key features such as outlining the various modes of communication available to clinicians, which ones are actually secure from an information security and patient privacy perspective, understanding the technical or infrastructure requirements for each mode of communication, and to educate around proper usage of ICTs. These ideas are explored earlier in Text Boxes 4.1–4.3.

ADDRESSING DIGITAL PROFESSIONALISM THROUGH FOCUSED FACULTY DEVELOPMENT

Some of the issues practicing clinicians who teach face are quite basic in nature. A faculty development model that asks and answers such questions as: What modes of electronic communication/social media exist? How do I separate my personal and professional identities on social media and create appropriate boundaries? What bodies provide guidance, and have legal jurisdiction, on how electronic communication and social media are used in the professional context? How should clinicians set up their privacy settings on social media?

Further teaching points may include: awareness and understanding of the potentials and pitfalls to using social media in medical practice; benefits of social media in one's own professional practice; a foundational overview of current media and how are they are being or may potentially be used. Focused faculty development curricula would ensure educators have a certain level of fluency and understanding around the use of social and other digital communications media. Thus, better equipping them for teaching in, with, or around these spaces.

ADDRESSING DIGITAL PROFESSIONALISM THROUGH THE FORMAL CURRICULUM

In the formal curriculum, digital professionalism can be explicitly taught as particular point within existing curricula already focused on the professionalism competency.

Fig. 4.2, below is what Kung et al. [22] refer to as *Focused reflective case based teaching* (journal club or academic half days).

TABLE 1. Digital Professionalism Presession Handout

Physician use of social media is on the rise and is evolving as a means of communication. The boundary between personal and public aspects of physicians' lives on the web can often be unclear. Even if inadvertent, breaches of patient privacy can occur via Facebook, Twitter, or other media despite using secure privacy settings. So how can physicians use social media safely?

In this session, we will discuss some of the recent cases surrounding social media as well as radiology-specific case scenarios to better understand the issues surrounding patient confidentiality, what constitutes acceptable online professional conduct, and how to safely incorporate social media into both personal and professional lives.

Clinical Cases:

Employees used a camera phone to take a picture of an x-ray and text it to one another. The image also was posted to one employee's personal Facebook page, but has since been removed and no information identifying the name or other information about the patient was displayed with the image.

Question: Is this a breach of patient privacy? Should these employees be disciplined?

Dr. T. communicated a few of her clinical experiences in the hospital's emergency department on her Facebook account but never specifically named the patients. However, one patient's injuries were such that an unidentified third party was able to identify the person.

Question: Should the doctor be found guilty of "unprofessional conduct" and should her clinical privileges be terminated?

Radiology department-specific scenarios:

1. You are on your Facebook account and a friend request pops up from your attending, Dr. S. Should you accept the invitation?
2. You are applying for fellowship training at a prestigious Institution. As part of the fellowship application, you are requested to allow the fellowship director to review your Facebook account. Is this request acceptable to you? Should you disclose your Facebook account?
3. One of the patients you recently imaged in the breast imaging section contacts you by e-mail as her recent breast biopsy was positive for malignancy. She tells you that her surgeon has recommended mastectomy. The patient wants to undergo bilateral mastectomy for definitive treatment and risk reduction but is concerned about the ultimate cosmetic result. She asks you to provide her with the name of the plastic surgeon you would recommend. How should you respond to this request? Does it matter if you performed any of her imaging studies and/or biopsy?

FIGURE 4.2 Digital professionalism precession handout. *Excerpted from Kung et al. © Elsevier.*

ADDRESSING DIGITAL PROFESSIONALISM THROUGH THE INFORMAL CURRICULUM

The informal curriculum can be viewed as a learning and teaching activity that occurs outside of the formally recognized and documented curriculum [19]. In the case of health professions it is best characterized by the teaching and learning that happens in the clinical setting. How this is approached, however, is tricky and largely depends on the academic healthcare setting of the learners. Very large schools will see residents covering rotations in multiple hospitals across multiple programs and departments, all with their own set of institutional norms. The most appropriate way in which to address the teaching of digital professionalism here is to role model professional and positive digital behavior.

One example of this might be to have learners and faculty follow a favorite Twitter #hashtag or podcast that is relevant to a particular rotation together. The success of this approach, however, is largely dependent on the faculty clinician's comfort level with the use of social media platforms. Other possible approaches might be present as a set of pre-orientation exercises including modules on appropriate and inappropriate uses of Social Media in practice or a short online quiz test knowledge on basic appropriate online behavior in professional settings. That said, there remain inherent challenges. While there are some digital modalities that are used more regularly in practice that can be effectively role-modeled (e.g., appropriate use of email, checking for email consent, etc.), most social media use is not currently part of regular clinical care, and much more difficult to effectively role model. This reality directly challenges our understanding of where to draw the line between personal and professional. These are clearly issues that contribute to the "content" of the hidden curriculum.

ADDRESSING DIGITAL PROFESSIONALISM THROUGH THE HIDDEN CURRICULUM

The hidden curriculum can be described as lessons that are learned but not explicitly intended [18]. It's "hidden" because it is usually unacknowledged or unexamined by students, educators, and the wider community [23]. More concretely, it is the lessons learned that are largely tied to institutional culture and or status quo behaviors. Often times these hidden lessons are at odds with what is expected or outlined in the formal curriculum. Social media platforms are ripe environments for modeling both positive and poor professional behavior. In the case of the Dalhousie Dentistry Gentlemen's club the response to remediate the students through a *restorative justice process* that was seen by many to be a small penalty not fitting to the deed [24], exemplifies how crucial it is to examine ways in which to effect positive changes that will carry forward to the hidden curriculum. Of course, one paragraph in a chapter is not enough to discuss institutional change. But there are small things that can be done beginning with eliciting pointed feedback from learners regarding their experiences.

The Indiana State University School of Medicine introduced a narrative teaching initiative wherein a team of volunteers conducted interviews with a range of individuals from the school seeking stories about the best prospects of the informal curriculum and aspirations for change. The stories and emerging themes were sent back to school members, discussed and debated and a process change was initiated [2]. This type of narrative initiative correlates well with Ellaway's call to address the negative aspects that a hidden curriculum may perpetuate around digital professionalism [25]. If we are going to buy into the digital native- digital immigrant argument then it goes without saying that a change in the way we teach professionalism in digitally mediated environments first begins with educators' understanding digital environments. If faculty are not comfortable in this space meaningful changes in any curriculum, be it formal or informal, will be a slow process. Only when faculty are truly comfortable with the digital can they begin to write the narratives in the curriculum that will effectively address the problem of unprofessional digital behavior.

The narratives around poor and inappropriate uses of social media by medical learners contribute to a collective development of negative viewpoints regarding social media uses in medical teaching and learning environments. By exploring current professionalism teaching models, and finding ways to integrate the concept of professional identity formation within the realm of social media, institutions can help shift the narrative into one that looks at ways in which to harness these technologies to positively influence professional practice.

DON'T FORGET ABOUT THE POTENTIAL FOR GOOD

So, after all the previously discussed challenges of digital health professionalism, is it really worth the effort? These are very personal decisions for health professionals. But we will end this chapter optimistically, by highlighting some of the opportunities of this brave new world of electronic communication between health professionals and patients. One use case we can all relate to, no matter if we are the health professional or the patient, involves end of life care. We have all experienced or know someone who has had a loved one suffer during the twilight moments of a fatal illness. At a certain point, a difficult decision is made to modify the goal of care from cure to comfort. This puts the closest relative to that patient in a new and sometimes unforgiving role—that of a caregiver. The caregiver may become overwhelmed by the endless work and emotional toll of the experience. A recent small study involved the establishment of an online closed Facebook group for bereaved caregivers from their hospice, using best practices previously known from in-person support groups [26]. Post-intervention survey comments by caregiver participants revealed that this platform led to the beneficial sharing of stories and of resources in the virtual presence of clinicians and other bereaved caregivers.

This accessible platform for peer support is but one example of a surge of positive use cases for electronic communication in healthcare within the last decade [27]. Other potential benefits include more convenient and therefore more frequent communication between clinicians, patients, and the public, which may improve the quality of and perceived satisfaction with medical care [28]. These electronic tools can also provide patients and the public with quicker and easier access to concise, up to date, and unambiguous medical expertise. For instance, posting (with copyright permission) comprehensible evidence-based medical information on social media sites may increase the volume of quality health information made available to patients and the public.

Even among clinicians, electronic communication can foster a culture of improved knowledge exchange and lifelong learning. Clinicians who subscribe to health-related email listservs and who follow health-related social media such as certain Twitter feeds, Facebook pages and "hashtags" can keep current with evidence-based daily clinical pearls. Similarly, those subscribed to platforms such as Sermo can request the almost instantaneous hive consultation of many online clinicians to assist with challenging clinical cases.

Moreover, electronic online tools may become reliably used for public health research, surveillance and interventions. Arguably, popular social media may aggregate individual health-related data that are more current than standard epidemiological surveys. While demonstrating the effectiveness of using online tools in these novel ways is still in its infancy, successes continue to push the field forward. For example, in April 2015, a team published an article that validated a novel data-mining tool of Facebook likes, fit to geographic locations, which adequately represented various health parameters of the populations in those locations [29]. Conversely, those same social media platforms can also be used by public health experts to communicate directly to a large target group about precautions. A yearly public health promotional campaign on Twitter by the Centre for Disease Control in the United States called #Vaxwithme successfully reminds millions to vaccinate themselves against influenza to reduce their risk of contracting the disease each year [30].

Online tools can even be used to influence health policy. Activist online voices can come together to pressure government policy makers to act. The "Coalition of Ontario Doctors," an unofficial group of Ontario physicians, was originally formed using email to connect like-minded physicians, who ultimately developed a group on Facebook called "Concerned Ontario Doctors." The group currently has over 20,000 [31] physicians, including over 10,000 on the Facebook page, and was a very effective platform for uniting many physicians in their common goal to improve healthcare in Ontario. The use of social media in this context has been extremely effective and led to ability for doctors to band together and support one another through social media [32].

As a driver for patient advocacy video platforms like YouTube and others allow clinicians and health professionals to create and publish important education content that can be made readily available to patients and the community. Two such examples: Dr. Mike Evans created a YouTube series "DocMikeEvans" [33] designed to bring health education to the public in an accessible manner. Similarly, Dr. Danny Sands has harnessed technology to engage patients to take active participation in their healthcare [34].

All these potential benefits inform what is called a "networked" model of patient care. The concept developed by cardiologists Bornkessel and Fornburg recommends four online strategies for clinicians to best make use of social media professionally [35]. First, as described earlier in this chapter, it is key for clinicians to become comfortable in writing effectively on health topics, in a comprehensible manner to the layperson. Second, clinicians must observe and listen to patients who are active on digital social networks. They should use strategies for engagement such as following comments under certain health-related social media hashtags to connect with relevant discussions. Third, clinicians should find the online information material that best delivers their intended clinical goal. A popular and effective way to do so is to post videos from a repository of evidence-based entertaining videos on selected health topics. Finally, clinicians working toward the networked model should seek opportunities to collaborate around and cocreate opportunities for the model. In other words, clinicians should champion this beneficial online dialogue among them, patients and other stakeholders on various medical issues. If done right, the "networked" model may have an enormous positive impact far beyond what the clinician would ever know. And so, despite the effort associated with being a digital health professional, using electronic communication tools effectively and pursuing the networked model of care may be the key to being a truly honorable clinician in the 21st century.

SUMMARY

A great many of the issues and challenges cited in this chapter reflect challenges related to the imminent change(s) brought on by new social technologies in a workforce where relationships and relationship building is paramount to becoming and *being* a professional. The tenets which characterize professionalism: *public good, ethical standards, values, integrity, honesty, altruism, humility, respect for diversity,* and *transparency* reflect an understanding of the weight placed on health care professionals. How do we ensure these high standards can be lived out in a digital world?

To summarize, we have looked at the many themes and issues with digital professionalism and leave you with more questions to complement the ones we started with in this chapter:

Given that electronic communications and social media are not only here to stay, but also rapidly expanding, clinicians need to find a way to meet society's expectations for communication in the 21st century while ensuring that standards of professionalism are maintained. Achieving this balance, without adversely affecting clinical workflow, and negating the very benefits achieved through electronic communication, is the key challenge. How can practicing clinicians, many of whom are generationally not adept with all electronic communications modalities, navigate the complex guidelines and standards to not only find this balance, but also then teach their more-adept students? Media representations of social media and its failings in healthcare settings tell us we need to do a better job at teaching professional behavior. How do we get faculty to become truly comfortable with technology when they are skeptical of it and often don't use it themselves? Where within formal and informal curricula can we teach and model professional digital behavior? What are the best uses of technologies and how do we weigh the risks and benefits in at a time where social and digital technology is evolving? Do we need to reevaluate our conceptions of privacy and security if we wish to make the best use of technologies to serve the greater good and uphold one of the most basic tenets of professionalism? Who are the change agents that will champion the goal toward creating a culture of change that will affect positive outcomes?

References

[1] Frank JR, Snell L, Sherbino J, editors. CanMEDS 2015 physician competency framework. Ottawa: Royal College of Physicians and Surgeons of Canada; 2015.
[2] Thistlethwaite J, et al. Professionalism in medicine. Oxford, England: Radcliffe Publishing; 2008.

[3] De M. Towards defining paternalism in medicine. AMA J Ethics 2004; [accessed 04.16] http://journalofethics.ama-assn.org/2004/02/fred1-0402.html.
[4] Birden H, Glass N, Wilson I, Harrison M, Usherwood T, Nass D. Defining professionalism in medical education: a systematic review. Med Teach 2014;36(1):47–61 2014 January.
[5] 2015 International Survey of Primary Care Doctors [Internet]. Originally published November 2015 [cited 2016 Aug 06]. Available from: http://www.commonwealthfund.org/interactives-and-data/surveys/international-health-policy-surveys/2015/2015-international-survey.
[6] Rayar M et al. Why can't you email your doctor? Healthy Debate, 2015. http://healthydebate.ca/2015/04/topic/innovation/email.
[7] Barlow et al. Unprofessional behaviour on social media by medical students. Med J Aust 2015;201(11):439.
[8] Chiose S. Dalhousie board calls emergency meeting; independent inquiry urged. The Globe and Mail [newspaper on the internet.] 2015 Jan 08 [cited 2016 Aug 06]. Available from: http://www.theglobeandmail.com/news/national/dalhousie-board-calls-emergency-meeting-independent-inquiry-urged/article22372362/.
[9] Ye et al. E-mail in patient-provider communication: a systematic review. Patient Educ Couns 2010 Aug;80 [2]:266-73. doi: 10.1016/j.pec.2009.09.038. Epub 2009 Nov 13.
[10] Using email communication with your patients: legal risks [Internet]. Originally published March 2005 / Revised May 2015. [cited 2016 Aug 06]. Available from: https://www.cmpa-acpm.ca/-/using-email-communication-with-your-patients-legal-ris-1.
[11] Canadian Medical Protective Association (CMPA), Consent to Use Electronic Communications [Internet]. CMPA. 2017 [cited 24 February 2017]. Available from: https://www.cmpa-acpm.ca/documents/10179/301287261/com_16_consent_to_use_electronic_communication_form-e.pdf.
[12] Top 10 tips for using social media in professional practice [Internet]. Originally published October 2014. [cited 2016 Aug 06]. Available from: https://www.cmpa-acpm.ca/-/top-10-tips-for-using-social-media-in-professional-practice.
[13] Sask. nurse charged with professional misconduct pleads not guilty. CBC News [newspaper on the internet]. 2016 Feb 15. [cited 2016 Aug 06] Available from: http://www.cbc.ca/news/canada/saskatoon/not-guilty-plea-from-sask-nurse-over-facebook-post-1.3449537.
[14] Farnan JM, Paro JAM, Higa JT, Edelson J, Arora VM. The YouTube generation: implications for medical professionalism. Perspect Biol Med 2008;51(4):517–24.
[15] Kaczmarczyk JM, Chuang A, Dugoff L, Abbott JF, Cullimore AJ, Dalrymple J, et al. e-Professionalism: a new frontier in medical education. Teach Learn Med 2013;25(2):165–170. April 2013.
[16] James R. Out of the box: the perils of professionalism in the digital age. Business Inform Rev 2012;29 (1):52–6.
[17] Prensky M. Digital natives, digital immigrants part 1. On the Horizon 2001;9(5):1–6.
[18] Cruess RL, Cruess SR. Teaching medical professionalism: supporting the development of a professional identity. Second ed, New York: Cambridge University Press; 2016.
[19] Levinson W, Ginsberg S, Hafferty FW, Lucey CR. Understanding medical professionalism. McGraw Hill; 2014.
[20] Future of Medical Education in Canada Postgraduate Project: A Collective Vision for Postgraduate Medical Education. [S.l.]: The Association of Faculties of Medicine of Canada; 2012.
[21] Ho K, Ellaway R, Littleford J, Hayward R, Hurley K. The CanMEDS 2015 eHealth Expert Working Group Report. Ottawa: The Royal College of Physicians and Surgeons of Canada; 2014.
[22] Kung JW, Eisenberg RL, Slanetz PJ. Reflective practice as a tool to teach digital professionalism. Acad Radiol 2012;19(11):1408–14 November 2012.
[23] Glossary of Education reform http://edglossary.org/hidden-curriculum/ [accessed 24.08.16].
[24] Dalhousie dentistry students break silence on 'Gentlemen's Club' Facebook scandal. Global News. 2015 May 21 Jennifer Tryon and Nick Logan http://globalnews.ca/news/2010585/dalhousie-dentistry-students-break-silence-on-gentlemens-club-facebook-scandal/.
[25] Ellaway RH, Coral J, Topps D, Topps M. Exploring digital professionalism. Med Teach 2015;37(9):844–9 09/02.
[26] Elaine Wittenberg-Lyles, et al. "It is the 'starting over' part that is so hard": using an online group to support hospice bereavement. Palliat Support Care 2015;13(02):351–7 doi: http://dx.doi.org/10.1017/S1478951513001235, Published online: 24 February 2014, Dalhousie University probes misogynistic student 'Gentlemen's Club'. *CBC News*. 2014 Dec 15.
[27] McCarthy D et al. Group Health Cooperative: reinventing primary care by connecting patients with a medical home, 3. New York, NY: Commonwealth Fund; 2009.

[28] Zhou Y et al. Improved quality at Kaiser Permanente through e-mail between physicians and patients. Health Affairs (Millwood) 2010;29(7):1370–5.

[29] Gittelman S, Lange V, Gotway Crawford C, Okoro C, Lieb E, Dhingra S, et al. A new source of data for public health surveillance: Facebook likes. J Med Internet Res 2015;17(4):e98. Available from: http://dx.doi.org/10.2196/jmir.3970.

[30] Join the effort by becoming a flu prevention partner. [Internet]. CDC. 2016 [cited 26 August 2016]. Available from: http://www.cdc.gov/flu/partners/success-vaxwithme.htm.

[31] Coalition of Ontario Doctors—A coalition of over 20,000 doctors in Ontario fighting for a fair PSA with the province. [Internet]. Coalitionofontariodoctors.ca. 2016 [cited 27 August 2016]. Available from: http://coalitionofontariodoctors.ca/.

[32] Collier R. Doctors v. Government: taking the fight online—Part III: Ontario Medical Association makes savvy use of social media to protest cuts to medical fees. CMAJ 2015;187(6) First published March 16, 2015, http://dx.doi.org/10.1503/cmaj.109-5015.

[33] Evans M. DocMikeEvans [Internet]. Date not available [cited 15 Feb. 2017] http://www.reframehealthlab.com.

[34] Sands D. Connecting Health, Healthcare, Technology, and Engagement [Internet]. 2013 [cited 15 Feb. 2017]. Available from: http://www.drdannysands.com/.

[35] Bornkessel A et al. Social media: opportunities for quality improvement and lessons for providers—a networked model for patient-centered care through digital engagement. Curr Cardiol Rep 2014;16(7):504. Available from: http://dx.doi.org/10.1007/s11886-014-0504-5.

CHAPTER

5

Privacy and the Hi-Tech Healthcare Professional

Paulette Lacroix[1] *and Seana-Lee Hamilton*[2]

[1]PC Lacroix Consulting, North Vancouver, British Columbia [2]Fraser Health Authority, Surrey, British Columbia

Throughout the ages, and well before the latest advancements in technology, healthcare professionals have maintained a principal role in the collection, use, and disclosure of a patient's/client's health information. Indeed, accurate and contemporaneous recording of personal health information is central to providing treatment and healthcare services. Healthcare professionals are responsible to ensure the confidentiality of health information in accordance with their codes of ethics and regulatory standards of practice. Digitalization of health information has added both new opportunities for care delivery and unique challenges in safeguarding information. Healthcare professionals can easily recognize the value of electronic data in practice, program evaluation, and research, but this technology also has a darker side that can place the privacy of patient/client information at risk. Therefore, it is important for the healthcare professional to understand legislative privacy compliance requirements that determine how an individual's personal health information will be collected, used, and disclosed. Private- and public-sector healthcare organizations are required to develop privacy policies and procedures and provide awareness training to employees and associated independent practitioners.

The privacy of patient/client health information will take on increasing importance as clinical information systems are integrated to encompass the continuum of care and new technologies enable the movement of information in different ways. More and more organizations are realizing the benefits of a privacy risk assessments early in the design phase of technical implementation projects when mitigation strategies are most cost effective.

Privacy, as a concept, was first considered as a social expectation, rooted in some of the oldest texts and cultures over millennia, such as the Bible, the Qur'an, Jewish law, and the cultural fabric of ancient China [1]. Privacy as a democratic right was first defined in 1890 by the US Supreme Court as "the right to be left alone" [2 p. 110]. Almost a century later in 1997 the United Kingdom's Calcutt Committee defined privacy as "the right of the

Health Professionals' Education.
DOI: http://dx.doi.org/10.1016/B978-0-12-805362-1.00005-X

individual to be protected against intrusion into his personal life or affairs, or those of his family, by direct physical means or by publication of information" [2 p. 11]. The Australian Privacy Charter in 1995 described privacy as "a free and democratic society [*that*] requires respect for the autonomy of individuals, and limits on the power of both state and private organizations to intrude on that autonomy" [2 p. 11]. The Canadian Charter of Rights and Freedoms (1982) does not specifically mention privacy; however, under Sections 7 and 8 legal rights extend to security of person and the right to be secure against unreasonable search or seizure [3].

Our contemporary view of information privacy emerged in the 1960s along with the increasing adoption of information technology by business, government, and academia. Privacy protection changed dramatically during the 1990s due to the forces of globalization, the rise of transnational corporations, the influence of the World Wide Web and a growing need for enforcement of data protection laws across jurisdictions. Personal information has now taken on significance as an exploitable resource in information economies leaving regulators attempting to deal with the vast range of technologies and processes not anticipated with current legislation. Generally stated, information privacy legislation establishes rules that govern the collection, use, disclosure, retention, and disposal of personal information. It is important to emphasize that privacy legislation extends to every individual; therefore, healthcare organizations have a duty not only to protect the healthcare consumer's personal information but also the healthcare professional's personal information collected for the management of an employee–employer relationship.

PERSONAL INFORMATION

The definition of personal information is specific to each privacy legislation and may be variable between jurisdictions. A generally accepted definition is considered to be any information of a type that is commonly used, alone or in combination with other information, to identify or purport to identify an individual. Personal information can include unique identifiers such as social insurance or security number, employee identification, driver's license and health insurance number; ethnicity including race, color, national origin; religion, marital status, blood type, fingerprints; employment, medical or criminal history; information on financial transactions; date of birth, street address, telephone number, and many other identifiable characteristics. As new technologies emerge, the concept of personal information has expanded to include biometric data, digital video footage, Internet Protocol (IP) address, and geo-location [4]. In the European Union, personal information is known as personal data and may be further classified as sensitive data that includes financial and medical information, political opinion, religious beliefs, racial or ethnic origins, and anyone's views or opinions about an individual [5].

Information privacy laws often exclude an individual's contact information as personal information, described as business contact information including first and last name, title, business email address, business phone number, and business fax number. Also, excluded from personal information is publicly available information that is published in a directory, phone book, or professional registry.

PERSONAL HEALTH INFORMATION

As a category of personal information, personal health information is considered highly sensitive even when compared to financial information, and some jurisdictions have legislated specific health information privacy laws at the federal and state/provincial levels [6,7]. Personal health information has been defined as identifying information in oral or recorded form about an individual that relates to the physical or mental health of the individual, including information respecting the individual's healthcare status and history and the health history of the individual's family; the provision of healthcare to the individual including information respecting the person providing healthcare (name, business title, address, phone number, license number, profession, job classification, and employment status); donations of body parts, registration information, payments or eligibility to benefits; and information collected in the course of and incidental to the provision of a healthcare program and a drug, healthcare aid, device, product, or equipment under prescription [8]. Identifying information is considered any information that identifies an individual or which is reasonably foreseeable in the circumstances that either alone or together with other information the individual would be identified [9].

The US Health Insurance Portability and Accountability Act of 1996 (HIPAA) identifies protected health information as all individually identifiable health information held or transmitted by a covered entity or its business associate in any form or media. Protected health information includes demographic data relating to an individual's past, present, or future physical or mental health condition, provision of healthcare to the individual, or the past, present, or future payment for the provision of health care to the individual, and that identifies the individual or for which there is a reasonable basis to believe it can be used to identify the individual. Individually identifiable health information includes many common identifiers such as name, address, date of birth, and social security number [10].

FAIR INFORMATION PRINCIPLES

For practical reasons privacy is described as a construct based on ten universal fair information principles that, together, provide a framework for legislation, policy, practice, and regulation. This framework is foundational in presenting a common understanding of how privacy is practiced and regulated across all industry sectors and jurisdictions. These principles are further supported by security requirements for electronic records that, when applied in data protection and regulation, allow for the safe transfer of personal information for individuals who work and live in different locations. In this era of globalization and interjurisdictional information transmission, many countries have adopted privacy frameworks that underpin the development of legislation, policy, and practice standards; assist in assessing information privacy risk; and enable consistent and fair regulatory oversight.

Privacy principles were first introduced In the United States in the 1970s through the development of the Code of Fair Information Practices, an endeavor for the safe transmission of information to support commerce [11]. In 1981, the Organization for Economic

Co-operation and Development (OECD), a 23-member body that includes the United States and Japan in addition to several European nations, published Guidelines Governing the Protection of Privacy and Trans border Data Flows of Personal Data that were updated in 2013 to recognize the growing importance of digital commerce [12]. In 1996, Canada adopted the Canadian Standards Association (CSA) Model Code for the Protection of Personal Information [13]. More recently, in 2004, the Asia-Pacific Economic Cooperation (APEC) with 21 Pacific-coast members in Asia and the Americas approved the APEC Privacy Framework [14]. While these frameworks may vary in degree, they all require that personal information be fairly and lawfully collected; used only for the original specified purpose; only disclosed lawfully; kept accurate and up to date; made accessible to the individual the information is about; kept reasonably secure; and destroyed after its purpose has been completed.

The CSA Model Code, based on the OECD guidelines and considered the most comprehensive framework, is presented for the purpose of demonstrating how a privacy framework is applied in the healthcare setting. These ten principles form a common foundation for developing legislation and policy, regulating public bodies and private enterprise, and in determining the level of privacy risk for emerging technologies [15].

Principle 1: Accountability. Accountability for the privacy and security of personal information is central to the governance of healthcare programs. An organization is responsible for personal information under its control and/or custody and must designate an individual to be accountable for the management of this information throughout its life cycle, most often to a Privacy Officer. The classification of personal information in terms of control or custody also has importance relative to an individual's right to access and validate the accuracy of their own information. Control of personal information is assigned to the organization that is the primary collector. Custody of the information may be granted to another entity, such as a referral to a practitioner, disclosure to a service provider, or mandatory reporting requirements to a healthcare institute, but only if the use of the information is consistent with the purpose for collection.

Principle 2: Identifying Purpose. The organization is required to identify the purpose for collecting personal information at or before its collection, and must be able to validate the information is necessary for a program or activity of the organization. Privacy legislation does not allow for frivolous collection of personal information. Use of personal information must always be consistent with the purpose for collection. In healthcare, the primary purpose for collecting personal information and/or personal health information from a patient/client is to provide treatment and care. Secondary use of this information includes education, service planning, and continuous quality improvement initiatives. Privacy legislation requires the purpose for collection be clearly stated and available to the individual at the time of collection, either in a consent format or a posted notification where allowed.

Principle 3: Consent. An underlying premise in privacy is the individual's right of ownership of personal information and consent is required prior to its collection. There are different types of consents. An individual may give an affirmative consent, for example, opt-in to receive information. Implied consent may reasonably be inferred from the action or inaction of the individual, for example, when an individual does not opt out. Explicit consent is a requirement for processing sensitive information and requires an active communication, most often a signed written and informed consent. Largely, explicit consent is required only where it is possible, practical, and required by law. Examples of where explicit consent is needed include law enforcement purposes, participation in exceptional collections such as for a novel program, for invasive treatment procedures, and to prevent fraud where the patient/client requires a legally authorized representative. Explicit consents must identify the organization's legal authority to collect, use and disclose personal information and other prescribed elements. More frequently used in healthcare is implied consent in conjunction with a posted notification explaining the organization's legal authority to collect, the purpose for collection, how the information will be used and disclosed, how long the information will be retained and who to contact for more information. The notification is highly visible at the point of collection and the patient/client implies consent to the collection of their information by registering at the healthcare facility. Of consequence is the need to obtain a new consent when there are new uses contemplated for the information.

Principle 4: Limiting Collection. An organization should limit collection of personal information to only what is required to fulfill the purpose(s) identified. The information should be collected directly from the individual and relate to a treatment or service specific to the collection. For example, collection of information for healthcare should not require an individual to provide additional information for use by the organization for marketing or fundraising. This is an important consideration for healthcare foundations, associate organizations, and when designing healthcare programs that use Internet platforms or social media.

Principle 5: Limiting Use, Disclosure, and Retention. An organization should only use or disclose personal information consistent with the purpose for collection, restrict retention to when this purpose is complete and securely destroy the information at the end of its life cycle. Health record retention guidelines are generally determined by the government and most professional regulatory bodies. Secure destruction of electronic records extends to personal information in the custody of third-party service providers. Healthcare professionals managing treatment and service delivery programs that contract with third-parties must be fully aware of what personal information is being collected by or shared with the service provider, how it is being used and disclosed, the record retention schedule, and when the information has been securely destroyed.

Principle 6: Accuracy. An organization should ensure the personal information it collects is accurate, up to date and complete. While information accuracy is critical to clinical decision making by healthcare professionals, this principle protects an individual's privacy rights by ensuring the most accurate information is available for any decision made about them. Healthcare organizations are required to have procedures in place for tracking updates to personal information they collect and corrections made to records.

Principle 7: Safeguards. Privacy legislation has provisions for "reasonable security" in safeguarding personal information in the custody and/or control of organizations. Reasonable security is expected to protect personal information from unauthorized collection, use, disclosure, or disposal and includes administrative, physical, and technical security. Healthcare security controls are guided by international standards and industry best practices that continually advance information confidentiality, integrity, and access. A security risk assessment is required prior to technical implementations, and security experts should be involved early in the project design phase to identify security concerns when they can be more easily addressed or avoided.

Principle 8: Openness. This principle requires organizations to have its personal information management policies and procedures available to the public upon request. In practice, these documents are written in language that is easy to understand, posted on websites, and/or made available in brochures and other media. New program initiatives must consider how the public will be informed of the appropriate management of their personal information and in a way that promotes trust.

Principle 9: Individual Access. Upon request, an individual shall be informed of the existence, use, and disclosure of his/her personal information and be given access to that information. The organization should have processes in place to allow an individual to request access to their personal information, and when the request is denied to be fully informed of the reason. An individual is also able to challenge the accuracy and completeness of their information and request to have it amended as appropriate. The health record may be updated, for example, when there is a change to verifiable information such as an individual's name or next of kin. Clinical notes in the records cannot be altered and any changes requested by the patient/client are made as an annotation. Both public and private sector privacy legislation allow for individual access to personal information, a requirement that must be taken into consideration when third-party service providers have custody and/or control of information for which individual access is the responsibility of the healthcare organization. Regulators expect an individual to have timely access to their personal information no matter where it is stored.

Principle 10: Challenging Compliance. An individual shall be able to express a challenge to the organization concerning its compliance with privacy legislation. An organization is required to have a complaint process in place and a responsible individual to oversee complaints and investigations, usually the Privacy Officer. Generally, healthcare organizations that comply with these privacy principles will experience fewer complaints from individuals challenging the protection of their personal information. All complaints must be fully addressed, and if not satisfied the individual has the right to file a complaint with the regulator asking that an investigation be initiated.

COMPLIANCE WITH PRIVACY LEGISLATION

Compliance activities in managing personal health information are similar to those for other personal information. In the United States, HIPAA regulators established national standards to protect personal health information that applies to all health plans, clearinghouses, and other healthcare providers conducting specific healthcare transactions electronically. The European Union has recently implemented Directive (EU) 2016/680 that provides interjurisdictional privacy compliance requirements for all categories of personal information including health. Canada does not have a national health information privacy law, leaving the legislative responsibility to provincial governments. In Australia personal health is protected by the Privacy Act and more recently new legislation, the Personally Controlled Electronic Health Records Act, was enacted to enable personal electronic health records [16].

The sensitivity level of personal information is a strong determination as to the type of safeguards required. Financial and health information are considered highly sensitive due in part to the risk of individual harm, such as identity theft, when information is breached. Information sensitivity can also be context-specific relating to age, religion, ethnicity, and socio-economics. An example of sensitive information in context is the individual who elects to password protect parts of their electronic health record (e.g., medications, lab results) while others do not. For the healthcare professional, password-protected records require extra diligence when accessing this information, and for privacy and security there is a higher threshold of reasonable protection. In certain situations, sensitive information may be adapted to decrease its level of risk, such as airport body scanners now represented as a caricature with no distinguishing features. Privacy invasive technologies, such as Closed Circuit TV (CCTV), GPS, and Radio-Frequency Identification (RFID), can often have their levels of risk mitigated through the use of privacy and security assessment tools.

The most effective privacy compliance strategy for an organization is a privacy management program that demonstrates accountability for the personal information within its custody and/or control. The program details accountabilities and responsibilities of information management within the organization, and includes governance, risk assessment, training, breach management, security controls, and service contracts. Operational accountability is designated to a Privacy Officer who oversees the development, implementation, and ongoing maintenance of the program, and is responsible to establish risk assessment tools and control mechanisms.

Governance within the privacy management program focuses on policies and procedures that guide employees and others within the organization. The Privacy Officer is responsible to develop these policies and monitor adherence through system audits and information breach investigations. Foundational to a privacy management program is a Personal Information Directory where all collections of personal information are recorded and proves to be a useful reference in the event of an information breach, defined as the unauthorized collection, use, disclosure, or disposal of personal information on any media. Privacy breaches can be very expensive on many fronts for the healthcare organization and, most importantly, can erode public trust. Protecting personal information really begins with healthcare professionals as is reflected in the reportedly high percentage of healthcare privacy breaches attributed to negligent or careless employees, findings that have remained unchanged from 2010 to 2015 [17]. Progressively more jurisdictions are enacting or strengthening privacy legislation to require mandatory breach reporting to regulators. While the risk of sanction and direct costs are real to both the organization and the healthcare professional, what must not be forgotten is the patient/client who bears the full impact of an information breach.

Privacy management programs utilize standardized privacy and security assessment tools to conduct risk assessments of all programs, systems, and technical projects before implementation. The Privacy Impact Assessment (PIA), based on the ten privacy principles, is used to assess what personal information is collected, the purpose for collection, how it will be used and disclosed, who will have access, what safeguards are in place and where the information will be stored. Another risk assessment tool, the Security Threat Risk Assessment (STRA) was developed from international information security standards to examine user access, audit capability, and network security. Compliance with privacy legislation requires demonstration of reasonable security, and PIAs and STRAs together can provide an organization with a comprehensive privacy and security posture [18].

An important privacy control of an electronic information system is the ability to limit access to only authorized users, most often expressed as a role-based access control system with a user authentication process and access privileges that are sufficiently granular to differentiate each role. For example, user access may be grouped by professional classification but different roles within that classification are assigned specific privileges based on their need to know. Systems must also have internal logs that record user access and activities performed by each user. Reasonable security also requires a robust user account approval and management process, and the ability to conduct access audits in the detection and investigation of privacy breaches.

Security of personal information includes administrative, physical, and technological security controls that range from locked filing cabinets to encrypted databases. Security experts have specialized certifications and technical expertise in applying international standards when assessing risks associated with electronic information systems. Designing network security protocols is guided by established best practices that include encryption, firewalls, secure transmissions, penetration testing, access and audit controls, and risk assessments. Security is continually challenged to look at new ways to harden the system network against external attack. Reported cyber threats facing healthcare organizations include ransomware, malware, and denial-of-service attacks. Security is also concerned about employee negligence leading to breaches, mobile device insecurity, use of cloud services that are insecure, employee-owned mobile devices, and mobile applications used in eHealth [19]. There is an underlying expectation by security that all healthcare professionals and third-party service providers will contribute to the security of personal information by adherence to information policies and procedures.

Privacy awareness training is a central component of a privacy management program and should be provided to all healthcare organization employees and associated independent practitioners. Awareness training promotes safe information management practices and informs healthcare professionals of privacy compliance requirements, such as when to obtain explicit consent from patients/clients. Responsibility for appropriately managing personal information extends to helping patients/clients understand their right to access personal information and the process to follow.

A substantial privacy risk to a healthcare organization is the information management practices of third-party service providers. Attempts to mitigate this risk is primarily through robust procurement practices and contractual agreements, legally requiring the service provider to comply with the same privacy requirements as the organization. Third-party service provider contracts must take into account the sensitivity of the personal information, any requirement for trans-border data flows and data retention, storage, and disposal obligations. A contract may also require a service provider to follow the healthcare organizations polices and provide timely notification in the event of a breach.

Advancing the use of technology in healthcare provides an interesting balancing act in supporting innovation while ensuring organizations comply with privacy legislation.

Below are examples of best practices used by healthcare organizations in minimizing unauthorized collection, use, and disclosure of personal health information:

- Document collection of personal health information for all programs and services, including online services, patient/client registration areas, clinics, and programs by assessing the types and sensitivity of personal information in the control or custody of the healthcare organization, so appropriate resources may be allocated to safeguard this information.
- Ensure users of personal health information understand the legally authorized uses such as direct provision of healthcare services, quality improvement studies, billing and accounts payable, and follow-up on patient/client complaints. Use of personal health information for research and evaluation generally has special legislative privacy provisions.
- Document all areas where personal information, including health information, is disclosed both internally and externally, and the legislative authority for disclosure.
- Provide data-retention policies to assist with records management and the secure destruction of personal health information when it is no longer required.
- Include a process in project management to conduct internal PIAs for every project, system, or initiative that is planned, existing or upgraded and to complete STRAs where personal information is involved.
- Use contracts such as Information Sharing Agreements, Data Access Agreements, Service Level Agreements, and other legal instruments to delineate legal responsibilities for information management by all parties.
- Provide information to the public in printed and electronic online format as to their privacy rights, including notification for collection and consent requirements, right of access and the complaint process.

QUALITY IMPROVEMENT, PROGRAM EVALUATION, AND RESEARCH

The collection, use, and disclosure of personal health information for quality improvement, program evaluation, and research can present healthcare professionals unique challenges in protecting personal information. Privacy legislation most often mandates restrictions on personal health information for research and program evaluation, but considers quality improvement as an extension of the direct provision of healthcare services and therefore is included in patient/client consent. The following definitions are helpful in understanding the purpose for collection and use of personal information, two essential considerations when meeting patient's/client's consent provisions [20–22].

Quality Improvement: The purpose of conducting quality improvement is to enhance internal processes, practices, and outcomes. This process is most often used to assess an existing practice or one proposed through literature reviews. Examiners use simple and easy evaluation methodologies and a large enough convenience sample to observe any change in outcomes with the new practice. Patient/client data is anonymized. Information is analyzed using descriptive statistics that help to demonstrate changes or trends. Quality improvement is conducted within a Plan, Do, Study, Act cycle where

findings can be quickly implemented and reevaluated. Results cannot be generalized outside of the existing practice. Healthcare organizations provide funding for ongoing quality improvement, and organizational approval is required before findings may be published.
Program Evaluation: The purpose of evaluating healthcare programs is to inform decisions, identify improvements, and provide information to support strategic planning and organizational effectiveness. Evaluators may use experimental or quasi-experimental designs with control groups of patients/clients, and controls are selected to minimize the influence of confounding or extraneous variables. A combination of valid and reliable instruments and program-specific data collection tools are used. Patient/client data must be deidentified or anonymized and risk of reidentification must be minimized. Inferential and descriptive statistical methods are applied to evaluate program design, implementation, and outcomes. Accountability-focused evaluations may be conducted to demonstrate the value of a program to external funders. Healthcare organizations provide funding for program evaluation as a way to test validity of organizational goals. Evaluation findings may be published with the approval of the organization.
Research: The purpose of research is to generate new knowledge consisting of facts, theories, principles, relationships, or the accumulation of information on which they exist. Research uses accepted scientific methods of observation. Patient/client data from existing databases must be deidentified or anonymized and risk of reidentification must be minimized. Primary research also involves collecting information directly from participants who will have consented to this new collection and use. Data are analyzed with inferential statistics to test for significant differences and descriptive statistics to compare and contrast qualitative information. Research can also include multisite trials involving patient/client information disclosures from other healthcare service providers. Research funding is generally received through grants or from an external agency, and research on human subjects requires approval from a Research Ethics Board. Findings may be published in peer-reviewed journals, contributing to a scientific body of knowledge that may inform policy and clinical practice.

Many healthcare organizations have well-established departments for research and program evaluation that oversee proposals to Research Ethics Boards, formal funding applications, cohort consent forms, Data Access Agreements, Information Sharing Agreements, PIAs, and STRAs. The main areas related to research and program evaluation that require privacy oversight are:

- Detailed list of data elements
- Data linkages
- Data deidentification methods/tools
- Encryption tools
- Cloud/server secure storage
- Retention requirements (e.g., data is often retained for as long as 25 years)
- Solicitation for research purposes
- Researcher collaboration
- Communication content and methods (e.g., email, text, and fax)

While privacy legislation has fewer restrictions on the use or disclosure of deidentified health information, it is important to note that reidentification of personal health information will constitute a breach. Deidentified health information must neither identify nor provide a reasonable basis to identify an individual. This may often be achieved by the removal of specified identifiers of the individual and any related information that may lead to reidentification. While research in this area initially established a common threshold for reporting deidentified data, further study has shown that removing unique identifiers and limiting cohort size may not always prevent reidentification in certain contexts. Care must be taken to assess level and type of disclosure controls in each situation where health information is used for analysis and reporting [23]. New guidelines being developed in this emerging field of inquiry include the requirement for ongoing risk assessments of reidentification of deidentified information. Some countries (Australia for example) are considering enacting legislation that makes it an offence to reidentify deidentified information.

SPECIAL PRIVACY CONSIDERATIONS

In privacy, there are special risk considerations for web-enabled technology, online programs for children, email and text messaging, online web-based services, use of social media, cloud computing, wearables, and mobile devices. Healthcare organizations are exploring new ways to assist in providing care delivery to patients/clients, enable augmented learning opportunities for students and residents, and provide outreach to marginalized populations that may only be able to be reached through technical options.

While advances in technology have outpaced privacy legislation, public privacy expectations have only strengthened with increased reporting of information breaches, investigations by regulators, and a growing realization that personal information has become a lucrative commodity in identity theft and targeted marketing.

One strategy gaining international acceptance is the concept of Privacy by Design, an approach that supports designing technology where information privacy is the default [24]. In practice, Privacy by Design builds the principles of privacy into technology early in the design phase and continues throughout the build, implementation, and management of the system. Early identification of privacy and security issues leads to more effective and cost-effective decisions. From a privacy perspective information and communication technologies (ICTs) are considered neutral, and places the emphasis on the choices that are made when designing and using them. Therefore, ICTs can be privacy invasive or privacy enhancing, depending on their design.

Web-Enabled Technology

Healthcare service delivery is no longer restricted by location thanks in part to web-enabled technology. While in the past clinical information systems were mainly hospital based, these systems can now digitally dialogue with other information systems and devices that provide healthcare professionals with timely access to the patient/client electronic record and an array of service options. For example, home health services utilize web-enabled applications and tablets to capture biometrics of patients/clients with chronic diseases that are transmitted to a central nursing station for triage and intervention. Mobile technology is used in complex wound care that quickly refers the most complex cases to specialists, decreasing patient/client wait times. Medical residents and nursing students are being taught to use web-enabled applications for practice-related content such as drug formularies, complex procedures, and educational resources.

Healthcare is also experiencing a growing trend in wireless digital health technology [25]. Consider implantable sensors that monitor the blood pressure of an individual with a recent cardiac arrest, or contact lenses that can detect a diabetic's level of glucose. Ingestible digital medicine, from camera capsules to thermometer pills, can record the internal information of an individual's body while traveling through the digestive tract. Technology is truly changing healthcare, but the shift toward greater dependency on technology brings with it security risks of cyber-attacks that can tamper, manipulate, or control the device and potentially cause harm to the patient/client. The challenge healthcare professionals face is not to deter or block technical innovation in treatment and service delivery, but to guide its evolution in a way that enables protection of the patient/client's personal health information.

Online Programs for Children

Privacy regulators in Europe, the United States, and Canada have highlighted the need for social responsibility when developing online environments targeted at children and youth, urging operators of websites created for this population to adopt the highest standard of privacy possible, ensuring privacy policies and user agreements are clear and simple to understand by the user. In the United States, the Children's Online Privacy Protection Act (COPPA) is regulated by the Federal Trade Commission and places parents in control over what information is collected online from their children under age thirteen. Verifiable parental consent is required before any collection, use or disclosure of personal information from children [26 p. 107–111].

There are many instances where healthcare professionals may choose to use the Internet to deliver programs benefiting young patients/clients, for example, creating online services and discussion forums targeting specialized groups in oncology and mental health. Increasingly young people only want to connect digitally and may respond more favorably to services accessible online. In an effort to engage children, service providers have applied game theory in health promotion activities, such as challenging participants to collect virtual tokens that could be exchanged for real-life rewards. In another instance, children were rewarded with virtual game points for completing physical exercise activities, thus closing the gap between physical and virtual reality.

While it is expected online marketing to children will continue to grow, this group is especially vulnerable when it comes to information privacy. Best practice requires an opt-in consent from the parent or guardian, limiting collection of personal information to what is necessary, using the information only for the purpose it was collected, ensuring the confidentiality of that information and, where consent is withdrawn, immediately deleting all such information from the database.

Email and Text Messaging

Email, as a primary communication tool, has become a societal norm. Most people send emails without even considering how vulnerable and nonsecure the system really is. An email system is dependent on three variables to operate: an email application, security software, and an Internet provider. Because the sender and receiver of an email may each use a different set of variables, this means that taken together the transmission of email can be varied, storage located anywhere and security questionable. Unencrypted emails are open to external penetration during transmission and can expose an electronic system to cyber-attacks. Anyone who has received spam will know email is a primary vehicle used for introducing malware and other viruses into computer networks. The sender of an email has no control over the content once sent since the message can be altered once received and forwarded without the knowledge of the original sender.

While an email system within a healthcare organization may be perceived as being safer than personal email, it is still not advisable to send personal information over email since the sender and receiver cannot be authenticated. This is particularly significant when a healthcare organization uses email to communicate outside of their network environment. For instance, if a healthcare organization is using email to communicate with a trusted entity, such as a government agency, the network systems will have ways to authenticate an email address through user directories and enrollment processes. But when emails are sent from a healthcare organization to a patient/client their authenticity can never be assured. The ability to validate a sender and a receiver of a communication is called nonrepudiation, and can only occur when a third party or process can independently verify the sender and receiver, an issue that is not easily resolved [27]. One way that healthcare organizations have used to authenticate the sender and receiver of an email is for the sender to use the telephone number on the patient/client email consent to call and validate the receiver before the email is sent and request a confirmation that the email was received. Healthcare professionals sending emails outside of a trusted network or forwarding work emails to their personal accounts are relinquishing control of the data and putting the organization at risk for noncompliance with privacy legislation.

Consideration must also be given to what healthcare information sent by email should also be filed in the health record to meet professional and legal standards. From a security perspective, healthcare organizations also need to ensure that calendar, task lists, and emails do not contain personal information that could be subject to Phishing attacks. For instance, using a desktop calendar as a scheduling tool that contains patient/client first and last name, date of birth, medical information, home phone number, and health/insurance number can lead to identity theft. Many organizations use cloud storage or sharing software such as Dropbox, and with mobile technology an email can be transmitted to multiple user devices which has significant privacy risks for personal health information. Other considerations include legal, regulatory and health record management requirements, retention and disposal timelines, and the risk of identity theft from cybersecurity attacks.

An increasingly common request from patients/clients is to communicate with healthcare professionals via email. This method of communication can be timely, effective, and convenient for both parties, and has contributed to the emerging practice of virtual clinics [28]. Best practice in email communication with patients/clients requires the healthcare professional to first obtain from the patient/client a written consent that advises the patient/client of inherent risks in sending personal information by email, and to agree as to what types of personal health information the healthcare professional is allowed to transmit. An alternative method of communication is to provide patients/clients with secure portal access to the healthcare organization's network; however, this is costly and user accounts are difficult to manage.

Texting has also become a preferred method of communication between healthcare professionals themselves as well as with patients/clients. The privacy risks of texting personal information are more significant than emailing, in that texts are bounced off of cell towers similar to telephone conversations.

Text messages are not encrypted in transit or storage, are subject to modification by the receiver, may be forwarded to others without the sender being aware, and can arrive hours or days later due to delays in cellphone communications. Texts can be misdirected if sent within a group and a receiver cannot be authenticated even if the text is sent to the correct phone number.

The limited number of characters in a text indicates it is intended for transitory messages of limited value and not for documenting patient/client care. Given the current limitations and privacy concerns of texting personal information, healthcare professionals should not use texting to disclose patient/client personal information, for example, medication or treatment orders, status updates, and end-of-shift reports. That said, texting does present healthcare professionals an opportunity for communicating with patients/clients that does not expose their personal information, for example broadcasting wellness events or health promotion activities, general health alerts and emergency response during disasters. Texting patient/client reminders of upcoming appointments is acceptable, providing the healthcare professional has first received written consent.

One of the main considerations in setting up email and text communication with patients/clients is to make sure they are aware healthcare professionals are not able to access and respond to emails or texts at all times, and that in an emergency situation the patient/client should attend in person at an urgent care clinic or hospital. Where professional regulatory and medical-legal restrictions dictate, such as personal health information that must only be documented on the legal health record, healthcare professionals may deny electronic communications with patients/clients. Another consideration when sending electronic communications is antispam legislation and, in the case of texting, "do not call lists." And in some jurisdictions privacy laws prohibit disclosure of personal information outside of the country.

Online Web-Based Services

Many online web-based services are offered at seemingly no cost to the user who registers their name and email address in order to download software applications, content, and other offerings. But behind the scenes digital networks collect personal information, device geo-location data, and communication metadata. What is not clearly evident in the user interaction is that personal information is now considered currency in a digital economy, with the highest value placed on personal health information.

The online economic model is opaque, predicated on the collection and use of personal information, and operates in ways that challenges the legal protections and traditions afforded to the offline world. As a consequence, it is difficult for the average user of information technologies to understand and meaningfully consent to these agreements or business practices [29]. Data about an individual can be collected, directly or indirectly, from social profiles, publicly available demographic data, web search logs, purchase histories, personal contacts, and lifestyle interests across a wide range of activity. From a privacy rights perspective, this covert information collection creates an imbalance as to the ownership and control of these data, and can lead to identity theft and fraudulent activity with serious consequences.

Online searches can heighten risk to a healthcare organization's electronic infrastructure, thus security generally limits the scope of online searches by healthcare professionals in an attempt to lessen the risk of introducing malware, ransomware or other viruses into its network. Most healthcare organizations now use Internet websites to support clinical programs. To avoid some of the inherent risks associated with Internet online searches healthcare organizations are advised to launch and support program websites within their internal networks, applying security measures to ward off cyber-attacks and unauthorized access.

Use of Social Media

Skype, Twitter, Facebook, Instagram, Blogs, Podcasts...the choices and alternatives to connect and communicate with each other are ever increasing. Social media can provide a platform for healthcare professionals to effectively reach their patients/clients. For example, some healthcare organizations have set up social groups by demographics, services or diagnosis to provide members a forum to share stories and similar experiences with the group.

Health information can be disseminated through blog posts or podcasts, and events can be broadcast or retweeted on Twitter. Skype can be used to engage patients/clients in face-to-face conversations, an especially effective medium in providing access to healthcare for patients/client in remote regions. Privacy regulators, governments, and healthcare organizations have recognized the growing use of social media in healthcare and have conducted privacy and security assessments for many of the more common platforms, applying information privacy risk as a way to determine which social media will best serve healthcare organizations.

With the public's rapid adoption of social media as a way to communicate, search for information and receive content, it is not surprising healthcare professionals are using these technologies to reach out to patients/clients. An accepted privacy-promoting approach in applying social medial in healthcare is to provide digital content to patients/clients through a secure portal or website account resulting in the dissemination of reliable healthcare information and a medium for the patient/client to communicate with a healthcare professional. A key privacy consideration when using a social media platform is to publish a clearly understandable terms of use policy and a privacy statement. A healthcare organization is also advised to continually monitor their digital presence to prevent misuse or external cyber intrusion.

Cloud Computing

Cloud computing involves storing data, temporarily or on a more permanent basis, over the Internet. In the last decade, many organizations have turned to cloud solutions for storage or back up to facilitate productivity and save money [30]. "The Cloud" is a concept used to describe the virtual nature of digital storage, which can mean the data are stored on servers physically placed in many geographical locations. Considering the proliferation of cloud storage as a cost-effective way to save large amounts of data, healthcare organizations are cautioned that not all cloud storage services are created equal. It is important to conduct a privacy risk assessment prior to signing on to a cloud computing service to confirm the healthcare organization will still be in compliance with privacy legislation. For instance, an organization may be required to store personal information within its jurisdiction and the cloud storage resides outside of this area.

A PIA and STRA are required to assess risk, and mitigation strategies may include data encryption in transit and storage, data segregation to ensure an organization retains custody and/or control of the personal information, strong authentication and access rules, vendor service levels that provide downtime procedures and data recovery timelines, and the ability to extract the organization's data at termination of the contract. Cloud service providers should be able to provide audit reports of user access and produce an audit log report if required during a privacy or security investigation. When a cloud service is used by a healthcare organization to collect personal information from a patient/client through an online process, this is considered a new collection and will require a consent mechanism.

Wearable Technology

The use of wearable technology for health and fitness is an internationally fast-paced consumer trend [31]. More people are using personal health and fitness devices that collect and store information such as heart rate, blood pressure, sleeping habits, exercise, and eating habits. Epidermal sensors are being used in sports to measure levels of lactic acid in an individual's sweat. Ear buds are not just for listening to music but can also be used to measure real-time biometric and physiological data, sending this information back to the user's smart phone. Wearables digitally connect an individual to their body, quantifying the human body as information.

It is not surprising these sensors and technologies can be privacy invasive [32]. Wearables are intended to merge with the body to extract very sensitive personal health information, but consumers may not fully understand when they consent to the terms of use. Many technologies require individuals to upload their information to a website for performance analytics and storage. Privacy concerns relate to the security of the information being collected and stored, and if it is being used as consented by the individual. Parental consent may be required for wearables targeted at children and youth where, for example, toy companies sell wristbands that can track and monitor fitness. The issue of nonrepudiation also applies to wearable technology and is related to the reliability of information used to identify and authenticate individuals online that can lead to exposure of one individual's personal information to others using the same service.

Healthcare professionals can expect to encounter the effect of patients/clients offering information results from wearables as part of a clinical assessment. The accuracy of the sensors in wearables is questionable and reliance on these data may not be appropriate in a clinical setting. However, given the high adoption rate of wearables and the range of devices available, healthcare professionals may consider incorporating their use as a general part of chronic disease management. Only devices and software tested and approved by health regulatory bodies can be relied upon for accurate and standardized health information used in treatment and care.

Mobile Devices

The use of mobile technology in the direct delivery of healthcare services continues to grow. Portable phones and tablets are being routinely deployed to healthcare professionals in acute and community care settings, allowing direct user access into digital health records from anywhere in the networked system. Mobiles may be used for communication between healthcare professionals for many reasons, one being organizational compliance with occupational working-alone laws. Healthcare professionals may utilize other types of mobile devices for care and treatment including bedside respirometers, ECG machines, MRI and CT scanners, patient location trackers, and many others that are resulting in operational efficiencies. Mobile technology may be used by patients/clients for quick and easy access to wellness advice, to book appointments, download resources, and access their own personal health information through a healthcare organization portal. Smartphones with advanced video capabilities and high-definition cameras make it increasingly possible for patients/clients to connect with healthcare providers outside of their region, a technology that can be promoted for providing rural and remote healthcare.

While mobile devices are becoming more compact, their capabilities are increasingly powerful, providing an ability to store and transmit massive amounts of personal data. There are many privacy risks with mobiles such as their small size and portability that increases risk of loss or theft. Communication by mobile phone may be intercepted and, like other computing devices, mobiles are vulnerable to cyber threats such as viruses and spyware, exposing organization networks to attack by malicious agents. These challenges are further complicated by the variety of devices on the market and poorly configured security settings. It is strongly advised that a healthcare organization conduct a PIA and STRA before implementing a mobile device platform to ensure compliance with privacy legislation.

An increasing number of healthcare professionals are requesting to use their own personal phones for work, driving the development of safer technologies for mobile Bring Your Own Device (BYOD) programs in healthcare. The benefit of BYOD to the healthcare professional is the convenience of using one mobile device to manage communications, while the organization benefits economically by not having to maintain an inventory of mobiles. The use of BYOD in healthcare is different from other industries and presents unique privacy challenges in maintaining control of personal health information that is used, disclosed and stored on mobile devices [33]. The technical solutions on the market propose to mitigate these risks through controlled registration, password-protected access, data encryption in transmission and storage, remote wiping of devices as needed, secure app downloads, and a secure digital vault to store personal information thus ensuring the organization retains legal control of the information. The healthcare organization is responsible to provide terms of use, a BYOD policy, and mobile device privacy awareness training for all users. The risk to patient/client personal information stored in mobile devices is very real and, once compromised, can lead to significant harm.

EDUCATIONAL IMPLICATIONS AND CHALLENGES FOR THE HEALTHCARE PROFESSIONAL

Healthcare professionals have access to an extensive amount of personal information and personal health information on a daily basis, much of which is very sensitive. It is also not uncommon for healthcare professionals to have access to multiple and integrated clinical information systems within one or more healthcare organizations, placing on them a significant responsibility to safeguard their patient's/client's personal health information. Armed with the knowledge of privacy and security requirements provided in this chapter and having insight into how each technology may be safely applied, the healthcare professional is in a privileged position as a front-line privacy advocate.

Technology will continue to advance and with it will emerge new opportunities in healthcare. It is imperative healthcare organizations and healthcare professions continue to exercise due diligence in assessing privacy and security risks to personal information. Respecting that all individuals have the right to know the collection, use, and disclosure of their personal information, healthcare professionals can help their patients/clients to understand ways to protect their information privacy. The use of privacy protective strategies by healthcare professionals will go a long way in building public trust in the healthcare system and in meeting the public's expectation of privacy.

References

[1] Swire PP, Bermann S. Information privacy. Official reference for the Certified Information Privacy Professional (CIPP). York, ME: International Association of Privacy Professionals; 2007.

[2] Klein K. Canadian privacy: data protection law and policy for the practitioner. York, ME: International Association of Privacy Professionals; 2009.

[3] Constitution Act 1982, Part I. Canadian Charter of Rights and Freedoms. Available from: <http://laws-lois.justice.gc.ca/eng/const/page-15.html> [accessed 06.12.16].

[4] Office of the Privacy Commissioner of Canada. A matter of trust: integrating privacy and public safety in the 21st century. Available from: <https://www.priv.gc.ca/en/privacy-topics/public-safety-and-law-enforcement/gd_sec_201011/#toc3>; November 2010 [accessed 06.12.16].

[5] Ustaran E, editor. European privacy: law and practice for data protection professionals. Portsmouth, NH: International Association of Privacy Professionals; 2012.

[6] McEwen J, Shapiro SS, Kosmala P, editors. U.S. Government Privacy. Essential policies and practices for privacy professionals. York, ME; 2009.

[7] Prosser S. Personal health information and the right to privacy in Canada. A FIPA Law Reform Report. British Columbia: Freedom of Information and Privacy Association; 2000. Available from: <http://fipa.bc.ca/library/Reports_and_Submissions/Personal_Health_Information_and_the_Right_to_Privacy_in_Canada-May_2000.pdf> [accessed 06.12.16].

[8] Cavoukian A, Grant D. A guide to the personal health information protection act. Information and Privacy Commissioner of Ontario; 2004. Available from: <https://www.ipc.on.ca/wp-content/uploads/Resources/hguide-e.pdf> [accessed 06.12.16].

[9] Data Protection Commissioner (Ireland). What is personal data? Available from: <https://www.dataprotection.ie/docs/What-is-Personal-Data-/210.htm> [accessed 06.12.16].

[10] U.S. Department of Health and Human Services, National Institutes of Health. What health information is protected by the privacy rule? HIPAA Privacy Rule Information for Researchers. Available from: <https://privacyruleandresearch.nih.gov/pr_07.asp> [accessed 06.12.16].

[11] Organisation for Economic Co-Operation and Development. Thirty years after the OECD privacy guidelines. OECD Publishing; 2011. Available from: <http://www.oecd.org/sti/ieconomy/49710223.pdf> [accessed 06.12.16].

[12] Organisation for Economic Co-Operation and Development. The OECD privacy framework. OECD Publishing; 2013. Available from: <https://www.oecd.org/sti/ieconomy/oecd_privacy_framework.pdf> [accessed 06.12.16].

[13] Holmes N. Canada's Federal privacy laws. Parliament of Canada. Law and Government Division; 2008. Revised September 25, 2008. Available from: <http://www.lop.parl.gc.ca/content/lop/researchpublications/prb0744-e.htm#appendixb> [accessed 06.12.16].

[14] Asia-Pacific Economic Cooperation. APEC privacy framework. Singapore: APEC Secretariat; 2005. Available from: <http://www.apec.org/Groups/Committee-on-Trade-and-Investment/~/media/Files/Groups/ECSG/05_ecsg_privacyframewk.ashx> [accessed 06.12.16].

[15] Canadian Standards Association. Mather D, Burford G, editors. Model code for the protection of personal information. Available from: <http://simson.net/ref/RSA/1996.CanadianStandardsAssociation.ModelCodeForProtectionOfPersonalInfo.pdf>; 1996 [accessed 06.12.16].

[16] Australian Government Department of Health. My health record. Available from: <http://www.health.gov.au/internet/main/publishing.nsf/content/ehealth-record> [accessed 06.12.16].

[17] Ponemon Institute Research Report. Fifth annual benchmark study of privacy & security of healthcare data. Available from: <https://media.scmagazine.com/documents/121/healthcare_privacy_security_be_30019.pdf>; 2015 [accessed 06.12.16].

[18] Government of British Columbia. Privacy impact assessments and security, threat and risk assessments. Available from: <http://www2.gov.bc.ca/gov/content/about-gov-bc-ca/citizen-centric/ux-toolbox/web-standards-and-guidelines/user-experience-standards-and-guidelines/privacy-impact-assessments-and-security-threat-and-risk-assessments> [accessed 06.12.16].

[19] Ponemon Institute Research Report. Sixth annual benchmark study on privacy & security of healthcare data. Available from: <http://www.ponemon.org/library/sixth-annual-benchmark-study-on-privacy-security-of-healthcare-data-1>; 2016 [accessed 06.12.16].

[20] Kring DL. Research and quality improvement: different processes, different evidence. MEDSURG Nurs 2008;17(3):162–9.

[21] Rozalis ML. Evaluation and research: differences and similarities. Can J Program Eval 2003; 18(2):1–31. Available from: <http://www.academia.edu/11679718/EVALUATION_AND_RESEARCH_DIFFERENCES_AND_SIMILARITIES> [accessed 06.12.16].

[22] Alberta Heritage Foundation for Medical Research: Alberta Research Ethics Community Consensus Initiative (ARECCI). ARECCI ethics decision-support tools for projects. Available from: <http://www.aihealthsolutions.ca/media/ARECCI-Ethics-Guidelines-2013_4a.pdf> [accessed 06.12.16].

[23] El Emam K, editor. Risky business. Sharing health data while protecting privacy. USA: Trafford Publishing; 2013.

[24] Cavoukian A. Privacy by design…take the challenge. Canada: Information and Privacy Commissioner of Ontario; 2009.

[25] Deloitte Centre for Health Solutions. Connected Health. How digital technology is transforming health and social care. Available from: <https://www2.deloitte.com/content/dam/Deloitte/uk/Documents/life-sciences-health-care/deloitte-uk-connected-health.pdf>; 2015 [accessed 06.12.16].

[26] Swire PP, Ahmad K. U.S. private-sector privacy. Law and practice for information privacy professionals. Portsmouth, NH: International Association of Privacy Professionals; 2012.

[27] International Association of Privacy Professionals. Glossary of privacy terms. Available from: <https://iapp.org/media/pdf/resource_center/IAPP_Privacy_Certification_Glossary_v2.0.0.2.pdf>; 2012 [accessed 06.12.16].

[28] Ellis E. The doctor is online, anytime—it's the freewheeling world of eHealth. Published online August 6, 2016. Available from: <http://vancouversun.com/news/local-news/the-doctor-is-online-anytime-its-the-freewheeling-world-of-ehealth> [accessed 06.12.16].

[29] Office of the Privacy Commissioner of Canada. Guidelines for online consent. Available from: <https://www.priv.gc.ca/en/privacy-topics/collecting-personal-information/consent/gl_oc_201405/>; 2014 [accessed 06.12.16].

[30] Office of the Privacy Commissioner of Canada. Fact sheet: Introduction to cloud computing. Available from: <https://www.priv.gc.ca/en/privacy-topics/technology-and-privacy/online-privacy/cloud-computing/02_05_d_51_cc/>; 2011 [accessed 06.12.16].

[31] McGrath MJ, Scanaill CN. Wellness, fitness, and lifestyle sensing applications. In: Sensor technologies. Healthcare, wellness and environmental applications. Apress Open Access. Available from: <http://link.springer.com/book/10.1007/978-1-4302-6014-1>; 2014. p. 217–48 [accessed 06.12.16].

[32] Thierer A. The internet of things and wearable technology: addressing privacy and security concerns without derailing innovation. Rich J Law Technol 2015;21(6). Available from: <https://www.researchgate.net/profile/Adam_Thierer/publication/268520567_The_Internet_of_Things_and_Wearable_Technology_Addressing_Privacy_and_Security_Concerns_without_Derailing_Innovation/links/54e5f8fe0cf2bff5a4f1e8e2.pdf> [accessed 06.12.16].

[33] Office of the Privacy Commissioner of Canada and Alberta and British Columbia Information Privacy Commissioners. Is a Bring Your Own Device (BYOD) program the right choice for your organization? Privacy and Security risks of a BYOD program. Available from: <https://www.oipc.bc.ca/guidance-documents/1827>; 2015 [accessed 06.12.16].

CHAPTER

6

Ethics, Obligations, and Health Informatics for Clinicians

David Wiljer[1,2] *and Anupam Thakur*[2,3]

[1]University Health Network, Toronto, ON, Canada [2]University of Toronto, Toronto, ON, Canada [3]Centre for Addiction and Mental Health, Toronto, ON, Canada

Case example: A third-year resident has a grand round coming up shortly. She had seen her former supervisor present a clinical case in a previous rotation at a different hospital site, which she thought was very relevant. She signs into the patient's electronic medical record for clinical information through her remote access and includes it in her presentation. She had thought of checking with the patient about using the information but was not able to get in touch. Nonetheless, she spoke about the case at the rounds. The patient was anonymous in the presentation but because of the rare nature of the clinical presentation and surgical success, some of the participants already knew about it. A first year medical student in the audience was intrigued by the case and tweets to his friends and the story reaches the local newspapers; the patient is upset about it. A resident is interested in researching this area and would like to look at other case records. However, the hospital Research Ethics Board [REB] refuses the application.

Questions: What are the key issues in this case? Can sharing of anonymized patient information be justified for educational purposes? What does the current legal framework of ethics have to say? What are the ethical issues around social media (SM) and medical education?

INTRODUCTION

This chapter aims to explore how ethical issues are important in the use of health informatics in day-to-day clinical practice and in health professions education. The chapter examines healthcare professional's (HCP) "digital responsibilities" in an educational

Health Professionals' Education.
DOI: http://dx.doi.org/10.1016/B978-0-12-805362-1.00006-1

context. This chapter also examines how social norms, service user's expectations, and culture can influence ethical dilemmas. As the case illustrates, the digitization of clinical data provides a rich tapestry for education experiences and learning opportunities. As intriguing and as enticing as this may be for improving the delivery of health care, the use of the data is complex. Even with altruistic purposes in mind, such as those that are predominant in medical education, the moral, ethical, and professional issues, and boundaries can quickly become complex and, at times, can lead to uncertainty. Uncertainty can lead to inappropriate action, inaction, or even failure to act. This chapter will explore the issues and address practical concepts that will allow HCP educators to meaningfully use the available data in a safe, ethical, and professional manner.

Why Do We Need Health Informatics Ethics for Health Professional Education?

Ethical code of practice has long been embedded into medical practice. The Hippocratic Oath is one of the most well-known codes from ancient times and more recently, there has been an emphasis on ethical principles such as autonomy, nonmaleficence, beneficence, and justice [1]. With the advent of the digital age and digital applications in healthcare, the role of ethics and obligations of HCP have become extremely important. Recording "interesting" cases in repositories and on mobile devices, sharing information in "closed groups" of social networking sites, communicating with patients using electronic media are a familiar sight. It is not only patients, but also healthcare professionals and institutions that face challenges of digital media due to its misuse, which includes legal challenges. Therefore, it is extremely important that the ethical frameworks around "digital professionalism" are clearly defined. A separate chapter (Chapter 4) in this book explores digital professionalism in more details. Training of HCPs and creating opportunities for lifelong learning are vital to spread knowledge and awareness of the ethical issues. Ethical compliance can be compromised due to poor understanding of a wide range of issues, including consent and lack of familiarity about relevant policies [2]. The resulting behavior can have significant impact for the patients and their families, students, teachers, the organization/institution and even the health system itself.

Ethical Challenges of Widespread Use of Information Technology

It is not uncommon to think about confidentiality issues of electronically stored patient information as the principal ethical challenges in the use of informatics in health care. Interestingly, other than data protection and privacy, important areas of application such as clinical decision-making support tools, big data research, telehealth, and healthcare-related information on the internet raise ethical challenges too. Incidents related to breaches in patient privacy and confidentiality have been reported time and again. Privacy refers to the individual's desire to control who has access to information about the individual and confidentiality indicates respecting service user's privacy, refers to safeguarding identifiable data and health information, and release it only with patient's consent. Even with the intention of respecting privacy and maintaining confidentiality, security must also be maintained. Security refers to ensuring that the safeguards are put in place that will protect the data either from accidental or mal-intentioned breaches. In the digital age, the importance of security increases because the magnitude and scope of a potential breach can be significantly increased. In addition, digital footprints are difficult to erase and once in the public realm, can cause significant impact—even through a simple tweet, as in the case example—for a long period of time.

The question then arises, is there a compelling argument for privacy to be inviolable or can it be breached if certain conditions demand disclosure in the best interests of the individual or for the wider public health interest?

Ethical Principles and Information Ethics

The patient in the case scenario was not informed that the contents of the clinical record were being accessed for academic activities. Further, the patient was not informed about the details of the case being shared in grand rounds. The general ethical principles of information ethics [3] suggest that all persons have a basic right to *privacy*, and therefore have control over the collection, storage, access, use, communication, and *disposition* of the data. Unfortunately, although the resident intended to discuss this with the patient, it never happened. Another general principle is that of *openness*. Information should be disclosed in an appropriate and timely fashion to the patient. It would have been a different story had the resident checked the information governance policy of the hospital or she had received the guidance and education on these types of issues throughout her professional training. At times, it might be difficult to understand how to approach these types of situations. Could the resident have spoken to an information governance specialist in the organization about data access and use? On hindsight, the resident could have spoken to a contact person in information governance for further advice.

The ethical issues may also extend beyond the rights of the individual and be influenced by context. In the case that has been presented, should the context of learning matter? Perhaps the resident thought that it was okay in the end to use the case because the patient would be fine with the sharing of the information for education purposes given that the patient was treated at a teaching hospital. Should patients who are being treated at a teaching hospital or an academic health sciences center have different expectations around the use of their data? In the modern world, clinical care is provided in a multitude of settings, including in a public funded health care system, private health care providers, or even online consultations. Clinical data can be used for a variety of reasons, depending on the setting. Academic centers may be more keen to utilize health care data for clinical research and systems medicine analysis (including big data approaches) whereas small scale private practices may be keen on quality improvement. Online practices can have an interest in contextual marketing apart from managing quality. Clinical care has now moved out of the consultation room to the homes of service users, especially with digital clinical monitoring tools. Examples include the "Diabeo system" which helps the diabetic patients calculate the dose of insulin required based on various parameters, automatically adjusts the dose and then sends the information from a smartphone to a remote computer, which facilitates further tele-consultation. Another such example is the use of "Cardiomobile," a real-time remote monitoring system for cardiac rehabilitation [4]. The opportunities for learning from digital data may be higher in settings where informatics has had the highest penetrance as in academic centers, hospital settings with a strategic focus on digital innovation and telemedicine. But this could imply a disparity in digital learning opportunities depending on the setting. Nevertheless, teaching hospitals should be at the forefront to harness learning opportunities from data in an ethical manner. Is this issue something that should be managed at the organizational level? For example, could a patient grant permission upon admission for a procedure for sharing their information for education purposes? It is a challenge and an important responsibility for teaching organizations to develop not only policies, but procedures and processes that promote safe and responsible educational practices and care in the context of health professional education.

A related question is about the use of data for healthcare systems, wider societal needs and the need for health information to be shared for training and improvement of HCP practices. If the case selected by the resident involved clinical manifestations of a rapidly spreading infectious disease, not previously known, would the application of ethical principles in the use of the information been any different? The principle of *legitimate infringement* suggests that there are situations when access, use, and communication of data can be valid and reasonable in the interests of wider societal needs [5]. An editorial in the widely respected journal Nature emphasized the need for rapid data sharing in public health emergencies; Ebola and Zika virus epidemics are important examples [6]. It is important that any legitimate infringement should be done in the least *intrusive* fashion and the accountability should be maintained at the highest level.

The ethical principles described above are mentioned in the International Medical Informatics Association's Code of Ethics [3]. To ensure that the above ethical principles are followed, the learner should be aware of policies and procedures in place. The learner should be aware of the obligations toward the persons whose healthcare records are being accessed. At the same time, it is the role of health professional educators to ensure that learners are digitally literate and have the skills and clinical judgment to function with confidence within the boundaries of the policies, procedures, and processes that are in place. Trust and advocacy play a significant role in patient–professional relationships in healthcare. The same is applicable in the use of health informatics by learners in healthcare. It is important that the learner is aware of and is educated about the major laws associated with use of electronic data. Familiarity with the appropriate legislation such as with Health Insurance Portability and Accountability Act [7,8] and organizational policies on data protection and sharing and use of digital information can help in a better understanding of the key issues. Privacy and confidentiality issues are extremely important in the healthcare system. The subsequent section will examine the key issues in more detail.

Data, Confidentiality, and the Learner

Confidentiality is integral to healthcare and trust is one of the cornerstones in patient–clinician relationships. Confidentiality in health care is a requirement for those who have access to patient information to hold it in confidence. It is important for healthcare professionals to be cautious about using digital media to ensure that patient-related data transmission and usage is secure. Use of digital media facilitates efficient ways of healthcare delivery. In day-to-day clinical care, electronic communication, use of personal devices, including mobile phones, and storage of patient-related data on devices is a common occurrence. Electronic communication in the form of emails has been used by clinicians to communicate about health care [9]. However, such communication has its own limitations. Every organization has its own policies, procedures on the use of digital tools such as email and text messaging for communication with patients. In addition, every organization has its own culture and the explicit and hidden curriculum [10] in this domain can be extremely powerful in shaping not only what is taught, but what is practiced and accepted as the norm.

In the example shared at the beginning of the chapter, let us think of a situation where the resident sends his presentation slides to the grand rounds coordinator in advance preparation. He uses the traditional email system in absence of a dedicated patient web portal or a secure data exchange system. Unfortunately, the coordinator's email is hacked and the information in the resident's email is accessed by unauthorized people. Clearly, this would have been in conflict with patient's right to privacy. Was there any way this situation could have been avoided? Perhaps, the data file containing slides could have been encrypted and sent out. Although it would not have been totally fail-safe, in absence of a secure exchange system, encryption could have provided some protection to the data being shared.

Similarly, security breaches in storage of data can compromise confidentiality of patient-related data if appropriate measures are not in place. This includes malware and viruses infecting an end user's device as well as unencrypted portable devices, which are lost or stolen. Crotty and Mostaghimi have enlisted a few practical steps to safeguard data [9]. Setting up passwords to access personal devices, disabling automatic photo sharing through services such as Dropbox and avoiding access to unsecured wireless networks in public places are some of them. Some health organizations host electronic records and other patient-related data in remote servers. This reduces the risks associated with use of portable devices mentioned earlier. However, it is important that systems are in place to secure cloud data.

The words privacy and confidentiality are used together quite often synonymously, but they do not mean the same thing and are important concepts in their own right. The term "confidentiality" is rooted in the patient–clinician relationship, the relationship of trust, as discussed earlier whereas "privacy" is the right of the patient to keep information about them and make decisions about how the information is shared [11]. The means by which privacy is protected and confidentiality ensured is called "security." Protection of privacy and confidentiality is important for safeguarding a person's rights. There are other benefits too. Patients who feel assured about their privacy might be more comfortable in disclosing information to the clinicians, which can help in their clinical care as well as help in public health-related issues. However, if the data are compromised, it can cause harm in many ways to the patient, including stigma, bias, and financial harm. The learner is always encouraged to verify organizational policies and procedures related to storage and transfer of patient-related information. Some organizations may allow use of specific encrypted flash drives or have agreements with external providers subject to adherence with data-protection regulations. Other organizations may put automatic systems in place to ensure that data must be encrypted before it can be loaded on external devices.

Various laws in different jurisdictions across the world have been enacted to protect the privacy of patients. Health Insurance and Accountability Act (HIPAA) [7,8] in the United States, Personal Information Protection and Electronic Documents Act (PIPEDA; 1999) in Canada [12,13], and various laws and codes of practice all over the world [14] help in safeguarding privacy rights and upholding confidentiality. Similarly, there are several health informatics codes in different countries. American Medical Informatics Association code of ethics [3], UK Council for Health Informatics Professions Code of Conduct, European Parliament Directive (95/46/EC), and Health Information Privacy Code [1998] in Canada [15] are some examples of how countries have developed codes of Ethics based on ethical principles [16,17]. Notably, medical students are usually bound by the Code of Ethics in the same way as they are bound by their national medical professional bodies [18]. The last couple of decades has seen wide ranging changes in the digital world and how it touches human lives. Perhaps revisiting the Health Information Privacy Code to understand how it reflects current understanding and expectations about privacy might be the next logical step.

Health organizations have designed local policies and regulations on security, access, and transfer of health data to uphold privacy rights of patients. At times, the protocols and guidelines may not be clear on handling of patient-related information, especially sharing of information and under what circumstances. For example, until a few years ago, there were not many guidelines about the use of social media (SM) for health care professionals. Now, many organizations have clear policies and guidelines for the use of SM and social networking sites. These guidelines often define what SM is, who can post to SM, what sites and channels can be used and a basic code of conduct. Some organizations will also outline clear procedures if staff or students run into challenges or difficult circumstances such as cyberstalking. At the end of the day, it is the responsibility of the end user to act with integrity to ensure the confidentiality of patient-related information. The gap that currently exists, however, is the effective role modeling and teaching of how to translate these policies and procedures into becoming a digitally proficient practitioner.

A related phenomenon is that of the "privacy paradox" where there is a constant tension between individual's fear of their autonomy because of personal and health-related data being shared and processed using information technologies, for example, genomic data; and, on the other hand, the same person might be sharing health information in health forums and social networking sites [18,19]. The important point in this apparent paradox is that the data are owned by the patient and the patient in almost all circumstances should make decisions about how and when their personal data are shared. HCP should also be trained and educated on to make best efforts to ensure that patients understand the consequences, risks, and benefits of sharing or not sharing personal health information.

Consent Issues

Going back to the case example at the start of the chapter, the resident tried to obtain consent from the patient but couldn't get in touch. There are several issues that are important here. What is meant by the informed consent for the patient and the resident? Who is responsible for getting the informed consent and are there any limits to the consent rights of the individual? While requesting consent, it is important to provide information that a "reasonable" patient would want to have, the patient should have the capacity to understand the information provided and give consent and thirdly, consent should be voluntary [5]. In some settings, patients might not have the capacity to consent. In those situations, ethical considerations would include capacity assessment and a process for substitute decision making. The healthcare professional carries the responsibility for obtaining informed consent. Had the resident been able to meet with the patient discussed earlier in this chapter, the individual should have been informed about the reason for accessing his electronic health records, the contents of the record being shared with attendees of the case conference, specify if the presentation was being webcast to a wider audience, and to whom (which audience) the information is being shared. Notably, the patient could have refused to consent to the information being shared. In the same example, if the researcher resident would have obtained the ethics board approval, the informed consent requires disclosure of the planned research work, who would be accessing data and for what purpose, duration for which the consent would be needed and the steps taken to ensure security of the data [5]. There can be situations when such procedures may not be possible, for example, unconscious or disorientated patients presenting in an emergency and do not have any substitute decision makers.

There are situations when consent is not necessary. These conditions hinge on the ethical principles of *impossibility* and *legitimate infringement*. Kluge [5,20] has outlined conditions such as healthcare professionals needing access for professional reasons, institutions and planners requiring access to data for planning and financing of health services and situations where access to specific data can further good of the society at large. However, Kluge [5,20] cautions that the *infringement* into patient's "informatic rights" should be demonstrated as *necessary*. For health professional educators and organizations, this also introduces the responsibility and obligation of supporting patients and their families in developing digital literacy or in ensuring that patients are aware of the full implications and risks of consenting to have even anonymized data shared on SM. In general terms, prior notification to the patient where possible to inform that data may be accessed by authorized persons on a need-to-know basis is best practice. Moreover, in such situations, the data should be used in an anonymized manner and systems should be in place to ensure that appropriate security measures are in place.

CLINICAL DATA AND EDUCATION: SECONDARY DATA USE

The use of data and cases that emerge from clinical practice are primary teaching texts for HCP educators. In the early part of the last century, William Osler articulated that "there should be no teaching without a patient for a text, and the best teaching is that taught by the patient himself" [21]. The emergence of electronic health records and clinical documentation over the last several decades has created numerous opportunities for educators to access digital records for teaching purposes. However, most clinical documentation systems have not been designed to promote teaching and very few organizations have created formal frameworks or structures to help HCP teachers to harness these data for teaching purposes. In the chapter case example, several potential issues and risks were identified. In fact, the use of clinical documentation for teaching purposes falls within a category of data use often referred to as secondary use of health information.

The American Medical Informatics Association (AMIA) has clearly outlined the benefits of health data for learning and practice improvement: "Secondary use of health data can enhance healthcare experiences for individuals, expand knowledge about disease and appropriate treatments, strengthen understanding about the effectiveness and efficiency of our healthcare systems, support public health and security goals, and aid businesses in meeting the needs of their customers" [22]. The concept of *secondary use* of data is an important framework for HCP education. It emphasizes that clinical data in fact was not intended for educational purposes and signals to both teacher and learner that there are boundaries and issues related to using the data for these purposes. Unfortunately, the use of clinical data for HCP education is an underdeveloped area of practice and study. There are some examples in the literature of using electronic health data for teaching purposes [23,24], but this is not a developed domain of practice with frameworks and guidelines to promote the adoption of secondary use of data in HCP education. There may in fact be a moral imperative to adopt better practices to learn from clinical data to improve the delivery of health care. However, substantial work needs to be done to assist leaners and educators in navigating the practical and ethical challenges of harnessing the potential power of clinical data for HCP formation and professional development [25].

THE EXPERT PATIENT AND HCP EDUCATION

The practice of health care has changed rapidly over the last several decades. This change in part is because of the emergence of role of the patient as some patients have access to more and more information and options for health care in the digital age. In fact, we have seen the emergence of expert patients who utilize and optimize digital tools to manage their personal health [26,27]. It has also led to the development of a digital divide where those who do not have access to these tools may be disadvantaged in their ability to access health information and health services [28,29]. For HCP education, this phenomenon has created some opportunities to improve the delivery of patient education and to harness the vast array of health information available online for teaching purposes. At the same time, HCP educators have a moral obligation to once again contribute to the formation of HCPs who have the digital skills and literacy to serve patients and families who are digitally literate, digitally disadvantaged, or somewhere along this vast spectrum. In order to deliver equitable and socially responsible care, HCP educators must prepare the next generation of professionals to work within this spectrum and navigate ethical issues of professionalism, professional boundaries, and equitable access to services.

Social Media, HCP Education, and Ethics

We saw in our case example how a tweet by a medical student about a patient had far reaching consequences. SM has transformed the way individuals interact at both personal and professional level. Various tools of SM including blogs, social networking sites, podcasts, and video-sharing sites [30] can enhance traditional teaching methods in HCP education. Many studies have explored the educational potential of SM tools [31–36]. Quick dissemination, ease of use and mass participation are some of the reasons for SMs popularity as a learning tool. SM plays an important role in engagement with community about public health issues and disseminating health-related information.

However, issues related to privacy and confidentiality, professional boundaries, integrity, accountability, and the blurring lines between personal and professional identities has led to concerns about ethical issues. A study [37] reported on issues related to professional boundaries including display of personal information and views of residents and fellows about patients sending friend requests on social networking sites. Eighty-five percent of the participants stated they would automatically decline the request and the rest stated they would decide on an individual basis. Privacy protection issues can have an impact on doctor–patient relationship [37] and this can be applicable to both the patient and the HCP.

We have touched on the issue of privacy and confidentiality several times in this chapter. Sharing health-related information in online forums for advice has been seen as an extension of the consulting room in the digital space. The growth of "*crowdhealth*," a term used to describe use of social networks to ask for information about diagnosis, treatment, or general advice, reinforces the debate about ethical issues related to privacy, confidentiality, consent, data ownership, and how the data are used in research [38]. Ethical and professionalism issues have led to development of formal guidelines by professional and regulatory bodies for use of SM in health care [39,40]. Awareness and application of ethical principles by *digital HCPs* can be further enhanced by inclusion of such topics in ethics curricula within HCP education. SM's promise as an educational tool for the future is destined to grow further. HCP educators can help harness its full potential by supporting use of SM tools within an ethical framework.

Mobile Computing Systems and the Learner

Mobile devices have become a part of our everyday life, patients, and healthcare professionals alike. They have transformed the way healthcare is provided because of their reach, access at point of care, convenience of use and multifunctionality. The personal digital assistants [PDAs] and later smartphones and tablets with a myriad of mobile medical apps have changed clinical practice in many ways [41]. A survey about the use of mobile devices by HCPs reported that 85% of a medical school faculty and medical students, and 90% of residents used mobile devices in a range of clinical settings including classrooms and hospitals [42]. The use of apps has revolutionized the way mobile devices have been used in healthcare. Mobile apps range from those used for diagnosis and treatment to drug reference guides, clinical guidelines, literature search portals, and e-learning [42]. HCPs use apps for clinical decision making [43–45], patient monitoring (especially remote monitoring) and collection of medical data [43,44,46–48], and information gathering [42,47,49–51]. Apps are making a significant mark in medical education and training. Within educational settings, apps are being used ever more to access information about diagnosis and treatment and to log educational experiences.

Apart from issues related to privacy and confidentiality, which have been discussed in detail elsewhere, concerns related to professional behavior have been raised. The same study [42] reported concerns related to blurring of personal and professional boundaries in use of mobile devices. The authors raise several pertinent questions: "What should you do if a friend accidentally sees a clinical photo or text relating to a patient on your phone?" Can a clinician receive emails from patients in their personal mobile devices or should separate devices be used for personal and professional purposes. Better security features and the organizational support with formal guidelines can help resolve some of the barriers in use of mobile devices in communication with patients and service users. A related issue to reflect on is about the use of mobile devices in the clinical space by patients and service users. What are the ethical implications of patients and families recording clinical sessions secretly by a patient? Recording of visits can possibly give opportunity to the patient to review information at their own convenience and comfort [52]. Recording without the HCPs consent can affect patient–clinician relationships, bringing issues related to trust to the fore.

HCP learners have reported advantages related to portability, quicker access, increased efficiency, and access to multimedia resources, thereby having a positive effect on their learning [42]. At the same time, mobile devices causing distraction in the clinical environment and those at a disadvantage not able to access the benefits as much as their better off peers calls for HCP educators and policy makers to make best use of mobile digital technology and lessen the potential risks.

Digital Images and Ethics

Digital photography and the development of teaching videos has rapidly become an important part of teaching and learning in health professions education [53]. Text Box 6.1 outlines practical suggestions about the use and storage of digital media for educational purposes [53].

TEXT BOX 6.1

PRACTICAL TIPS FOR TAKING AND STORING DIGITAL IMAGES FOR EDUCATIONAL PURPOSES

1. Written, informed consent from patient; make sure that the individual is aware of the purpose for taking digital images.
2. Keep an audit trail by documenting the consent process, storage of the image, and procedures to follow up on withdrawal of consent.
3. Store images in a secure environment, check with organizational policies for transportation and storage of data, avoid portable unencrypted devices.
4. When in doubt, seek advice about local policies and procedures.

Adapted from Hubbard V, Goddard D, Walker S. An online survey of the use of digital cameras by members of the British Association of Dermatologists. Clinical and experimental. Exp Dermatol 2009;34(4):492–494.

The advent of smartphones and similar devices has transformed the way digital images are used in clinical care, teaching, and research in healthcare settings. The ease of use, high quality of images, minimal costs, and convenience in storage and distribution has contributed to their increased use [2]. However, this has raised issues about a lack of attention to ethical standards of the use of digital images in HCP education. A study on ethical implications of digital images for teaching and learning in healthcare settings identified issues related to poor understanding of consent when taking digital images and lack of familiarity with policies related to ethical issues [2]. This can easily lead to unsafe practice and increasing the risk of breach in patient's privacy. Notably, in a study involving dermatological trainees, consent was sought from consultants rather than patients for taking digital images [54]. The same study also reported that nearly one-third of the participants were not aware of guidelines around digital photography in the clinical space. A lack of awareness about the idea of a written, valid consent for taking photos and using digital images for educational purposes suggests the need for developing policies and guidelines to educate HCPs about the ethical issues. As digital technologies are ever evolving, it is important that continuous professional development activities are developed for HCPs to keep them up-to-date. The General Medical Council, UK (2015) and other bodies have developed guidelines on use of audio and visual recordings of patients [55]. Interestingly, in situations where consent was obtained, it was usually verbal, infrequently informed, lacked specificity about educational purposes, and unclear about storage or with whom the pictures could be shared [3]. Another point of note is the emerging evidence of digital images being used for purposes other than clinical education. Kornhaber et al. (2015) recommend educational workshops and further training within clinical settings to address knowledge and practice gaps related to use of digital images in teaching and learning, primarily ethical issues [3].

The use of video in HCP education [56] also extends beyond patients to students themselves. As approaches such as simulation become more widespread, organizations capture videos of learners in order to help them develop skills [57]. Most studies have focused on learning skills and competencies. As the use of digital technology increases in HCP education, ethical issues related to data privacy, access and control, as well as digital professionalism as a competency, will gain more significance. HCP educators and organizations must develop clear policies and procedures around how these digital tools will be used, stored, and disposed of in an appropriate manner.

Ethical Aspects of Telemedicine and Tele-Education

With the advent of the digital age, healthcare delivery has moved from clinical settings to ubiquitous ones such as homes and communities. Telemedicine is the use of information technology in healthcare provision with a clinician as one of its participants. Telehealth involves broader applications including health promotion. Technology bridges the gap between the users. The applications range from video-link and telephone, to remotely run clinics and telesurgeries done distantly [58]. This has helped in increasing access to health care providers, efficient service delivery with fewer resources, quality improvement, and decreased costs. Researchers have noted that this also led to a personalized approach to health care, which can be considered empowering [58]. Telemedicine plays an important role in equitable distribution of healthcare resources.

One of the ethical issues related to telemedicine applications is that of their design, and especially their user-friendliness. The traditional model of care relies on face-to-face contact between the patient and the clinician whereas in telemedicine, the clinical encounter is in virtual space. The risk of missing important subtle visual cues during a clinical encounter, the individual's familiarity (or lack of) with new technology influencing the activity, and reliance on data rather than *'living and breathing patients'* as sources of information for the healthcare professional have emerged as ethical concerns when it comes to making autonomous choices about their healthcare. At the same time, huge strides made by telemedicine in accessing quality healthcare has promoted "informed autonomy," where increased access to information can enable decision making by patients [58].

From the perspective of HCP education, the issue of telemedicine once again points to the importance of developing the professional identity of students who have the competencies and capabilities to practice not only in person, but also in digital realms. Does one have the same identity and approaches as a "digital" HCP? What are the issues related to practice in the digital realm without physical boarders and how does one manage the issues that surface. For example, issues related to cross-border involvement, jurisdictions of practice, who is liable, which codes of ethics are applicable, and what is permissible by the regulatory and professional bodies that oversee different professions.

In HCP education, it is important to consider ethical issues of telemedicine and other digital realms of practice. Digital and online consultations are emerging in many domains of HCP practice. Students should be introduced to key concepts as part of their training and professional identity formation. One important example is the issue about confidentiality in the virtual environment. Patients may be reluctant to give information for good clinical care unless they are assured of its confidentiality. Sharing of information about the technology being used is vital so that the patient is fully aware and understands the functions and usage. Particular attention should also be paid to design issues to minimize chances of design flaws affecting assessment, interventions, and clinical outcomes. Another important area of development is training and mastery in communication techniques used in the digital world. For example, telemedicine consultations in psychiatry involve working with patients, carers, and health teams in the digital space. Mental state examination during an assessment on a screen is not the same as a traditional face-to-face clinical interview. A study identified specific domains of competency including technical skills, assessment skills; relational skills and communication; collaborative and interprofessional skills among other competencies [59].

HCP Education, Digital Data and Research

Throughout the chapter, we have explored how a digital world has revolutionized healthcare. HCP education has a pivotal role in shaping knowledge and practices of health professionals to harness the full potential of the opportunities. In the newly revised CanMEDS 2016 framework by the Royal College of Physicians and Surgeons of Canada, the "Manager" role has been replaced by the "Leader" role [60]. One of the key competencies in the role is that of using health informatics to improve quality of patient care and optimize patient safety [60]. This ranges from describing data in health information systems, through use of data for clinical performance measures, to participation in systems-based informatics development and improvement. Competency in health informatics is emerging as an important topic in medical education literature [61,62] and ethics is an important part of the proposed curriculum (60).

Secondly, within a Kirkpatrick model of evaluation framework [63], assessment of impact of HCP education in digital healthcare is crucial. The model recommends evaluation of training programs' effectiveness at four levels, starting from reaction at level 1, learning at level 2, changes in behavior in level 3 and systemic changes in level 4. However, to assess changes in behavior and wider system level changes in patient-related outcomes, electronic health records, and other digital source of information remain key sources of information. Access to patient data for evaluation purposes calls for an ethical approach to use of clinical data. One of the primary ethical questions is about consent from service users about use of data for educational and research purposes at the outset e.g, how informed would the consent be in such situations? Kaplan [64] recommends transparency, flexibility in approach and accountability for ethical use of health information data in research.

An additional issue related to teaching of *digital ethics* in healthcare curricula is about its relevance in research. This is important because the lines between healthcare databases and other linked databases can easily get blurred in the digital world [65]. Web browsing data, which is mostly unregulated and yet has a wealth of health-related information raises issues related to consent, ownership, and control of data. The potential of electronic health records, big data, and data mining in advancing the cause of patient care improvement is well accepted. From an ethical standpoint, what constitutes use of data in the public interest, combining public and private data, "commodifying data for sale" emerge as important issues. In a UK study, it has been observed that individuals have concerns about use of personal data, privacy, and control over access but are in favor of using them for public health research [66]. Another study reported willingness to share anonymous personal health data to advance research for the good of the public [67]. Use of anonymized data, defining informed consent from data-centric perspective for research purposes, and strict governance mechanisms are central to the ethical debate to ensure health informatics continues to play an important role in advancement of healthcare research.

CONCLUSIONS

This chapter started by asking the question around why we need health informatics ethics in medical education and has explored a number of facets to this question. The context of the digital age has increased the complexity of HCP education and is raising a number of challenges that need to be addressed as the opportunities of practicing digitally emerge. The issues raised, however, point to a challenge for health professions education. The formation of professional identity must also include the formation of the virtual or the digital professional identity. Health professionals that are being trained must be prepared to practice in the digital world of health care. From an ethical perspective, there are many challenges facing educators of health professionals: (1) how can digital tools be ethically integrated into the formation and education of health professionals? (2) how are health professional learners prepared to manage the uncertainty and ethical dilemmas that may present themselves? and (3) how does health professional education contribute to the development of professional identity of digitally competent and capable professionals who can make the most of and sense of the opportunities in dynamic, digital practices in a safe, responsible and meaningful way?

The formation of health professionals has to include the ability to rapidly adapt to changing practice, which includes both physical and digital contexts. In the case of our health professional students, digital competency and capability must be a shared responsibility of the learners and the educators. It is not enough to assume that growing up in a digitally enabled environment ensures digital literacy and the competencies and capabilities to practice health care in this environment. Educators themselves must become comfortable addressing these challenges and fluidly moving between the digital and physical realities of education and care. In the circumstances of our case of the student sharing information on SM, educators must make an important decision: should students be taught not to have a digital professional identity or should they be encouraged to develop this aspect of their professional identity and help students navigate the digital pitfalls as they emerge in practice. Health professional educators themselves have a moral and ethical responsibility and accountability to foster and develop the next generation of health professionals for the challenges and context in which they will work.

References

[1] Beauchamp TL, Childress JF. Principles of biomedical ethics. USA: Oxford University Press; 2001.

[2] Kornhaber R, Betihavas V, Baber RJ. Ethical implications of digital images for teaching and learning purposes: an integrative review. J Multidiscip Healthc 2015;8:299.

[3] International Medical Informatics Association (IMIA). IMIA code of ethics for health information professionals. Available at <www.imia-medinfo.org>; 2002 [accessed 13.01.17].

[4] Mosa AS, Yoo I, Sheets L. A systematic review of healthcare applications for smartphones. BMC Med Inform Decis Mak 2012;10(12):67.

[5] Kluge E. Informed consent to the secondary use of EHRs: informatics rights and their limitations. Medinfo 2004;11(Pt 1):635–8.

[6] Anonymous. Benefits of sharing. Nature 2016;530(7589):129.

[7] Annas GJ. HIPAA regulations—a new era of medical-record privacy? New Engl J Med 2003; 348(15):1486–90.

[8] Centers for Disease Control and Prevention. HIPAA privacy rule and public health. Guidance from CDC and the US Department of Health and Human Services. MMWR: Morbid Mortal Wkly Rep 2003;52(Suppl. 1):1–17, 9.

[9] Crotty BH, Mostaghimi A. Confidentiality in the digital age. BMJ 2014;348:g2943.

[10] Hafferty F, Gaufberg E, O'Donnell J. The role of the hidden curriculum in "on doctoring" courses. AMA J Ethics 2015;17(2):130.

[11] Erickson J, Millar S. Caring for patients while respecting their privacy: renewing our commitment. Online J Issues Nurs 2005;10(2).

[12] Austin LM. Is consent the foundation of fair information practices? Canada's experience under PIPEDA. Univ Tor Law J 2006;56(2):181–215.

[13] Austin LM. Reviewing pipeda: control, privacy and the limits of fair information practices. Can Bus Law J 2006;44:21.

[14] Knoppers BM. Confidentiality of health information: international comparative approaches. A health information network for Australia 2000.

[15] Willison DJ. Health services research and personal health information: privacy concerns, new legislation and beyond. CMAJ 1998;159(11):1378.
[16] de Lusignan S, Chan T, Theadom A, Dhoul N. The roles of policy and professionalism in the protection of processed clinical data: a literature review. Int J Med Inform 2007;76(4):261–8.
[17] Claerhout B, DeMoor G. Privacy protection for clinical and genomic data: The use of privacy-enhancing techniques in medicine. Int J Med Inform 2005;74(2):257–65.
[18] Masters K. Health Informatics Ethics. In: Hoyt RE, Yoshihashi AK, editors. Health Informatics: practical guide for healthcare and information technology professionals. Lulu.com; 2014. p. 195–215.
[19] Büschel I, Mehdi R, Cammilleri A, Marzouki Y, Elger B. Protecting human health and security in digital Europe: how to deal with the "Privacy Paradox"? Sci Eng Ethics 2014;20(3):639–58.
[20] Kluge E-HW. Informed consent and the security of the electronic health record (EHR): some policy considerations. Int J Med Inform 2004;73(3):229–34.
[21] Roter D, Hall JA. Doctors talking with patients/patients talking with doctors: improving communication in medical visits. Westport, CT: Praeger; 2006.
[22] Botsis T, Hartvigsen G, Chen F, Weng C. Secondary use of EHR: data quality issues and informatics opportunities. AMIA Summits Transl Sci Proc 2010;2010:1–5.
[23] Ellaway RH. Medium, message, panopticon: the electronic health record in residency education. J Grad Med Educ 2016;8(1):104–5.
[24] Milano CE, Hardman JA, Plesiu A, Rdesinski MRE, Biagioli FE. Simulated electronic health record (Sim-EHR) curriculum: teaching EHR skills and use of the EHR for disease management and prevention. Acad Med 2014;89(3):399.
[25] Ellaway RH, Coral J, Topps D, Topps M. Exploring digital professionalism. Med Teach 2015;37(9):844–9.
[26] Cordier J-F. The expert patient: towards a novel definition. Eur Respir J 2014;44(4):853–7.
[27] Wardrope J, Ravichandran G, Locker T. "Expert patient"—dream or nightmare? Group 2000;343:94–9.
[28] Sarkar U, Karter AJ, Liu JY, Adler NE, Nguyen R, López A, et al. Social disparities in internet patient portal use in diabetes: evidence that the digital divide extends beyond access. J Am Med Inform Assoc 2011;18 (3):318–21.
[29] Chang BL, Bakken S, Brown SS, Houston TK, Kreps GL, Kukafka R, et al. Bridging the digital divide: reaching vulnerable populations. J Am Med Inform Assoc 2004;11(6):448–57.
[30] Peters ME, Uible E, Chisolm MS. A Twitter education: why psychiatrists should tweet. Curr Psychiatry Rep 2015;17(12):1–6.
[31] Bahner DP, Adkins E, Patel N, Donley C, Nagel R, Kman NE. How we use social media to supplement a novel curriculum in medical education. Med Teach 2012;34(6):439–44.
[32] Wang AT, Sandhu NP, Wittich CM, Mandrekar JN, Beckman TJ. Using social media to improve continuing medical education: a survey of course participants. Mayo Clinic Proc 2012;87(12):1162–70.
[33] Langenfeld SJ, Vargo DJ, Schenarts PJ. Balancing privacy and professionalism: a survey of general surgery program directors on social media and surgical education. J Surg Educ 2016.
[34] Guraya SY. The usage of social networking sites by medical students for educational purposes: a meta-analysis and systematic review. N Am J Med Sci 2016;8(7):268.
[35] Nwosu AC, Monnery D, Reid VL, Chapman L. Use of podcast technology to facilitate education, communication and dissemination in palliative care: the development of the AmiPal podcast. BMJ Support Palliat Care 2016; bmjspcare-2016-001140.
[36] Rosenkrantz AB, Won E, Doshi AM. Assessing the content of YouTube videos in educating patients regarding common imaging examinations. J Am Coll Radiol 2016;13(12):1509–13.
[37] Moubarak G, Guiot A, Benhamou Y, Benhamou A, Hariri S. Facebook activity of residents and fellows and its impact on the doctor–patient relationship. J Med Ethics 2010;37(2):101–4.
[38] Denecke K, Bamidis P, Bond C, Gabarron E, Househ M, Lau A, et al. Ethical issues of social media usage in healthcare. Yearb Med Informa 2015;10(1):137.
[39] Kind T. Professional guidelines for social media use: a starting point. AMA J Ethics 2015;17(5):441.
[40] Kind T, Patel PD, Lie D, Chretien KC. Twelve tips for using social media as a medical educator. Med Teach 2014;36(4):284–90.
[41] Ventola CL. Mobile devices and apps for health care professionals: uses and benefits. PT 2014;39(5):356.
[42] Wallace S, Clark M, White J. 'It's on my iPhone': attitudes to the use of mobile computing devices in medical education, a mixed-methods study. BMJ Open 2012;2(4):e001099.

[43] Mickan S, Tilson JK, Atherton H, Roberts NW, Heneghan C. Evidence of effectiveness of health care professionals using handheld computers: a scoping review of systematic reviews. J Med Internet Res 2013;15(10): e212.
[44] Divall P, Camosso-Stefinovic J, Baker R. The use of personal digital assistants in clinical decision making by health care professionals: a systematic review. Health Inform J 2013;19(1):16–28.
[45] Aungst TD. Medical applications for pharmacists using mobile devices. Ann Pharmacother 2013;47 (7–8):1088–95.
[46] Mosa ASM, Yoo I, Sheets L. A systematic review of healthcare applications for smartphones. BMC Med Inform Decis Mak 2012;12(1):1.
[47] Ozdalga E, Ozdalga A, Ahuja N. The smartphone in medicine: a review of current and potential use among physicians and students. J Med Internet Res 2012;14(5):e128.
[48] Boulos MNK, Wheeler S, Tavares C, Jones R. How smartphones are changing the face of mobile and participatory healthcare: an overview, with example from eCAALYX. Biomed Eng Online 2011;10(1):1.
[49] O'Neill K, Holmer H, Greenberg S, Meara JG. Applying surgical apps: smartphone and tablet apps prove useful in clinical practice. Bull Am Coll Surg 2013;98(11):10–18.
[50] Payne KFB, Wharrad H, Watts K. Smartphone and medical related App use among medical students and junior doctors in the United Kingdom (UK): a regional survey. BMC Med Inform Decis Mak 2012;12(1):1.
[51] Murfin M. Know your apps: an evidence-based approach to evaluation of mobile clinical applications. J Physician Assist Educ 2013;24(3):38–40.
[52] Rodriguez M, Morrow J, Seifi A. Ethical implications of patients and families secretly recording conversations with physicians. JAMA 2015;313(16):1615–16.
[53] Harting M, DeWees J, Vela K, Khirallah R. Medical photography: current technology, evolving issues and legal perspectives. Int J Clin Pract 2015;69(4):401–9.
[54] Hubbard V, Goddard D, Walker S. An online survey of the use of digital cameras by members of the British Association of Dermatologists. Clin Exp Dermatol 2009;34(4):492–4.
[55] GM Council. Making and using visual and audio recording of patients. London: General Medical Council; 2013.
[56] Williams DM, Fisicaro T, Veloski JJ, Berg D. Development and evaluation of a program to strengthen first year residents' proficiency in leading end-of-life discussions. Am J Hosp Palliat Care 2011;28(5):328–34.
[57] Wouda JC, van de Wiel HB. The effects of self-assessment and supervisor feedback on residents' patient-education competency using videoed outpatient consultations. Patient Educ Couns 2014;97(1):59–66.
[58] Kaplan B, Litewka S. Ethical challenges of telemedicine and telehealth. Camb Q Healthc Ethics 2008;17 (04):401–16.
[59] Crawford A, Sunderji N, López J, Soklaridis S. Defining competencies for the practice of telepsychiatry through an assessment of resident learning needs. BMC Med Educ 2016;16:28.
[60] Frank J, Snell L, Sherbino J. The draft CanMEDS 2015 physician competency framework–series IV. Ottawa: The Royal College of Physicians and Surgeons of Canada; 2014.
[61] Fetter MS. Improving information technology competencies: implications for psychiatric mental health nursing. Issues Ment Health Nurs 2009;30(1):3–13.
[62] Brixey JJ. Health informatics competencies, workforce and the DNP: why connect these 'dots'? 750. Nurs Inform 2016; 2016.
[63] Kirkpatrick DL. Evaluating training programs. San Francisco: Tata McGraw-Hill Education; 1975.
[64] Kaplan B. How should health data be used? Privacy, secondary use, and big data sales. Privacy, secondary use, and big data sales (August 1, 2014). Yale University Institute for Social and Policy Studies Working Paper No. 14-025. 2014.
[65] Aicardi C, Del Savio L, Dove ES, Lucivero F, Tempini N, Prainsack B. Emerging ethical issues regarding digital health data. On the World Medical Association Draft Declaration on Ethical Considerations Regarding Health Databases and Biobanks. Croat Med J 2016;57(2):207.
[66] Luchenski S, Balasanthiran A, Marston C, Sasaki K, Majeed A, Bell D, et al. Survey of patient and public perceptions of electronic health records for healthcare, policy and research: Study protocol. BMC Med Inform Decis Mak 2012;12(1):40.
[67] Bietz MJ, Bloss CS, Calvert S, Godino JG, Gregory J, Claffey MP, et al. Opportunities and challenges in the use of personal health data for health research. J Am Med Inform Assoc 2016;23(e1):e42–8.

Further Reading

Hersh WR, Gorman PN, Biagioli FE, Mohan V, Gold JA, Mejicano GC. Beyond information retrieval and electronic health record use: competencies in clinical informatics for medical education. Adv Med Educ Pract 2014;5:205.

Norberg PA, Horne DR, Horne DA. The privacy paradox: personal information disclosure intentions versus behaviors. J Consum Affairs 2007;41(1):100–26.

SECTION III

PATIENT SAFETY AND QUALITY ASSURANCE THRUSTS IN DIGITAL HEALTHCARE AND THEIR INFLUENCE ON CLINICIANS AND PATIENTS

CHAPTER

7

Transformative Technology: What Accounts for the Limited Use of Clinical Decision Support Systems in Nursing Practice When Compared to Medicine?

Antonia Arnaert[1], *Norma Ponzoni*[1], *John Liebert*[2] *and Zoumanan Debe*[3]

[1]McGill University, Montreal, QC, Canada [2]Private Practice of Psychiatry, Scottsdale, AZ, United States [3]Consultant, Ontario, QC, Canada

INTRODUCTION

Healthcare organizations are complex environments where components of the systems are interrelated and interdependent [1], making it difficult to introduce change and modifying healthcare practices. Current governmental austerity measures pressuring public Canadian healthcare organizations to be more cost effective have a direct impact on the resources available within the institution to innovate or to invest in addressing emerging concerns such as the current widespread epidemic of Zika fever in mid-2016 [2]. Furthermore, the characteristics of the patient population served by these institutions has increased in acuity overtime due to the shift toward ambulatory care, the increasing number of elderly patients, the prevalence of chronic diseases, advancing technologies, and the availability of an array of new treatment options, etc., further stressing an already overstretched system.

Working in such challenging, complex environments where numerous demands are placed on clinicians drives the need for functional, efficient clinical performance; on the other hand, it also significantly increases clinicians' cognitive load, or their ability to

Health Professionals' Education.
DOI: http://dx.doi.org/10.1016/B978-0-12-805362-1.00007-3

process information and perform expected tasks. For example, a nurse that is mentally preoccupied by complex mathematical calculation will not have the ability to perform another task simultaneously, that is, the brain can only process a certain amount of incoming information at a time. The strain of performance can lead to breaches in patient safety and medical errors as the thinking process involved in clinical environment requires practitioners to integrate and apply information that is retrieved from multiple sources, such as the patient him or herself, the medical records, previous experience, etc. [3]. Furthermore, there is the expectation that clinicians will remain competent by regularly engaging in professional development and that their practice conforms with recognized guidelines and evidence-based practice. This is no different for nurses, who have also experienced increased professional expectations on performance.

In the era of emerging health information technologies, healthcare organizations are increasingly implementing clinical tools that can benefit nurses with daily tasks, such as computerized nursing documentation, electronic health records (EHRs), mobile health applications, etc. Invariably, this shift to incorporate technology will eventually transform healthcare, changing the practice of nursing over the coming years and forcing them to build new skills in this area [4]. Given the current complexities in the practice context, one of these electronic tools that has been increasingly explored particularly within the practice of medicine is the use of clinical decision support systems (CDSSs), which assists clinicians with the process of clinical reasoning and decision-making. CDSSs thereby help clinicians process the vast amount of information they are presented with and provide a mechanism for insuring the integration of evidence-based practice into care delivery.

Despite the fact that CDSSs are increasingly used to support physician practice and the existent body of nursing literature making reference to CDSSs, there are little to no examples of their use in the nursing domain [5]. This is true even for advanced practice nurses, such as nurse practitioners who, despite being nurses at the core, have similar scopes of practice when compared to physicians. This chapter will provide an overview of the burgeoning evolution of CDSSs in medicine and the state of its current use in nursing practice and education.

DEFINITION, FUNCTIONALITY, AND ARCHITECTURE OF CDSSs

Over time, CDSS has been defined in a number of ways [6]; typically, these definitions conceptualize a CDSS as electronic "support" tools that aids clinical practice. The word "support" in the term CDSS, which points to the fundamental goal of these systems, is often interpreted and concretized by scientists in different ways and consequently the products that are developed can be quite varied. Initially, the goals and architecture of CDSSs in medicine, also called "expert systems," were focused on the diagnostic decision-making process. These early CDSSs, also recognized as diagnostic support systems (DSSs), relied on clinicians entering patient data for the system to generate a "diagnosis" or treatment plan in order to independently validate or replace the clinicians' reasoning process. Later on, scientists felt that computer-assisted decision-making should aid practitioners, the recognized experts, in working through the process. In this circumstance, CDSSs do not make the decision for the clinician, but rather optimize the process of making reasoning easier and more transparent by offering information based on current research, recognized best-practice standards, etc. Spooner supports this categorization whereby CDSSs range from "the passive display of information to intense computation designed to model complex clinical reasoning" [7].

The technical architecture of CDSSs is engineered with a common structure to receive clinical data from clinicians manually and/or automatically populated by drawing from EHRs as input and to produce information or even clinical knowledge as output such as reminders, alerts, recommendations for patient diagnosis, treatments plans, etc. In order to produce these outputs, the CDSS needs to interact with the incoming clinical data and information from, for example, evidence-based guidelines, research on signs and symptoms of diseases, reference information, similar cases, etc. that are stored in various databases, however all terms and concepts used need to be standardized and clearly defined. Subsequently, data mining reasoning engines or inferential tools combine the information from the different databases, which are integrated and interoperable, and communicate them to the user as an output in the form of integrated displays and clinical reports [8]. More specifically, the inferencing methods are able to transform, combine, and detect associations or recognize patterns from different pieces of information [9] and make a differential diagnosis according to rules and algorithms that are based on current, a priori literature-driven medical knowledge and expert opinion [10,11].

The functionality of CDSSs can be deployed on a variety of platforms such as mobile [12–14], cloud-based or locally installed on a workplace computer by the healthcare professional [15,16]. The implementation of the CDSS varies across healthcare organizations and depends on the type of clinical systems, workflow and security systems in those organizations, vendor offerings, and governmental constraints. In addition, they can be utilized strategically to reinforce quality improvement and patient safety initiatives within organizations to reduce cost and improve care delivery [17].

CDSSs AND CLINICAL REASONING SUPPORT

Due to increasing patient acuity, today's clinical environments are hectic, fast-paced, and rapidly changing forcing clinicians to develop highly adaptable and honed reasoning skills. These thinking skills, often referred to as "clinical reasoning," are the foundations of clinicians' practice and are required by all healthcare professionals. Clinical reasoning is described by Simmons as "a complex process that uses cognition, metacognition (reflection) and discipline-specific knowledge to gather and analyze patient information, weigh alternatives and evaluate the best possible treatment regimen" [18]. Every healthcare profession has a unique, but quickly growing, body of discipline-specific knowledge which healthcare professionals are expected to master. This challenging practice context serves to increase the clinicians' cognitive load, or the intensity of mental effort required to complete a task. Paas, Tuovinen, Tabbers, and van Gerven define this mental effort as "the aspect of cognitive load that refers to the cognitive capacity that is actually allocated to accommodate the demands imposed by the task" [19]. For example, when one is faced with a highly complex problem to solve, the mental effort required to work through the problem can be said to be a measure of one's cognitive load; in other words, the more complex the problem, the higher the cognitive load. It would therefore be logical to assume that computerized clinical systems such as CDSSs with fast processing speed and the ability to store vast amounts of knowledge within its memory can aid clinicians with these complex tasks.

The thinking process required in healthcare is challenging, and specifically in the context of nursing practice, Tanner [20] has elaborated and tested an integrative model of clinical judgment over a number of years in an attempt to explicate the nuances and complexity of clinical reasoning in nursing practice. More specifically, a nurse must gather facts or information from a variety of sources in order to formulate a pertinent diagnosis or clinical impression. For example, when a nurse practitioner practicing in Quebec (Canada) is asked to describe how he/she came to her clinical impression, the explanation would likely be as follows: (1) information is obtained from the health history, physical exam, and laboratory tests; (2) the relative importance of various signs and symptoms expressed are evaluated; and (3) a list of all possible diseases/diagnoses pertinent to the situation are elaborated then ruled-out one by one; this is referred to as the process of differential diagnosis. This is a rather superficial description of the process, as expert reflection is much more than the processing of clinical information. Making inferences relies not only on declarative knowledge, but also on drawing from past experience, reading and interpreting a patient's nonverbal cues, managing uncertainty, and reflective practice.

In comparison to the expert clinician's reasoning abilities, today's CDSSs are not yet designed intelligently and are only capable of addressing lower level cognitive tasks with functions such as reminders and alerts [21]. Historically, CDSSs have focused on the diagnostic process, as described in the more simplistic, linear fashion above. One may ask oneself, how this type of "computer assistance" can work in the airline industry, where pilots' tasks and reasoning are equally as complex; however, scientists considering this question found the comparison to be unrelated, as autopilot operations are successful because the procedures and operations of normal flights are highly predicable based on data that can be gathered objectively. In aviation, the rules for decision-making can be fully specified and implemented, barring exceptional circumstances, which is not the case in healthcare [22]. In their current form, CDSSs can be used to assist practice [23], especially for novice clinicians [9], but are not tailored to "replace" expert clinician reasoning for the purposes of validation [24].

EVOLUTION OF CDSSs IN MEDICINE

Over the course of the last 45 years, a number of CDSSs have been developed and evaluated [25,26]. Initially, CDSSs were independent programs designed to provide diagnostic and prescriptive support; yet these programs were time consuming as they required physicians to manually enter pertinent patient information [6]. Early discussions around diagnostic support started with the 1959 paper of Ledley and Lusted, "Reasoning foundations of medical diagnosis" [27], where the authors propose a probabilistic model to achieve a differential diagnosis using symptom cards. Over the next 15 years, a number of scientists innovated various standalone CDSSs that targeted specific diseases or symptomology. A few examples are offered, however the ensemble of examples cited do not constitute an exhaustive listing. A few years after Ledley and Lusted's ground-breaking paper, Warner produced a highly accurate probabilistic model for identifying congenital heart defects, using a matrix to map signs and symptoms to a number of congenital heart diseases [28,29]. This model used contingency tables to map symptoms and signs to diagnoses, based on the frequency of manifestation for each symptom or sign given an underlying diagnosis. A similar system was developed by Collen in the early 60s and implemented at Kaiser Permanente in the United States, where patients were asked to sort cards to indicate the symptoms they were experiencing in order for a computer to generate an initial differential diagnosis [30]. By the end of the 1960s, Bleich developed a unique algorithm-based system that automated the required calculations of blood test results in order to propose an appropriate intervention, in addition to a diagnosis, for acid-base imbalances [31,32]. The creation of de Dombal's highly accurate system (91.8%) for identifying the underlying cause for acute-onset abdominal pain, paved the way in the early 1970s for clinicians to demonstrate that these systems had practical value [33]. With the emergence of computers and artificial intelligence in the 1970s, systems started to incorporate rule-based processing. For example, Shortliffe created MYCIN, an expert system that suggested appropriate antibiotic treatment working backward from the clinician's observations of the infectious process [34]. In addition to entering clinical data, the ATTENDING system developed by Perry Miller, asked clinicians to elaborate a treatment plan, after which time, the system would offer suggestions [35,36]. The next phase in the development of standalone CDSSs in the 1980s saw attempts at more inclusive programs that focused on a broader array of diagnosis in a specialty area. Randolph Miller and Harry Pople from the University of Pittsburg were the major collaborators on project INTERNIST-I, one of the first programs that covered the majority of possible diagnosis in internal medicine and performed as well as the average physician [37]. Another example of this is the current, web-based DXplain system, which is still commercially available and has been in use for over 25 years; this system was different in that it explained its reasoning to justify the diagnostic conclusion [38,39].

The progressive development of CDSSs over the years was driven by the notion that medicine had become increasingly complex and that the knowledge required for competent practice was extensive, therefore computer-aided technology was required to help humans provide safe care. However, to overcome the difficulties arising from reentering data, limited access to patient-specific information such as lab results and to reduce medical errors, scientists began developing various clinical systems such as HELP [40] and RMRS [41], which coupled standalone CDSSs with computer-based physician order entry (CPOE) systems and EHRs [42]. Beyond the integration with other clinical systems, there was a need to have a standardized interface and query language in order to access and share decision support content [43]. The Arden Syntax is one example that combined the syntaxes used by the HELP system and the RMRS system to increase interoperability of clinical decision support knowledge.

Gradually initial optimism that CDSSs could serve to assist clinicians in their daily work waned over time from the 1960s to late 1990s when faced with the complexity involved in creating these programs. Systems gradually moved from programs that attempted to provide diagnostic support to simpler applications that serve to provide checks, warnings, and reminders to physicians [44]. Despite renewed passion for CDSSs in the new millennium due to quality and safety concerns, they are not yet commonplace in medicine [45], and even less so in nursing [46].

CDSSs IN CLINICAL NURSING PRACTICE

The nursing profession's body of scientific knowledge is growing exponentially making it challenging for nurses to remain up-to-date by reading current literature and to integrate and access all of this information within the context of daily practice [47,48]. It is the expectation of professional practice that nurses base their care decisions on current research, however according to Pravikoff [49], nurses feel more comfortable referring to peers or looking online rather than using recognized databases such as CINAHL. In parallel, nurses, generally speaking, struggle to make sense of scientific literature and integrate findings into their clinical practice, especially when conflicting recommendations are proposed. Furthermore, on the care units, nurses often do not have open and free access to the journal articles if the healthcare organizations do not have annual subscriptions with multiple publishers through their libraries. Failure to consult the scientific literature contributes to a stagnation of practice, impacting the quality of care that patients receive; this leaves the nurse dependent on his/her clinical experience and accumulated knowledge base to serve as a basis for making healthcare decisions. This is not optimal practice because failure to consult contributes to higher error rates and may lead to patient harm [50]. In contrast, while novice nurses have more "up to date" knowledge, they lack the clinical experience and expertise to back their decisions. Yet regardless of clinical experience, one can quickly become obsolete if there is no investment in remaining current on new developments [51]. Ideally, in order to make good care decisions, nurses' judgment, and evidence-based best practices should be combined with the use of CDSSs.

Progress in CDSS development is lagging behind in nursing when compared to other disciplines such as medicine [52]. And even when developed, they are not specifically engineered with nursing care in mind, but rather are an extension of existing CDSSs used for medicine [53]. Despite the developments in nursing informatics, a subspecialty that has existed since the 1980s [11], developments in innovation have been hindered primarily due to a lack of standardization in the terminology used to describe the "nursing process" or the method that nurses use to think through a clinical problem [52]. Furthermore, the early systems were designed as stand-alone programs and not integrated in the workflow [54], burdening nurses with the requirement of data input thereby increasing the effort required in using these programs and consequently, unnecessarily increasing the cognitive load rather than decreasing it as intended [55]. Repeated data entry, without the understanding or sense of the recognized benefit, quickly becomes onerous on highly acute and stressful units as individual nurses may not value the potential that data offers and "just want to get through their day" [56]. Knowing that nurses are the primary users of EHRs because they document tremendously in comparison to other healthcare professions, it is logical to assume that the development of integrated systems, that combine EHRs with CDSSs, would be user-friendly and tailored to their needs. However, today's EHRs are complex and create an additional hurdle to the delivery of highly reliable evidence-based nursing care [56]. In addition to more systemic challenges, nurses tend to lack informatics competencies and comfort with technology [57,58,24]; this discomfort often creates fears and resistance of relinquishing control of their clinical expertise to a machine or automated CDSS [59]. In addition to the fear created, the move to more automated systems can serve to threaten nurses who, as other healthcare professionals, value their autonomy, clinical experience, and acquired intuitive reflexes.

While various standalone CDSS programs have been developed for very specific, focused uses to screen for certain diseases such as cancer [48], to prevent certain complications such as falls [60] or pressure ulcers [61], or with certain nursing specialties in mind, such as school nursing [58], overall there is a scarcity of studies evaluating CDSS-use for general nursing care [5]. Generally speaking, findings suggest that in order to significantly improve patient safety and clinical practice [57,62,63], CDSSs must be embedded in an organization's EHR, seamlessly integrated with the nurses' clinical workflow and available at the point of care [16,64]. The integration of systems, into one place of reference, will consequently improve communication and allow for a certain level of oversight on care and outcomes among administrators [56], and as such may result in a more cost-effective system for patients and healthcare organizations [65]. In other words, these integrated systems can provide clinicians and administrators with a "bird's eye view" of all aspects of care in a timely manner to ensure quality of care within their institution [66].

On an individual level, nurses follow the steps of the "nursing process" (assessment, hypothesis-generation, diagnosis, goal setting, interventions, and evaluation) to address patient problems and to promote health. CDSSs support this process by making the steps involved explicit, forcing nurses to think through the process systematically [59]. In addition, steps in the nursing process can be informed by quick access to evidence-based best practices [24,58,67] that serve to help nurses consider additional information in order to avoid jumping to ill-supported conclusions prematurely and to consider a wider array of treatment options and strategies [68]. Moreover, beyond providing easy access to evidence, CDSSs have influenced nurses' adherence to recommended guidelines [69,70]. Furthermore, when the CDSS is able to validate a nurse's thought process [71], this instills confidence, especially in novice nurses, and may serve to reassure clinicians that they are moving forward in a competent and safe way. By combining the nurse's expertise with the theoretical knowledge a CDSS can provide, better clinical decisions can be made then either what the nurse or computer could make alone [72]. An additional benefit of the CDSS is that the entire process of decision-making is well documented [73], including the justification for care decisions. Overall, a systematic review on CDSSs used within the context of various disease processes indicated that they improve provider performance in 64% of the 97 reviewed studies [74]. When clinicians observe direct care benefits to patients and added value to their practice [75], they are motivated and believe that CDSSs can improve efficiency [76]; but often indicate the lack of training and technology-savviness impedes their ability to use these systems [71]. Something which could be overcome in the long run through the provision of professional development on CDSSs for practicing nurses and their systematic integration in curricula for nursing students.

USE OF CDSS IN CLINICAL NURSING EDUCATION

It is not just the practicing nurses that were trained before the age of electronic documentation that lack informatics literacy; newly graduated nurses and nursing students maybe techno-savvy with the use of mobile health applications and social media, but often lack the critical informatics knowledge and skills required to use CDSSs embedded in EHRs effectively. According to Miller et al. [77], it can take up to six months for novice nurses to become comfortable using EHRs; in this study, beyond the consideration of age and previous clinical experience, the nursing school from which one graduates seems to have a major impact on this level of comfort. In this day and age, it seems essential for schools of nursing to incorporate nursing informatics in their respective curricula [24]; however, in order to prepare graduates to "hit the ground running" with the knowledge and skills expected of novice practitioners, educational institutions need to have access to the equipment and computer programs to train them. Many colleges and universities, despite having state-of-the-art simulation centers, cannot afford to purchase or keep up with the equipment that replicates the clinical environment. Especially with the recent evolution of EHRs, where there is broad institutional variation in use of commercially available EHRs [78,79], hospitals are establishing restrictive policies that limit students access and use of their EHRs due to concerns surrounding ownership of documentation, patient safety and confidentiality, interference with billing, etc. [79]. This has a significant impact on the students' ability to prepare for their clinical practicum as it has been demonstrated that students that have the opportunity to document in EHRs resulted in greater subjective responsibility for their patients [80]. One possible solution could be to create distinct sections in EHR for students in nursing and other disciplines. Another suggestion would be to create close strategic partnerships between clinical and academic institutions that would delineate the conditions by which students can be granted access to the hospital or clinic's clinical information systems onsite and remotely [80]. An example of such a strategic partnership is the collaboration between the Cleveland Clinic and their affiliated hospitals where a student nurse portal was created to allow students to learn how to enter data and how the resulting information or knowledge can enable them to provide better care. The student nurse portal is hosted on the Cleveland Clinic servers and is available to students anywhere, anytime. Students have reported that access to this portal is a valuable tool in preparing for the use of the EHR during their clinical experiences [80]. While this revolutionary vision is promising, it will be key for educators to be prepared adequately to guide students in the use of these systems. Currently, most educators, not being hospital based, do not have expertise [81] and may feel unprepared to guide students in the use of this emerging technology [82,83]. In order to facilitate the inclusion of information technology content in nursing curricula and to prepare educators to teach this content, the global T.I.G.E.R. initiative was established in North America, to promote these competencies for nursing students, practicing nurses, and nursing organizations [24].

In the meantime, a few academic institutions have invested in the purchase of simulated EHRs (S/EHRs) which are internet-based programs that contain simulation scenarios and accompanying EHRs that document patients' previous health information, assessments, and interventions. Overall students perceived that the S/EHR was realistic, easy to navigate and increased their confidence and ability to provide patient care and administer medications safely [84]. In parallel to these trends, "stand-alone" mobile applications (apps) such as electronic drug guides, medical laboratory references, and dictionaries [85], are becoming more popular to provide evidence-based information to students as a reference in order to support student learning and decision-making [86]. While these stand-alone mobile apps are useful as educational tools and promote the provision of evidence-based care [87,88], they do not effectively support the development of clinical reasoning as the reference information they provide is not patient specific [89]; something that can only occur when the mobile device is connected directly with both a CDSS and an EHR [90]. Currently, research is exploring the use of CDSSs installed on mobile devices and some institutions are pioneering in this area, such as the Columbia University School of Nursing that recently added decision support features on their mobile platform used within the context of their nurse practitioner curriculum [46]. In medicine, Johnston et al. indicated that CDSSs on a mobile device provide medical students with better access to high quality information, knowledge of evidence-based medicine and increased their adoption of point-of-care tools [91]. Along the same lines, a study of a web-based CDSS created for dental students reported the technology was easy to use and saved time, but more importantly, it increased students' adherence to treatment guidelines [92]. In addition, the University of Utah School of Medicine developed an expert system, ILIAD, that is used as a teaching tool for medical students; this tool allows for simulation to help students sharpen their diagnostic skills [93]. In nursing education, prototype expert systems as a tool for clinical decision support, has been the subject of much discussion and developmental activity [94]. Examples of two expert systems that are used in nursing education are EXTEND [95] and FLORENCE [96], which are computer-assisted case-based reasoning systems that guide students through the diagnostic process. In order for these expert systems to take-off on a large scale and reach their full potential in education and clinical practice, limitations to their implementation and sustainability will need to be addressed.

SUCCESSFUL IMPLEMENTATION OF A NURSING CDSS

Once a healthcare organization has decided to employ the use of a CDSS to enhance nursing practice, many factors need to be considered for the purposes of successful implementation and sustainability. It is paramount that large-scale change that has systemic influence must include potential users and exploit their motivation, knowledge, and skills in the domain [97]. Resistance to change is an expected phenomena [69]; however to overcome this obstacle, nurses must see the potential benefits of using CDSSs for themselves and for the patients in their care. More specifically, the tool must meet their needs at the point of care, be user-friendly and nonobstructive [98], be transparent regarding the justification behind their suggestions [16], and the payoffs of using the system must outweigh the effort invested in its use. Essential to the initial success of change initiatives is to have positive early adopters or champions that can drive the process forward; studies [99] have suggested that in the case of CDSSs, advance practice nurses working as knowledge brokers should be the ones to take on this role. They are well placed for playing this part within a healthcare organization as they are familiar with practice policy and protocols, can identify potential pitfalls or bottlenecks, are familiar with clinical workflows, but most importantly, they are expert clinicians that can translate the clinical reasoning process involved in decision-making to the hospital-based information technology (IT) experts [16]. This being said, the implementation team needs to be broader than the advanced practice nurse and the IT expert; it needs to incorporate representatives from nursing administration, bedside users, etc. [100,101]. Furthermore, from a technological perspective, the CDSS implementation team needs to work in close collaboration with the selected CDSS vendor in order for the product to be adapted to the healthcare milieu as the CDSS needs to be integrated within the hospital's EHR. Unit-specific variability in terms of language and clinical parameters is an ever-present reality; however, a CDSS can be a useful tool in working toward the goal of institutional standardization where nursing practice is consistent from unit to unit [16]. When the CDSS is successfully integrated from a clinical point of view and is interoperable from a system's perspective [97], this tends to ensure sustainability and user appreciation for these systems.

DISCUSSION AND RECOMMENDATIONS

In our opinion, this chapter addressing the use and challenges of CDSSs in nursing practice and education warrants three points of discussion: (1) the new landscape of healthcare delivery—an inevitable course; (2) lack of consensus regarding the purpose of CDSSs; and (3) a sustainable CDSS requires long-term investment and commitment.

The New Landscape of Healthcare Delivery—An Inevitable Course

Over the last decade, healthcare has undergone tremendous change and evolution with the arrival of technological advancement, which has resulted in numerous challenges, but also many opportunities for improvement and optimization. While pervasive change can be threatening, it is essential that healthcare administrators and clinicians keep an open mind and remind themselves of the bigger picture contextually. Technology has been an integral part of our daily lives for some time, but it is only now that the digital age of healthcare seems to be upon us, inevitably changing nursing practice. For example, healthcare clinicians are using social networking as a way to engage patients who are "connected" technologically. In addition, the use of mobile technology as a support to healthcare practice is widespread and growing exponentially [87]; however, we still see certain organizations react conservatively by adopting policies that limit their use. As one can imagine, this causes considerable frustration for younger generations of nurses and nursing students that feel they could benefit from the use of the many commercially available healthcare apps in existence [86]. In spite of this looming technology tsunami, many hospitals are not yet equipped with EHR, much less with CDSSs. As we know, medicine is making progress in this domain, contrary to nursing, where there are limited commercially viable products currently available. This discrepancy becomes blatant when even nurse practitioners, who have very similar scopes of practice to medicine [73], do not have access to a CDSS tailored to their practice. CDSSs in nursing education are virtually nonexistent, which is a real issue for nursing students in terms of their ability to prepare for practice. In addition, the nursing academics that should be pioneering the use of CDSSs have a limited understanding of its potential benefits and usefulness. In this vein, the Canadian Association of Schools of Nursing (CASN) have published entry-to-practice nursing informatics competencies and developed a toolkit with teaching strategies to support the uptake and integration of these competencies. In addition, CASN supported the creation of a Digital Health Nursing Faculty Peer Network in early 2015 to further guide educators in the domain of digital health.

Lack of Consensus Regarding the Purpose of CDSSs

Among nurses and scholars that have begun to reflect on CDSS integration in healthcare, there seems to be a general lack of understanding regarding the ultimate purpose of these applications and how they should be used practically. For example, Are CDSSs to provide evidence-based information during the decision-making process?, Are they meant to support the clinical reasoning process in steps from assessment to evaluation?, Are they meant to provide the 'answer' independently so that the practitioner can validate their own reasoning?, Are they meant to be used before, during or after patient encounters?, etc. Because these questions remain unanswered, there seems to be general confusion within the community of practitioners and in an attempt to compensate, there has been a proliferation of apps, which only serve the more limited purpose of generating information for users. Previous negative experiences with new software systems, such as CPOEs, which aim to support practice and the pervasive confusion that remains, may have contributed to the delayed progression of CDSS implementation [102].

A Sustainable CDSS Requires Long-Term Investment and Commitment

The hospitals and clinics that have decided to pioneer in the use of this technology must be aware that beyond initial efforts, significant, and sustained commitment to the process is required if the endeavor is to be successful in the long run. Beyond adequate training of employees in the use of CDSSs, clinical experts must be liberated to work closely with IT specialists for the purposes of customizing the commercial products to meet their needs and to ensure adequate integration into the clinical workflow. It is of the utmost importance that new working relationships, between professionals that are not used to working together, are valued and sustained. Change is never easy and institutions must patiently support and document the process of evolving to new practices. Beyond researching patient and provider outcomes, evidence is necessary on the structural and process components vital to the widespread implementation of CDSSs. If success stories are made public, the current confusion and reticence may fade, making way for healthcare practices that are consistent with our vision of care for the 21st century.

References

[1] Kannampallil TG, Schauer GF, Cohen T, Patel VL. Considering complexity in healthcare systems. J Biomed Inform 2011;44(6):943–7.

[2] European Centre for Disease Prevention and Control. Zika outbreak in the Americas and the Pacific [Internet]. Sweden: ECDC. Available from: <http://ecdc.europa.eu/en/healthtopics/zika_virus_infection/zika-outbreak/Pages/zika-outbreak.aspx> [cited 09.08.16].

[3] Leppink J, van den Heuvel A. The evolution of cognitive load theory and its application in medical education. Perspect Med Educ 2015;4(3):119–27.

[4] Huston C. The impact of emerging technology on nursing care: warp speed ahead. Online J Issues Nurs [Internet] 2013;18(2):108. Available from: <http://nursingworld.org/MainMenuCategories/ANAMarketplace/ANAPeriodicals/OJIN/TableofContents/Vol-18-2013/No2-May-2013/Impact-of-Emerging-Technology.html> [cited 24.08.16].
[5] Mitchell N, Randell R, Foster R, Dowding D, Lattimer V, Thompson V, et al. A national survey of computerized decision support systems available to nurses in England. J of Nurs Manag 2009;17(7):772–80.
[6] Wright A, Sittig DE. A four-phase model of the evolution of clinical decision support architectures. Int J Med Inform 2008;77(10):641–9.
[7] Spooner SA. Mathematical foundations of Decision Support Systems. In: Bernier ES, editor. Clinical Decision Support Systems. 3rd ed. Switzerland: Springer International Publishing; 2007. p. 19–44.
[8] El-Sappagh SH, El-Masri S. A distributed clinical decision support system architecture. J King Saud Univ Comp Inf Sci 2014;26(1):69–78.
[9] Clemmer TP, Gardner RM, Shabot MM. Medical informatics and decision support systems in the intensive care unit: state of the art. In: Shabot MM, Gardner RM, editors. Decision Support Systems in Critical Care. New York: Springer-Verlag; 1994. p. 3–24.
[10] Osheroff JA, Teich JM, Middleton B, Steen EB, Wright A, Detmer DE. A roadmap for national action on clinical decision support. J Am Med Inform Assoc 2007;14(2):141–5.
[11] Englebardt SP, Nelson R. Health care informatics: an interdisciplinary approach. St Louis: Mosby; 2002.
[12] Mickan S, Atherton H, Roberts NW, Heneghan C, Tilson JK. Use of handheld computers in clinical practice: a systematic review. BMC Med Inform Decis Mak 2014;14:56.
[13] Martinez-Perez B, de la Torre-Diez I, López-Coronado M, Sainz-de-Abajo B, Robles M, Garcia-Gómez JM. Mobile clinical decision support systems and applications: a literature and commercial review. J Med Syst 2014;38(1):4.
[14] Mopsa AS, Yoo I, Sheets L. A systematic review of healthcare applications for smartphones. BMC Med Inform Decis Mak 2012;12(1):67.
[15] Ketcham M, McBride S, Tietze M, Padden J. Clinical decision support system. In: McBride S, Tietze M, editors. Nursing informatics for the advance practice nurse. New York: Springer; 2016. p. 461–92.
[16] Berner ES. (Department of Health Services Administration, University of Alabama at Birmingham). Clinical decision support systems: state of the art. Report. AHRQ Agency for Healthcare Reform and Quality (U.S.); 2009 June. Report No.: 09-0069-EF.
[17] Kuperman GJ, Bobb A, Payne TH, Avery AJ, Gandhi TK, Burns G, et al. Medication-related clinical decision support in computerized provider order entry systems: a review. J Am Med Inform Assoc 2007;14(1):29–40.
[18] Simmons B. Clinical reasoning: concept analysis. J Adv Nurs 2010;66(5):1151–8.
[19] Paas F, Tuovinen JE, Tabbers H, van Gerven PW. Cognitive load measurement as a means to advance cognitive load theory. Educat Psychol 2003;38:63–71.
[20] Tanner C. Thinking like a nurse: a research-based model of clinical judgment in nursing. J Nurs Educ 2006;45(6):204–11.
[21] Islam R, Weir CR, Jones M, Del Fiol G, Samore MH. Understanding complex clinical reasoning in infectious diseases for improving clinical decision support design. BMC Med Inform Decis Mak 2015;15:101.
[22] Miliard M, editor. What clinical decision support can learn from a helicopter cockpit [Internet]. Healthcare IT News; 2015. Available from: <http://www.healthcareitnews.com/news/what-clinical-decision-support-can-learn-helicopter-cockpit> [cited 23.08.16].
[23] Aleksovska L, Loskovka, S, ed. Review of reasoning methods in clinical decision support systems. 18th Telecommunications forum TELFOR; 2010 November 23–25. Belgrade, Serbia.
[24] Byrne MD, Columnist G. Nursing informatics and ASPAN: clinical decision support through the perianesthesia data elements. J PeriAnesth Nurs 2010;25(2):108–11.
[25] Greenes RA. Clinical decision support: the road ahead. Boston: Elsevier; 2007.
[26] Power DJ, editor. A brief history of decision support systems [Internet]. DSSResources.COM, World Wide Web; 2007. Available from: <http://DSSResources.COM/history/dsshistory.html> [cited 23.08.16].
[27] Ledley RS, Lusted LB. Reasoning foundations of medical diagnosis; symbolic reasoning foundations of medical diagnosis; symbolic logic, probability, and value theory aid our understanding of how physicians reason. Science 1959;130(3366):9–21.

[28] Warner HR, Toronto AF, Veasey LG, Stephenson A. A mathematical approach to medical diagnosis: application to congenital heart disease. JAMA 1961;177(3):177–83.

[29] Ramnarayan P, Britto J. Paediatric clinical decision support systems. Arch Dis Child 2002;87:361–2.

[30] Collen MF, Rubin L, Neyman J, Dantzig GB, Baer RM, Siegelaub AB. Automated multiphasic screening and diagnosis. Am J Public Health Nations 1964;54:741–50.

[31] Bleich HL. Computer evaluation of acid-base disorders. J Clin Invest 1969;48(9):1689–96.

[32] McCallie DP. Clinical decision support: history and basic concepts. In: Weaver CA, Ball MJ, Kim GR, Kiel JM, editors. Healthcare Inform Manag Sys. New York: Sprinter Cham Heidelberg; 2016. p. 3–19.

[33] de Dombal FT, Horrocks JC, Staniland JR, Guillou PJ. Construction and uses of a "data-base" of clinical information concerning 600 patients with acute abdominal pain. Proc R Soc Med 1971;64(9):978.

[34] Shortliffe EH, Davis R, Axline SG, Buchanan BG, Green CC, Cohen SN. Computer-based consultations in clinical therapeutics: explanation and rule acquisition capabilities of the MYCIN system. Comput Biomed Res 1975;8(4):303–20.

[35] Miller PL. Critiquing anesthetic management: the "ATTENDING" computer system. Anesthesiology 1983;58 (4):362–9.

[36] Miller PL. Extending computer-based critiquing to a new domain: ATTENDING, ESSENTIAL-ATTENDING, and VQ-ATTENDING. Int J Clin Monit Comput 1986;2(3):135–42.

[37] Miller RA, Pople HE, Myers JD. Internist-1, an experimental computer-based diagnostic consultant for general internal medicine. N Engl J Med 1982;307(8):468–76.

[38] Barnett GO, Cimino JJ, Hupp JA, Hoffer EP. DXplain, an evolving diagnostic decision-support system. JAMA 1987;258(1):468–76.

[39] mghlcs.org [Internet]. Using decision support to help explain clinical manifestations of disease. Boston: The Massachusetts General Hospital Laboratory of Computer Science. Available from: <http://www.mghlcs.org/projects/dxplain/> [cited 26.08.16].

[40] Openclinical.org [Internet]. HELP Health evaluation through logical processing. Utah: University of Utah; 2004. Available from: <www.openclinical.org/aisp_help.html> [cited 25.8.16].

[41] McDonald CJ, Murray R, Jeris D, Bhargava B, Seeger J, Blevins L. A computer-based record and clinical monitoring system for ambulatory care. Am J Public Health 1977;67(3):240–5.

[42] Berner ES, La Lande TJ. Overview of clinical decision support systems. In: Berner ES, editor. Clinical decision support systems: theory and Practice. 2nd ed New York: Springer; 2007. p. 3–22.

[43] Samwald M, Fehre K, de Bruin J, Adlassnig KP. The Arden Syntax standard for clinical decision support: experiences and directions. J Biomed Inform 2012;45(4):711–18.

[44] Carter JH. Design and implementation issues. In: Berner ES, editor. Clinical decision support systems: theory and practice. 2nd ed New York: Springer; 2007. p. 64–98.

[45] Lutgenberg M, Weenink JW, van der Weijden T, Westert GP, Kool RB. Implementation of multiple-domain covering computerized decision support systems in primary care: a focus group study on perceived barriers. BMC Med Inform Decis Mak 2015;12(15):82.

[46] Bakken JS, Jia H, Chen ES, Choi J, John RM, Lee NJ, et al. The effect of a mobile health decision support system on diagnosis and management of obesity, tobacco use, and depression in adults and children. J Nurse Practitioner 2014;10(10):774–80.

[47] Lewis A, Belmonte-Mann F. Decision support tools aid school nursing practice. J School Nurs 2002;18 (3):170–3.

[48] Gottesman O, Scott SA, Ellis SB, Overby CL, Ludtke A, Hulot JS, et al. The CLIPMERGE PGx program: clinical implementation of personalized medicine through electronic health records and genomics – pharmagenomics. Clin Pharmacol Ther 2013;92:214–17.

[49] Pravikoff DS, Tanner AB, Pierce ST. Readiness of U.S. nurses for evidence-based practice. Am J Nurs 2005;105:40–51.

[50] Henneman EA. Unreported errors in the intensive care unit: a case study of the way we work. Crit Care Nurse 2007;27(5):27.

[51] Ebright PR, Urden L, Patterson E, Chalko B. Themes surrounding novice nurse near-miss and adverse-event situations. J Nurs Adm 2007;34(11):531–8.

[52] Anderson J, Wilson P. Clinical decision support in nursing. CIN 2008;26(3):151–8.

[53] Brokel JM. Infusing clinical decision support interventions into electronic health records. Urol Nurs 2009;29 (5):345–52.
[54] Barton AJ. Health information technology safety: implications for the clinical nurse specialist. Clin Nurse Spec 2012;26(4):198–9.
[55] Institute of Medicine. Health IT and patient safety: building safer systems for better care. Washington (DC): The National Academies Press; 2012. 234 p.
[56] O'Brien A, Weaver C, Settergren T, Hook ML, Ivory CH. EHR documentation: the hype and the hope for improving nursing satisfaction and quality outcomes. Nurs Admin Q. 2015;39(4):333–9.
[57] Bakken S, Currie LM, Lee NJ, Roberts WD, Collins SA, Cimino JJ. Integrating evidence into clinical information systems for nursing decision support. Intern J Med Inform 2008;77:413–20.
[58] Omachonu VK, Einspruch NG. Innovation in healthcare delivery systems: a conceptual framework. Innovat J 2010;15(1):1–20.
[59] Ryan SA. An expert system for nursing practice. J Med Syst 1985;9:29–41.
[60] Carroll DL, Dykes PC, Hurley AC. An electronic fall prevention toolkit: effect on documentation quality. Nurs Res 2012;61:309–13.
[61] Beeckman D, Clays E, Van Hecke A, Vanderwee K, Schoonhoven L, Verhaeghe S. A multi-faceted tailored strategy to implement an electronic clinical decision support system for pressure ulcer prevention in nursing homes: a two-armed randomized controlled trial. Intern J Nurs Stud 2013;50:475–86.
[62] Collins SA. Nurses spurring innovation. Nurs Manag 2014;45(10):17–19.
[63] Anderson JA, Willson P, Peterson NJ, Murphy C, Kent TA. Prototype to practice: developing and testing a clinical decision support system for secondary stroke prevention in a veteran's healthcare facility. CIN 2010;28(6):353–63.
[64] Kawamoto K, Houlihan CA, Balas EA, Lobach DF. Improving clinical practice using clinical decision support systems: a systematic review of trial to identify features critical to success. Brit Med J 2005;330(7494):765.
[65] Peterson JJ, White KW, Westra BL, Monson KA. Anesthesia information management systems: imperatives for nurse anesthetists. AANA J 2014;82(5):346–51.
[66] Fathauer L, Meek J. Initial implementation and evaluation of a hepatitis C treatment clinical decision support system. Appl Clin Inform 2012;3(3):337–44.
[67] Bright TJ, Wong A, Dhurjati R, Bristow E, Bastian L, Coeytaus RR, et al. Effect of clinical decision-support systems: a systematic review. Ann Intern Med 2012;157:29–43.
[68] Ozbolt JG. Developing decision support systems for nursing: theoretical bases for advanced computer systems. Comput Nurs 1987;5(3):105–11.
[69] Rincon TA. Integration of evidence-based knowledge management in microsystems: a tele-ICU experience. Crit Care Nurse Q 2012;35(4):335–40.
[70] Lyerla F, LeRouge C, Cooke DA, Turpin D, Wilson L. The impact of a nursing clinical decision support system and potential predictors on head of bed positioning for mechanically ventilated patients: final results. Am J Crit Care 2010;19(1):39–47.
[71] Weber SA. Qualitative analysis of how advanced practice nurses use clinical decision support systems. J Am Acad Nurse Pract 2007;19:652–67.
[72] Sinclair VG. Potential effects of decision support systems on the role of the nurse. Comput Nurs 1990;8 (2):60–5.
[73] Vetter MJ. The influence of clinical decision support on diagnostic accuracy in nurse practitioners. Worldviews Evid Based Nurs 2015;12(6):355–63.
[74] Garg AX, Adhikari NK, McDonald H, Rosas-Arellano MP, Devereaux PJ, Beyene J, et al. Effects of computerized clinical decision support systems on practitioner performance and patient outcomes: a systematic review. JAMA 2005;293:1223–38.
[75] Randell R, Dowding D. Organizational influences on nurses' use of clinical decision support systems. Intern J Med Inform 2010;79(6):412–21.
[76] Garrett B, Klein G. Value of wireless personal digital assistants for practice: perceptions of advanced practice nurses. J Clin Nurs 2008;17(16):2146–54.
[77] Miller L, Stimely M, Matheny P, Pope M, McAtee R, Miller K. Novice nurse preparedness to effectively use electronic health records in acute care settings: critical informatics knowledge and skill gaps. OJNI 2014;18 (2):1–24.

[78] Friedman E, Sainte M, Fallar R. Taking note of the perceived value and impact of medical student chart documentation on education and patient care. Acad Med 2010;85(9):1440–4.
[79] Heiman HL, Rasminsky S, Bierman JA, Evans DB, Kinner KG, Stamos J, et al. Medical students' observations, practices, and attitudes regarding electronic health record documentation. Teach Learn Med 2014;26(1):49–55.
[80] Carroll K, Parker C, editors. The invisible impact: Students' usage of the EHR. Chicago: HIMSS News; 2011. Available from <http://www.himss.org/news/invisible-impact-students-usage-ehr> [cited 26.08.16].
[81] McNeil BJ, Elfrink VL, Bickford CJ, Pierce ST, Beyea SC, Averill C, et al. Nursing information technology knowledge, skills, and preparation of student nurses, nursing faculty, and clinician: a U.S survey. J Nurs Educ 2003;42(8):341–9.
[82] Malloch K, Porter-O'Grady T. Introduction to evidence-based practice in nursing and health care. 2nd ed. Sudbury Massachusetts: Jones and Bartlett Publishers; 2010.
[83] Graber ML, Tompkins D, Holland JJ. Resources medical students use to derive a differential diagnosis. Med Teach 2009;31:522–7.
[84] Jansen DA. Student perceptions of electronic health record use in simulation. J Nurs Educ Pract 2014;4 (9):163–72.
[85] Staudinger B, HöB V, Ostermann H. Nursing and clinical informatics: socio-technical approaches. New York: Medical Information Science Reference; 2009.
[86] Beauregard P, Arnaert A, Ponzoni N. Nursing students' perceptions of using smartphones in the community practicum. Computers, Informatics, Nursing (in review).
[87] Lee NJ, Bakken S. Development of a prototype personal digital assistant-decision support system for the management of adult obesity. Intern J Med Inform 2007;76:S281–92.
[88] Cato K, Hyun S, Bakken S. Response to a mobile health decision-support system for screening and management of tobacco use. Oncol Nurs Forum 2014;41(2):145–52.
[89] Jenkins ML, Hewitt C, Bakken S. Women's health nursing in the context of the national health information infrastructure. JOGNN 2006;35(1):141–50.
[90] Curran C, Sheets D, Kirkpatrick B, Bauldoff GS. Virtual patients support point-of-care nursing education. Nurs Manag 2007;38(12):27–8.
[91] Johnston JM, Leung GM, Tin KY, Ho LM, Lam W, Fielding R. Evaluation of a handheld clinical decision support tool for evidence-based learning and practice in medical undergraduates. Med Educ 2004;38 (6):628–37.
[92] Montini T, Schenkel AB, Shelley DR. Feasibility of a computerized clinical decision support system for treating tobacco use in dental clinics. J Dent Educ 2012;77(4):458–62.
[93] Openclincial.org [Internet]. Iliad. Expert system for internal medical diagnosis. Utah: University of Utah; 1995. Available from: <http://www.openclinical.org/aisp_iliad.html> [cited 26.08.16].
[94] Hanson AC, Foster SM, Nasseh B, Hodson KE, Dillard N. Design and development of an expert system for student use in a school of nursing. Comput Nurs 1994;12(1):29–34.
[95] Koch B, McGovern J. EXTEND: a prototype expert system for teaching nursing diagnosis. Comput Nurs 1993;11(1):35–41.
[96] Bradburn C, Zeleznikow J, Adams A. FLORENCE: synthesis of case-based and model-based reasoning in a nursing care planning system. Comput Nurs 1993;11(1):20–4.
[97] Trivedi MH, Daly EJ, Kern JK, Grannemann BD, Sudnerajan P, Claassen CA. Barriers to implementation of a computerized decision support system for depression: a observational report on lessons learned on "real world" clinical settings. BMC Med Inform Decis Mak 2009;9:1–9.
[98] Settig DF, Wright A, Osheroff JA, Middleton B, Teich JM, Ash JS, et al. Grand challenges in clinical decision support v10. J Biomed Inform 2008;41(2):387–92.
[99] Barton AJ, Thompson TL. Decision support and the clinical nurse specialist. Clin Nurse Special 2009;23 (1):9–10.
[100] McCool C. A current review of the benefits, barriers, and considerations for implementing decision support systems. Online J Nurs Inform [Internet] 2013;17(2). Available from: <http://ojni.org/issues/?p = 2673>.
[101] Carney PH. Information technology and precision medicine. Semin Oncol Nurs 2014;30(2):124–9.
[102] Greenes RA. Definition, scope, and challenges. In: Greenes RA, editor. Clinical decision support: the road to broad adoption. 2nd ed. Waltham: Elsevier; 2014. p. 3–29.

SECTION IV

HEALTH INFORMATION LITERACY AND CREDIBILITY ASSESSMENT

CHAPTER

8

Developing Digital Literacies in Undergraduate Nursing Studies: From Research to the Classroom

Maggie Theron[1], *Elizabeth M. Borycki*[2] *and Anne Redmond*[1]

[1]Trinity Western University, Langley, BC, Canada [2]University of Victoria, Victoria, BC, Canada

INTRODUCTION

Online sources, including postings or blogs, websites, Facebook, and Twitter, are used by the general public to seek information on health topics, and by health professionals to provide education and advocacy. [1] Healthcare professionals are challenged to develop and incorporate ways in which social media can be utilized to improve patient outcomes in a complex multilevel, healthcare system of delivery. Increasingly, digital technologies and social media applications are used to expand access to healthcare services, develop and apply digital interventions, and monitor patient outcomes. Diviani and colleagues [2] have expressed concerns regarding the quality of online health information and the ability of online users to accurately evaluate media. These concerns include the lack of available and consistent evaluation criteria and definitions for the general public and health professionals to use in appraising the quality of online health information. In a recent systematic review of 38 articles in 2014, Diviani and colleagues identified that people with low literacy (pertaining to online health information) were not able to apply evaluation criteria appropriately and effectively. [2] The findings support the need for further research into the relationship between low literacy rates and patient assessment of the trustworthiness of online health information. In addition to this, the continuous development and refinement of measurable evaluation criteria published in open access journals focusing on health information are emphasized as being important. [2] Currently, the internet and social media are mostly accessed by undergraduate students and other young people, although use by those 65 and older has increase threefold since 2010. [3,4]

Health Professionals' Education.
DOI: http://dx.doi.org/10.1016/B978-0-12-805362-1.00008-5

Different cohorts or populations may demonstrate variability in their purpose and use of social media. Even though there is a belief that undergraduate students possess expert Internet search skills, they still seem to struggle with locating and accurately appraising the trustworthiness of the health information they find online. Therefore, it is important to educate undergraduate students about how to critically appraise the quality of health information available on the Internet and social media. Health professional students, have a further responsibility to educate the general public about health, internet or social media information. The mere fact that internet usage is growing in all aspects of the population escalates the importance of instilling strong information literacy practices into nursing education.

To better understand how this can be done, the authors of this chapter will:

Outline how the digitalization of health information on the Internet and social media has impacted health education.
Describe some of the current concepts of health informatics literacy.
Discuss the relation of these digital literacies to informatics competency development including competencies for undergraduate nursing students.
Review the current literature focusing on teaching digital literacy as part of informatics, specifically related to undergraduate students and young people.
Identify evaluation criteria for the assessment of health information found on the Internet and social media, as well as provide strategies for engaging student health professionals in improving their health literacy competencies.
Identify current barriers and challenges to developing these competencies.
Outline current informatics competencies, strategies, and tools that build digital literacy skills in undergraduate nursing students in the context of social media (e.g., Facebook, Twitter), and discuss ways in which informatics competencies are integrated into undergraduate nursing curricula.

DIGITIZATION OF HEALTH BY THE INTERNET AND SOCIAL MEDIA

The modern era of computing has revolutionized patient care and digitalized health care, but this would not have been possible without the introduction of the Internet and social media technologies over the past six decades. Both the Internet and social media have become important tools that the general public, patients, and health professionals (including health professional students) use to obtain information used in their decision-making surrounding health care.

The Internet and the Use of Online Health Information

In the late 1960s, the computers became a general tool that could be used to support education, research, and government organizations, and evolved through the 1980s to equipment that had ability to allow for access to distributed database systems. [5] Hence the birth of the "inter" net. Today, 87% of American adults use the Internet to obtain information they need for work and personal use. [6] Statistics indicate 66% of healthy adults search for health information on the Internet, and 51% of chronically ill adults also seek health information on the Internet. [7] In 2013, 72% of Internet users accessed health information online. [8] Most of these online health information seekers (77%) initiated their search for health information using a search engine such as Google, Bing and Yahoo. Thirteen percent of health information seekers used an online website for specialized health information such as WebMD. Two percent of health information seekers used Wikipedia to start their search, and the remaining group of individuals used a social media site such as Facebook (i.e., 1%). [6]

The Internet is an important source of health information for the general public, patients and health professionals. Many individuals are able to find evidence-based, accurate, and current information on the Internet, yet this is not the case for all those who use the Internet. For example, some individuals have found that the quality of the information found on the Internet can be harmful to the individual if used in health related decision-making. [9,10] Therefore, it is important for health professional and health professional students to become well versed in understanding how to assess the quality of information found on the World Wide Web, and to help others wading through the poor quality information that is also posted on the web.

Social Media and Health Care

Social media technologies were first introduced in the late 1970s. Today, social media has changed the way we gather, communicate, and share information globally. [11] Early forms of social media consisted of newsgroups, list serves, and chat rooms, but it was not until the introduction of Wikipedia, an online encyclopedia, that we saw the first crowd sourced approach to creating shareable information internationally. Today, 65% of adults use social media in North America [12] Twitter, Instagram, Facebook, Pinterest, Snapchat, and WhatsApp have become popular social media used by the general public and patients to exchange information about everyday life and health. The information communicated via social media continues to grow and influence health professional and patient decision making. [11] Of note, 90% of 18 to 29 year olds use social media with 35% of those over the age of 65 now using social media to communicate with the general public [6]. Differing groups, especially students are using information posted via social media to support every day decision-making. Some researchers believe that social media has become a ubiquitous aspect of everyday life in North America with its influence being expected to continually grow. [12] This influence is expanding in varying ways. Social media is being used as a means by well and chronically ill individuals as well as health professionals and health care organizations to communicate information. Patients and family members who have rare diseases are searching for information using social media sites (e.g., PatientsLikeMe) in hopes of networking with others who have the same disease to learn about treatments that have been effective in disease management and to provide social support. [10] Yet, the quality of information provided by social media is also variable. Some users of social media link to posts outlining the latest evidence based research, while others post information that, if used by a patient, may be harmful. Moorhead et al. discusses two such potentially harmful examples that occur in social media. They describe the potential of adverse health consequences resulting from information found on social media sites, such as pro-smoking imagery, and negative health risk behaviors displayed online, such as unsafe sexual behavior. [13] Ineffective or unethical social media use can lead to privacy breaches (as has been show in media reports where nurses' have posted confidential patient information online). [10] Health professionals and health professional students need to be technology savvy (i.e., they can use social media), use technology ethically, and effectively evaluate the quality of information provided via social media tools. [10]

Health Professionals Remain an Important Source of Health Information

Among the general public, 70% of citizens in our society continue to seek out health information through personal interaction with doctors, nurses and other health professionals. [8] The research by Gutierrez and colleagues emphasize the fact that patients rely on healthcare providers to gain health information, irrespective of the patient's health literacy. Their results reiterate the significance of effective patient-provider communication in reaching health care decisions. [14] Individuals also seek out personal connection as they expand their reach for health information. When we look at where individuals look for other sources of health information, they report obtaining information and support from family and friends (i.e., 60%) and from others who have had a similar condition (i.e., 24%). [8] This information suggests that individuals do look beyond professionals for information about their health concerns. Health professionals have a role not only in being self-consumers of online health care information, but also as appraisers of the quality of information their patients' access.

There are a number of reasons why patients and caregivers continue to prefer seeking out health information from health professionals. [14] Health professionals are well educated in their disciplines and are skilled in assessing and evaluating the quality of health information. In addition to this, many health professionals are able to individualize and tailor information to patients based on their advanced knowledge of an individual's disease, life style habits, and socio-economic status. Patients and families are also becoming more critical of the health information they read on the Internet and social media. Patients and family members are looking toward health professionals to help them understand, interpret, and assess the quality of information found on the World Wide Web. There is an increasing recognition by the general public that the quality of the information may be variable, poor or incorrect. Health professionals have a role in educating patients and families and increasing their awareness of the variable nature, and in some cases poor quality, of information that may be found on the Internet and in social media. Health professionals need to assess the health and eHealth literacy of patients and families seeking information. They must also be able to tailor their interactions to patients and families who have looked for such information, recognizing that patients and families have been empowered by the Internet and social media and that empowerment may be greater/optimized if the right information is available to support individual decision making when and information needs arises.

In summary, both the internet and social media represent two important tools that the general public, patients, and their families use to learn about how to promote their health and address acute and chronic illnesses. Health professionals use the Internet to learn about new evidence based approaches to manage wellness and chronic illness. Health professionals need to educate the public about how they can best evaluate and use health information gathered from the Internet and social media posts to effectively promote their health rather than harm it.

HEALTH INFORMATICS LITERACY

Several definitions have emerged over time to describe processes for both the public and professionals to locate and use social media and online information sources in a proficient manner. Informatics is defined by McGonigle and Mastrian [15] as a field that integrates specialty science, computer science, cognitive science, and information science to communicate data, information, knowledge and wisdom in a specialty's practice. Information literacy is defined as the students' ability to recognize when information is needed and having the ability to locate, evaluate and effectively use the needed information. Information literacy in general represents an intellectual framework for finding, understanding, evaluating, and using information, including information found from online sources.

According to Diviani et al., [2] health literacy is the "the cognitive and social skills that determine motivation and ability to gain access, understand and use information to promote and maintain good health."(np) The definition framed by Baker (based on the Institute of Medicine Health Literacy, 2004, and US Department of Health and Human Services document, Healthy People, (2010) is similar: "the degree to which individuals have the capacity to obtain, process, and understand basic health information and services needed to make appropriate health decisions." [16] eHealth refers to healthcare initiatives and practice supported by electronic or digital media with a typical focus to provide patient and family education where information is communicated electronically. Stellefson and colleagues [17] suggest "eHealth literacy" involves "six core skills or types of literacy: traditional literacy, health literacy, information literacy, scientific literacy, media literacy, and computer literacy" based on Norman and Skinner's Lily model. A more concise definition of eHealth literacy is later proposed by the same authors as the "ability to obtain, process, evaluate, and use information acquired through electronic resources" [17](p 467). The concept of eHealth literacy is used by many authors to describe foundational skills required by students and other health professionals as they search for and use online health information in their education and their day to day work. We know that eHealth literacy is influenced by a person's presenting health issue, educational background, and health status at the time of the eHealth encounter. It is important for health professionals to know the patient's motivation for seeking information, and the technologies that are used. Like other literacies, eHealth literacy is not static; rather, it is a process-oriented skill that evolves over time as new technologies are introduced and the personal, social, and environmental contexts change. Like other literacy types, eHealth literacy is a discursive practice that endeavors to uncover the ways in which meaning is produced and inherently organizes the individual's ways of thinking and acting. Norman and Skinner [18] state that, "eHealth literacy aims to empower individuals and enable them to fully participate in health decisions informed by eHealth resources" (p 69). Building on their definition, eHealth literacy can be further expanded or narrowed.

Stellefson and colleagues [19] suggest the "ability to seek, find, understand, and appraise information from electronic resources and apply such knowledge to solve health problems [is part of eHealth literacy]" (np). For nursing, this ability is a required practice competency and an essential part of nursing education.

The Canadian Association of Schools of Nursing have adopted the 2009 International Medical Informatics Association's definition of Nursing informatics, defined as, "A science and practice [which] integrates nursing, its information and knowledge, and their management, with information and communication technologies to promote the health of people, families and communities worldwide." [20] Essentially, nurses use information and knowledge to inform practice and to educate individuals, families and communities with information that will assist them in making healthcare decisions that will positively impact their quality of life. To extend this concept to information that occurs online or digitally, nurses require digital literacy competencies. These are defined and discussed in the following section. It is important to attend to the fact that digital information includes social media.

O'Grady and colleagues call attention to the main aims of collaborative, adaptive, and interactive technologies (social media) for health as providing information or tools for decision support, social support, self-management, or self-management support [21]. Kostik and peers amalgamate a definition of social media as a group of Internet-based applications that permit users to create and engage their generated contents, acquire real time feedback and to continuously adjust content. They describe how social media assumes the characteristics of information openness, participation, interaction, sharing, connectedness, creativity, autonomy, collaboration, and reciprocity [22]. Social media sites hold great promise for information sharing and patient support, but should be used in conjunction with patient-provider dialogue. Conversely, social media sites require online presence of scholars and healthcare providers. These types of sites that patients access for health information provide health professionals with challenges to evaluate the quality, trustworthiness, and credibility of the information. Kostik and colleagues state that the sources of social media patients may use are in need of analysis, as no systematic criteria for measuring quality of health related social media currently exist. [20]

Clearly, health literacy and eHealth literacy are still evolving as concepts as we learn more about how information on the internet is being created and disseminated and what it is used for in healthcare by organizations, patients, the general public, health care professionals, and undergraduate students in the health professions. Jakubec and Astle mention some of these foundational literacy skills for health professionals as the searching, interpreting, and appraising evidence for use in practice. There have been several attempts by academics and professional organizations to build these literacies. For example, several courses and assignments in health literacy and informatics have been developed to try and address these necessary skill sets in a continuous, rapid, evolving digital age. [23,24]

DEVELOPING DIGITAL LITERACY AND INFORMATICS COMPETENCY FOR HEALTH PROFESSIONALS

The definition of digital literacy is "Those capabilities that mean an individual is fit for living, learning, and working in a digital society." [25] Martin [17] suggest that digital literacy is an umbrella that has a broad range of interrelated skills, which traditionally fall under other literacy abilities, including literacy in technology, information, communication, social, and visual arenas. While digital literacy is sometimes assumed for today's youth, because of the ease with which they engage with their devices, they still must develop knowledge, values, and skills to avoid remaining amateur users of information and communication technology (ICT). [19] This increased level of knowledge and understanding of digital information has produced the development of professional literacy and informatics competency standards.

Today, both general public and health care employers want health professionals to have health informatics competencies as applied to their profession. [26–28] Increasingly, health care employers are expecting health professionals, for example, nurses, to be knowledgeable about the assessment and evaluation of the quality of health information that is presented online in addition to more traditional informatics competencies such as evaluating the quality of information found in research publications and in peer reviewed journals. There is an expectation that student nurses are knowledgeable and incorporate research into their practice, [29] but nursing informatics is not integrated in all nursing curricula. [30]

There has also emerged a health care employer and professional association expectation that nurses be able to use technologies being introduced in health care organizations as part of the modernization of health care. For example, many employers in North America, Europe, and Asia have the expectation that nurses are able to review information in an electronic health record as well as use decision support tools within the electronic health record, such as body mass index calculators, care guidelines, patient education resources, and drug reference information. Employer expectations for nurses include that nurses are able to contribute to health information by providing evidence-based and professional content. These expectations are often clearly outlined in professional practice standards. [30,31]

Yet, even as these health care employer and professional organization expectations have grown, many nurses graduate without having these competencies. Few student nurses learn about the electronic health record in the classroom context due to the lack of availability of electronic health records and qualified faculty that understand the technology from a conceptual, technical, and practice-based point of view. [32] Internationally, this has led national nursing organizations to develop nursing informatics competencies, resources, and approaches for educating nursing students about nursing informatics to support faculty to educate students to work effectively in health care settings upon graduation [31,33,34]

Much of this work has involved identifying nursing informatics competencies that are needed for nurses to work in a modern health care system. [31,33,34] This is the case around the world. In the United States, nursing practice, research and educational experts have come together to develop nursing informatics competencies (i.e., the Technology Informatics Guiding Educational Reform (TIGER) competencies). [31] In Canada, CASN brought together nursing informatics experts who work in education and in health care to identify the key competencies expected of student nurses after they have graduated from a 4-year undergraduate program. [30,31] In Australia, educators surveyed nursing informatics experts across the country to develop a comprehensive list of nursing informatics competencies expected of student nurses following graduation from an undergraduate nursing program. [31,33,34] These works have led to the development of faculty resources in the form of lectures, readings, and digital tools (e.g., online websites, YouTube videos) that can be used in a classroom context to help nursing students develop nursing informatics competencies. [34]The term competency as described in the document is the ability of nursing students to draw upon various types of knowledge, skills, and other internal and external resources to address specific problems in nursing practice. Specific areas of focus to nursing informatics competency development have been the assessment of health information found online, including information from social media. The CASN competencies include the ability of entry-level nurses to "assist patients and their families to access, review and evaluate information they retrieve using ICTs (i.e., current, credible, and relevant) and with leveraging ICTs to manage their health (e.g., social media sites, smart phone applications, online support groups, etc.)." [28] The result has been the development of competencies/guidelines specific to social media posts. [31] This has led to an increased urgency and focus upon developing informatics competencies among nursing students in an effort to address a public health educational need and to prevent patient privacy breaches as described above. We begin by reviewing the current state of appraisal of the quality of health information online and how these informatics literacies can be taught in the context of undergraduate educational programs to health care providers. [35]

Review of the Literature Focusing on Teaching Digital Literacy to Undergraduate Students

Most student nurses enter undergraduate nursing programs as lay individuals undoubtedly having the same abilities of the general public when assessing and reviewing health information online. Over time health, eHealth and digital literacy grows as the student nurse learns about health, disease, communication skills, and the technologies that can be used to support wellness and self-management of chronic illness. The health information they retrieve may influence their health care choices and personal health decisions. [36] Many social media sites and some internet websites allow for open access pertaining to the editing of healthcare information, whether it be through formal publications or sharing of health-related experiences. This type of engagement with the editing processes results in the possibility of deceptive and confusing health information. For their own purposes and the purpose of patient education, there is a need for all health professional students to identify available and high quality health information. [36] The general public frequently encounters inaccuracies in online health information. There is a need for the general public to increase their health literacy and for health care providers who are in the process of advancing their competencies to do so also [2,36].

This is best illustrated by Diviani and colleagues [2] who completed a systematic review of the literature where they explored how consumers with low health literacy evaluated online health information. The researchers concluded that people with low health literacy searched fewer sources of information to find health information, chose different sources of information than those health care consumers with high literacy levels, and displayed poor interpretation ability of the sources they located. The authors elaborated on the inability of the general public and specific patient populations to use evaluation criteria for online health information. [2] When people have low health literacy, they seemed to rate high-quality websites as low in quality and to rate low quality websites as high in quality. These ratings may be due to the fact that people with low literacy use criteria that are not well established to evaluate the websites. Those with low health literacy also used celebrity endorsement, position in search results, quality of images, and website authorship as evaluation criteria. The lack of confidence in government and religious sites as sources of health information was obvious, although university researchers seemed to be believed.

The general public lacked the skills to navigate vast amounts of information, lacked knowledge of evaluation criteria to appraise the quality of information and trusted the information that was presented based on their own subjective judgments. Based on the findings of this research, Diviani and colleagues [2] were clear that low health literacy plays a role in the appraisal of the quality of online health information. In a secondary analysis of the 2003 National Assessment of Health Literacy United data, researchers stated that household survey data indicated only 12% of their sample size seemed to be proficient in health literacy [37]. The researchers suggested a number of factors may potentially influence the level of health literacy; for example, older age, belonging to a minority group, having less education and lower socio-economic status, being single, and being a recent immigrant. [37] Level of education, however, was found to be a factor that had varying levels of influence in terms of predicting proficiency in health literacy. [2],[35], [22]

The study of the predictive factors of health literacy is ongoing. We live in a society where many factors that influence health literacy exist. As well, health care consumer and health professional student exposure to inaccurate or incomplete health information may leave individuals unable to identify health issues and to display behaviors that may influence one's health outcomes negatively. [2] This is especially the case when one considers vulnerable populations at risk for disparities. The impact of low health literacy may also intensify if people with higher socio economic status have speedier access to high quality health information. [2,38,39] The emergence of such digital inequalities may contribute to "increasing knowledge gaps and health disparities." [38][(p 13)] There is a need for further exploration of how people appraise online health information and the development of shared definitions and appraisal criteria for online health information. [2] More importantly, there is a need to educate student health professionals who may have low health literacy when they begin their program of study.

Internet use by both the general public and health professionals has improved access to health information. In addition to this, the millennial generation prefers to use the internet to locate health information as it provides rapid access to a vast amount of health related information. The variety of mobile applications that are continuously emerging, and the low cost of conducting information searches for the purpose of health promotion also makes both high and poor quality information more available. [40,41] As Lu [42] explains, the Internet provides a platform for anonymous engagement in health discussions, especially around topics of a sensitive nature like sexually transmitted diseases. The Internet was already ranked by college students as the third source for obtaining information regarding HIV/AIDS in Taiwan. [42] Since then, several publications highlight the use of the internet by most young people, including undergraduate nursing students, as the number one source for accessing health information. [35,40,43–46]

Undergraduate students in the health professions are no exception, when it comes to navigating and searching for information online for themselves, course work, or for assisting them in health care related decisions to be used in the process of providing patient care. [40,47,48] In a survey of 1202 Canadian college students, Kwan and partners [49] found that 79 % of survey respondents used the Internet as the most common source of information when seeking health information. [49] Students frequently sought online information to help them decide whether or not to consult a health professional. [50] Undergraduate nursing students (89%) were found to use the internet often to access health information for themselves and for their nursing education. [43] Accessing the Internet often did not exclude the use of other sources of information. Many students in the health professions do seek health information from multiple sources to help them make health care decisions. To date, not much is known about the impact of the disparities in both access and skill as part of the dissemination process of evidence found online and the most effective ways to develop best practices in seeking health information to be used in the process of health care. [39] Research suggests that undergraduate students search for health information online because they perceive that they have a lack of knowledge or they are experiencing fear-related emotions that drive the health information seeking behaviors. [50] A small portion of students are motivated by curiosity or their values of living healthier lives when they search for online health information. Students in the health professions use health informatics and technologies to collect, manage, and transfer health information. [35] If these students and other health professionals do not achieve adequate levels of health literacy, inaccurate or incomplete information use may influence the quality of patient care and ultimately patient safety. [35]

Evaluating Online Health Information

Steffelson and colleagues [35] in their systematic review suggested that eHealth literacy be developed among college students. Their message was clear: many students lack skills to seek and evaluate online health information. Their work was consistent with Zhang's [51] findings that students seem to make quick and superficial judgments of the quality of online health information. As one participant stated "if it's on the internet, I just take what they tell me." [44] (p 192) Even though undergraduate students are deemed to be "savvy" internet users, their knowledge of eHealth is limited. Undergraduate students require formal informatics training to develop health and eHealth competencies and this is critical to their development of nursing informatics competencies. [30,43] As Robb and Shellenbarger's [40] work conducted with 59 college students using the eHealth Literacy Scale (eHeals) developed by Norman and Skinner [18] in 2006 suggests, there is a need to explore the perceptions of nursing students regarding their eHealth literacy after completing an introductory course about health and wellness. As their findings propose, there is a need to evaluate students' abilities to use the internet to locate pertinent health information, and to assess the level of student confidence in using this information in making nursing practice decisions. [40] Yet, this represents only a limited view of health and eHealth literacy. There is also a need to develop competencies that focus on searching for high quality information.

Searching for High Quality Information

Health literacy skills include the ability of students to effectively and efficiently search for relevant health information using electronic resources. Students in various studies seem to focus on finding health information versus appraising the information that is found. Such student behavior has led researchers to conclude that undergraduate students lack advanced and appropriate search skills. [35,40,51–53] Additionally, researchers observed that two factors, namely: students' familiarity with websites and their confidence to skillfully use search strategies affected their search results and evaluation abilities of their selected online health information. [44,46] Many students experience low self-efficacy regarding their searching abilities and struggle to distinguish fact from fiction. As a result, these students feel uncertain about the quality and usability of the information. [50] Generally, students seem to search for health information superficially by search query or general browsing. [44] Often only one set of key words is used, and information is selected using the position in listing of sources, believing a Google list has the most trustworthy site at the top of the list. [44,46,52] Google is the search engine that is used most often by students, followed by other sites such as Yahoo Answers and WebMD. [46,50] The "organization of information" is a website evaluation criterion described by Zhang to address the interface aspects of the website and include the student's navigation ability and procedural knowledge regarding searching for online health information. [41] Websites from organizational domain names like .gov and .edu and .org. are largely trusted by students' more than individual websites. [46] As well, most students are aware of the Honcode© criteria. Although Google is often mentioned by students as their number one search engine, it is surprising how many students still use Wikipedia as a source of information. In a study by Menchen-Trevino and Hargattai, [54] 77% of students accessed Wikipedia during their searches, 19 % went to Wikipedia directly, and only 47% searched Wikipedia through other search engines. One of the students participating in their study said: "everyone uses it although it was not a legitimate resource, particularly for coursework." [54] (p 30) Awareness that Wikipedia can be edited by virtually anyone was not collective common knowledge among students who participated in this study. The students were unaware of Wikipedia's editing and technology norms. It can be said that these students lacked the basic knowledge that is crucial for site navigation underpinning information searches. [50] A further issue for students that arose is a lack of interest in investigating the authors of the online health information. [52] Few students, even though they knew they should, checked the currency of information. [46] Factors influencing the selection of a website for obtaining health information included the site's navigation ability, usability, look, and the kinesthetic experience. [44] Various studies have confirmed students are challenged to differentiate between various types of information sources and did not understand the difference between scholarly sources and sales pitches. [35,44,46,52], The above summary regarding student's abilities to search and identify adequate health information for evidence-based practice clearly reinforces the message that access does not guarantee discernment between quality and credibility of health information available online. The need for teaching these skills in the context of an undergraduate nursing program is evident. There is still much to learn about how young people and undergraduate students in the health professions, as well as health professionals, interpret health information they find through electronic sources including social media.

CRITERIA USED BY STUDENTS TO APPRAISE ONLINE HEALTH INFORMATION

In an extension of this, many authors suggest there is a need to educate students about how to appraise the credibility of the content of health information found online, and to develop students' ability to use the information as part of their work in health care. In research where undergraduate students were asked to judge the trustworthiness of health-related websites or articles, about 40%–50% of students' judgments were incorrect when compared to those of experts. [52,55] It is becoming increasingly more evident to educators that undergraduate students are utilizing evaluation criteria either inappropriately or using different sets of criteria to evaluate the credibility and quality of health information from the internet and social media. To illustrate, Zhang [51] constructed a mental model for 38 students who were given an assignment and then monitored for five minutes as they used Medline Plus to find and appraise health information. The mental model Zhang constructed from the data includes three categories of criteria students use to evaluate online health information. Students judged the quality of sites according to esthetics, including sophisticated presentation of the site, and nonessential hints of source and message credibility, whereas experts judged these same websites to be unsuitable and not trustworthy. [51] Noam Tractinsky and colleagues [56] concur with Zang, their study demonstrated evidence that a secure relationship exists between student perceptions of systems usability and esthetics. The uptake of information from esthetically pleasing social media sites is therefore not an unusual practice among students. The most commonly used criteria to appraise credibility of online health information by students were related to source credibility referring to the expertise and trustworthiness of the website's sponsor influenced by several features such as the name of the sponsor, privacy policy, endorsements by third parties, and message features. [57] Commonly used criteria by students are the domain name, infomercialism, currency of information, and appearance. [44] In addition, students are influenced by their emotions as they evaluate online health information and were found to make quick judgments on quality of information. [51] This statement is supported by Senkowski and Branscrum [46] in that students often relate their experiences and knowledge, and apply their beliefs when applying online information to health scenarios. Students are more reluctant to choose or believe health information that is contradictory to their personal beliefs. [46] (p 236) Students more likely guide searches toward websites or answers matching their experiences, current knowledge base, or predetermined beliefs. [46] The perceived relevance and importance of the health information guided the students' selection of certain sources for health information. [50,54] An interesting hypothesis is posed that emotional motivation plays a role in how students search for and select health information. The search of information is a process to regulate an emotion hence the selective attention they display to the types of information. [50] Two other factors that influenced the students' searches for online health information were when healthy living was part of their core values and when certain topics were highlighted by the media. [50]

The discrepancy between what students think the source of health information represents versus the application of credibility criteria for sources of health information are obvious throughout several studies. Menchen-Trevino and Hargattai [54] define credibility in three ways: "message credibility" originates from the trustworthiness of the individual; "sponsor credibility" represents the trustworthiness of site presented; and "site credibility" refers to the reliability of the site as a whole. Most students declared their absence of paying attention to the reliability and credibility of online information. [44] According to Laura O'Grady [58], since websites are not research articles, credibility criteria for evaluation of websites are needed. O'Grady discusses the need for common terminology and researching of a theoretical framework for depicting credibility in health care websites. Undergraduate students tend to evaluate the actual content of online health information according to perceived subjectivity of the content and format. [51] Format criteria include the amount, usefulness, helpfulness, reliability, believability, accuracy, readability, currency, clarity of content, and how informative the content is. [44,51] The presence of large amounts of information on selected sites constituted comprehensiveness of information to many students. [44] Other credibility criteria often employed by students are the presence of quotations, references (list of source citations), and statistics. Students believe the quality of the site is higher if a list of sources is cited. [44,54,57] Although students realize the value of verification of facts by other credible sources, very few students actually complete the process of verification of the sources listed. In a study by Menchen-Trevino and Hargattai, [54] 2011, only 23% of students verified the information they found online. Information sources need to be critically evaluated for the quality of the sources they use, as many websites like Wikipedia make use of mostly government websites and news articles as sources instead of academic articles. As demonstrated by Luyt's [57] 2010 research, Wikipedia is unsuitable for reference work. In addition, many students, "even though they know they should, do not check the currency of information" [46] (p 238) or look for the author's intentions. [46,57]

Beyond most students' inability to apply high standards of evaluation criteria to online health information, several studies highlight the tendency of students to present online health information as their own ideas without citing the online source/s. [44,46,54] "Students thought it okay to present other's ideas as their own without citing conflicting views about the appropriateness of "copying and pasting" answers directly from webpages. This concern may stem from students performing similar searches in an academic setting, where this practice is often viewed as an act of plagiarism." [46] (p 238)

In studies where students' evaluations of online health information were compared to those of experts, it was clear that the experts were mostly unsure about the credibility of the content and used some surface criteria in their evaluation more prominently such as the author's credentials and sources of information. [44,55] Whereas, students' uncertainty in evaluation is related to presentation and esthetics as described, and with less concern regarding the appropriateness of using and failing to cite data from other sources. O'Grady, in clarifying the terminology associated with health care information appraisal, would describe students' evaluations as being more about the evaluation of "surface credibility." Surface credibility is based on the components, impression, and structure of a site. O'Grady suggests that the distinction between what is perceived credible by an individual and what is known credible by an expert is important. [58]

The notion that students overall view text as information sources only, limits their ability to situate the health information into a social context and enforces a tendency to cite resources more frequently than providing actual answers to more complex health scenarios. [46,57] It is therefore not surprising that students express a misunderstanding regarding the scenario or feeling unsure about what was asked or expected of them in response to a given assignment requiring an academic response to a health scenario. [46] More research needs to be done regarding students' performance in more difficult scenarios as they currently display poor judgment on the credibility of online health information and an inclination to be led by these sources. [44,54] In summary, the suppositions of these studies are that college students lack basic credibility evaluation skills of online health information, and when attempting to develop health informatics competencies, their proficiency self-reports are overestimated. [44] Social media, however, may constitute different challenges than the use of the traditional search engines, which are discussed in detail below.

CRITERIA USED BY STUDENTS TO APPRAISE THE CREDIBILITY OF SOCIAL MEDIA INFORMATION

Social media and social networking sites may provide a platform for patient education, psycho-social support, shared decision making and follow up services that are cost effective, save time, and promote access to health information, especially for patients who are unable to travel. [45] Social media and networking sites have the potential to increase people's health-related knowledge, serve as a support network or increase coping and self-efficacy abilities in their users. Across many health sectors, social media are also used to try and enhance workflow and extend the relationship between the patient and health care professionals. Horgan and Sweeney [41] affirm the value of social media to support young people with mental health struggles. Interestingly, 68% of young people ($n = 922$) indicated they will make use of online support; however, 79.4% still prefer support via face-to-face connection. The need for health professionals to develop their knowledge of the current online health services, including via social media, was reiterated/reinforced for the benefit of providing patients with accurate and reliable information and support. [41]

Several cross-sectional surveys in Canada and the US indicate high rates of use of online health information including social media (blogs, podcasts, videos, and wikis). Gustafson and Woodworth [59] define social media as "an electronic medium that allows interactive comment, as in the case of Twitter and YouTube, or collaboration and information sharing as in Pinterest and Wikipedia." [59] (p 3) Social media is also defined as a "continually changing set of tools that facilitate online relationships which enable users to participate in online networking." [60] (p 21)

Social networking sites such as Facebook are commonly used to seek and share online health information, but a knowledge gap exists around the use of health information on these social networking sites. [38] Facebook was invented for networking among undergraduate students, but remains the most commonly accessed social media site for specific heath information, followed by YouTube, blogs, and Twitter. [45] Social media acts primarily as informational and support resources. Sixty five percent of social media sites are used by patients undergoing a health decision-making process. When this occurs, visual media are mostly accessed. Seventy nine percent of social media sites invite feedback. Ninety-seven percent of sites are easy to navigate, and the majority of sites are presented at fourth and fifth grade readability levels. Although 70% of the sites provide accurate information, a third of the sites contained inaccuracies and or inconsistencies in the information provided. [45] Although social media can be informational and serve as a support system, incorrect health information has elicited further urgency and has led to the development of interventions to regulate and detect low quality information. [45] Large numbers of irrelevant pages regularly pop-up when general health searches are performed, with most pages devoted to marketing and promotion sites and only a few for support purposes. Therefore, further research is needed to explore how health information can be accurately streamlined for intent to support and educate. [45]

According to Hale et al., [38] there is a need for more research surrounding social media in health information searches. Due to the nature of social media, mostly unregulated content is represented and the content can be altered or edited by one or several authors. [54] The idea that knowledge can be created through a social process by several stakeholders poses several challenges for the evaluation of credibility obtained from social media using traditional academic standards. [54] Knowledge regarding the opportunities for the advantage of social media is evolving together with several ethical issues around boundaries of public and private communications. [59] These ethical issues are being addressed by authors Wiljer and Thakur as well as Bahr, Crampton and Domb in previous chapters of this book. Therefore, it is important for health professionals to consider the possible inaccuracies of health information communicated through social networks as perceived credibility is an important factor in the health professional's decision to use the information as part of the provision of health care. [22]

Although there remains some uncertainty about the influence of social media and quality of the content found, a few tools have been developed to assess the quality of blogs and podcast use in education. [46,61] In an attempt to create credibility criteria, several websites have developed policies for website authors and designers. Many researchers and policy makers equate credibility with believability, or a newer concept of trust in online health information. [61] As Ivanitskaya and colleagues [47] state, it is quite clear in their argument that website evaluation criteria do not ensure the accuracy of health information. The challenge continues. To date, website criteria developed by Hoosuite, Bitly, Symphur, and Google analytics are not sufficient for appraising the quality of health information found on social media. [45]

Numerous researchers have proposed the further development of shared definitions and consensus criteria surrounding online health appraisal skills and criteria for appraisal of electronic sources of information. Researchers have advocated for the need to have measures that reflect people's actual evaluation ability of online health information, including social media. Researchers have also suggested that there is a need for more research focused on understanding health information seeking behaviors [2]. In regards to social media, the characteristics of the message and the source of the message are part of the credibility of online health information and should be appraised [2]. Supporting this direction of thought, Robb and Shellernarger [40] have advised health professionals to pay attention to who created the message, the currency of the information and the specific evidence cited as part of the message or information. Generally, the more users generate information, the lower the quality of the information. Health professionals generally create higher quality health information. [45]

Criteria specific to the quality of information traditionally include whether the content reflects academic and well-researched facts in addition to the content being clear, consistent, thorough, reputable, reliable, and trustworthy. Researchers have further suggested there is a need to pose questions specific to the credibility of each social networking site and to examine the authorship of each social networking site. [51] Kammerer and Gerjets [62] recommend health professionals examine information sources and the primary purpose of a web page or social media post. In addition, health professionals should examine whether the social media posts provide factual neutral information based on research and are backed up with references. Blogs should exchange personal experiences and display author characteristics (author role, expertise, and motives). The authors further remind health professionals to ascertain whether the social media page reflects first or second hand knowledge and to check document characteristics such as the date of publication, type of document, and publisher information. [62]

A Delphi process was followed by Thoma and colleagues [61] to develop quality indicators for bloggers and podcasters that can also be used as evaluation criteria for the credibility of the content of these sites. Thoma et al., [61] via surveys, identified a list of 14 and 26 quality indicators for bloggers and podcasters. Quality indicators focused on aspects of credibility, content and design and included: resource credibility; an evaluation of editorial process (independent from sponsors); references; good quality information that is accurate; information that is presented in a logical, clear, coherent, transparent manner with information about the author identification, author information, and information about financial conflicts of interest.

Several rating scales are available to assess or evaluate quality of online health information. For example, the DISCERN (non- acronym) Instrument was developed in the United Kingdom in the 1990s to assist consumers in rating the quality of health information. [63] The instrument provides a 16-question Likert scale with a handbook to help consumers judge the quality of health information, particularly with a focus on treatment and to assist them in identifying evidence based information. [63] Quickness in health information seeking is often important to users. Khazaal and partners [64] used the DISCERN instrument to develop a Brief DISCERN tool using just six questions from the original DISCERN to appraise the quality of online health information (questions, 4, 5, 9, 10, 11, and 13 of the original DISCERN questions). In another rating scale, Kostick and colleagues [45] provide a newer framework for analyzing health related social media since the lack of systemic guidelines for evaluation remains a concern. Kostick et al., [45] propose the following seven parameters in an evaluation of several social media sites (Facebook, YouTube, Blog, Twitter, Pinterest, and Yahoo Answers in a variety of mediums). The seven parameters are summarized as the interactivity and user-friendliness of the site, readability of information, medium used (visual, audio, text or a combination of any of these), the purpose of the site, the intended audience, any inaccuracies and inconsistencies, and lastly, the tone of the page and information. [45] These parameters reflect the searching and engagement aspects of social media in accessing health information; hence, the parameters still lack the criteria to appraise the actual content of information properly. [45] The framework highlights the need to become literate in finding relevant information rather than the appraisal of information. In addition to the above evaluation tools, Rowley et al., [3] developed a trust in online health information tool specific for students to use to determine whether the information they find online can be trusted. The Trust in Online Health Information Tool comprises the subsequent dimensions: "authority, style, content, usefulness, and brand, ease of use, recommendation, credibility, and verification." [3] (p 316, 322)

Representation of diverse criteria for the evaluation of health information found online and in social media enforces the need for further research and development in this area. The increasing use of social media as part health education poses several challenges. Criteria for social media based scholarship are largely underdeveloped. Key suggestions for this type of scholarship should be "original, advance education by building on theory, research or best practices, be archived and disseminated, provide the education community the ability to comment, provide feedback and be transparent." [64] (p 552) Although accessible interactive learning can be facilitated via the social construction of knowledge, what types of evidence will support transparent critical appraisal of the different types of knowledge on social media? Navigating ethical challenges such as freedom of expression and privacy in social media, as part of the critical appraisal process may present increased complexity in the evaluation of the quality of information as part of undergraduate education, especially since several organizations use social networking sites for education, support and marketing purposes. [38] In summary, it is pertinent that undergraduate students exercise foundational appraisal skills to evaluate the credibility and quality of information relevant to their own health and the health of their patients. Undergraduate students need to develop a skill set, knowledge, and confidence to collect and critically appraise online health information. [26]

STRATEGIES FOR IMPROVING HEALTH LITERACY COMPETENCIES

Several countries have developed competencies to create an educated health informatics workforce. [28] These competencies extend beyond finding and appraising relevant quality information online or through social media. Competences include basic computer literacy, understanding, and managing data and clinical information systems, navigating the health service sector, ethics and security, implementation of evidence-based practices, and understanding the relationship between the health care systems and evolving technologies. [28]

The health professional workforce is comprised of various health care professionals (e.g., physicians and nurses) and other employees (regulated and nonregulated health professionals). Researchers have identified that there is a need for all health professionals to be proficient in informatics. [43] The use of information and communication technologies is obvious as part of the day-to-day life and education of undergraduate students in the health professions. These competencies are described as the ability to locate quality information, increase understanding of various online and social media sites and how these sites portray health information, the development of awareness of their literacy skill set, and the development of use of relevant evidence as part of health care decisions. [43,47]. Harper suggests the development of "comprehension, numeracy, media literacy, and computer literacy" [53] (p 125) as part of informatics literacy competencies in young college students. The idea that students need to pay attention to authorship and the art of information seeking has also surfaced in the literature. [53] The American Medical Association started a redesign of a new curriculum for medical students, which incorporate new content in health care delivery and health system science. The Association of Faculties of Medicine (AFMC) developed eHealth initiatives for clinical information management, personal and shared information management, health information management and clinical decision management. [26] In addition, The Association of Faculties of Pharmacy of Canada (AFPC), the national non-profit organization advocating the interests of pharmacy education and educators, also developed entry to practice pharmacy informatics competencies. [27] Canada Health Infoway helps realize the vision of healthier Canadians through innovative digital health solutions and provides guidelines for appraisal of online health information [65].

Health care organizations and regulatory bodies of most health professions expect students to graduate with informatics competencies in order to use healthcare systems effectively and use online health information in health service delivery. [31] Since Registered Nurses make up the largest portion of the workforce in healthcare, it is pertinent that they can function optimally as part of the system. [31] This leaves the education system responsible for ensuring baccalaureate nursing students graduate with adequate entry to practice level informatics competencies.

BARRIERS IN COMPETENCY DEVELOPMENT

There are several barriers to building nursing informatics capacity in nursing students. In the face of the swift development of technology, smartphones, and several applications, access has rapidly increased to a surge of online health information. [31,35,40,53] Many countries around the world have not been able to develop informatics competencies, yet there has emerged a need to educate nursing students to adequately search, find, appraise, and incorporate online health information as part of best practices. [31] In addition to this, there is a lack of credibility criteria and search skills. [44,66] Students are indeed more savvy in the use of technology but less savvy regarding the appraisal and use of the online health information. [47,53]

Students have grown to use Google as their homepage, and display a sense of the research process to be simply a compilation of facts. [66] When books and academic literature are presented online, it blends with the idea of "just another website." Students exhibit the skill set to look up a topic in depth, but are unable to address the issue adequately. [66] Within the scarcity of adequate criteria available to students, they lack the ability to adequately articulate the evaluation criteria they use to critically appraise online health information. [55,66] The absence of the fundamental understanding of the purpose and procedures of doing research, results in superficial searches based on the visual quality of a website or existence of a source list, and students struggle to make sense of expectations. [66] Students tend to explore only the first few links and do not study the authorship of the online health information they select. [52]

The influences of a fatalistic world view and level of education on the need to search and appraise online health information is emerging as themes from recent work done by Myrick and colleagues. Several students believed their higher level of education absolved them from adequately seeking health information even if these students recommend other students should gather information from online sources. [50]

Perceptions of Students Regarding Their Digital Literacy Appraisal Skills

In a research study of 1914 college students, Ivanitskaya and colleagues [47] determined students were unable to critically evaluate online health information and the actual health literacy abilities of these students were much lower than the students own perceived abilities. [47] In fact, only 31% of college students were able to accurately estimate their true health literacy level. [23,47] Similar findings occurred in a study by Brown and Dickson [67] with occupational therapy students and several other research studies with undergraduate students. [35,46,52], The tendency for students to perceive their informatics competency level generally higher than their actual abilities creates significant hesitation to use self-rating scales as a means to evaluate the attainment of informatics competencies.

CURRENT INFORMATICS COMPETENCIES FOR NURSING STUDENTS

CASN commenced a rigorous process to create national nursing informatics competencies for entry level nurses in Canada [31,68] In 2014 CASN published a document outlining these entry-level-to-practice nursing Informatics competencies. The predominant informatics competency described by CASN for entry-level nurses is the ability to use "information and communication technologies to support information synthesis in accordance with professional and regulatory standards in the delivery of patient care." [68] (p 4) Three competencies were identified to provide direction for undergraduate nursing curriculum development of this overarching competency, consisting of "information and knowledge management, professional and regulatory accountability, and use of ICT's." [68](p 3,4) Students are clearly expected to perform high level searches and critical appraisal of online health information and other resources to provide evidence informed nursing care upon graduation. In addition, students are to provide patients and their families with assistance in processes of access, selection, and use of relevant health information and information technology, including social networking sites. Mentoring nursing students to attain these informatics competencies may require a shift in teaching philosophy and the educator role whereby the process of gaining the competency becomes as important as teaching the content. [47,66] Nursing specifically lends itself to problem based learning and with a strong formative approach, including guidance and feedback, students can successfully attain these competencies.

Integration of Digital Literacy Principles into Nursing Curricula: Tools and Strategy

Accompanying the list of CASN informatics competencies for entry- to –practice nurses, the goal to build competence and confidence in nurse educators to teach nursing informatics throughout nursing curricula is suggested. [35,66,68,69] Several regulatory bodies such as the College of Registered Nurses in British Columbia (CRNBC) have incorporated some of the CASN informatics competencies as part of their scope of entry-level nursing practice. More research is required as validation processes are created and questions emerge as part of informatics competency development in undergraduate curriculums. [31] The information seeking behaviors of graduate nurses still requires exploring, and learning experiences and tools to assist new graduates need to be developed, validated, implemented and evaluated using different research methods.

Drawing upon social learning theories Jakubec and Astle [24] suggest relational learning practices and the social construction of knowledge in small groups may help students learn foundational literacy skills and cultivate confidence in the attainment of these literacy competencies. The incorporation of planned instructional experiences that move students from developing intentional processing of informatics skills to automatic processing of these skills is encouraged throughout their undergraduate degree. [35,47] Students need to understand one portion of an assignment and practice the foundational skill required for that specific portion, before moving to the next skill in order to build capacity in informatics literacy. [66] The importance of clarity of the purpose of informatics assignments, guided practice learning activities with feedback and examples, and space and time for discussions around types of knowledge are emphasized by experts. [40,66] How scholars in these undergraduate programs attend to practice problems may strongly influence the development of informatics competencies in students as they move evidence into practice [66].

Searching for Health Information Online

Activities to help undergraduate students cultivate the skills, knowledge and confidence to gather and appraise online health information, in fact, inspiring students to move past internet searches are proposed. [40] The millennial generation generally knows how to work with technology, but needs assistance to navigate different information platforms and sources to acquire the information that may help them solve practice problems. In higher education, students should be guided in the efficient use of social media platforms as a method for collecting quality health information when other sources are inaccessible. For example, guidance sessions regarding the appropriate use of keywords and processes for adequate evaluation of search results are proposed. Furthermore, students should be taught not to solely trust Google when gathering health information and rather how to identify and select credible and reputable websites. [46] Students are encouraged to allocate enough time to their search queries, and faculty are encouraged to invest time with the students to solve search challenges when teaching informatics [44].

Emotional and cognitive factors as well as the epistemic beliefs that underpin the concept of knowledge are all factors that influence students' abilities to conduct quality online health information searches. [50,62] Generally, students who view knowledge as evolving and tentative, place more value in establishing the trustworthiness of sources and demonstrate increased source evaluations, better searches, and distinguish higher quality from low quality information than students who view self as a constructor of knowledge or knowledge as absolute. [62] Creating space to elicit self-awareness and reflections from students upon their worldviews and influencing factors may assist them to identify biases and evoke curiosity to explore other ways of knowing or select other sources of knowledge. A practical strategy to provide students with a tabular format of the ranking and source list of the search interface may move them to some degree to distinguish different types of sources and become more confident with searches and knowledge about the search interface [62].

These suggestions may prove useful to build competency in students to adequately navigate advance searches, use Boolean operators to control searches, and discern quality sources. [35,54] Adequate skill in searching for health information though does not guarantee students' proficiency in the appraisal or use of the online health information.

Critical Appraisal Skills

Menchen-Trevino and Hargittal [54] view many of the tools students use to evaluate online health information as impractical and rather suggest the development of heuristics that assist students to judge content quickly. The level of debate, the prominence of the topic, adequate citations, and the intensity of dispute are a few indicators of the quality of content students may consider. These criteria may in fact, still not be enough or used effectively by students in the application process and several challenges can sprout from the application of these criteria. Rather educators should motivate students to explore and examine the discussion and history of the specific information piece to highlight the complexity of determining quality and credibility of information. [54] Educators also deal with the challenge that students may view text as knowledge and language as information, instead of understanding knowledge may be generated by the community in specific socio-cultural, historical, ideological, and political processes. [57] The idea that knowledge is part of social construction and will evolve over time needs to be taught as part of informatics literacy versus a generic skills-based approach to enhance critical analytical processes of information appraisal. The co-construction of knowledge as part of social media is still an emerging field for scholarly research and worth discussion time in class. Specific criteria to appraise health information on social media is also still evolving, but may serve as important conversation points as part of practical quality assessments in class [61].

Collaboration with Librarians and Practice Stakeholders

Collaboration between educators and librarians provides students with resources to find and evaluate online health information. [44,47,66,70] A partnership model between educators and librarians results in education sessions and assessable means to change students' information seeking and information evaluation behaviors. Students who have practice in health organizations have access to interdisciplinary mentorship or other partnerships in applying informatics to actual practice problems and not simulated problems. [47,66] Student peer mentorship is encouraged where students with high competency in informatics can support other students [50].

CONCLUSION

Expectancy does not ensure or produce competency. This statement highlights the active role required in developing digital competencies in undergraduate nursing students. The internet and social media have significantly changed our society. They have also changed how we receive information and have made us more conscious of the variability of health information online. Social media and the World Wide Web have also made the general public more cautious about the health information they locate on the internet and in social media. Nurses can educate the general public about health information found on the internet and the World Wide Web, but in order to do this we need to educate individuals who are embarking on nursing careers during their undergraduate studies about the variability of health information found on the internet and social media. This begins with understanding the underlying motivations and ways in which students and, more specifically the health professional student, seeks health information. In addition to this, we must educate our educators about the tools that can be used to critically appraise online information and information obtained through social media. Lastly, there is a need to integrate nursing informatics competency development into undergraduate nursing education to advance the development of digital literacies

References

[1] Peek HS, Richards M, Muir O, Chan SR, Caton M, MacMillan C. Blogging and social media for mental health education and advocacy: a review for psychiatrists. Curr Psychiatry Rep 2015;17(11):1–8. doi 10.1007/s11920-015-0629.

[2] Diviani N, van den Putte B, Giani S, van Weert JC. Low health literacy and evaluation of online health information: a systematic review of the literature. JMIR 2015;17(5):e112. Available from: <http://dx.doi.org/10.2196/jmir.4018>.

[3] Rowley J, Johnson F, Sbaffi L. Students' trust judgements in online health information seeking. Health Informatics J 2015;2(4):316–27. Available from: http://dx.doi.org/10.1177/1460458214546772.

[4] Fergie G, Hunt K, Hilton S. What young people want from health-related online resources: a focus group study. J Youth Studies 2013;16(5):579–96. Available from: http://dx.doi.org/10.1080/13676261.2012.744811.

[5] Collen Morris F. In: Marion J Ball, editor. A history of medical informatics in the United States. New York: Springer; 2015.

[6] Pew Research Centre (2015). Social media usage: 2005-2015. Available from: <http://www.perinternet.org/2015/10/08/social-networking-usage>.

[7] Fox S, Purcell K. Chronic disease and the Internet. Washington, DC: Pew Internet & American Life Project; 2010.

[8] Pew Research Centre (2013). Health fact sheet. Available from: <http://www.pewinternet.org/health-fact-sheet/>.

[9] Crocco AG, Villasis-Keever M, Jadad AR. Analysis of cases of harm associated with use of health information on the internet. JAMA. 2002;287(21):2869−71.

[10] Ventola CL. Social media and health care professionals: benefits, risks, and best practices. PT 2014;39(7):491. Available from: <http://www.ncbi.nlm.nih.gov/pmc/articles/PMC4103576/>.

[11] Morrison K. The Evolution of Social Media [Infographic] | SocialTimes [Internet]. Adweek.com. 2016. Available from: <http://www.adweek.com/socialtimes/the-evolution-of-social-media-infographic/620911> [cited 26.8.16].

[12] Perrin A. Social media usage: 2005-2015. Washington, D.C.: Pew Internet & American Life Project. Available from: <http://www.pewinternet.org/2015/10/08/social-networking-usage-2005-2015/> [retrieved 12.10.15].

[13] Moorhead SA, Hazlett DE, Harrison L, Carroll JK, Irwin A, Hoving C. A new dimension of health care: systematic review of the uses, benefits, and limitations of social media for health communication. J Med Internet Res 2013;15(4):e85.

[14] Gutierrez N, Kindratt TB, Pagels P, Foster B, Gimpel NE. Health literacy, health information seeking behaviors and internet use among patients attending a private and public clinic in the same geographic area. J Community Health 2014;39(1):83−9.

[15] McGonigle D, Mastrian K. Nursing informatics and the foundation of knowledge. (3rd Edition) Jones & Bartlett Publishers: Sudbury, MA; 2014.

[16] Baker DW. The meaning and the measure of health literacy. J General Intern Med 2006;21(8):878−83.

[17] Martin A. Digital literacy and the 'digital society. Digit Lit Concepts, Policies Pract 2008;30:151−76.

[18] Norman CD, Skinner HA. eHealth literacy: essential skills for consumer health in a networked world. J Med Internet Res 2006;8(2):e9.

[19] Digital Literacy Fundamentals | MediaSmarts [Internet]. Mediasmarts.ca. 2017. Available from: <http://mediasmarts.ca/digital-media-literacy-fundamentals/digital-literacy-fundamentals> [cited 25.2.1017].

[20] Kostick KM, Blumenthal-Barby JS, Wilhelms LA, Delgado ED, Bruce CR. Content analysis of social media related to left ventricular assist devices. Circ Cardiovasc Qual Outcomes 2015;8(5):517−23.

[21] O'Grady L. Accessibility compliance rates of consumer-oriented Canadian health care Web sites. Med Inform Internet Med 2005;30(4):287−95.

[22] Paige SR, Stellefson M, Chaney BH, Alber JM. Pinterest as a resource for health information on Chronic Obstructive Pulmonary Disease (COPD): a social media content analysis. Am J Health Educ 2015;46 (4):241−51. Available from: http://dx.doi.org/10.1080/19325037.2015.1044586.

[23] Theron M, Redmond A, Borycki EM. Nursing students' perceived learning from a Digital health Assignment as part of the nursing care for the childbearing family course. Stud Health Technol Inform 2017;234:328.

[24] Jakubec SL, Astle BJ. Students connecting critical appraisal to evidence-based practice: a teaching−learning activity for research literacy. J Nurs Educ 2012;52(1):56−8.

[25] Developing digital literacies [Internet]. Web.archive.org. 2017. Available from: <http://web.archive.org/web/20141011143516/http://www.jiscinfonet.ac.uk/infokits/digital-literacies/> [cited 25.2.2017].

[26] ehealth Initiatives - The Association of Faculties of Medicine of Canada [Internet]. AFMC.ca 2017. Available from: <https://afmc.ca/medical-education/ehealth-initiatives> [cited 25.2.2017].

[27] Pharmacy informatics- Entry to practice competencies for pharmacist. Association of Faculties of Pharmacy of Canada 2013. Available from: <https://www.afpc.info/system/files/public/AFPC> [cited 25.2.2017].

[28] Parry D, Hunter I, Honey M, Holt A, Day K, Kirk R, et al. Building an educated health informatics workforce—the New Zealand experience. Stud Health Technol Inform 2013;188:86−90.

[29] Mallidou AA, Converse M, Randhawa GK, Atherton P, MacPhee M, Bryant LA, et al. Health services researcher pathway for registered nurses: an integrative literature review. Health Care; Curr Rev 2013;2 (1):1−6.

[30] Nagle LM, Crosby K, Frisch N, Borycki EM, Donelle L, Hannah KJ, et al. Developing entry-to-practice nursing informatics competencies for registered nurses. Nurs Inform 2014;356−63.

[31] Borycki EM, Foster J, Sahama T, Frisch N, Kushniruk AW. Developing national level informatics competencies for undergraduate nurses: methodological approaches from Australia and Canada. Stud Health Technol Inform 2013;183:345−9.

[32] Borycki EM, Lemieux-Charles L, Nagle L, Eysenbach G. Evaluating the impact of hybrid electronic-paper environments upon novice nurse information seeking. Methods Inf Med 2009;48(2):137−43.

[33] Borycki EM, Foster J. A comparison of Australian and Canadian informatics competencies for undergraduate nurses. Stud Health Technol Inform 2014;201:;349–55.
[34] Cummings E, Borycki EM, Roehrer E. Consumers using mobile applications. Enabling health and healthcare through ICT: available. Tailored Closer 2013;183–277.
[35] Stellefson M, Hanik B, Chaney B, Chaney D, Tennant B, Chavarria EA. eHealth literacy among college students: a systematic review with implications for eHealth education. J Med Internet Res 2011;13(4):e102.
[36] Kammerer Y., Gerjets P. Effects of search interface and Internet-specific epistemic beliefs on source evaluations during Web search for medical information: an eye-tracking study. Behav Inform Technol [serial on the Internet]. January 2012 [August 12, 2016];31(1):83-97.
[37] Martin LT, Ruder T, Escarce JJ, Ghosh-Dastidar B, Sherman D, Elliott M, et al. Developing predictive models of health literacy. J Gen Intern Med 2009;24(11):1211–16.
[38] Hale TM, Pathipati AS, Zan S, Jethwani K. Representation of health conditions on Facebook: content analysis and evaluation of user engagement. JMIR 2014;16(8):e182..
[39] Percheski C, Hargittai E. Health information-seeking in the digital age. J Am College Health 2011;59 (5):379–86.
[40] Robb M, Shellenbarger T. Influential factors and perceptions of eHealth literacy among undergraduate college students. Online J Nurs Inform (OJNI) 2014;18(3):18–29 [cited August 12, 2016].
[41] Horgan A, Sweeney J. Young students' use of the Internet for mental health information and support. J Psychiatr Ment Health Nurs [serial on the Internet] 2010;17(2):117–23 [cited August 12, 2016].
[42] Lu HY. Source preferences and the displacement/supplement effect between Internet and traditional sources of sexually transmitted disease and HIV/AIDS information. Sex Educ 2009;9(1):81–92.
[43] Edirippulige S, Smith AC, Beattie H, Davies E, Wootton R. Pre-registration nurses: an investigation of knowledge, experience and comprehension of e-health. Aust J Adv Nurs 2007;25(2) 78.
[44] Kim H, Park SY, Bozeman I. Online health information search and evaluation: observations and semi structured interviews with college students and maternal health experts. Health Info Libr J. 2011;28(3):188–99. Available from: http://dx.doi.org/10.1111/j.1471-1842.2011.00948.x.
[45] Kostick KM, Blumenthal-Barby JS, Wilhelms LA, Delgado ED, Bruce CR. Content analysis of social media related to left ventricular assist devices. Circ Cardiovasc Qual Outcomes 2015;8(5):517–23.
[46] Senkowski V, Branscum P. How college students search the internet for weight control and weight management information: an observational study. Am J Health Educ 2015;46(4):231–40. Available from: http://dx.doi.org/10.1080/19325037.2015.1044139.
[47] Ivanitskaya L, Boyle IO, Casey AM. Health information literacy and competencies of information age students: results from the interactive online Research Readiness Self-Assessment (RRSA). JMIR 2006;8(2):e6.
[48] Heuberger RA, Ivanitskaya L. Preferred sources of nutrition information: contrasts between younger and older adults. JIR 2011;9(2):176–90.
[49] Kwan MY, Arbour-Nicitopoulos KP, Lowe D, Taman S, Faulkner GE. Student reception, sources, and believability of health-related information. J Am Coll Health 2010;58(6):555–62.
[50] Myrick JG, Willoughby JF, Verghese RS. How and why young adults do and do not search for health information: Cognitive and affective factors. Health Educ J 2016;75(2):208–19.
[51] Zhang Y. Dimensions and elements of people's mental models of an information-rich Web space. J Assoc Inf Sci Technol 2010;61(11):2206–18.
[52] Eysenbach G, Köhler C. How do consumers search for and appraise health information on the world wide web? Qualitative study using focus groups, usability tests, and in-depth interviews. BMJ. 2002;324 (7337):573–7.
[53] Harper R. Development of a health literacy assessment for young adult college students: A pilot study. J ACH 2014;62(2):125–34.
[54] Menchen-Trevino E, Hargittai E. Young adults' credibility assessment of Wikipedia. Inf Commun Soc 2011;14(1):24–51.
[55] Theron M, Redmond A, Borycki EM. Baccalaureate Nursing Students' Abilities in Critically Identifying and Evaluating the Quality of Online Health Information. Stud Health Technol Inform 2017;234:321.
[56] Tractinsky N, Katz AS, Ikar D. What is beautiful is usable. Interact Comput 2000;13(2):127–45.
[57] Luyt B, Tan D. Improving Wikipedia's credibility: references and citations in a sample of history articles. J Am Soc Inf Sci Technol 2010;61(4):715–22 doi: 10.1002/asi.21304. Available from: <http://onlinelibrary.wiley.com/doi/10.1002/asi.21304/full>.

[58] O'Grady L. Future directions for depicting credibility in health care web sites. Int J Med Inform 2006;75 (1):58–65.

[59] Gustafson DL, Woodworth CF. Methodological and ethical issues in research using social media: a metamethod of Human Papillomavirus vaccine studies. BMC Med Res Methodol 2014;14(1):1.

[60] Golden, M. Social Media: Strategies for professionals and their firms. John Wi. 2011. As cited p.21in CRNBC, Competencies in the Context of entry level registered Nurse practice in British Columbia, 2013. Retrieved from <https://www.crnbc.ca/Registration/Lists/RegistrationResources/375CompetenciesEntrylevelRN.pdf>.

[61] Thoma B, Chan TM, Paterson QS, Milne WK, Sanders JL, Lin M. Emergency medicine and critical care blogs and podcasts: establishing an international consensus on quality. Ann Emerg Med 2015;66(4):396–402.

[62] Kammerer Y., Gerjets P. Effects of search interface and Internet-specific epistemic beliefs on source evaluations during Web search for medical information: an eye-tracking study. Behav Inform Technol [serial on the Internet]. (2012), [cited August 12, 2016];31(1):83-97. Available from: CINAHL Complete.

[63] Charnock D, Shepperd S, Needham G, Gann R. DISCERN: an instrument for judging the quality of written consumer health information on treatment choices. J Epidemiol Commun Health 1999;53(2):105–11.

[64] Sherbino J, Arora VM, Van Melle E, Rogers R, Frank JR, Holmboe ES. Criteria for social media-based scholarship in health professions education. Postgraduate Med J 2015;91(1080):551–5. doi: 10.1136/postgradmedj-2015-13300. Available from: <https://scholar.google.ca/scholar?hl = en&q = sherbino + arora&btnG = &as_sdt = 1%2C5&as_sdtp >.

[65] Canada Health Infoway: Digital Health in Canada [Internet]. Infoway.ca. 2017. Available from: <https://www.infoway-inforoute.ca/en/> [cited 25.2.2017].

[66] Badke W. A letter from a librarian to professors everywhere. Online searcher [serial on the Internet]. (2016), [cited August 12, 2016];40(4):63-65. Available from: Library, Information Science & Technology Abstracts.

[67] Brown CA, Dickson R. Healthcare students' e-literacy skills. J Allied Health 2010;39(3):179–84.

[68] Nagle L, Borycki E, Donelle L, Frisch N, Hannah K, Harris A, et al. Nursing informatics: entry to practice competencies for registered nurses. Ottawa: Canadian Association of Schools of Nursing; 2012.

[69] Sand-Jecklin K, Murray B, Summers B, Watson J. Educating nursing students about health literacy: from the classroom to the patient bedside. OJIN 2010;;15(3). Available from: http://dx.doi.org/10.3912/OJIN.Vol15No03PPT02.

[70] Hallyburton A, Kolenbrander N, Robertson C. College health professionals and academic librarians: collaboration for student health. J ACH 2008;56(4):395–400.

PART II

EXPERIENCES FROM THE FIELD

SECTION V

TRAINING CLINICIANS IN INFORMATICS AND PRACTICING IN IT-ENABLED SETTINGS

SUBSECTION V.A

CURRICULUM DESIGN AND IMPLEMENTATION STRATEGIES

CHAPTER

9

Lessons Learned and Looking Forward With Pharmacy Education: Informatics and Digital Health

Kevin A. Clauson[1], Timothy D. Aungst[2], Roger Simard[3], Brent I. Fox[4] and Elizabeth A. Breeden[5]

[1]Lipscomb University College of Pharmacy and Health Sciences, Nashville, TN, United States [2]MCPHS University, Worcester, MA, United States [3]Uniprix, Inc., Montreal, QC, Canada [4]Auburn University, Harrison School of Pharmacy, Auburn, AL, United States [5]Lipscomb University College of Pharmacy and Health Sciences, Nashville, TN, United States

BACKGROUND

As the practice of pharmacy continues to change from a product-focused to a patient-focused model, the roles for information and communication technologies (ICT) in pharmacy are of increasing importance. The integration of clinical information systems and other health information technologies (HIT) has already impacted pharmacy operations and care delivery, whereas the potential of digital health and MISST technologies (Mobile Imaging, pervasive Sensing, Social media, and Tracking) offer the promise of harnessing patient-generated data and advancing toward a participatory medicine model [1]. In order for this potential to be realized, pharmacist educators need to recognize the current role of informatics across multiple settings, embrace an innovative approach to revising curricular goals and content delivery, and in some cases adopt the unconventional role of the futurist to best serve students—and ultimately—their patients. This chapter will provide an overview of offerings and approaches to date surrounding informatics in pharmacy education, explore what can be done to integrate ICT into the learning experience, and suggest methods to pair emerging digital health technologies with experiential and interprofessional learning environs.

Health Professionals' Education.
DOI: http://dx.doi.org/10.1016/B978-0-12-805362-1.00009-7

CHANGING WORLD OF THE PRACTICE OF PHARMACY: FOCUS ON TECHNOLOGY

The past 20–30 years have seen a concerted effort directed toward moving the practice of pharmacy from a product-focused model to a patient-focused model [2–5]. This call to action for a new paradigm for pharmacy practice originates from a European perspective [5] and is highlighted in a joint global effort [3] between the World Health Organization (WHO) and International Pharmaceutical Federation/Federation Internationale Pharmaceutique (FIP), which represents over 120 Member Organizations worldwide [6].

> Pharmacists should move from behind the counter and start serving the public by providing care instead of pills only. There is no future in the mere act of dispensing. That activity can and will be taken over by the internet, machines, and/or hardly trained technicians. The fact that pharmacists have an academic training and act as health care professionals puts a burden upon them to better serve the community than they currently do. [5]

Similarly, processes and models ranging from White's office-based pharmacy model to Helper and Strand's pharmaceutical care model to medication therapy management, have been adopted, adapted, and revised in North America. This has been done in an effort to move toward a practice-based model [7,8]. In the United States (US), there are hopes that these efforts and more can be accelerated by the recent passage of legislation including the Affordable Care Act (ACA) and the Health Information Technology for Economic and Clinical Health (HITECH) Act. To this end, informatics (Fig. 9.1), pharmacy informatics (Fig. 9.2) and information and communication technologies (ICT) in pharmacy (Appendix 9.1: Glossary) are increasingly viewed as likely catalysts for this change in approach [8]. The integration of clinical information systems and other health information technologies (HIT) have already impacted pharmacy operations and care delivery, whereas the potential of digital health and MISST technologies offer the promise of harnessing patient-generated data and advancing toward a participatory medicine model. Pharmacist educators need to recognize the current role of informatics across multiple settings in order for this potential to be realized. They must also embrace an innovative approach to revising curricular goals and content delivery, and in some cases, adopt the unconventional role of the futurist to best serve students—and ultimately—their patients. This chapter will provide an overview of offerings and approaches to date surrounding informatics in pharmacy education, explore what can be

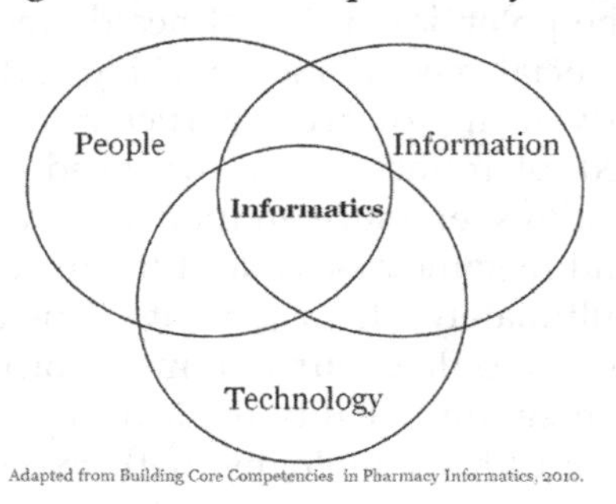

FIGURE 9.1 Informatics as the nexus of people, information, and technology. *Adapted from American Pharmacists Association (APhA), Building Core Competencies in Pharmacy Informatics (2010). Used with permission.*

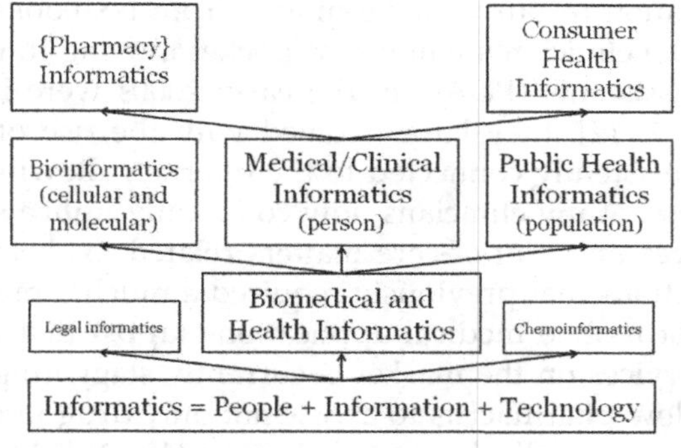

FIGURE 9.2 Informatics and selected health-related subspecialties. *Adapted from Hersh W. A stimulus to define informatics and health information technology. BMC Med Inform Dec Mak 2009;9:24. Used under CC License Atribution 4.0 (CC BY 4.0) https://creativecommons.org/licenses/by/4.0/.*

done to integrate ICT into the learning experience, and suggest methods to pair emerging digital health technologies with experiential and interprofessional learning environs.

WHY PHARMACY EDUCATION MUST PIVOT TO BENEFIT STUDENTS AND PATIENTS

The landscape of medical practice has seen greatly disruptive technology emerge in the past two decades, with substantial ramifications for clinical practice. The rapid rise and implementation of the Internet and wireless technology has played a significant role in creating an environment where information can be shared seamlessly and has integrated multiple devices and infrastructures. The Internet of Things (IoT) reflects numerous devices and technology interlinked, with pharmacy practice challenges to adapt to a broader utilization of technology and the tremendous amount of data that is generated. Electronic prescribing (e-prescribing), robotic filling machines, automated dispensing cabinets, and computerized physician order entry (CPOE) are just a few examples of technology that is quickly finding a place in the hospital and community pharmacy.

While this technology may have improved the workflow and productivity of pharmacists, the integration of the technology takes significant work and planning. Not only do the infrastructure and materials need to be incorporated into the pharmacy, but the staff must be trained on how to implement, utilize, and adapt the technology which is fundamental to the practice of pharmacy. The rise of informaticians (see Section 3) and associated information technology (IT) pharmacy positions has been linked closely to the utilization of electronic resources in the pharmacy environment and space, necessitating the role of a pharmacist with clinical knowledge and skills of their practice that provide a merging of clinical and technical abilities.

Moving beyond integrated health systems bound by computerized hardware, the realm of mobile technology has created new ways for pharmacists to engage in daily practice.

Many of today's information resources have moved from textbooks and handheld guides traditionally trusted to the clinician's white coat pocket moving to the palm of their hand. While personal digital assistants (PDAs) in the early 2000s were initial forays for health information companies [9–12], they have evolved with the rise of tablet computers and smartphones that are consistently connected to the Internet, allowing data to be constantly updated for ease of access. Many clinicians now commonly utilize smartphones to consult traditional drug references on point-of-care matters related to checking drug information, dosing, and drug interactions that previously required a much more laborious process.

Indeed, the number of mobile medical applications (apps) that help provide clinicians with information and services on the market is currently staggering and constantly developing, with apps that allow pharmacists to access not only drug information, but also clinical studies, medical papers, medical news, and more [13–15]. Mobile devices have also been demonstrated as beneficial for pharmacists to help track their daily activities and assist with order-entry in the hospital environment [16–20]. While such technological advances in informatics and the transition of traditional resources to a more accessible format has no doubt been a boon for pharmacists in their daily lives to enhance productivity, it has also been a benefit for patients.

The rapid rise of mobile technology has contributed to the rise of MISST technologies and digital health: the use of technology to help personalize the level of care for patients and enhance health professionals' interaction. Digital health includes the use of health information technology (HIT), wearable devices, mobile health (mHealth), and telehealth. Health professionals recognize the potential implications of digital health on patient care, and many health researchers and organizations have expanded to pilot or integrate such novel technology [21]. Examples include the use of patient portals; the ability to refill medications, or to serve as reminders to take medications from a mobile app on the patient's phone, as these provide services to patients that can help them keep abreast of their health data [22–24]. The development of wearable technology that provides direct patient data to health care professionals so as to make therapeutic interventions provide additional examples. The ability to assess a patient's medication adherence or response to medications to guide pharmacotherapy is also pertinent to the practice of pharmacy.

Digital health is still relatively new and lacks a requirement in the education of future health care practitioners. While pharmacy education does require exposure to and education in informatics, the components of digital health are largely absent. This is a conundrum for pharmacy educators, considering that health researchers and organizations are currently investing time and effort into the future implications of digital health infrastructure. The realm of pharmacy has been targeted by health industries and companies traditionally unaffiliated with health care and serves as a fertile area for investment and the introduction of technological innovations into practice. These areas include traditional community- and hospital-based pharmacy operations and clinical pharmacy.

In mobile platforms, apps available on smartphones offer many services that may help inform patients about their health on a daily basis. They offer a mix of user engagement features to help patients manage their conditions and address lifestyle habits that may contribute to the development of harmful comorbidities. The concept of creating prescribable apps (e.g., Welldoc BlueStar) to recommend for patients based on their disease states has also garnered some attention. For pharmacists, apps that help patients manage their medications and serve as improved adherence reminders are of particular interest [23,24]. Alternatively, the rise of digital health devices, such as Bluetooth-enabled medication vials or ingestible biosensors that can track when patient take their medications, may offer enhanced solutions. Proteus Digital Health has helped pioneer the "smart pill" that integrates such a sensor to track medication intake. One study of their smart pill technology was conducted in Europe to evaluate the impact of tracking patients' medication adherence and the management of blood pressure via pharmacist assessment [25]. This work is culminating in a push for the creation of a category of "digital medicines" with integrated sensors (e.g., Otsuka Pharmaceuticals' Abilify product, currently seeking Food and Drug Administration (FDA) approval [26]).

Other examples of digital health products include myriad devices developed for health, fitness, and patient-specific vital-sign tracking [27]. Many fitness trackers help count steps and track caloric expenditure throughout the day, with more advanced devices containing sensors that can track sleep cycles and heart rates. These devices are attractive to consumers who want to receive feedback on their daily activities, and appeal to health care communities advocating their utilization for patients to be more engaged with their personal health [28]. More advanced digital health devices incorporate sensors that can perform actions previously attainable only in a health care setting under the purview of medical staff. A recently FDA-approved sensor that fits on the back of a smartphone allows users to measure their heart rhythm, with the ability to track arrhythmias [29]. This device was utilized by community pharmacies in Australia to help identify patients in the community with atrial fibrillation [30,31]. The use of these devices may represent a boon to pharmacist clinical activities (i.e., as a means of helping educate and evaluate patient health and medication compliance) especially in chronic disease management - utilizing telemonitoring services [32].

The large-scale integration of technology into society and the health care sphere has also led organizations and researchers to determine new avenues for medication monitoring and postmarket data collection. Social media has been seen as one form of mass patient participation, with many patients using online forums and services to discuss their health and outcomes. Perhaps one of the best examples is Dave deBronkart (a.k.a. e-Patient Dave), who has given many widely acclaimed presentations about his process of surviving cancer, which is attributed to finding more information online and partnering with open-minded clinicians [33]. As the Internet offers many online information services and opportunities for patients to connect with one another, the ability to collect postmarket information on the effects of medications represents an exciting new stream of research. This includes platforms such as Twitter, Facebook, and Instagram where patients who mention side-effects of their medication may lead to new versatile ways of identifying issues with medications not seen in clinical trials [34,35]. The online community platform PatientsLikeMe, which started with a focus for those with rare diseases to foster camaraderie and share experiences, has proven one example of impact from such data tracking. The data aggregated from PatientsLikeMe has now been integrated into the Walgreens Pharmacy patient portal, giving users information gathered from the patient community about potential side-effects and time to benefit [36].

The implication of these technological innovations is that the practice of pharmacy and the veritable landscape of pharmacy itself is witnessing drastic developments that future pharmacists must be prepared for. Pharmacists are quickly being thrust into a digital health world where technology plays a pertinent role in patient care, and also have the opportunity to leverage such tools in their practice and serve as guides for their patients [37]. New material for the pharmacy curriculum that prepares students for a digital health future will require much innovation and additions to the current curriculum. Many of these new tools entering the market may help augment current modalities of pharmacy interventions and roles. The integration of informatics, MISST technologies, and digital health tools into pharmacy education can take many forms and opens up new possibilities for education. It will also help foster more interprofessional opportunities for educators.

TRANSITIONING TO INCLUDE CORE INFORMATION AND COMMUNICATION TECHNOLOGIES (ICT), HEALTH INFORMATION TECHNOLOGIES (HIT), AND INFORMATICS COMPETENCIES

Fundamentally, a "*transition*" occurs when something changes from one state or condition to another. This term is, arguably, an appropriate characterization of what is needed in areas of pharmacy education that are related to clinical informatics and digital health topics. The current state of pharmacy informatics education is that it is largely inconsistent across Doctor of Pharmacy (PharmD) programs, and in some cases, off the mark [38,39]. Despite the increased importance of informatics in pharmacy practice, the incorporation of informatics in pharmacy education has remained flat by a certain measure when compared to that of 10 years ago. This reality is less than desirable, and poses even greater concern given that pharmacy operations and health care delivery continue to be driven daily by health information technologies. In addition, the relatively recent emergence of pharmacy informatics and digital health topics is an important contributor to the need for robust informatics educational efforts in professional programs. The rapid rate of advancement and change that is inherent in the digital domain is also an important contributor to the barriers that pharmacy educators face.

The resulting challenge for pharmacy education is to develop salient learning experiences that reflect the current state of pharmacy informatics and digital health. Educational competencies can serve as a useful starting point in development of these learning experiences. The International Medical Informatics Association (IMIA) published the first set of health and medical informatics education recommendations in 2000, [40] then a revised set in 2010 [41]. While not explicitly labeled as "competencies," the intent of the published recommendations is to support development of educational efforts (i.e., courses, tracks, and programs). In the United States (US), the Institute of Medicine (IOM) identified informatics as one of five core competencies for all health professions education in 2003 [42].

The Accreditation Council for Pharmacy Education (ACPE) promulgates educational standards for PharmD programs. Pharmacy informatics first appeared in the ACPE Standards in 2007 [43]. Initial and continued inclusion in the Standards [44,45] is certainly important to the advancement of pharmacy informatics education. However, increased granularity in guidance documents may help PharmD programs develop educational experiences for a domain in which there is a limited number of faculty trained in informatics and digital health-related topics. The Center for Advancement of Pharmacy Education

(CAPE) has also published suggested educational outcomes for PharmD programs, dating back to 1993 [46]. Moreover, the three subsequent CAPE Outcomes have included those that seem to suggest a focus on informatics in [47–49].

Returning to the need for granularity, the continued challenge is that pharmacy informatics and digital health topics are relatively new, rapidly changing, and faculty with expertise in these topics are lacking in many PharmD programs. Relevant competencies—albeit arguably in need of review with an eye toward enhancement that reflects recent advancements specific to pharmacy informatics—have been published [50]. Developed through a modified Delphi technique, these competencies largely focus on pharmacy practice and management.

In addition, there is a need for attention to more recently emerging topics such as digital health and MISST technologies, reflecting the reality in the current educational domain. The rapidly changing nature of technology and its equally rapid incorporation into professional practice as well as adoption by consumers has created an environment where education is reacting to practice. Practice, in turn, is increasingly relying on automation for efficiency and safety gains [51]. Governmental interventions continues to drive electronic health record (EHR) adoption, although much work to optimize their use remains undone [52–55]. Patients are finally getting the opportunity to engage and become equal participants in their own care decisions, [56,57] even though they should always have been the center of care. All of these realities result in a dynamic informatics and digital health landscape, creating an almost constantly moving target for educational experiences.

Fortunately, opportunities exist to utilize currently available resources in transitioning to an educational experience that incorporates informatics and to some extent, digital health competencies. "Partners in E" is a freely available source of informatics educational materials designed to be integrated into PharmD programs [58]. This resource focuses largely on informatics topics. Similarly, the Association of Faculties of Pharmacy of Canada (AFPC) and Canada Health Infoway offer Canadian PharmD programs a competency-based informatics education resource that includes digital health topics [59]. These programs offer the advantage of a plug-and-play, modular resource that can be integrated into existing programs, or used as complete programs.

The potential inability to modify resources as needs develop over time is an important limitation of the use of externally developed educational resources. In fact, existing resources may not meet important needs upon initial use. For these and other reasons, the option to create informatics and digital health educational experiences from scratch may be an appealing venture. In either case, the starting point is to identify the desired outcomes of target learners. At this point, competencies like those discussed above can be a valuable resource [50]. Educators can also turn to the practice setting for insight into competency development.

While the focus here is on PharmD education, there are many pharmacists (and other health care providers) whose daily practice *is* informatics. These practitioners are often hands-on with the HIT that their professional colleagues rely upon to support their practice. Known as "informaticians," these individuals often blend their understanding of a specific health care discipline with delivery, administrative, and financial models found in the health care ecosystem [60]. Like other specialty areas, informaticians possess a unique set of knowledge and skills that allows them to positively impact practice and, ultimately, patient outcomes. Accordingly, efforts have led to the identification of core content for specialty training [61] and informatics certification opportunities [62–64]. A complementary type of expertise that could be leveraged to optimize processes with an eye toward improved patient care is the health knowledge broker (e.g., role of medical librarian), which is distinctly different from a health informatician. These individuals, also known as health infomediaries, occupy more of an interpretive managerial role rather than that of a tool operator [65]. Health infomediaries require judgment and wisdom beyond simple technical proficiency and familiarity with emerging technological innovations. This is more in keeping with the role of a health professional [65].

An important distinction to note before hastily adopting specialty training content for use in PharmD programs is to recognize that the goal is not merely a graduate who has expertise in informatics and digital health topics. Instead, the goal is to produce qualified PharmD graduates who understand how informatics and digital health tools fit their practice and how these tools can support patient activities that contribute to better health outcomes. Stated another way, the present focus is the pharmacist's use of these tools to *support their practice*, not the use of these tools *as their practice*. The intersection of providers and patients' use of these tools should also be a central theme of educational efforts.

Bounded by the use of informatics and digital health as supportive tools for pharmacists' practice, a common framework can be helpful to guide transitioning to include these domains in pharmacy education. The medication use system (Fig. 9.3) provides a useful framework that is largely universal across the primary pharmacy practice settings (i.e., institutional, ambulatory, and community). Pharmacy educators will want to create

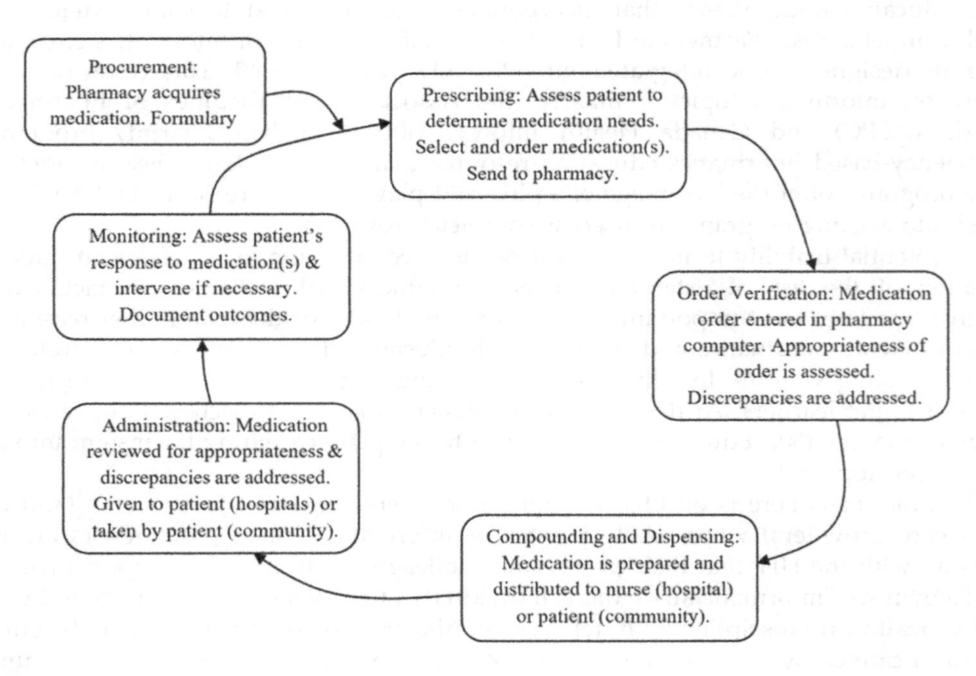

FIGURE 9.3 The medication use system

engaging learning experiences that apply to the eventual practice settings of their students. And while a majority of graduates from a particular PharmD program may migrate to a specific type of practice, many may also change jobs throughout the course of their professional careers and others may work part-time in a setting that is different than their primary setting.

Another tangible benefit of framing informatics and digital health education in the medication use system is that it illustrates the interprofessional nature of pharmacy practice [66]. While widespread interest in interprofessional education has somewhat recently become a central focus in health care, [67,68] its importance is clear and is increasingly becoming a focal point within pharmacy education [69–72]. Digital tools that span practice settings, connect patients and providers, and allow patients to engage the health care system in new ways illustrate the interconnectedness that is the reality in today's pharmacy practice settings. Additionally, discussions framed by the medication use system can identify the tools, information, and people engaged at each point in the system, allowing learners to identify inputs, processes, and outputs relevant to informatics, as described below.

First, consider the Procurement step in which medications are ordered and received. Inputs to consider include institutional formularies, wholesaler data, medication availability information, and cost data. Processes central to procurement include identification of medications to order and the actual method by which those medications are ordered, usually through direct connections with wholesaler information systems. The output of procurement is a medication delivery to the pharmacy. The challenge of counterfeit and unapproved medications entering the medication supply chain is of current importance [73,74]. This is an area in which pharmacists and informatics play important roles [75].

From the patient's perspective, the medication use system begins with a prescription. The patient's diagnosis, medical and medication histories, insurance status, intolerances and allergies, and treatment and pharmacy preferences are all important inputs. In both institutional and community settings, the prescribing process has increasingly shifted to an electronic exchange of information between the prescriber, third party payer (when appropriate), and the pharmacy. Clinical decision support systems (CDSS) bring an additional input to the prescribing process by presenting prescribers with real-time information about the patient and the medication [76–78] to ensure optimal medication therapy.

Order verification is closely tied to prescribing. At this step, pharmacy staff members determine the appropriateness of a prescription in the context of the patient's medical history and current medication profile. Beyond the received prescription, inputs include the patient's medical and medication histories. As in the Prescribing stage, CDSS provides an important input to support the process of comparing the ordered medication to what is known about the patient (e.g., current medications, previous experiences with medications, diagnoses, insurance status, etc.) and what is known about appropriate medication therapy, ultimately leading to a prescription that is ready for preparation. In the community setting, pharmacy staff must also address issues related to payment, which often involves a third party payer. Refill and prior authorization requests are increasingly being performed through electronic communication.

Prescription preparation is the act of compounding and dispensing. Automated devices that support safe and efficient compounding and dispensing are some of the most commonly found technologies in pharmacies [51,79]. The array of technologies is quite broad and is one of the most commonly found areas where pharmacy informaticians practice in institutional settings. For the typical pharmacist whose practice is supported by informatics, it is critical to understand the advantages and limitations of automated compounding and dispensing devices, as well as the information that feeds in and out of these devices. Medication errors can occur despite the use of automation, so all pharmacists should be able to articulate where and how these errors are most likely to occur, and more importantly, how to prevent them.

Administration is in fact the last step to prevent medication errors before they reach the patient. Accordingly, institutional settings have made barcode medication administration systems, electronic medication administration records, and intelligent infusion devices ("smart pumps") top priorities. These systems share important information with pharmacy systems and are critical to safe medication use. While the community pharmacist is largely disconnected from patients' medication administration data, the pharmacist is well positioned to help patients identify digital tools to support appropriate medication-taking behaviors. Often known as "adherence devices," these tools run the gamut from simple alarm reminders on smartphones to Web-connected digital pill boxes that use multiple reminder prompts, engage the patient through questions and answers to assess well-being, and present all gathered data to an authorized caregiver for review and follow-up. Regardless of the device's capabilities, future pharmacists should have an appreciation for the appropriate role of these devices as well as the critical decision points that result in the selection of one device over another.

When an adherence device—or something similar—includes the ability to view longitudinal patient data, it is functioning as a remote monitoring tool. Today, patients have the ability to monitor much more than their medications. They can monitor their physical activity, sleep patterns, vital signs, environmental influences on their health, dietary behaviors, finances, and virtually anything else that can be measured [80,81]. This domain, known as the "quantified self," [82] is arguably the most explosive in the digital health arena and is enabled by the widespread availability of relatively inexpensive devices that can record and share longitudinal data created throughout a patient's daily activities. When the data a patient supplies is insightful to understanding how they take their medication—or why a medication is or is not effective—pharmacists should be prepared to use data visualization tools to interpret the data made available to them. They must also understand the limitations of the data as well as what additional data they need to have a complete picture of the patient.

Considering the discussion above, informatics and digital health educational experiences should reflect the current practice of pharmacy as well as the rapidly advancing world of tools available to patients. It is challenging to incorporate this dynamic field into current professional life; on the other hand, pre-existing resources are available [83–85].

DIGITAL HEALTH AND EMERGING TEACHING MODALITIES TO PREPARE HEALTH PROFESSIONAL STUDENTS

With the advent of digital health technology and informatics creating a larger role in patient care, students will need to be prepared to work in a health care landscape that will be unlike that of the past, and most likely change rapidly over the course of their professional lives. For that reason, health care education must not only teach future practitioners about current and foreseeable technological innovations and implications, but also how to adjust and be malleable for an unknown future. This is analogous to preparing pharmacy students to keep current with the medical literature. Just as students are taught how to navigate the medical literature and interpret it for themselves, we must teach students how current technology impacts patient care, and the basic concepts to be aware of for the future [86]. Such education modalities can be approached in varying ways, with some traditional and some very novel.

Didactic lecture-based presentations to students outlining the current technological health care landscape, including topics surrounding digital health and informatics in practice, may be the simplest way. This method is also most concurrent to traditional teaching modalities. Expanding beyond lectures, hands-on exploration of informatics tools and digital health devices can be pragmatically introduced into physical or clinical assessment courses. There is also value in presenting students with emerging conceptual technologies with potential applications to pharmacy. One recent example is blockchain technology, which is a technology solution that serves as a shared, immutable ledger that can be used to record transactions [87]. Blockchain has been proposed as a means to help secure the drug supply chain and even comply with country-specific legislation such as the Drug Supply Chain Security Act (DSCSA) in the US as well as a vehicle to enable pharmaceutical companies to securely share clinical trial data with research participants [88]. Examples on topics and cases for use in education are presented in Table 9.1.

The largest barrier to these proposals relating to pharmacy programs (and any other health care educational program) will be the associated costs of creating and sustaining a technologically intensive course that will require frequent updates with regards to hardware and software over time. With that in mind, faculty seeking to engage students in the use of this type of technology, and its impact on current and future clinical practice, should focus on the mechanics of said technology. University faculties will quickly recognize that technology flow changes greatly, and that preparing students to recognize and then learn this concept is essential. With limited tools, faculty can still teach and train students the basic concepts needed for using digital health, MISST, and associated technologies. As drastic changes occur (e.g., that advent of virtual reality (VR) or augmented reality- AR), then a budgetary change or purchase would likely be required to be responsive. Faculty should also invest time and effort to educate their departments on new tools and trends to keep abreast of major health care developments in the realm of pharmacy practice.

Several examples of novel ways in which digital health has entered the pharmacy classroom are now present. The topic of pharmacogenomics has been increasingly discussed; with recent changes to CAPE outcomes, it is now mandatory in the pharmacy curriculum. While educational needs can be met by simple didactic lectures, the University of Pittsburgh Pharmacy School has created an advanced program utilizing 23andMe, Inc. personal genomic testing to teach students about pharmacogenomics with their own personal genomes [89,90]. Such novel ways to teach students through hands-on experiences can be equally expanded to other areas of digital health, such as health devices and informatics, including the integration of these tools into laboratory experiences rather than lecture-based content [91,92].

TABLE 9.1 Case Studies for Incorporating Informatics Topics into Pharmacy Teaching

Blockchain Technology in Health Care	Facilitate pharmacy students' exploration of documented, ongoing efforts to use blockchain in pharmacy by organizations (e.g., Hashed Health, IEEE, Hyperledger Project) and companies (e.g., Blockchain Health, Boehringer Ingelheim, Eli Lilly, Merck, Pfizer, YouBase.io) to foster discussion and problem solving using real-world examples of technology.
Digital Health Devices Assessment	Many connected devices are now available to aid patients in self-monitoring (e.g., heart rate, blood glucose, or blood pressure); therefore, training students to understand how to use these devices is pertinent. These devices can be demonstrated and operated in a clinical practice lab or similar setting.
Mobile Health Apps Use	Students can be introduced and educated on how to utilize clinical applications (e.g., drug information, clinical information databases) for practice as well as how to assess and recommend apps for patients in their own health management.
Pharmacogenomics and Personalized Medicine	Utilizing commercial products (e.g., 23andMe) to give personal demonstrations on the role of pharmacogenomics in patient care via real-life experience.
Practice Use of Electronic Health Records (EHRs)	With most health care systems and small group practices adopting the use of EHRs into the workflow, training students in their use is becoming the baseline for education. Some EHR vendors make academic licenses available for this purpose.
Virtual Reality (VR) and Augmented Reality (AR)	The use of VR/AR is quickly becoming more accessible as the technology has seen meaningful development in the past 5 years for health care. Medical education companies are seeking to create products for the health care education field that pharmacy educators may capitalize upon in small practice settings. These may range from disease and pharmacological education to simulations of patient care and settings.
Wearable Devices Practice	Students practice using wearable health devices and apps to monitor their own health outcomes. The assessment of these activities can include student's self-actualization of their own health and patients' health, and possible uses of these devices in patient care. These activities could easily be added into classes early in the curriculum.

THE NEXT GENERATION OF INTERPROFESSIONAL EDUCATION

Interprofessional education (IPE) has become a key focal point in medical education across all health care environments during the last decade. As a demonstrative factor, the recent ACPE Standards 2016 [93] identifies Standard 11, Interprofessional Education, to be one of the keystones in pharmacy education. Based on the core competencies for interprofessional collaborative practice from 2011 [94], the ACPE identifies three key elements, including (1) interprofessional team dynamics, (2) interprofessional team education, and (3) interprofessional team practice as the areas to be fulfilled. Health knowledge brokerage is another fundamental example of interprofessional collaboration. Furthermore, health infomediary activities can be taught from a patient-centered perspective [65].

Universities may struggle with fulfilling these IPE activities over the course of education for pharmacy and allied health programs. Several reasons include logistical barriers, identifying and creating events in line with their student's curriculum, and faculty understanding. Many IPE activities focus on clinical knowledge or practice, which precludes early years of practice-based or knowledge gaps in the student's education. Nonetheless, technology presents a novel opportunity to engage students in IPE activities. With increasingly technologically savvy students entering pharmacy, a greater percentage may gravitate toward technological advances in their daily lives for their education or personal needs. By leveraging this technological familiarity, many educators are seeking to apply their students' innate interest in technology to clinical practice.

Some pharmacy schools and health care institutions are creating innovation labs to spur new disruptive services across the health care environment, by capitalizing on their staff and student body to foster new ideas to put into practice. Innovative practice areas are now appearing across the US, such as the University of Pittsburgh's PillLab and the Thomas Jefferson University's JeffDesign program, where pharmacy students can collaborate with others to design and rethink current practice. While many of these programs involve practicing clinicians, students are also taking advantage of these programs for their own professional growth. This has been spurred on by the role of events like "Hackathons" whereby individuals across many disciplines (e.g., design, business, and health care) assemble to create new solutions for a problem as an interdisciplinary team. Examples include the MIT Grand Hack sponsored by the Hacking Medicine Institute and Stanford's Health Hackathon (Health++), which has gone on to create new products and ideas in the pharmacy environment.

These Hackathons and similar events offer a new opportunity for IPE activities, as they assemble students from health care backgrounds to work together to address problems from the start of their education. As their clinical knowledge develops over the course of their education, they may be better able to refine their ideas, and perhaps in the end culminate in a viable product or service for patient care. Several examples currently exist where pharmacy schools have introduced hackathons or "Shark Tanks" into their curriculum, such as at the Keck Graduate Institute School of Pharmacy, which has started to integrate shark tanks into their curriculum to meet CAPE Standard 4 (Innovation and Entrepreneurship).

While these innovative ideas may help foster new ways of approaching pharmacy education and IPE activities, they will require much consideration. Technology in itself, and even business knowledge, may be outside the scope of most pharmacy (and health care educator) faculty acumen. As such, partnering with other individuals at the university level or the community would be beneficial. The university may benefit from business possibilities based on these relationships, and students will develop more knowledge than current faculty may be able to offer. Other considerations will be teaching space and gathering required material or support required.

LESSONS LEARNED AND MOVING FORWARD IN THE DAYS AHEAD

The roles that informatics and digital health technologies can play in future pharmacy practice demonstrate rapid changes and advancements at a pace necessitating adaptation by academics. Integration of these topics into the curricula still lags behind market demands and can be expected to play a significant role in education in the near future. Implications on interprofessional duties and education can be utilized to meet other related goals of academic outcomes, and help further utilize students' clinical knowledge and developing skills in practice.

Future pharmacy practice must demonstrate the quick accession to the technological expectations of society (i.e., that health care services adapt to a wireless age). This is implied by the integration of community pharmacy activities into larger health care networks. As informatics becomes increasingly involved in all health data systems, pharmacists may need to incorporate knowledge of its use and its limitations in their practice. Likewise, understanding digital health and its implications for patient care, as well as the rise of personalized health, also necessitates improved and continually developing knowledge amongst future practitioners.

We can expect that in the near future traditional drug delivery and dispensing services will be usurped in some ways by growing technological integration (e.g., automation). However, digital health tools and MISST technologies open up more methods for pharmacists to care for their patients in a clinical role, especially at the community level where medication therapy management (MTM) services could realize great benefit. This will be especially important for chronic disease management, thanks in part to devices and mobile apps targeting such issues as hypertension, diet, exercise, diabetes, and mental health. As pharmacies are the bulwark for many patients to purchase products related to their health, the pharmacy could also serve as the literal and figurative provider of digital health devices. Ultimately, living in a connected world will require that we have connected pharmacists.

References

[1] Nebeker C, Lagare T, Takemoto M, Lewars B, Crist K, Bloss CS, et al. Engaging research participants to inform the ethical conduct of mobile imaging, pervasive sensing, and location tracking research. Transl Behav Med 2016;6(4):577–86.

[2] American College of Clinical Pharmacy. A vision of pharmacy's future roles, responsibilities, and manpower needs in the United States. Pharmacotherapy 2000;20(8):991–1047.

[3] Wiedenmayer K, Summers RS, Mackie CA, Gous AGS, Everard M, Tromp D, et al. Developing pharmacy practice: a focus on patient care/Elargir la pratique pharmaceutique: recentrer les soins sur les patients. 2006 ed. Geneva: World Health Organization; 2006.

[4] Dowse R. Reflecting on patient-centred care in pharmacy through an illness narrative. Int J Clin Pharm 2015;37(4):551–4.

[5] van Mil JW, Schulz M, Tromp TF. Pharmaceutical care, European developments in concepts, implementation, teaching, and research: a review. Pharm World Sci 2004;26(6):303–11.

[6] Federation Internationale Pharmaceutique (FIP). Global Pharmacy Workforce and Migration Report. Federation Internationale Pharmaceutique, 2008.

[7] Bluml BM. Definition of medication therapy management: development of professionwide consensus. J Am Pharm Assoc 2005;45(5):566–72.

[8] Latif A, Waring J, Watmough D, Barber N, Chuter A, Davies J, et al. Examination of England's New Medicine Service (NMS) of complex health care interventions in community pharmacy. Res Social Adm Pharm 2016;12(6):966–89.

[9] McCreadie SR, Stevenson JG, Sweet BV, Kramer M. Using personal digital assistants to access drug information. Am J Health Syst Pharm 2002;59(14):1340–3.

[10] Honeybourne C, Sutton S, Ward L. Knowledge in the palm of your hands: PDAs in the clinical setting. Health Info Libr J 2006;23(1):51–9.

[11] Baumgart DC. Personal digital assistants in health care: experienced clinicians in the palm of your hand? Lancet (London, England) 2005;366(9492):1210–22.

[12] Felkey B, Fox BI. Emerging technology at the point of care. J Am Pharm Assoc 2003;43(5 Suppl 1): S50-S51.

[13] Mosa AS, Yoo I, Sheets L. A systematic review of healthcare applications for smartphones. BMC Md Inform Decis Mak 2012;12:67.

[14] Aungst TD. Medical applications for pharmacists using mobile devices. Ann Pharmacother 2013;47 (7-8):1088–95.

[15] Kullar R, Goff DA. Transformation of antimicrobial stewardship programs through technology and informatics. Infect Dis Clin North Am 2014;28(2):291–300.

[16] Patel RJ, Lyman Jr. AE, Clark DR, Hartman TJ, Chester EA, Kicklighter CE. Personal digital assistants for documenting primary care clinical pharmacy services in a health maintenance organization. Am J Health Syst Pharm 2006;63(3):258–61.

[17] Collins MF. Measuring performance indicators in clinical pharmacy services with a personal digital assistant. Am J Health Syst Pharm 2004;61(5):498–501.

[18] Ford S, Illich S, Smith L, Franklin A. Implementing personal digital assistant documentation of pharmacist interventions in a military treatment facility. J Am Pharm Assoc 2006;46(5):589–93.

[19] Raybardhan S, Balen RM, Partovi N, Loewen P, Liu G, Jewesson PJ. Documenting drug-related problems with personal digital assistants in a multisite health system. Am J Health Syst Pharm 2005;62(17):1782–7.

[20] Ray SM, Clark S, Jeter JW, Treadway SA. Assessing the impact of mobile technology on order verification during pharmacist participation in patient rounds. Am J Health Syst Pharm 2013;70(7):633–6.

[21] Bhavnani SP, Narula J, Sengupta PP. Mobile technology and the digitization of healthcare. Eur Heart J 2016;37(18):1428–38.

[22] DiDonato KL, Liu Y, Lindsey CC, Hartwig DM, Stoner SC. Community pharmacy patient perceptions of a pharmacy-initiated mobile technology app to improve adherence. Int J Pharm Pract 2015;23(5):309–19.

[23] Dayer L, Heldenbrand S, Anderson P, Gubbins PO, Martin BC. Smartphone medication adherence apps: potential benefits to patients and providers. J Am Pharm Assoc 2013;53(2):172–81.

[24] Choi A, Lovett AW, Kang J, Lee K, Choi L. Mobile Applications to improve medication adherence: existing apps, quality of life and future directions. Adv Pharmacol Pharm 2015;3(3):64–74.

[25] Noble K, Brown K, Medina M, Alvarez F, Young J, Leadley S, et al. Medication adherence and activity patterns underlying uncontrolled hypertension: assessment and recommendations by practicing pharmacists using digital health care. J Am Pharm Assoc 2016;56(3):310–15.

[26] Kane JM, Perlis RH, DiCarlo LA, Au-Yeung K, Duong J, Petrides G. First experience with a wireless system incorporating physiologic assessments and direct confirmation of digital tablet ingestions in ambulatory patients with schizophrenia or bipolar disorder. J Clin Psychiatr 2013;74(6):e533–40.

[27] Topol EJ, Steinhubl SR, Torkamani A. Digital medical tools and sensors. JAMA. 2015;313(4):353–4.

[28] Patel MS, Asch DA, Volpp KG. Wearable devices as facilitators, not drivers, of health behavior change. JAMA 2015;313(5):459–60.

[29] Baquero GA, Banchs JE, Ahmed S, Naccarelli GV, Luck JC. Surface 12 lead electrocardiogram recordings using smart phone technology. J Electrocardiol 2015;48(1):1–7.

[30] Lowres N, Krass I, Neubeck L, Redfern J, McLachlan AJ, Bennett AA, et al. Atrial fibrillation screening in pharmacies using an iPhone ECG: a qualitative review of implementation. Int J Clin Pharm 2015;37 (6):1111–20.

[31] Lowres N, Neubeck L, Salkeld G, Krass I, McLachlan AJ, Redfern J, et al. Feasibility and cost-effectiveness of stroke prevention through community screening for atrial fibrillation using iPhone ECG in pharmacies. The SEARCH-AF study. Thromb Haemost 2014;111(6):1167–76.

[32] Margolis KL, Asche SE, Bergdall AR, Dehmer SP, Groen SE, Kadrmas HM, et al. Effect of home blood pressure telemonitoring and pharmacist management on blood pressure control: a cluster randomized clinical trial. JAMA. 2013;310(1):46–56.

[33] deBronkart D. Meet e-Patient Dave https://www.ted.com/talks/dave_debronkart_meet_e_patient_dave: TEDxMaastricht; 2011 Available from: https://www.ted.com/talks/dave_debronkart_meet_e_patient_dave.

[34] Powell GE, Seifert HA, Reblin T, Burstein PJ, Blowers J, Menius JA, et al. Social media listening for routine post-marketing safety surveillance. Drug Safety 2016;39(5):443–54.
[35] Freifeld CC, Brownstein JS, Menone CM, Bao W, Filice R, Kass-Hout T, et al. Digital drug safety surveillance: monitoring pharmaceutical products in Twitter. Drug Saf 2014;37(5):343–50.
[36] PatientsLikeMe. Patientslikeme Adds Information About Patient Experiences With Medications To Walgreens Pharmacy Website 2005. Available from: <http://news.patientslikeme.com/press-release/patientslikeme-adds-information-about-patient-experiences-medications-walgreens-pharma>.
[37] Clauson KA, Elrod S, Fox BI, Hajar Z, Dzenowagis JH. Opportunities for pharmacists in mobile health. Am J Health Syst Pharm 2013;70(15):1348–52.
[38] Flynn AJ. The current state of pharmacy informatics education in professional programs at US colleges of pharmacy. Am J Pharmaceut Educ 2005;69(4):66.
[39] Fox BI, Karcher RB, Flynn AJ, Mitchell SH. Analysis of pharmacy informatics syllabi in professional programs at US colleges of pharmacy. Am J Pharm Educ 2008;74(4): Article 89.
[40] Anonymous. Recommendations of the International Medical Informatics Association (IMIA) on education in health and medical informatics. Methods Inf Med 2000;39(3):267–77.
[41] Mantas J, Ammenwerth E, Demiris G, Hasman A, Haux R, Hersh W, et al. Recommendations of the International Medical Informatics Association (IMIA) on education in biomedical and health informatics: first revision. Methods Inf Med 2010;49:105–20.
[42] Institute of Medicine. Health professions education: a bridge to quality. Washington, DC: National Academies Press; 2003.
[43] Accreditation Council for Pharmacy Education. Accreditation standards and guidelines for the professional degree program in pharmacy leading to the doctor of pharmacy degree. 2006.
[44] Accreditation Council for Pharmacy Education. Accreditation standards and guidelines for the professional degree program in pharmacy leading to the doctor of pharmacy degree, version 2.0. 2011.
[45] Accreditation Council for Pharmacy Education. Accreditation standards and key elements for the professional program in pharmacy leading to the doctor of pharmacy degree. 2015.
[46] Education CtICiP. Background paper II: Entry-level, curricular outcomes, curricular content and educational process. Am J Pharmaceut Educ 1993;57:377–85.
[47] American Association of Colleges of Pharmacy. Center for the Advancement of Pharmacy Education Educational Outcomes 1998. 1998.
[48] American Association of Colleges of Pharmacy. Center for the Advancement of Pharmacy Education Educational Outcomes 2004. 2004.
[49] Medina MS, Plaza CM, Stowe CD, Robinson ET, DeLander G, Beck DE, et al. Center for the advancement of pharmacy education 2013 educational outcomes. Am J Pharmaceut Educ 2013;77:162.
[50] Seaton TL. Setting the stage: consensus-based development of pharmacy informatics competencies. In: Fox BI, Felkey BG, Thrower M, editors. Building core competencies in pharmacy informatics. Washington, DC: American Pharmacists Association; 2010.
[51] Fox BI, Pedersen CA, Gumpper KF. ASHP national survey on informatics: assessment of the adoption and use of pharmacy informatics in U.S. hospitals—2013. Am J Health-Syst Pharm 2015;72(8):636–55.
[52] Payne T.H. The electronic health record as a catalyst for quality improvement in patient care. Heart (British Cardiac Society). 2016.
[53] Middleton B, Bloomrosen M, Dente MA, Hashmat B, Koppel R, Overhage JM, et al. Enhancing patient safety and quality of care by improving the usability of electronic health record systems: recommendations from AMIA. J Am Med Inform Assoc 2013;20(e1):e2–8.
[54] Charles D, Gabriel M, Furukawa MF. Adoption of electronic health record systems among U.S. non-federal acute care hospitals: 2008-2013. Office of the National Coordinator for Health Information Technology, Technology OotNCfHI; 2013.
[55] Technology TOotNCfHI. Report to Congress: Update on the adoption of health information technology and related efforts to facilitate the electronic use and exchange of health information. 2014.
[56] Delbanco T, Walker J, Bell SK, Darer JD, Elmore JG, Farag N, et al. Inviting patients to read their doctors' notes: a quasi-experimental study and a look ahead. Ann Intern Med 2012;157(7):461–70.
[57] Ricciardi L, Mostashari F, Murphy J, Daniel JG, Siminerio EP. A national action plan to support consumer engagement via e-health. Health Aff (Millwood) 2013;32(2):376–84.

[58] Healthcare Information and Management Systems Society (HIMSS). Partners in E 2015. Available from: <https://www.himss.org/library/pharmacy-informatics/partners-in-e>
[59] Association of Faculties of Pharmacy of Canada (AFPC) and Canada Infoway. Informatics for Pharmacy Students E-RESOURCE 2015. Available from: <http://afpc-education.info/moodle/index.php>.
[60] Hersh W. Who are the informaticians? What we know and should know. J Am Med Inform Assoc 2006;13 (2):166–70.
[61] Gardner RM, Overhage JM, Steen EB, Munger BS, Holmes JH, Williamson JJ, et al. Core content for the subspecialty of clinical informatics. J Am Med Inform Assoc 2009;16(2):153–7.
[62] Fridsma DB. The scope of health informatics and the Advanced Health Informatics Certification. J Am Med Inform Assoc 2016;23(4):855–6.
[63] Gadd CS, Williamson JJ, Steen EB, Andriole KP, Delaney C, Gumpper K, et al. Eligibility requirements for advanced health informatics certification. J Am Med Inform Assoc 2016;23(4):851–4.
[64] Gadd CS, Williamson JJ, Steen EB, Fridsma DB. Creating advanced health informatics certification. J Am Med Inform Assoc 2016;23(4):848–50.
[65] Van Eerd D, Newman K, DeForge R, Urquhart R, Cornelissen E, Dainty KN. Knowledge brokering for healthy aging: a scoping review of potential approaches. Implementation Sci: IS. 2016;11(1):140.
[66] Institute of Medicine. Health professions education: a bridge to quality. Washington, DC: The National Academies Press; 2003. p. 192.
[67] Interprofessional Education Collaborative Expert Panel. Core competencies for interprofessional collaborative practice: Report of an expert panel. Washington, DC, 2011.
[68] Johnson AW, Potthoff SJ, Carranza L, Swenson HM, Platt CR, Rathbun JR. CLARION: a novel interprofessional approach to health care education. Acad Med 2006;81(3):252–6.
[69] Dobson RT, Stevenson K, Busch A, Scott DJ, Henry C, Wall PA. A quality improvement activity to promote interprofessional collaboration among health professions students. Am J Pharm Educ 2009;73(4):64.
[70] Buring SM, Bhushan A, Brazeau G, Conway S, Hansen L, Westberg S. Keys to successful implementation of interprofessional education: learning location, faculty development, and curricular themes. Am J Pharm Educ 2009;73(4):60.
[71] Buring SM, Bhushan A, Broeseker A, Conway S, Duncan-Hewitt W, Hansen L, et al. Interprofessional education: definitions, student competencies, and guidelines for implementation. Am J Pharm Educ 2009;73(4):59.
[72] Jones KM, Blumenthal DK, Burke JM, Condren M, Hansen R, Holiday-Goodman M, et al. Interprofessional education in introductory pharmacy practice experiences at US colleges and schools of pharmacy. Am J Pharm Educ 2012;76(5):80.
[73] Barlas S. Track-and-trace drug verification: FDA plans new national standards, pharmacies tread with trepidation. P&T 2011;36(4):203–31.
[74] Jackson G, Patel S, Khan S. Assessing the problem of counterfeit medications in the United Kingdom. Int J Clin Pract 2012;66(3):241–50.
[75] Chambliss WG, Carroll WA, Kennedy D, Levine D, Mone MA, Ried LD, et al. Role of the pharmacist in preventing distribution of counterfeit medications. J Am Pharm Assoc 2012;52(2):195–9.
[76] Berner ES. Clinical decision support systems. New York: Springer; 2007.
[77] Kawamoto K, Houlihan CA, Balas EA, Lobach DF. Improving clinical practice using clinical decision support systems: a systematic review of trials to identify features critical to success. BMJ 2005;330(7494):765.
[78] Bright TJ, Wong A, Dhurjati R, Bristow E, Bastian L, Coeytaux RR, et al. Effect of clinical decision-support systems: a systematic review. Ann Int Med 2012;157(1):29–43.
[79] Pedersen CA, Gumpper KF. ASHP national survey on informatics: assessment of the adoption and use of pharmacy informatics in U.S. hospitals--2007. Am J Health Syst Pharm 2008;65(23):2244–64.
[80] Swan M. Emerging patient-driven health care models: an examination of health social networks, consumer personalized medicine and quantified self-tracking. Int J Environ Res Public Health 2009;6(2):492–525.
[81] Barrett MA, Humblet O, Hiatt RA, Adler NE. Big data and disease prevention: from quantified self to quantified communities. Big Data 2013;1(3):168–75.
[82] Swan M. The quantified self: fundamental disruption in big data science and biological discovery. Big Data 2013;1(2):85–99.
[83] Cain J, Fox BI. Web 2.0 and pharmacy education. Am J Pharm Educ 2009;73(7): Article 120.

[84] Fox BI, Flynn AJ, Fortier CR, Clauson KA. Knowledge, skills, and resources for pharmacy informatics education. Am J Pharm Educ 2011;75(5):93.
[85] Fox BI, Thrower MR, Felkey BG, editors. Building core competencies in pharmacy informatics. Washington, DC: American Pharmacists Association; 2010.
[86] Aungst TD. Integrating mHealth and mobile technology education into the pharmacy curriculum. Am J Pharm Educ 2014;78(1):19.
[87] Yli-Huumo J, Ko D, Choi S, Park S, Smolander K. Where is current research on blockchain technology?—a systematic review. PLoS One 2016;11(10):e0163477.
[88] Mackey T, editor. Blockchain: the new frontier in the fight against counterfeit medicines? Stanford Medicine X: Stanford University; September 17, 2016.
[89] Weitzel KW, Aquilante CL, Johnson S, Kisor DF, Empey PE. Educational strategies to enable expansion of pharmacogenomics-based care. Am J Health Syst Pharm 2016;73(23):1986–98.
[90] Adams SM, Anderson KB, Coons JC, Smith RB, Meyer SM, Parker LS, et al. Advancing pharmacogenomics education in the core PharmD curriculum through student personal genomic testing. Am J Pharm Educ 2016;80(1):3.
[91] Miranda AC, Serag-Bolos ES, Aungst TD, Chowdhury R. A mobile health technology workshop to evaluate available technologies and their potential use in pharmacy practice. BMJ Simul Technol Enhanced Learn 2016;2(1):23–6.
[92] Rodis J, Aungst TD, Brown NV, Cui Y, Tam L. Enhancing pharmacy student learning and perceptions of medical apps. JMIR mHealth and uHealth 2016;4(2):e55.
[93] Accreditation Council for Pharmacy Education. Accreditation standards and key elements for the professional program in pharmacy leading to the doctor of pharmacy degree. 2016.
[94] IPEC. Core Competencies for Interprofessional Collaborative Practice. Washington, D.C: American Association of Colleges of Nursing, American Association of Colleges of Osteopathic Medicine, American Association of Colleges of Pharmacy, American Dental Education Association, Association of American Medical Colleges, and Association of Schools of Public Health, 2011.
[95] Gibbons MC, Wilson RF, Samal L, Lehmann C, Dickersin K, Lehmann H, et al. Impact of Consumer Health Informatics Applications. Rockville, MD.: Agency for Healthcare Research and Quality, Services USDoHaH; 2009 October. Report No.: Contract No.: 09(10)-E019.
[96] Kostkova P. Grand challenges in digital health. Front Public Health 2015;3:134.
[97] Hersh W. A stimulus to define informatics and health information technology. BMC Med Inform Decis Mak 2009;9(1):24.
[98] SPM. Definition of participatory medicine. Newburyport, MA: Society for Participatory Medicine; 2014. Available from: <http://participatorymedicine.org/>.

APPENDIX 9.1 GLOSSARY OF INFORMATICS-RELATED TERMINOLOGY USED IN THIS CHAPTER

Term	Definition
Automation	The use of automated machinery or equipment in manufacturing or other services
Blockchain	A decentralized transaction and data management database solution that maintains a continuously growing list of data records (i.e., ledger) confirmed by participating nodes; first developed for the cryptocurrency Bitcoin [87].
Consumer health informatics	"Any electronic tool, technology, or electronic application that is designed to interact directly with consumers, with or without the presence of a health care professional that provides or uses or personal information and provides the consumer with individualized assistance, to help the patient better manage their health or health care [95]."
Digital health	"The use of information and communications technologies to improve human health, healthcare services, and wellness for individuals and across populations." [96]
Health information technology	"The use of information and communication technology in health care settings." [97]
Informatics	The intersection of people, information, and technology
Information and communication technologies	Information technology with a focus on communication and networking
Information technology (IT)	"Activities and tools used to locate, manipulate, store, and disseminate information." [85]
MISST	"Mobile Imaging, pervasive Sensing, Social media, and Tracking." [1]
Mobile health (mHealth)	"The use of mobile devices and global networks to deliver health services and information." [37]
Participatory medicine	"A model of cooperative health care that seeks to achieve active involvement by patients, professionals, caregivers, and others across the continuum of care on all issues related to an individual's health." [98]
Pharmacy informatician	"Pharmacists whose practice is devoted to the development, implementation, management, and support of HIT systems are pharmacy informaticians." [85]
Pharmacy informatics	"The use and integration of data, information, knowledge, and technology involved with medication use processes to improve outcomes." [85]
Wearables/wearable technology	Accessories and/or sensors that enable personalized collection and transmission of mobile generated data

CHAPTER

10

Patient-Centered Technology Use: Best Practices and Curricular Strategies

Maria L. Alkureishi[1], *Wei Wei Lee*[1] *and Richard M. Frankel*[2,3,4]

[1]University of Chicago, Chicago, IL, United States [2]Indiana University School of Medicine, Indianapolis, IN, United States [3]Richard L. Roudebush Veterans Administration Medical Center, Indianapolis, IN, United States [4]Cleveland Clinic, Cleveland, OH, United States

INTRODUCTION

Electronic medical record (EMR) use at the point of care in the clinic or hospital room is quickly becoming the norm world-wide as a result of health care organization and governmental mandates and incentives to promote their adoption [1–3]. There is a concurrent global movement toward patient-centered care, with the World Health Organization's (WHO) Global Strategy on People-Centered and Integrated Health Services calling for improved people-centered care that empowers, educates and engages individuals and incorporates technology in an efficient and effective manner [4].

The necessity for clinicians to be both clinically efficient with the EMR while maintaining meaningful interactions with patients can be difficult to achieve in the context of the clinical interaction [5,6]. As a result, it is not uncommon for unintended adverse communication behaviors to arise, which can have downstream effects on communication and the patient–doctor relationship [7,8]. While it is important for clinicians to be aware of potential communication challenges, it is also important for clinicians to recognize opportunities to use technology as a positive patient education and engagement tool [9]. In order to provide efficient humanistic patient-centered care, clinicians need to be trained in the practical integration of patient-centered communication strategies while attending to the demands of exam room computing activities.

Health Professionals' Education.
DOI: http://dx.doi.org/10.1016/B978-0-12-805362-1.00010-3

Recent calls to action have highlighted the importance of educating medical trainees in patient-centered EMR communication strategies. In the United States, the Liaison Committee on Medical Education (LCME) specifically states that schools must "prepare medical students for entry into any residency program and for the subsequent contemporary practice of medicine" with "specific instruction in communication skills as they relate to communication with patients and their families, colleagues, and other health professionals." [10] As a result, the Alliance for Clinical Education (ACE) recommends that medical schools formally train students on a "clear set of competencies" related to EMR use to ensure preparation for clinical practice [11]. Similarly, the Accreditation Council for Graduate Medical Education (ACGME) Milestones requires residents and fellows to be competent in "incorporating patient-specific preferences into plans of care" and "role modeling effective communication and development of therapeutic relationships" while "utilizing information technology with sophistication." [12] Lastly, the American Medical Association (AMA) highlights the need to educate attending physicians in how to use and teach EMR skills as well [13].

The United States is not alone in advocating for the need to educate trainees and practicing physicians in best communication practices as it relates to the EMR. For example, the Royal College of Physicians and Surgeons of Canada developed the CanMeds framework for improving patient care by enhancing physician training [14]. This framework highlights the need to communicate effectively through the use of the EMR, sharing information with patients in a manner that enhances understanding. However, despite the existence of United States mandates, models such as CanMeds, and similar recommendations abroad, most medical trainees and practicing physicians receive no formal training in patient-centered EMR communication strategies [15–17].

Below, we describe best practices and strategies for patient-centered computer use and curricular approaches targeting medical trainees and practicing physicians. Challenges to curricular implementation will be discussed with a specific focus on the importance of training learners across the continuum to promote sustained behavioral change.

BEST PRACTICES: PROMISES AND PITFALLS

Strategies to improve clinician communication behaviors and promote opportunities for patient-centered EMR use are summarized below. The recommendations are based on recent empirical studies and literature reviews examining the impact of EMR use on the patient–doctor relationship and communication [5,7,8,18]. Research on the effects of computers in the clinical interaction is relatively recent and the level of evidence, methods, and theoretical frameworks is variable. As a result, we have included informal input from experts in the field as well as our own collective experience in adult and pediatric care at academic medical centers in inpatient and outpatient settings as well as private and community settings. Of note, while the best practices discussed here initially seem most applicable to outpatient settings, these patient-centered strategies apply equally well to inpatient, hospital-based care environments. Any situation in which a provider is interacting with the computer or mobile technology in front of a patient is an opportunity to use that technology to help facilitate discussion, increase understanding and promote its positive collaborative use as a communication tool.

ACQUAINT YOURSELF WITH THE CHART

One of the most important steps to facilitate a successful clinical interaction is the preparation that occurs prior to the actual clinical interaction itself [5,19,20]. At a minimum, taking a minute to briefly scan the electronic chart for the last note, recent labs and studies, and preventative health services enables the clinician to catch up on interval events in their patient's care and contextualizes the encounter. This process is also critical to ensuring key clinical items are not overlooked, particularly when multiple clinicians are involved in a patient's care.

From the perspective of the patient, advanced preparation decreases distractions and awkward silences that can arise when a clinician attempts to piece together information in the EMR during the clinical interaction. The bottom line is that adequate preparation promotes efficient use of the limited face-to-face time during the encounter [8,15,21,22]. It is important to note that while pre-encounter chart review can serve to create the clinician's agenda for the interaction, the patient's concerns, and priorities must then be integrated into the agenda. Advance preparation, even if done minutes prior to the interaction, fosters a sense of familiarity and helps clinicians address and anticipate their patients' healthcare needs. In terms of encounter organization and its relationship to efficiency, having a tentative agenda at the beginning of the interaction allows for greater efficiency of data entry into the electronic record and results in fewer revisions of the history of present illness, symptoms, and diagnostic possibilities.

HONOR THE GOLDEN MINUTE

The beginning of the clinical interaction is critical in establishing the personal connection between clinician and patient and sets the tone for the remainder of the encounter in terms of determining priorities [23]. When a computer is present, the first minute of the clinical interaction is often devoted to the physician interacting with the technology rather than the patient and his or her concerns [23]. It is essential to begin encounters technology-free with an undivided focus on the patient [8,21,23–25]. First impressions matter, and this can be particularly true during new patient interactions as both parties are setting expectations for future patient-doctor-computer encounters.

How this is practically carried out depends on the individual physician's workflow. For example, in the United States outpatient setting, it is commonly necessary for the physician to enter the exam room and login to the computer to access the patient's medical chart. Rather than immediately engaging with the computer, the clinician should instead enter the room and make a human-to-human connection by shaking their patient's hand as culturally appropriate. Engaging in brief non-medical talk, such as commenting on the weather or inquiring about the patient's family also helps to establish an interpersonal as well as transactional dimension to the visit.

These simple yet meaningful acts are part of a trust-building process and signal the start of a caring relationship between equals [26]. They convey sincerity, warmth and call attention to the patient, not the computer, as the most important focus in the room [26]. In outpatient situations where it is the patient that must enter the provider's consultation room, the same rules of conduct apply. The provider should take time to detach from the computer to make a personal connection and convey that their mental and physical focus has shifted from the computer screen to the patient.

The next step in "Honoring the Golden Minute" is to elicit the patient's concerns and goals for the interaction prior to involving the computer [5]. It is crucial to attend to the patient first before engaging the EMR, particularly for worried or anxious patients who may be more likely to ignore information if their main anxieties have not been addressed appropriately [27]. Furthermore, patients' concerns are often not known to the clinician ahead of time or are listed inadequately in the "Chief Complaint" field of the EMR, so providers cannot rely on the computer as a proxy to properly relay a patient's worries. It is recommended that clinicians start the interaction in an open-ended manner and determine directly from the patient what their needs and priorities are prior to engaging with the technology in the room.

INTRODUCE THE TECHNOLOGY

As the clinical interaction unfolds, there will come a point when the natural extension of the conversation is to transition to opening and reviewing the patient's chart in the EMR. Recapping and recording key points of information, conducting a review of systems, and reviewing medications, previous medical history or results together are examples of transitions to using the computer. When the appropriate time to engage the computer arises, a conscious effort should be made to integrate the computer while remaining patient centered. The computer should be introduced verbally to the patient in terms of its role and purpose and the provider should discuss when they will likely use the EMR in the interaction [15,19,25]. This helps to guide patients in the next phase of the interaction and reinforces the computer as a collaborative and educational tool to further their care [28]. Statements such as, "I will be using the computer from time to time in our visit to make sure that your concerns are accurately recorded in the EMR. If you have any questions while I'm working on the computer, don't hesitate to interrupt me and ask," can be useful in providing a rationale for the attention being paid to the computer during the interaction.

AIM FOR A CONVERSATIONAL STYLE

It is important to maintain a conversational style during the encounter. The use of computers can alter the exchange and result in long periods of silence as physicians type, relegate patient speech to short responses timed to avoid interruption of clinician computer use, and unnatural template-led questioning [29–31]. In order to maintain a conversational flow, clinicians must be facile with the EMR by mastering the ability to type and talk simultaneously and efficiently [5,15,22,24,32]. Familiarity with the specific EMR being used is necessary to support the clinician and patients' needs. This includes working knowledge of the various EMR screens, templates, order entry fields, and customization options to suit the encounter [19,22,33]. Often, individual screens or tabs on the EMR interface will be used out of order, so fluency with the EMR layout allows ease of navigation and minimizes how much the computer dictates the flow of conversation.

It is important to approach technology use from the patient's perspective. Clinicians should utilize tools to promote patient-shared decision making, empowerment and education by visually representing disease-specific data, graphing lab trends to counsel on behavior change, using images to explain conditions or procedures, and conducting internet searches for evidence-based resources when appropriate [5,22,28]. Investing time to learn how to use the EMR to enhance the patient's understanding of health conditions and treatments can add to the conversational feel and augment the clinical interaction.

EMPLOY THE "TRIANGLE OF TRUST"

In most instances, the physical orientation of record keeping technology in the examination room was designed prior to the integration of computers into clinical practice. As such, the location of the computer is often an after-thought, placed close to electrical outlet rather than in a purposeful location that will facilitate patient-centered use. Where this is the case, it is critical for the clinician to create a "Triangle of Trust" configuration by which the clinicians place themselves, the patient, and the computer screen at each of the three corners of a triangle (Fig. 10.1) to allow for screen sharing [25]. Doing so physically sets the stage for shared viewing of the screen and creates an environment for a transparent

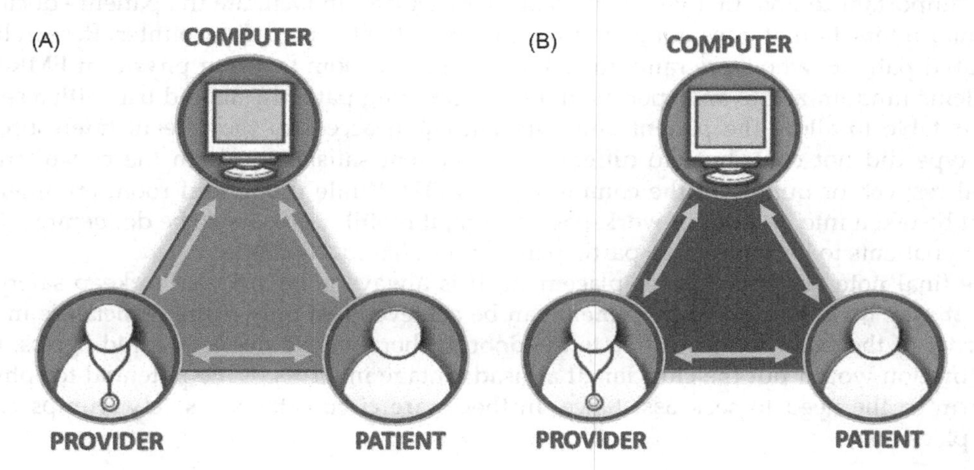

FIGURE 10.1 Illustration of the "Triangle of Trust" configuration in which the clinician purposefully positions themselves, the patient, and the computer screen at each of the three corners of a triangle in order to achieve shared viewing [25,34]. *Adapted from: M.L. Alkureishi WW. Lee, J. Farnan, V. Arora, Breaking Away from the iPatient to Care for the Real Patient: Implementing a Patient-Centered EMR Use Curriculum. MedEdPORTAL: 2014. Available at: www.mededportal.org/publication/9953. Accessed 2016, July 19. Used under Creative Commons License (copyright: the authors)*

and collaborative experience [8,21,22,25,35]. Furthermore, research confirms that the simple but important act of showing the screen to a patient increases their satisfaction with the computer and allows them to gain a better understanding of their care [22,30].

The "Triangle of Trust" configuration can be achieved regardless of the original orientation of the room and its components. When a computer or workstation on wheels (COW, or WOW) is used, it generally takes little effort on part of the provider to move the computer into position when ready to share the screen, and push it aside when it is no longer needed or helpful. Studies have shown however that in certain circumstances COWs can be cumbersome and difficult to navigate around and between patients, thus they should only be used in situations where they can be easily maneuvered to streamline patient care for the patient and provider perspective [36]. These issues can be easily addressed by using tablet computers, which are small and offer the ultimate benefit of mobility. Tablets can be brought out and shared with ease despite any space restrictions of the room. Downsides of a tablet are small screen size that may be a barrier to patients with vision impairments, as well as the lack of a keyboard for large amounts of data entry; however, a separate keyboard can often be added if needed. In addition, since tablets are small and portable, they can be easily lost or stolen and physicians must ensure the device is password protected, no patient information is stored on the device, and providers need to have the ability to remotely wipe the data remotely. If a desktop computer is used, some may offer movement in terms of positioning of the screen, and should be adjusted to create the triangle configuration. Even in rooms where the screen is fixed into position, for example, mounted on a wall, the "Triangle of Trust" can still be achieved by repositioning the furniture in the room that is modifiable, for example moving the patient or provider's chairs to create the triangle.

It is important to note that while an ideal room set up can facilitate the patient–doctor–computer interaction, it does not guarantee a more patient-centered encounter. Researchers compared patients who were randomized to a standard room favoring physician EMR use to patients randomized to an experimental room favoring patient-centered use with a semicircular table to allow the patient and clinician equal access to the screen. Interestingly, room type did not contribute to differences in patient satisfaction with the consultation, mutual respect, or quality of the communication [37]. While thoughtful room ergonomics should be taken into account in workspace design, it is still necessary to be deliberate about inviting patients to become active participants in the shared "Triangle."

One final note about computer placement. It is always good practice to keep safety in mind. It may be that an ideal placement can be achieved but only if the clinician is in the far corner of the exam room relative to the door. If there were a need for rapid egress, this configuration would put the clinician at a disadvantage in terms of the potential for physical harm or the need to seek assistance. In these rare circumstances, safety trumps computer placement.

DON'T FALL VICTIM TO THE MYTH OF MULTITASKING

It has been shown in a number of studies that it is virtually impossible to engage in complex, non-automatic parallel actions with equal focus and attention [38,39]. Individuals can rapidly shift attention from one focus to the next under executive function control of the frontal cortex, thereby determining which tasks to concentrate on and what information is seen as irrelevant and is ignored [39–41]. The ability to jump from one focus to another can create the illusion of attending to multiple tasks simultaneously, however, what is actually occurring is rapid switching between tasks, each switch at the expense of the other [39,40].

While experience and practice can transform a function from being deliberate to one that is more automatic, true multitasking is only possible when both tasks are entirely automatic and can occupy an individual's working memory simultaneously [39]. In addition, a lag of up to several tenths of a second occurs each time the brain handles a switch between two foci, which has cumulative negative effects on efficiency [42]. Attempting similar tasks at once creates an additional level of difficulty, as they utilize similar regions of the brain [41]. Just as writing a detailed email and talking on the phone is impossible to do well at the same time, documenting one issue in the EMR and talking to a patient about another is impractical as both actions involve communication and conflict is created between the two processes.

Attempts to multitask are made worse by the inescapable distractions inherent in the clinical care setting (a knock on the door, receiving a page, or seeing a pop-up alert in the EMR) and often result in what has become known as "distracted doctoring" [43]. Much like distracted driving, distracted doctoring can have similarly devastating consequences on a clinician's efficiency and accuracy [39,44–47]. From the patient's perspective, clinicians can become so noticeably distracted that they may seem to overtly ignore them, focusing on the technology rather than the human interaction [31,48,49].

It is important that clinicians accept the fact that one cannot multitask complex and parallel acts without considerable detriment to at least one if not both foci. Granted, there will be times when focused attention to the technology is needed; however, instead of trying to maintain a distinctly different conversation with the patient while using the EMR for an unrelated function, clinicians should instead signpost and say in so many words what they are doing and that concentrated attention to a task is needed [48].

In order to avoid long silences during computer use, clinicians should be transparent in their actions and explain what they are doing as they are doing it. Transparency in action, also known as the "talk aloud method" of verbalizing one's actions as they are doing them, is a common teaching method used by clinicians while performing tasks in front of trainees [50]. By keeping dialogue related to the activity at hand, as opposed to having a separate and unrelated conversation accompany an activity, one avoids the risk of multitasking and its associated negative consequences. More importantly, this practice involves the other person by explaining actions, thereby elucidating the otherwise "hidden" elements of their clinical reasoning and promoting transparency.

When used with trainees, the "talk aloud" method is an efficient means of achieving a shared understanding of what is occurring and why. With patients, it helps foster a natural conversational tone rather than one that is punctuated by long silences while the computer is in use [5,8,22,24]. More importantly, this technique is educational in that it provides an opportunity for the clinician to explain their actions and thought processes to their patients along the way [22,25,28]. Research has shown that if a patient understands the computer's functions and what their clinician is using it for, they feel more positively about it and better understand their care [8,28,30,51–55]. Knowing what a clinician is doing in real-time also provides an opportunity for patients to ask questions, thereby encouraging discussion, clarification and increased collaboration and shared understanding.

THE IMPORTANCE OF REAL-TIME CHARTING

It is unlikely that documentation of an entire clinical interaction will be captured in full during the course of the encounter, nor should this always be the goal. Conversely, it should not be the objective of the interaction to ignore the EMR and intentionally avoid using it in the room with the patient. Relying on scraps of notes or memory alone is not an efficient means of documenting a clinical encounter. Doing so creates a "cognitive load" on one's working memory and forces the clinician to have to accurately recall what was discussed including pertinent positive and negative historical points, physical findings, clinical reasoning, assessments, and plans [56]. Research on human memory has demonstrated that, in contrast to long-term memory which is practically unlimited, working memory is a limited and fragile resource that can easily reach storage capacity [57,58]. Failure to document key information in the EMR in real time can lead to high working memory load and result in errors in recall and a decline in overall performance and efficiency [59].

Another reason to avoid leaving a large majority of chart documentation until after the encounter is the undue burden that charting after-hours places on the clinician's well-being. Studies examining clinician work-life balance and wellness in the setting of increased EMR use demonstrate disturbing rates of physician burn out and stress secondary to documentation pressures and time spent outside of work (i.e., home) charting [60–64]. Although some clinicians may feel that they are doing their patient a service by completely ignoring the computer during the interaction, doing so comes at the expense of the provider's well-being and satisfaction. Leaving the bulk of charting until after the workday, otherwise known as "pajama time" charting, is neither a sustainable practice nor one that is conducive to protecting one's emotional health or nourishing the joy that brought physicians to the medical profession in the first place.

An equally important reason not to ignore the EMR is that it eliminates opportunities to use technology to enhance communication, education, and understanding between clinician and patient. From the patients' viewpoint, failing to document concerns in real time may also raise questions as to how seriously they and their concerns are being taken [21]. Avoidance of the EMR discounts its potential to enhance the clinical interaction and further reinforces its perception as a barrier to, rather than a facilitator, of care.

A practical balance must be struck in terms of how much information is recorded, and at what point in the conversation. For example, after "honoring the golden minute" and eliciting a patient's concerns for the day, the clinician can then open the chart and jointly review the patient's past medical history, active problem list, medications, and allergies and update the EMR with changes and current concerns. Documentation of a patient's history of present illness can also be done together as a recap of what was just discussed. This allows the patients to feel that their concerns were heard, and also provides an opportunity for collaborative documentation of their narrative with clarification if there was a misunderstanding. Pertinent findings from a review of systems or physical exam as well as updated problems, orders and referrals should be documented in real-time to minimize delays in patient care and ensure that key information is accurate and not forgotten after the encounter. Lastly, the after-visit or discharge summary can be created together as a tailored review of the encounter including the plan and goals for the next appointment. This process permits further clarification of remaining questions, and ensures both parties are on the same page in terms of next steps.

Concurrent documentation during the encounter allows the essential components of the interaction to be captured in a natural manner, and the remaining documentation can be quickly completed afterward. Of course, clinical interactions, particularly those with a psychiatric focus, may deviate from the approach described above and the patients' needs must always supersede the goal of real-time documentation. For the majority of encounters, however, the workflow described can help further the dual purposes of accurately capturing the details of the interaction in the moment, while capitalizing on the communication opportunities that the EMR offers.

ACTIVELY ENCOURAGE PATIENT ENGAGEMENT WITH TECHNOLOGY

In order to capitalize on the educational and communicative benefits of the computer, clinicians should be mindful of, and proactive in, seeking opportunities to use it collaboratively with patients [5]. Beyond patient inclusion in real-time documentation, clinicians can thoughtfully encourage patients to engage with the technology and seek ways to use it to augment involvement in their care. A very simple but effective means of doing so is by graphing and demonstrating trends in clinical data over time (i.e., vital signs such as body mass index and lab results such as hemoglobin A1c) [28,65]. A graphic easily and quickly conveys information in a powerful manner, particularly for non-native language speaking patients or when health literacy level is a concern. Graphic representation serves as the starting point by which to explain a disease process, understand disease progression over time, reinforce healthy lifestyle changes or begin a conversation about how to best address the issue together [19,21,28]. Graphs can also be printed and given to patients to refer to at a later time and can be used by the patient to include other family members who were not present for the discussion during the clinical encounter [5,66]. Showing patients their computer-based radiographic images is another very helpful visual tool that promotes engagement. Researchers found that demonstrating radiographic images on a tablet resulted in significantly improved involvement in care decisions, helped patients understand what was being explained to them, and resulted in an overall positive effect on their overall hospital experience [67].

Beyond the tools of the EMR itself, it is important to capitalize on other resources that an exam room computer offers [28]. For example, if discussing a medical diagnosis with a patient that involves reference to the anatomy of the body, clinicians can use websites or other online resources to bring up pictures or images to demonstrate the body part and visually illustrate what is being discussed. Providing patients links to recommended videos can greatly aid in reinforcing information discussed in an encounter. There are a number of websites that host a wide range of free and reputable videos covering an array of medical topics such as how to use medical devices like an inhaler or blood sugar monitor, demonstrations of physical therapy exercises or workout regimens, and educational videos on medical procedures and tests.

Patients with chronic diseases can benefit from ongoing support resources beyond those of their healthcare home, and it is important for clinicians to use Internet tools to augment counseling activities. For example, patients with obesity can benefit from meal planners and low calorie recipe websites, links to support group forums and blogs, and website searches to tap into different community fitness resources online. Additionally, the health app technology sector is rapidly expanding and these tools can also serve as a valuable means for engaging and motivating patients in their care plans. Clinicians can use these additional technology-based resources to provide further guidance and ongoing support to patients who may find utility in using apps on their mobile devices.

A word of caution, however, as not all apps are created equally. Likewise not everything on the Internet is reliable, so patients will increasingly look to clinicians for help in determining which sources provide credible information and guidance. Clinicians should challenge themselves to become more aware of the variety and quality of Internet resources in order to properly advise patients on their use. In so doing, today's physicians build and maintain their toolkit of technology resources not just for their own medical learning but for patient education as well. This step is key to recognize that the computer in the room is not just an electronic "notes to self," as paper records tended to be, but rather a portal to a wide variety of patient teaching tools that can augment discussion, enhance self-care, and promote improved understanding of overall health and well-being.

ACCENTUATE THE POSITIVE, ELIMINATE THE NEGATIVE

The EMR is an imperfect but useful tool, and there will be times (i.e., software errors, slow login times, and interrupted Internet connectivity) when its use makes for a frustrating experience for the clinician and patient. Although it is tempting to focus on the negative, forgetting the positive benefits of the EMR is ultimately unproductive and can have negative unintended consequences on patient care and perceptions. Research shows that a clinician's perceptions of the EMR affect their patients' perceptions not only of the EMR, but of their overall satisfaction and opinion of the quality of care they receive [25,68,69]. Patients can sense clinician aggravation while trying to navigate specific EMR functions or documentation tasks. Those who sense provider frustration are more likely to feel lost during the clinical interaction as well as to view the EMR as a barrier to communication [65]. A clinician's positive EMR attitude instills confidence in the patient and promotes a positive tone to the encounter and the medical care in general [70].

It is important, particularly during periods of stress and dissatisfaction, to employ positive rather than potentially harmful communication behaviors with patients. Even if internal annoyance with the EMR is present, clinicians should attempt to avoid outwardly expressing and thus transferring their aggravation and negativity to patients. As end users of the technology, clinicians should proactively inform their information technology staff and administrative support of the issues they are encountering so that constructive troubleshooting of potential solutions can begin. While this requires an extra step on part of the busy clinician, it is critical in helping advocate for future positive change in how EMRs function from the perspective of the end user.

MAXIMIZE USE OF VERBAL AND NONVERBAL COMMUNICATION BEHAVIORS

Face-to-face communication is thought to consist of three separate but very important elements; words and their content, tone of voice, and body language [71,72]. Although all three of these elements can be used at the same time to convey an overall message, the most powerful components of communication are the tone of voice and body language [71,72]. As such, it is important to actively engage in positive nonverbal communication behaviors throughout the course of the clinical encounter in order to facilitate a more connected and patient-centered interaction.

Researchers have found that while employing appropriate amounts of meaningful social touch (i.e., a hand shake or pat on the back) is positively related to a patient's assessment of a clinician's communication skills, the amount of eye contact a clinician displays is the most important determinant of their perceptions of clinician connectedness and empathy [73]. Clinician empathy is a key component of building a trusting patient–provider relationship. It encourages information sharing, improves patient satisfaction and adherence, and has been found to have an intrinsically therapeutic effect on patients. As a result, it is important that clinicians remain especially mindful of maximizing eye contact with patients [15,22,24,25,74,75]. In order to sustain significant eye contact during a clinical interaction, clinicians in practice today must learn to type effectively [5,24,48,76,77]. Being able to type quickly, softly, and touch type without needing to look at the keys helps make computer use less distracting and in turn helps promote a more conversational interaction [15,29,32].

While eye contact is likely the most critical single nonverbal behavior in the patient-clinician interaction, it is also important to be aware of, and purposefully employ, other helpful nonverbal skills as well. Meaningful actions such as affirmative head nodding can augment the quality of the conversation in a simple but significant way. Likewise maintaining an overall open body position as much as possible (i.e., facing the patient with head, upper, and lower body oriented toward the patient) promotes unspoken and continued engagement and minimizes negative perceptions related to computer use [8,66,78–80]. Even while using the computer, it is still possible to maintain elements of positive open-body language which helps support the message of the clinician and also the manner in which it is delivered.

Verbal communication behaviors can also help convey active listening and reinforce engagement. Continuers ("uh-huh," "go on," and "I see"), echoing statements, short requests ("tell me more") and brief summarizing statements are all useful and easy strategies to demonstrate involvement with, and understanding of, the patient experience [15,66]. A caution here. There may be situations where it is not appropriate, based on a patient or clinician's culture or beliefs, to have physical contact outside of actual direct medical care, or when sustained eye contact may be interpreted negatively [81,82]. As such, one must always seek to provide patient care that is respectful and responsive of individual patient preferences, needs, and values while employing appropriate communication behaviors that facilitate a sense of emotional investment and connectivity [83].

RECOGNIZE WHEN TO DISENGAGE FROM TECHNOLOGY

Inevitably during the course of clinical interactions, sensitive or particularly important topics will arise. Patient cues in this realm may be subtle (i.e., a slight gesture or facial expression, a small change in speech inflection or intonation); as a result, vigilance is required on part of the clinician to be receptive and attuned and to respond appropriately. Likewise, it may be the clinician who wants to discuss sensitive information with the patient, and this requires a noticeable change in tone to "signpost" the shift in the topic.

During these instances, it is necessary for the clinician to focus entirely on the patient and their needs and ensure that the patient is reciprocally engaged [5,24]. This is best accomplished when the clinician discontinues use of the computer [5,22,66,84]. All typing should cease, and nonverbal communication behaviors should support continued focus on the patient without interference [5,22]. Clinicians should avoid even having their hands on the keyboard or mouse, because even though it is not in use, it can convey a message akin to that of the clinician who talks to their patient whilst standing with their hand on the door handle. The message is, "My attention is elsewhere and I'm not really listening" [85].

Clinicians should completely disengage from the keyboard, move the screen aside if possible, and physically orient their head, upper torso, hips, and legs such that their gaze and entire body is directed toward the patient in an open, receptive orientation [5,22]. It is important that one's physical positioning mirrors the emotional intent and verbal behaviors that occur in order to impart a sense of special attention to the topic.

MEANINGFUL ENCOUNTER CLOSURE

It is important to be as thoughtful in closing the clinical encounter as it is in opening it, since the end often creates a lasting and meaningful impression on the patient [86]. An open-ended question (i.e., "What questions or concerns do you have about what we discussed today?") is important to provide a final but important opportunity to share additional issues or ask for further clarification. Conclusion of the encounter should include a recap of the plan and verification of understanding, as well as providing any last patient-related educational materials such as handouts, community support resources or after-visit summaries [15,21,66]. It is also important for the clinician to ensure that key history is documented (i.e., significant physical findings, history of present illness) and crucial tasks (i.e., orders, prescriptions, and referrals) are completed prior to moving on to the next patient [15].

In closing the in-person encounter, clinicians can also extend an invitation to patients to continue future dialogue virtually. Promoting use of online patient portals for example is an excellent means of using a "cold" technology to enable "warm" relationships, allowing for ongoing updates regarding treatment or clarification of care-related questions [5,82,87]. While virtual portals and electronic communication are not a substitute for in-person communication, they are an important and useful tool to augment patient care overall and provide extended access to the entire care team beyond their limited face-to-face availability [87,88]. By encouraging interaction with these platforms, patients can become invested in self-management of their care and wellness beyond the confines of the clinic or hospital [5,89].

Lastly, clinicians should log off of the EMR while explaining their actions and assuring the patient that their information is protected [25]. This is particularly important if an individual has fears about information safety, confidentiality, and others' access to their medical records [25,66]. Once the EMR is secured and the interaction has reached its natural close, the physician should end the encounter with a simple but meaningful touch, as appropriate, such as a hand shake or gentle pat on the back. This serves as a warm reaffirmation of the human exchange that has just occurred, and also of the clinician's connectedness to the patient and the human dimension of the clinician–patient relationship [26]. Having the last physical contact be between clinician and patient, not between clinician and computer, helps convey the patient-centered focus at the end of the interaction.

CHALLENGES TO PRACTICAL INTEGRATION

There are a number of barriers to successful implementation of the best practice strategies enumerated above. The first is the traditional mindset of what patient-centered care is, and recognition that technology use can in fact be patient centered. Modern medicine is practiced in a world of increasing technological innovation, and these advancements stand to further care and promote the manner in which it is delivered. Today's clinician must be encouraged to view the EMR and computer resources as tools by which to encourage understanding and empower patients to improve their health status and care [5]. This change in thinking is likely the most challenging mental block for providers to overcome, particularly for those that are wary of computers. It is however the most important barrier to overcome as it influences both the clinician's daily work-life perspective and also the experience of patients and their perceptions of care.

Practically, the time required to navigate the EMR and use it with patients during clinical interactions can be a major challenge during early phase-in of EMR adoption. It is important for the encounter to flow naturally, with time allowances for pauses, questions, and clarification. At the same time, productivity demands make it unfeasible to allow for extra time for each clinical interaction simply because an EMR is present. It is necessary to maximize use of limited face-to-face time to help promote patient-centered communication while integrating the EMR. The provider can document key points in real time during the interaction and capitalize on the time prior to and after the encounter to review the chart and allow for completion of documentation after its conclusion. The more facile a provider is with the technology, the more seamless their EMR use will become.

Clinician EMR skills and overall computer use can be a significant barrier to providing effective patient-centered care. Clinicians must recognize when their skills are lacking and have appropriate organizational resources with which they can augment their technical proficiency. Few organizations provide communication and training in patient-centered care strategies, much less afford opportunities for self-observation or feedback on one's actual skills [15]. Doing so is an important part of the self-reflective process in medicine, and is a foundational tool in establishing lifelong learning habits that will improve one's overall professional competence [90].

Lastly, many EMRs are user unfriendly, created by the need to capture and run data inquires to fulfill data aggregation and analysis (e.g., Meaningful Use in the United States) and billing criteria rather than promote seamless workflow integration for the provider and patient-centered communication. As such, clinicians must seek avenues within their organizations and at a policy and product development level to advocate for platforms that facilitate ease of use and features specifically designed for patient education and engagement.

CURRICULAR INTERVENTIONS

Although patient-centered EMR use training for providers of any level is a rarity, there are a variety of promising curricular innovations to teach best practices. These curricular models serve as examples, which organizations can use to provide similar training at various levels of trainees and skills.

MEDICAL STUDENT CURRICULA

Recommendations have highlighted the need to introduce EMRs into curricula early on, with the Alliance for Clinical Education (ACE) suggesting that this should start during the first year of Undergraduate Medical Education (UME) [11]. Targeting learners early in their training allows for gradual introduction of the EMR beginning with instruction on its functionality and implications for health care delivery, institutional documentation expectations, and subsequent hands-on practice [91]. Early exposure can help foster positive EMR perceptions and behaviors prior to any potential negative role modeling they may encounter during their clerkships [15]. An additional benefit of progressive undergraduate training is that it provides students multiple opportunities to become comfortable with the EMR and its direct incorporation into clinical care, thereby cultivating their EMR literacy and competency [11]. The practical implementation of these recommendations however is quite variable, especially with respect to the access (or lack thereof) that medical students are given to the EMR [92].

The Alpert Medical School of Brown University took an interesting approach to UME, which is modeled on relevant conceptual frameworks of reflective practice and narrative medicine [93]. Their curriculum formally introduces medical students to the computer and its role in the physician-patient relationship as part of a third year 'Doctoring' course, with further skill development in a second advanced EMR training module later in the third year. Curriculum components include a lecture on effective communication with the EMR followed by a standardized patient (SP) encounter with direct observation by faculty. Students then participate in a reflection-on-action exercise in which feedback is given by a co-teaching faculty team consisting of a physician and behavioral science specialist who directly observe students' EMR use and communication skills. In addition, there is a narrative medicine component using selected reflective readings to optimize effective integration of concepts. Lastly, students are asked to identify learning needs for preserving patient-centered care behaviors while using the EMR. Further learning opportunities for their third and fourth year are then mapped to behavior-focused grids, highlighting opportunities in which students can gain added practice, insight and feedback on their EMR use skills.

The University of Arizona College of Medicine—Phoenix introduced an EMR curriculum even earlier in the UME process, and provided basic training on navigation of the EMR to all second year students [94]. A subset of students were selected to receive additional training which included practice with an SP and/or specific EMR ergonomic training during their Doctoring clinical skills course sessions. The investigators found a significant positive effect of EMR ergonomics training on students' relationship-centered EMR use, with trained students reporting improvements in a variety of domains including ability to use the EMR to effectively engage patients and integrate the EMR into patient encounters. Student self-assessments were strongly corroborated by SP and faculty assessments, and illustrated how simple instruction on EMR ergonomics can result in improved utilization of the EMR in a relationship-enhancing fashion. It is important to note, a minimum of three ergonomic training sessions were needed to see overall improvement in EMR use, underscoring the importance of repeated exposure to training content over time to ensure successful outcomes.

A similar approach is ongoing at the University of Chicago for both second and third year medical students, and involves a mixture of interactive didactic lectures on barriers and best practices combined with simulation exercises [34,95,96]. The University of Chicago experience is discussed in further detail below, however, it and each of the examples already discussed highlight key components of an effective model for undergraduate curricula; EMR access, early and repeated introduction of core concepts, a safe environment in which EMR skills can be applied, practice with an SP, and opportunities for feedback and self-reflection. These models also illustrate how patient-centered EMR use training can be seen as a natural extension of a student's written and oral communication skills, and how it can logically and practically be incorporated into existing clinical skills curricula [11].

RESIDENT CURRICULA

There are considerable difficulties in implementing widespread resident curricula in patient-centered EMR use, most notably access and space to teach a large number of residents at one time, ability to require training, methods by which to evaluate performance, and availability of trained clinician teachers. As a result, there are few ongoing curricular examples of resident education in patient-centered EMR use. Despite these barriers, attempts at implementing various approaches are promising and can be used to help guide future efforts.

Reis et al describe a study in which 36 volunteer Family Medicine residents from a large Israeli health maintenance organization were divided into two study groups [84]. The control group was introduced to various tips for enhancing the computerized clinical encounter via three separate passive didactic lectures lasting 75–90 minute each. Topics covered included: doctor-patient communication; doctor-patient-computer communication (theoretical background and practical tools); and EMR use for advanced users. The intervention group was introduced to the same tips; however, they participated in hands-on practice. Simulation consisted of three SP encounters lasting 12 minutes each, followed by SP feedback. A physician observer viewed the encounters behind a one-way mirror, and gave video-based feedback for all three initial encounters. Following observer input, another three encounters were conducted in the same format with video-based feedback.

Results from this work are encouraging as they demonstrated improvement in performance, attitudes, and competence levels of residents in both the intervention and control groups. Although there was no significant difference between simulation or lecture groups on physician observer scores, residents in the experimental group rated the contribution of simulation to their learning higher than the control group and also resulted in higher satisfaction scores. Interestingly, most participants in the lecture group noted that they were interested in further training and emphasized it should be simulation-based, highlighting the role of simulation as an effective and well-received means for educating resident trainees.

A follow-up to Reis's initial work included development and pilot testing of a low-cost computer-based simulation for family medicine residents, EMR-sim, in order to assess the usability and impact on self-reported competencies and attitudes [97]. Simulation scenarios covered a range of topics including privacy and documentation, safety concerns, communicating with a patient who is distracted by the computer, and use of the EMR for patient education. Following each scenario, residents were given feedback by the program relative to their choices. Sixteen Family Medicine residents piloted the simulation and demonstrated improved competency and attitude scores. Mean scores for perceived usefulness and ease of use of the simulation were good; however, residents cited issues with usability aspects of the program preferring a more interactive representation of the EMR and increased options for shared decision making.

Although there are limitations to these initial projects, they are promising steps toward introducing the concept to residents and also studying the various means by which training can be offered in a graduate medical education environment. Lecture, interaction with an SP, and computer simulation all hold promise as effective and acceptable tools for teaching resident trainees how to better use EMRs in clinical encounters. Future resident training should aim to incorporate patient feedback in addition to direct observation by supervising clinicians since patient experiences will better prepare residents to practice post residency when expectations to see more patients and adhere to quality measures will be more pressing.

PRACTICING PHYSICIAN CURRICULA

The barriers encountered in implementing curricula for practicing physicians are perhaps the most difficult to overcome. It is a challenge to require training since it is not currently mandated or integrated into conditions for licensure, Board Certification or

Continuing Medical Education (CME) requirements [16]. Until there is nationalized support for these requirements, formal education is highly dependent on the availability of organizational support to provide the training itself and also the protected time during which the training can be delivered. With few exceptions, education in this area is most often left to the discretion of the individual practicing physician. Some large organizations such as Cleveland Clinic and Kaiser Permanente have made communication skills mandatory for all physicians, though requirements such as these are few and far between at this point in time [98]. While there are promising CME activities available nationally and internationally on patient-centered EMR use, practicing physicians undergo an inherent self-selection by nature of their chosen participation and as a result those that may need the training the most may not be aware of their limitations, the availability of coaching and feedback, or have the time or resources to actively seek it out [99–102].

One such voluntary CME activity worth highlighting, however, involved a joint partnership between the Cleveland Clinic (CC) and University of Chicago (UC) in which two types of training were explored [103]. A 4-hour training for CC primary care faculty and a shorter condensed 90-minute training for UC general internal medicine faculty was developed. Both sessions included a lecture highlighting barriers and best practices for patient-centered EMR use as well as a Group Objective Structured Clinical Examination (Group-OSCE; GOSCE) to practice skills with an SP and mock patient chart in an EMR training environment. Thirty-two academic primary care faculty in total took part, and the training was highly rated, with 100% of faculty agreeing training was important and relevant to their practice, enabling them to better teach and role model patient-centered care for trainees. There were no differences in mean ratings between the shorter or longer training groups. Interestingly, faculty participating in longer training reported higher GOSCE efficacy, however the shorter workshop appeared more informative and effective with higher reported rates of new knowledge gained [103]. Such training, although voluntary, is promising particularly given busy primary care clinician schedules and demands, and illustrates that a relatively short intervention may be a feasible and effective way to spread best practices to faculty. Future education, however, should require training in all departments, so that all practicing physicians are grounded in best practices.

A promising approach to provide required training was undertaken by the Johns Hopkins Miller-Coulson Academy of Clinical Excellence with the implementation of a single EMR system across all ambulatory and inpatient care settings. This development presented a unique educational opportunity to improve communication between clinicians and patients [104]. To optimize patient-centered use of the EMR, the Academy created a 4.5-minute video that all clinicians (residents, fellows, and practicing physicians) were required to view as part of their EMR training. The video and accompanying pocket cards reminded users of the do's and don'ts of using the EMR while interacting with patients with examples drawn from Kahn's paper on etiquette-based medicine [105]. These Miller-Coulson Academy "ABC" guidelines highlight many of the best practice strategies outlined above, including reminders to involve the patient and maximize eye contact as much as possible while navigating the demands of the EMR [104]. Training was well received and while brief and not ongoing in nature, it is a promising model to build upon and offers an easily reproducible format in which to provide much needed education during a critical transitional time. It is recommended that organizations find ways to similarly take advantage of existing resources and novel methods to require and support practicing physicians' training in order to ensure personal learning of best practices and for use in their teaching and role modeling responsibilities.

EVALUATION METHODS

In the United States, the Liaison Committee on Medical Education (LCME) mandates that a "medical education program must include ongoing assessment activities that ensure that medical students have acquired and can demonstrate on direct observation the core clinical skills, behaviors, and attitudes that have been specified in the program's educational objectives." [10] Similarly, the Accreditation Council for Graduate Medical Education (ACGME) Milestones for assessment of resident and fellowship clinical competency includes proficiencies such as being able to "incorporate patient-specific preferences into plans of care" and "role model effective communication and development of therapeutic relationships in both routine and challenging situations," while being able to "utilize information technology with sophistication." [12] While other countries have similar guidelines, proper assessment of one's communicative behaviors and ability requires the availability of reliable and validated evaluation tools that are practical to use.

There has been considerable development of communication skills assessment tools and methods in the literature to evaluate a trainee or clinician's performance; however, most of these are prior to large scale introduction of EMRs into healthcare and therefore do not take into account the effects of EMR use while interacting with a patient [106]. Despite evidence demonstrating both positive and negative effects of healthcare computerization on a clinician's ability to communicate with patients, there is a paucity of formally validated evaluation tools that incorporate items that are relevant for communication tasks during EMR use.

One such tool was developed by the chapter authors, Alkureishi and Lee, and is known as the electronic-Clinical Evaluation Exercise (e-CEX) to assess patient-centered EMR use [107,108]. The e-CEX was created to assess EMR specific communication skills and consists of a 10-item, 90-point tool for use in direct or video-taped observation assessment by trained observers (Fig. 10.2). The content of the tool was based on best EMR communication practices identified by the authors' recent published literature review, as well as our collective clinical experience, which includes inpatient and outpatient pediatric and internal medicine in academic, community, private and Federally Qualified Health Center settings.

The tool consists of 10 items central to patient-centered EMR use: (1) prior review and familiarity with the patient's chart, (2) facilitated shared screen viewing, (3) allowed the patient to start with their concerns and did not prematurely engage the EMR, (4) introduced and explained the role of the EMR in the visit, (5) integrated EMR use around the patient's needs, (6) aware of, responded to, and used nonverbal communication cues to convey listening, (7) encouraged patient interaction with the EMR, (8) proficient in EMR use and navigation, (9) able to document key points in real-time, and (10) overall ability to utilize EMR to promote individualized and collaborative care in a respectful, humanistic environment. Each item is scored on 1–9 point scale with behavioral anchors

e-CEX: Patient-Centered Use of EMR rating tool

Provider's First & Last Name:______________________ **Date:**______________

Observer:______________ **Problem/Dx:** ______________________

Area of Focus: ☐ Encounter Preparation ☐ Communication Skills ☐ Technology Skills

1. **Preparation: Acquaint yourself with chart before visit (*A – HUMAN*[94])** Prior to entering room, review previous visits, pertinent imaging / labs to prepare and anticipate needs. (☐ N/A*)

1 2 3	4 5 6	7 8 9
Unsatisfactory	**Satisfactory**	**Superior**
No review or minimal review	*Acquainted with some elements of chart*	*Reviewed all pertinent parts chart and prior visits*

2. **Preparation: Set the stage, Use the Triangle of Trust (*U – HUMAN*[94]).** Arrange provider, patient, and computer screen in a triangle configuration to allow shared viewing and collaboration. (☐ N/A*)

1 2 3	4 5 6	7 8 9
Unsatisfactory	**Satisfactory**	**Superior**
Screen not visible to pt, provider's back to pt	*Screen partly visible, provider occasionally with back to pt*	*Triangle setup optimal, verifies pt can see screen, faces pt*

3. **Communication: Honor the Golden Minute, Allow patient to start with their concerns (*H – HUMAN*[94])** Greets patient, elicits patient's concerns / goals for the visit, then engages technology. (☐ N/A*)

1 2 3	4 5 6	7 8 9
Unsatisfactory	**Satisfactory**	**Superior**
No greeting or goals asked, Immediate technology use	*Greets patient, elicits some goals, Immediate technology use*	*Greets pt, elicits concerns. Avoids technology at visit start*

4. **Communication: Introduce and explain the technology.** While maintaining conversational flow, explain actions with EMR (***E – LEVEL***[94]), and value / benefit to patient (***V – LEVEL***[25]). (☐ N/A*)

1 2 3	4 5 6	7 8 9
Unsatisfactory	**Satisfactory**	**Superior**
No explanation and long silences with EMR-use	*Some conversation with EMR-use. Doesn't explain EMR actions/benefits*	*Conversational flow. Explains EMR actions / benefits*

5. **Communication: Integrate technology in patient-centered manner.** Integrate technology in patient-centered manner. Use EMR in natural flow of visit and integrate patient needs (i.e. typing). Nix the screen and disengage from EMR when discussing sensitive topics (***N – HUMAN***[94]). (☐ N/A*)

1 2 3	4 5 6	7 8 9
Unsatisfactory	**Satisfactory**	**Superior**
Unable to integrate EMR, disengages inappropriately	*Integrates some EMR use, usually disengages appropriately*	*Uses EMR seamlessly around pt needs Always disengages appropriately*

***N/A – Not Applicable** HUMAN - Alkureishi et al, 2013 [94]; LEVEL - Mann et al, 2004 [25]

6. **Communication: Awareness of non-verbal cues.** Maximize eye contact (***E – LEVEL***[25]), open body language, and other nonverbal actions to convey listening and understanding. Attuned to patient's nonverbal cues and responds appropriately. (☐ N/A*)

1 2 3	4 5 6	7 8 9
Unsatisfactory	**Satisfactory**	**Superior**
Misses pt cues, nearly absent nonverbal actions	*Catches some pt cues, adequate nonverbal actions*	*Attuned to pt cues, excellent nonverbal actions*

7. **Communication: Encourage patient interaction with technology.** Invites and encourages patient to interact with EMR, lets them look on (***L – LEVEL***[25]). Maximizes patient interaction (***M – HUMAN***[94]); shows results, imaging, graphs in explaining & discussing care / treatment plan. (☐ N/A*)

1 2 3	4 5 6	7 8 9
Unsatisfactory	**Satisfactory**	**Superior**
No invitation, patient not engaged	*Invites patient to engage, patient somewhat engaged*	*Patient actively engaged, EMR used as collaborative tool*

8. **Technology Skills: Proficient in technology use.** Adept typist, easily navigates EMR screens and tabs to facilitate flow of visit. Logs off at end of visit (***L – LEVEL***[25]). (☐ N/A*)

1 2 3	4 5 6	7 8 9
Unsatisfactory	**Satisfactory**	**Superior**
Clumsy use	*Somewhat proficient*	*Adept, easy EMR use. Logs off at end of visit*

9. **Technology Skills: Minimizes cognitive overload.** While integrating EMR into clinic visit, effectively uses time to document note in EMR. (☐ N/A*)

1 2 3	4 5 6	7 8 9
Unsatisfactory	**Satisfactory**	**Superior**
Inefficient during visit, most of document done post-visit	*Relatively efficient during visit, some documentation post-visit*	*Efficient during visit, minimal documentation post-visit*

10. **Overall Professionalism, Humanism & Patient-Centered Care:** Utilizes EMR to promote individualized and collaborative care in respectful, humanistic environment. (☐ N/A*)

1 2 3	4 5 6	7 8 9
Unsatisfactory	**Satisfactory**	**Superior**
Ineffective or cold collaboration, scant communication	*More effective utilization, some collaboration and communication*	*Seamless and effective collaboration, respectful and communication comprehensive communication*

Evaluation Time: **Observing:** ________min **Providing Feedback:** ________min

Observer satisfaction with evaluation - Strongly Disagree 1 2 3 4 5 Strongly Agree

Provider satisfaction with evaluation - Strongly Disagree 1 2 3 4 5 Strongly Agree

Comments: ______________________________

FIGURE 10.2 10-item, 90-point electronic-Clinical Evaluation Exercise Tool (e-CEX) [107,108]. Each item is scored on a 1–9 point scale and is based on best practices identified by literature review and the authors' collective clinical experience. To be used during observation of patient–provider-computer interaction for performance feedback and/or evaluation. *Copyright: the authors.*

(1–3 = unsatisfactory, 4–6 = satisfactory; and 7–9 = superior), and is modeled on the widely used Mini-Clinical Evaluation Exercise for direct observation of clinical history and exam skills in internal medicine clerkships and residencies and similar to the ACGME tool to evaluate of core competencies [109,110].

A trained SP and volunteer clinical faculty reviewed the tool for readability and usability. Testing of the tool was conducted with second year medical students at the University of Chicago's Pritzker School of Medicine who had participated in a patient-centered EMR use curriculum consisting of a 1-hour lecture and Objective Structured Clinical Examination (OSCE). Concurrent third year students who received no formal education on patient-centered EMR use participated in an identical OSCE and served as the control group. Trained clinician observers evaluated videotaped OSCE performance using the e-CEX, and the tool was found to be both reliable and valid. [107,108]

Additionally, faculty who used the tool reported that it was easy to use and were satisfied with amount of time required to complete the evaluation. For evaluators who have limited time to compete a direct observation, a short 3-item version of the e-CEX was developed and found to have similar explanatory power as the longer tool, which holds promise for its use in busy academic settings. [107,108] The e-CEX has subsequently been used with residents and attending physicians and has been instrumental in helping to provide efficient, structured feedback using both standardized and real patients.

A similar tool, e-SEGUE, was developed by Assis-Hassid et al. and is modeled after SEGUE (Set the stage, Elicits information, Give information, Understand the patient's perceptive, and End the encounter), one of the most widely used assessment tools for physician's communication skills teaching and assessment in North America [66,111]. Their comprehensive literature review identified 27 computer-related communication skills that facilitated evaluation of a physician's ability to provide patient-centered care while using the computer (PDCC) during a medical encounter. These behaviors were used to develop the e-SEGUE framework, and highlight specific behaviors such as introducing the computer and its role to the patient, using verbal and nonverbal behaviors to demonstrate active listening, and teaching the patient about their situation by showing test results on the screen. A two-phased content validity analysis of the PDCC behaviors addressed in e-SEGUE was conducted using a general panel of participants familiar with EMRs and patient-doctor communication (i.e., medical students, residents, primary care, and physicians) as well as an 8-person panel of leaders in national healthcare decision making. Panels rated the 27 behaviors and 23 of those showed high levels of agreement.

While assessments in this area are in its early stages, these tools are a promising start to being able to objectively observe, assess, and provide structured feedback on a user's patient-centered EMR use skills. As evaluation efforts in this area continue to develop, it will be critical for patient centered computer skills to become routinely integrated into required simulation training and evaluation, for example, in the United States by including them in the Clinical Skills component of the United States Medical Licensing Examination (USMLE) for medical students. By making patient-centered technology use a standard part of simulation and evaluation, it is likely more medical schools will move toward including and addressing this important competency in their curricula.

A LONGITUDINAL EXAMPLE: THE UNIVERSITY OF CHICAGO

While it is important to ensure early exposure to patient-centered EMR use, equally essential is the need to provide continued training and reinforcement throughout the remainder of one's medical career. Medical students, residents, and fellows may witness negative EMR communication behaviors of peers and supervisors, so it is important to emphasize best practices throughout the course of undergraduate and graduate medical education. Likewise, attending physicians need training in how to employ and role-model best practices given their busy clinical demands, and be given the tools to teach and give feedback on these skills to trainees. Providing multiple opportunities for continued training, regardless of trainee level, promotes the maintenance of positive and sustainable communication behaviors, reinforces the importance of reflection in medicine, and helps nurture an atmosphere of life-long learning and improvement.

The University of Chicago has implemented a longitudinal approach that incorporates various types of learners. To begin with, students are trained in basic EMR functionality and are given full reading access to the EMR beginning in their first year of medical school. In the middle of their second year, students receive a 1-hour interactive lecture embedded into the Advanced Topics in Communication component of their Clinical Skills Course. The lecture is preceded by an individual reflective exercise in which they are asked to recall a time when they or a loved one was in a clinical setting and an EMR was used in their care. Students are asked to reflect on how that altered or affected the patient–doctor relationship and communication. Following an introduction to Kolb's Experiential Learning Cycle, the exercise helps to jump start students' own personal experience relating to the EMR and grounds their learning in the perspective of a patient or family member rather than that of a healthcare provider [112,113]. A didactic lecture follows which provides an abstract conceptualization of the EMR's role in clinical care, its potential benefits and challenges, and a step-by-step skills-based approach for best practices to patient-centered EMR use as framed by the mnemonic HUMAN LEVEL (Fig. 10.3) [34].

Within two weeks of the lecture, students actively experiment by participating in a Group OSCE (GOSCE, with 3–4 participants per group) during which one student uses the EMR in a patient-centered manner to educate and engage a standardized patient (SP) to discuss lab results and lifestyle modification. The students use a mock patient chart in the training version of the EMR, which includes prior clinic visit notes, labs, and studies. The GOSCE takes place in a simulated clinic environment with unobtrusive video recording capability to allow for later viewing without compromising the high fidelity nature of the simulation. Students receive feedback immediately after their performance by the SP who uses a standardized checklist of EMR-behaviors developed from a systematic literature review, a faculty facilitator using the e-CEX tool as described above, and the remainder of the students in the group who observed the encounter [34]. After the GOSCE, students have the opportunity to view their videotaped performance and reflect on their encounter individually as well.

Student reception of the curriculum is extremely high. In a follow-up assessment survey, the majority of respondents agreed that the topic was important to their training and should be required for all medical students [95,96]. Interestingly, during the first year of curriculum

HUMAN LEVEL - 10 Tips to Enhance Patient-Centered EMR Use [25,94]		
H	Honor the "Golden Minute"	Make the start of the visit completely **technology free**. Greet the patient, start with **their** concerns and establish an **agenda** for the visit *before* engaging technology.
U	Use the "Triangle of Trust"	Create a **triangle configuration** that puts you, the patient and the computer screen at each of the three corners. This allows you to look at both the patient and screen without shifting your body position, and also enables **shared** screen viewing.
M	Maximize patient interaction	Encourage patient **interaction**. Pause for questions and clarification. Allow time for questions and to verify understanding.
A	Acquaint yourself with chart	Review the chart **before** you enter the room to prepare, inform and **contextualize** your visit.
N	Nix the screen	When discussing **sensitive** information, **completely disengage** from the EMR (look at the patient, turn away from screen, take hands off keyboard, etc.)
L	Let the patient look on	**Share** things on the screen with your patients.
E	Eye contact	Maintain **eye contact** with patients as much as possible. Treat patient encounters as you would a conversation with friends or family members.
V	Value the computer	Praise the **benefits** of the EMR and take advantage of opportunities to use technology as a tool to **engage** patients (pull up lab result to review together, utilize graphics, etc.).
E	Explain what you're doing	Be **transparent** about everything you do. Avoid long silences, aim for conversational EMR use by explaining what you are doing as you are doing it.
L	Log off	At the end of the visit, **log off** of the patient's chart while they are still in the exam room. This reassures the patient that their medical information is **secure**.

FIGURE 10.3 HUMAN LEVEL mnemonic summarizing best practices for patient-centered EMR use. The HUMAN portion of the mnemonic is the author's original work [34]. The LEVEL portion of the mnemonic is from the Kaiser Permanente Group, published by W.R. Mann and J. Slaboch, Computers in the exam room—Friend or foe? Perm J. 2004 Fall;8(4):49–51 [25] and is used with permission.

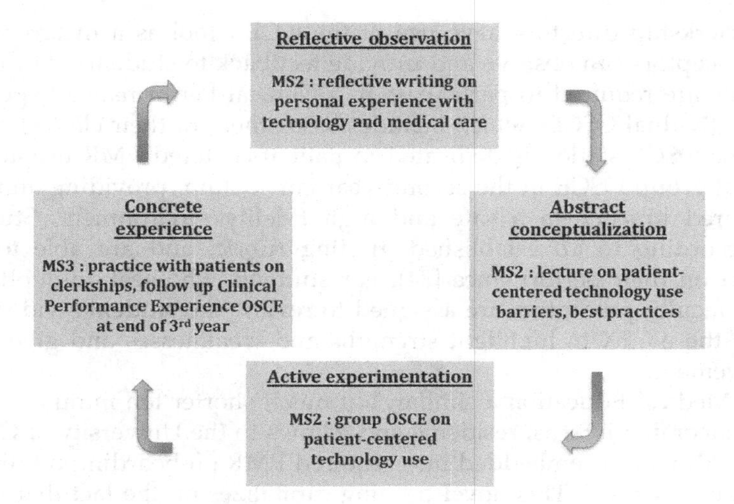

FIGURE 10.4 University of Chicago second (Ms2) and third-year (Ms3) medical student curricular components modeled after Kolb's Experiential Learning Cycle [112,113]. Abbreviations: Ms2, second year medical student; Ms3, third year medical student; OSCE, Observed Structured Clinical Examination.

implementation in 2013, second-year students who received the one-hour lecture performed significantly better on SP evaluations of their OSCE performance than concurrent third year peers who had not received the lecture but did take part in the same OSCE [95,96]. This result was unexpected given the lack of hands-on patient interactions for mid second-year students compared to their end of third-year colleagues. The finding highlights the importance and impact of providing a structured curriculum on best practices early on in medical training. Of note, when the same second-year students who participated in the GOSCE were asked to complete a nearly identical OSCE on patient-centered EMR use at the end of their third year, they scored significantly lower than their original performance [114]. This erosion of skills may be related to negative role modeling encountered during their clinical rotations, underscoring the importance of follow-up training to reinforce and allow feedback on their patient-centered EMR use skills during third year clerkships.

In response to the erosion of performance, an additional reinforcement in the curriculum for second-year students was added (Fig. 10.4). Immediately prior to the transition to third year clerkships, all students now participate in a 3-day Clinical Biennium which trains rising third year students in hands-on skills that they will need during their third and fourth years of medical school. Included in this transition is a lecture on "Technology Skills for the Wards," which reviews the practical and etiquette-based aspects of using computers with patients, and provides a review of documentation expectations and patient-centered EMR use best practices, utilizing the HUMAN LEVEL mnemonic [34].

Moving into third year, students are encouraged to use the EMR with patients throughout their clinical clerkships, thereby giving them the opportunity to refine their ability through concrete experiences and further practice. Additionally, third-year students are granted full EMR access, which includes the ability to place, and pend orders to residents or attendings for cosignature as well as write medical student admission and progress notes. Faculty clerkship directors have use of the e-CEX tool as a means by which they and clinician preceptors can observe and provide feedback to students. At the end of third year, all students are required to participate in a Clinical Performance Experience, a day-long series of individual OSCEs which includes assessment of their clinical and communication skills. One OSCE station is dedicated to patient-centered EMR use and is modeled after the original group OSCE in the second-year curriculum, providing another opportunity for structured practice in a safe and high fidelity environment. Students receive SP feedback according to an established grading rubric, and are able to individually view and reflect on their performance [34]. For students who seek in-depth faculty-level review, trained faculty preceptors are assigned to review the student's videotaped performance and use the e-CEX to highlight strengths and weaknesses and give specific guidance for improvement.

In Graduate Medical Education a similar, but much shorter ten minute curriculum was adapted for all incoming interns, residents and fellows to the University of Chicago, which was adopted in 2015 and is embedded into required EMR on-boarding training at the start of each academic year [115]. This novel training capitalizes on the fact that incoming residents and fellows from all specialties are available during this time, and utilizes the skill set of University of Chicago EMR trainers to deliver the content and highlight the importance of learning both the functionality of the EMR, and the human dimension of its use with patients. The training introduces the best practices using the HUMAN LEVEL mnemonic, identifies barriers to patient-centered EMR use, and reviews institutional documentation requirements including the unsafe cut-and-paste practices and authorship concerns [34]. A longer 1-hour curriculum, expanding on practical strategies, is also given to all Internal Medicine and Pediatrics residents annually, with plans to modify and expand this to other specialties.

Lastly, practicing clinicians have multiple opportunities to receive training in patient-centered EMR use via Faculty Advancing in Medical Education (FAME), a Continuing Medical Education program sponsored by the University of Chicago Academy of Distinguished Medical Educators and the MERITS Program (Medical Education Research, Innovation, Teaching, and Scholarship). Through FAME, faculty participate in a lecture highlighting barriers and best practices and are introduced to the e-CEX both as a means for assessing their own performance and as a tool that can be used to teach and evaluate trainees. Following the lecture, clinicians participate in GOSCE with a standardized patient (SP) in order to try their hand at incorporating best practices and using the e-CEX as an evaluation tool. Each faculty member has a chance to interact one-on-one with the SP during the GOSCE, and observes their peers engaging with the SP. A clinician facilitator guides an immediate feedback and debriefing session after each scenario, and faculty observers use the e-CEX tool to structure feedback for their colleagues. Faculty participants reported that a highlight of the GOSCE experience is the opportunity to observe and learn from peers and share practical strategies with one another. Dedicated patient-centered training has also been provided to the Section of General Internal Medicine, and utilizes a short lecture and 1-hour GOSCE format [103].

Reception of the resident and faculty components of the longitudinal curriculum has been extremely high, with most reporting training is effective and increases knowledge of barriers and best practices. Participants have significantly increased comfort levels in employing the strategies for best practice post training and state they will likely change their future practice as a result [103,115]. All faculty respondents felt the curriculum enabled them to better teach and role model patient-centered care for trainees [103].

The second-year medical student curriculum serves as the foundation for each of these progressive training components, and has been modified, in turn, to suit the needs of each learner level. The student curriculum has been peer-reviewed and published on MedEdPORTAL, a free open-access publication service provided by the American Association of Medical Colleges (AAMC), and provides a stand-alone complete teaching resource that allows for similar adoption and implementation at other institutions [34].

CHALLENGES TO IMPLEMENTATION

There are a number of challenges to patient-centered EMR use curricula implementation, the most basic being the limitation some institutions place on student access to medical records. The expectation that students will be proficient users of the EMR in their clinical and postgraduate years, despite lacking access and training during their preclinical years, is an untenable one. For decades, there has been no physical barrier to students' access to patient records as they simply accessed the physical paper chart [11]. However, the need for permission to access the EMR and concerns regarding billing and documentation requirements have resulted in great variability in medical students' access with some institutions barring students from using the EMR at all, to others granting full but supervised access, and still others allowing a middle ground of partial EMR access such as viewing only without charting or order entry privileges [11,92].

The legality of student access and documentation has been addressed by the Alliance for Clinical Education, with the most often cited reason for restricted access being the guidelines imposed by The Center for Medicare and Medicaid Services (CMS) [11]. CMS governs what specific elements of a student's documentation a teaching attending may refer to; review of systems and/or past family and social history with attending verification and independent re-documentation of the history of present illness, physical examination and medical decision making activities [116].

While institutions and individual teaching physicians must become familiar with and observe the CMS limitations, there is no reason to exclude medical student access to the EMR. Concerns may be alleviated by proper designation in the EMR that a student's note is precisely that, and will not be used for billing purposes. Doing so minimizes risk of documentation or billing errors, and more importantly affords students the educational benefit of documenting a clinical interaction, synthesizing one's clinical reasoning into words and receiving feedback, as well as the practical experience of using a computer with patients. These activities are central to training physicians today and are an essential part of caring for patients [11].

A potential workaround for institutions, if they are unable to provide students with sufficient access to the live EMR, is to use of EMR simulators. Such dedicated simulators, or even providing students access to the training versions for the institution's live EMR, allows students to practice without risk of interfering with the actual EMR environment.

There are a few examples of educational EMR simulators successfully being used with trainees; however, their use requires an investment of effort and resources [117–119]. The ultimate goal of using an EMR simulator should be to prepare trainees for actual practice and allow them to become familiar and have supervised access to the actual EMR they are likely to use as practicing physicians.

An interesting challenge of implementing technology-related training is the fundamental shift that has occurred in the background subject knowledge on part of the teacher and trainee. For the most part, trainees have been raised with the tools of the digital age: smart phones, computers, and the Internet. While their teachers have adopted most or all of these new technologies, unlike their trainees they have not spent their entire lives using these tools and thus their familiarity and approach to learning related to technology may be different. Given these two potential groups of health care practitioners, the "digital natives" and "digital immigrants," it is important to take baseline technology expertise and preferred methods of learning into account when designing curricula [120]. Training must be given to digital immigrants in order to promote efficient individual use of technology as well as enable adequate teaching and evaluation of trainees. Additionally, the important skill set that digital natives have by virtue of their long exposure to a variety of technology tools is one that needs to be acknowledged and capitalized on. As familiar users, digital natives are in the ideal position to help advocate for and contribute toward quality improvement, product enhancement, and usability development of the EMR and related patient-education resources.

Perhaps the most difficult aspect of incorporating longitudinal training in patient-centered EMR use is the challenge of resident and faculty daily workflows. As mentioned earlier, EMRs in their current state are often user-unfriendly, making seamless integration into the encounter inherently difficult. Compounded by time constraints and productivity demands, it is not surprising that stress can result in adoption of negative behaviors. It is important that medical providers at all levels recognize patient-centered EMR use as an extension of their overall professionalism and aim to employ and role-model positive communication behaviors despite the challenges of time and throughput. Equally important is the amount of support organizational leadership provides to their clinicians, trainees, and staff. From offering protected time for ongoing longitudinal training in best practices to eliciting and responding to user input to improve quality and delivery of care, organizational commitment to creating a positive culture of healthcare and overall satisfaction for patients and clinicians alike is a critical determinant of successful use of the EMR.

CONCLUSION

While there are inherent challenges in using technology with patients, the benefits of the EMR and computer use at the point of care are great. Strategies for patient-centered use and best practices exist and are teachable. Organizational support and implementation of a longitudinal curriculum to promote formal training and structured hands-on learning opportunities with the EMR will be essential to develop maintain, and role model positive computer-based communication behaviors. In so doing, the computer can become an extension of the clinician, and one of the many tools used to promote efficient, comprehensive, and compassionate personalized care.

References

[1] Blumenthal D. Launching HITECH. N Engl J Med 2010;362 382–38.

[2] National Physician Survey. Available from: <https://www.nationalphysiciansurvey.ca/> [accessed 25.3.16].

[3] Schoen C, Osborn R, Squires D, et al. A survey of primary care doctors in ten countries shows progress in use of health information technology, less in other areas. Health Aff 2012;31(12):2805–2816.

[4] World Health Organization (WHO), WHO Global Strategy on Integrated People-Centered Health Services 2016-2026. Available from: <http://www.who.int/servicedeliverysafety/areas/people-centred-care/en/> [accessed 25.3.16].

[5] Shachak A, Reis S. The impact of electronic medical records on patient–doctor communication during consultation: a narrative literature review. J Eval Clin Pract 2009;15(4):641–9.

[6] Barrett A, Stephens K. Making electronic health records (EHRs) work: informal talk and workarounds in healthcare organizations. Health Commun 2016;32(8):1004–13.

[7] Alkureishi MA, Lee WW, Lyons M, Press VG, Imam S, Nkansah-Amankra A, et al. Impact of electronic medical record use on the patient-doctor relationship and communication: a systematic review. J Gen Intern Med 2016;31(5):548–60.

[8] Crampton NH, Reis S, Shachak A. Computers in the clinical encounter: a scoping review and thematic analysis. J Am Med Inform Assoc 2016;23(3):654–65.

[9] Ventres WB, Frankel RM. Patient-centered care and electronic health records: it's still about the relationship. Fam Med 2010;42(5):364–6.

[10] Liaison Committee on Medical Education (LCME). Functions and structure of a medical school: standards for accreditation of medical education programs leading to the M.D. degree. June 2015. Available from: <http://lcme.org/wp-content/uploads/filebase/standards/2016-17_Functions-and-Structure_2016-07-22.docx> [accessed 29.4.16].

[11] Hammoud MM, Dalymple JL, Christner JG, Stewart RA, Fisher J, Margo K, et al. Medical student documentation in electronic health records: a collaborative statement from the Alliance for Clinical Education (ACE). Teach Learn Med 2012;24(3):257–266.

[12] Accreditation Council for Graduate Medical Education (ACGME) Internal Medicine Milestone Project. July 2015. Available from: <http://www.acgme.org/portals/0/pdfs/milestones/internalmedicinemilestones.pdf> [accessed 29.4.16].

[13] American Medical Association. 2012 Report of the Board of Trustees. Exam Room Computing and Patient-Physician Interaction. 2012 Available from: <http://www.ama-assn.org/assets/meeting/2013a/a13-bot-21.pdf> [accessed 29.4.16].

[14] CanMEDS: Better Standards, Better Physicians, Better Care. 2015. Royal College of Physicians and Surgeons of Canada. Available from: <http://www.royalcollege.ca/rcsite/canmeds/framework/canmeds-role-communicator-e> [accessed 26.11.16].

[15] Duke P, Frankel RM, Reis S. How to integrate the electronic health record and patient-centered communication into the medical visit: a skills-based approach. Teach Learn Med 2013;25(4):358–65.

[16] Graham-Jones P, Jain SH, Friedman CP, Marcotte L, Blumenthal D. The need to incorporate health information technology into physicians' education and professional development. Health Aff (Millwood) 2012;31 (3):481–7.

[17] Cooke M, Irby DM, O'Briend BD. Educating physicians: a call for re-form of medical school and residency. Carnegie Foundation for the Advancement of Teaching. San Francisco: Jossey-Bass; 2010.

[18] Kazmi Z. Effects of exam room EHR use on doctor-patient communication: a systematic literature review. Inform Prim Care 2013;21(1):30–9.

[19] Zhang J, Chen Y, Ashfaq S, Bell K, Calvitti A, Farber NJ, et al. Strategizing EHR use to achieve patient-centered care in exam rooms: a qualitative study on primary care providers. J Am Med Inform Assoc 2016;23 (1):137–43.

[20] Frankel RM. The effects of exam room computing on the doctor patient relationship: a human factors approach to electronic health Records and physician-patient communication. In: Agrawal A, editor. Safety of Health IT: Clinical case studies. New York: Springer; 2016.

[21] Frankel RM. Computers in the examination room. JAMA Intern Med 2016;176(1):128–9.

[22] Wuerth R, Campbell C, King WJ. Top 10 tips for effective use of electronic health records. Paediatr Child Health 2014;19(3):138.

[23] Pearce C, Trumble S, Arnold M, Dwan K, Phillips C. Computers in the new consultation: within the first minute. Fam Pract 2008;25:202–208.

[24] Ventres W, Kooienga S, Marlin R. EHRs in the exam room: tips on patient-centered care. Fam Pract Manag 2006;13(3):45–7.

[25] Mann WR, Slaboch J. Computers in the exam room—friend or foe? Perm J 2004;8(4):49–51.

[26] Bedell S, Graboys T. Hand to hand. J Gen Intern Med 2002;17(8):654–6.

[27] Kessels R. Patients' memory for medical information. J R Soc Med 2003;96(5):219–22.

[28] White A, Danis M. Enhancing patient-centered communication and collaboration by using the electronic health record in the examination room. JAMA 2013;309(22):2327–8.

[29] Greatbatch D, Heath C, Campion P, et al. How do desk-top computers affect the doctor–patient interaction? Fam Pract 1995;12(1):32–6.

[30] Als AB. The desk-top computer as a magic box: patterns of behaviour connected with the desk-top computer; GPs' and patients' perceptions. Fam Pract 1997;14(1):17–23.

[31] Booth A, Lecouteur A, Chur-Hansen A. The impact of the desktop computer on rheumatologist-patient consultations. Clin Rheumatol 2013;32(3):391–3.

[32] Margalit RS, Roter D, Dunevant MA, Larson S, Reis S. Electronic medical record use and physician-patient communication: an observational study of Israeli primary care encounters. Patient Educ Couns 2006;61 (1):134–141.

[33] Saleem JJ, Flanagan ME, Russ AL, McMullen CK, Elli L, Russell SA, et al. You and me and the computer makes three: variations in exam room use of the electronic health record. J Am Med Inform Assoc 2014;21 (e1):e147–51.

[34] Alkureishi ML, Lee WW, Farnan J, Arora V. Breaking away from the iPatient to care for the real patient: implementing a patient-centered EMR use curriculum. MedEdPORTAL: 2014. Available from: <www.mededportal.org/publication/9953> [accessed 19.7.16].

[35] Kumarapeli P, de Lusignan S. Using the computer in the clinical consultation; Setting the stage, reviewing, recording, and taking actions: multi-channel video study. J Am Med Inform Assoc 2013;20(e1):e67–75.

[36] Alsos OA, Das A, Svanaes D. Mobile health IT: the effect of user interface and form factor on doctor-patient communication. Int J Med Inform 2012;81(1):12–28.

[37] Almquist JR, Kelly C, Bromberg J, et al. Consultation room design and the clinical encounter: the space and interaction randomized trial. HERD 2009;3(1):41–78.

[38] Chneider W, Shriffin RM. Controlled and automatic human information processing: I. detection, search, and attention. Psychol Rev 1977;84:1–66.

[39] Skaugset LM, Farrell S, Carney M, Wolff M, Santen SA, Perry M, et al. Can you multitask? Evidence and limitations of task Switching and multitasking in emergency medicine. Ann Emerg Med 2016;68(2):189–95.

[40] Buschman TJ, Miller EK. Shifting the spotlight of attention: evidence for discrete computations in cognition. Front Hum Neurosci 2010;4:194.

[41] Kuchinskas S. Why Multitasking Isn't Efficient. WebMD. August 2008. Available from: <http://www.webmd.com/mental-health/features/why-multitasking-isnt-efficient> [accessed 15.5.16].

[42] Rubinstein JS, Meyer DE, Evans JE. Executive control of cognitive processes in task switching. J ExpPsychol Hum Percept Perform 2001;27(4):763–97.

[43] Richtel M. As Doctors Use More Devices, Potential for Distraction Grows. New York Times. Dec 2011. Available from: <http://www.nytimes.com/2011/12/15/health/as-doctors-use-more-devices-potential-for-distraction-grows.html?_r=0> [accessed 15.5.16].

[44] To Err is Human: Building a Safer Health System. In: Kohn LT, Corrigan JM, Donaldson MS, editors. Institute of Medicine (US) Committee on Quality of Health Care in America. Washington, DC: National Academies Press; 2000.

[45] Grundgeiger T, Sanderson P. Interruptions in healthcare: theoretical views. Int J Med Inform 2009;78 (5):293–307.

[46] Halamka J. Order Interrupted by Text: Multitasking Mishap. Agency for Healthcare Research and Quality (AHRQ) Patient Safety Network (PSNet). December 2011. Available from: <https://psnet.ahrq.gov/webmm/case/257> [accessed 15.5.16].
[47] Frankel RM, Saleem JJ. "Attention on the flight deck": what ambulatory care providers can learn from pilots about complex coordinated actions. Patient Educ Couns 2013;93(3):367–72.
[48] Booth N, Robinson P, Kohannejad J. Identification of high-quality consultation practice in primary care: the effects of computer use on doctor-patient rapport. Inform Prim Care 2004;12(2):75–83.
[49] Dowell A, Stubbe M, Scott-Dowell K, et al. Talking with the alien: interaction with computers in the GP consultation. Aust J Prim Health 2013;19(4):275–282.
[50] Delany C, Golding C. Teaching clinical reasoning by making thinking visible: an action research project with allied health clinical educators. BMC Med Educ 2014;14:20.
[51] Rosen P, Spalding SJ, Hannon MJ, et al. Parent satisfaction with the electronic medical record in an academic pediatric rheumatology practice. J Med Internet Res 2011;13(2):e40.
[52] Hsu J, Huang J, Fung V, et al. Health information technology and physician-patient reactions: impact of computers on communication during outpatient primary care visits. J Am Med Inform Assoc 2005;12:474–480.
[53] Weaver RR. Informatics tools and medical communication: patient perspectives of "knowledge coupling" in primary care. Health Commun 2003;15(1):59–78.
[54] McCord G, Pendleton BF, Schrop SL, Weiss L, Stockton L, Hamrich LM. Assessing the impact on patient-physician interaction when physicians use personal digital assistants: a Northeastern Ohio Network (NEON) study. J Am Board Fam Med 2009;22(4):353–9.
[55] Hirsch O, Keller H, Krones T, Donner-Banzhoff N. Arriba-lib: association of an evidence-based electronic library of decision aids with communication and decision-making in patients and primary care physicians. Int J Evid Based Healthc 2012;10(1):68–76.
[56] Chandler P, Sweller J. Cognitive load theory and the format of instruction. Cogn Instr 1991;8(4):293–332.
[57] Simon HA, Chase WG. Skill in chess. Am Sci 1973;61(4):394–403.
[58] Eysenck MW. Psychology: An International Perspective. New York: Psychology Press; 2004.
[59] Sweller J. Cognitive load during problem-solving: effects on learning. Cogn Sci 1988;12(2):257–85.
[60] Linzer M, Poplau S, Babbott S, Collins T, Guzman-Corrales L, Menk J, et al. Worklife and wellness in academic general internal medicine: results from a national survey. J Gen Intern Med 2016;31(9):1004–10.
[61] Babbott S, Manwell LB, Brown R, Montague E, Williams E, Schwartz M, et al. Electronic medical records and physician stress in primary care: results from the MEMO study. J Am Med Inform Assoc 2014;21(e1):e100–6.
[62] Friedberg MW, Chen PG, Van Busum KR, et al. Factors Affecting Physician Professional Satisfaction. Rand Corporation, 2013. Available from: <http://www.rand.org/pubs/research_briefs/RB9740.html> [accessed 20.6.16].
[63] Sinsky C, Colligan L, Li L, Prgomet M, Reynolds S, Goeders L, et al. Allocation of physician time in ambulatory practice: a time and motion study in 4 specialties. Ann Intern Med 2016;165(11):753–60.
[64] Shanafelt TD, Dyrbye LN, Sinsky C, Hasan O, Satele D, Sloan J, et al. Relationship between clerical burden and characteristics of the electronic environment with physician burnout and professional satisfaction. Mayo Clin Proc 2016;91(7):836–48.
[65] Rose D, Richter L, Kapustin J. Patient experiences with electronic medical records: lessons learned. J Am Assoc Nurse Pract 2014;26(12):674–80.
[66] Assis-Hassid S, Reychav I, Heart T, Pliskin J, Reis S. Enhancing patient-doctor-computer communication in primary care: towards measurement construction. Isr J Health Policy Res 2015;4:4.
[67] Furness ND, Bradford OJ, Paterson MP. Tablets in trauma: using mobile computing platforms to improve patient understanding and experience. Orthopedics 2013;36(3):205–8.
[68] Ratanawongsa N, Barton JL, Lyles CR, Wu M, Yelin EH, Martinez D, et al. Association between clinician computer use and communication with patients in safety-net clinics.. JAMA Intern Med 2016;176(1):125–8.
[69] Lelievre S, Schultz K. Does computer use in patient–physician encounters influence patient satisfaction? Can Fam Physician 2010;56(1):e6–e12.
[70] Gibson S. Three ways to ease patient fears about health IT. Health Care Tech Review. February 24, 2012. Available from: <http://healthcaretechreview.com/3-waysto-ease-patient-fears-about-health-it/> [accessed 20.6.16].
[71] Mehrabian A, Wiener M. Decoding of inconsistent communications. J Person Social Psychol 1967;6:109–14.

[72] Mehrabian A, Ferris SR. Inference of attitudes from nonverbal communication in two channels. J Consul Psychol 1967;31:248–52.
[73] Montague E, Chen P, Xu J, Chewning B, Barrett B. Nonverbal interpersonal interactions in clinical encounters and patient perceptions of empathy. J Participat Med 2013;5:e33.
[74] Hojat M, Louis D, Maxwell K, Markham F, Wender R, Gonnella J. Patient perceptions of physician empathy, satisfaction with physician, interpersonal trust, and compliance. Int J Med Educ 2010;1:83–7.
[75] Rakel D, Barrett B, Zhang Z, Hoeft T, Chewning B, Marchand L, et al. Perception of empathy in the therapeutic encounter: effects on the common cold. Patient Educ Couns 2011;85(3):390–7.
[76] Makoul G, Curry RH, Tang PC. The use of electronic medical records: communication patterns in outpatient encounters. J Am Med Inform Assoc 2001;8(6):610–15.
[77] Frankel R, Altschuler A, George S, et al. Effects of exam-room computing on clinician–patient communication: a longitudinal qualitative study. J Gen Intern Med 2005;20(8):677–682.
[78] Noordman J, Verhaak P, van Beljouw I, et al. Consulting room computers and their effect on general practitioner-patient communication. Fam Pract 2010;27(6):644–651.
[79] McGrath JM, Arar NH, Pugh JA. The influence of electronic medical record usage on nonverbal communication in the medical interview. Health Informatics J 2007;13(2):105–18.
[80] Lee WW, Alkureishi MA, Ukabiala O, Venable LR, Ngooi SS, Staisiunas DD, et al. Patient perceptions of electronic medical record use by faculty and resident physicians: a mixed methods study. J Gen Intern Med 2016;31(11):1315–22.
[81] Morris D. Bodytalk: The Meaning of Human Gestures. New York: Crown Trade Paperbacks; 1994.
[82] Argyle M, Cook M. Gaze and Mutual Gaze. Cambridge: Cambridge University Press; 1976.
[83] Institute of Medicine. Crossing the Quality Chasm. Washington: National Academies Press; 2001.
[84] Reis S, Sagi D, Eisenberg O, Kuchnir Y, Azuri J, Shalev V, et al. The impact of residents' training in Electronic Medical Record (EMR) use on their competence: report of a pragmatic trial. Patient Educ Couns 2013;93(3):515–21.
[85] Mehta P. Communication skills—talking to parents. Indian Pediatr 2008;45(4):300–4.
[86] White J, Levinson W, Roter D. "Oh, by the way . . .": the closing moments of the medical visit. J Gen Intern Med 1994;9(1):24–8.
[87] Reis S, Visser A, Frankel R. Health information and communication technology in healthcare communication: the good, the bad, and the transformative. Patient Educ Couns 2013;93(3):359–62.
[88] van Gurp J, Hasselaar J, van Leeuwen E, Hoek P, Vissers K, van Selm M. Connecting with patients and instilling realism in an era of emerging communication possibilities: a review on palliative care communication heading to telecare practice. Patient Educ Couns 2013;93(3):504–14.
[89] Rubinelli S, Collm A, Glässel A, Diesner F, Kinast J, Stucki G, et al. Designing interactivity on consumer health websites: PARAFORUM for spinal cord injury. Patient Educ Couns 2013;93(3):459–63.
[90] Epstein RM, Hundert EM. Defining and assessing professional competence. JAMA 2002;287(2):226–235.
[91] Hersh WR, Gorman PN, Biagioli FE, Mohan V, Gold JA, Mejicano GC. Beyond information retrieval and electronic health record use: competencies in clinical informatics for medical education. Adv Med Educ Pract 2014;5:205–12.
[92] Mintz M, Narvarte HJ, O'Brien KE, Papp KK, Thomas M, Durning SJ. Use of electronic medical records by physicians and students in academic internal medicine settings. Acad Med 2009;84(12):1698–704.
[93] Wald HS, George P, Reis SP, Taylor JS. Electronic health record training in undergraduate medical education: bridging theory to practice with curricula for empowering patient- and relationship-centered care in the computerized setting. Acad Med 2014;89(3):380–6.
[94] Silverman H, Ho YX, Kaib S, Ellis WD, Moffitt MP, Chen Q, et al. A novel approach to supporting relationship-centered care through electronic health record ergonomic training in preclerkship medical education. Acad Med 2014;89(9):1230–4.
[95] Lee WW, Alkureishi MA, Farnan J, Arora, VA. A novel curriculum to teach patient-centered use of the electronic medical record. Plenary Presentation, Clerkship Directors of Internal Medicine (CDIM) National Meeting, held as part of Academic Internal Medicine Week (AIMW). New Orleans, LA. September 2014.
[96] Alkureishi MA, Lee WW. Patient centered EMR use. Platform Presentation at Council on Medical Student Education in Pediatrics (COMSEP) & Association of Pediatric Program Directors (APPD). Nashville, TN. March 2013.

[97] Shachak A, Domb S, Borycki E, Fong N, Skyrme A, Kushniruk A, et al. A pilot study of computer-based simulation training for enhancing family medicine residents' competence in computerized settings. Stud Health Technol Inform 2015;216:506–10.
[98] Boissy A, Windover AK, Bokar D, Karafa M, Neuendorf K, Frankel RM, et al. Communication skills training for physicians improves patient satisfaction. J Gen Intern Med 2016;31(7):755–61.
[99] Lewis B, Ratanawongsa N, Reis S, Chisolm M, Alkureishi M. You, me and the computer makes three: how to integrate the computer in the exam room for optimal patient experience & what learners should know. American Academy on Communication in Healthcare (AACH) ENRICH Forum Workshop. New Haven, CT. June 2016.
[100] Lee WW, Alkureishi ML, Chi J, Arora VA. Breaking away from the iPatient to care for the real patient. Academic Internal Medicine Week (AIMW). New Orleans, LA. October 2013.
[101] Lee WW, Alkureishi ML, Arora, VA. "Teaching patient-centered electronic medical record (EMR) use to millennial learners: are we preaching to the choir?" AAMC National Integrating Quality Meeting Workshop. Chicago, IL. June 2014.
[102] Alkureishi ML, Frankel R, Chisolm M, Arora VA. Teaching Patient-Centered Communication Skills in the Digital Age. AAMC Annual Medical Education Meeting, Emerging Solutions Workshop, Baltimore, MD. November 2015.
[103] Alkureishi L, Farnan J, Arora V, M. Mayer, Isaacson J, Windover A, et al. Teaching the Teachers: Adapting a Faculty Development Workshop on Patient-Centered Electronic Medical Records Use for Busy Clinicians. Research Oral Abstract. AAMC Annual Meeting. Seattle, WA. November 2016.
[104] Shapiro S. Charting a Mindful Approach. Dome, March 2013. Available from: <http://www.hopkinsmedicine.org/news/publications/dome/dome_march_2013/charting_a_mindful_approach> [accessed 19.7.16].
[105] Kahn MW. Etiquette-based medicine. N Engl J Med 2008;358(19):1988–9.
[106] Assis-Hassid S, Heart T, Reychav I, Pliskin JS, Reis S. Existing instruments for assessing physician communication skills: are they valid in a computerized setting? Patient Educ Couns 2013;93(3):363–6.
[107] Alkureishi L, Lee WW, Lyons M, Wroblewski K, Farnan J, Arora V. Development and Validation of e-Clinical Evaluation Exercise (e-CEX) Tool to Assess Patient-Centered Electronic Medical Record Use. Research Oral Abstract. AAMC Annual Meeting. Seattle, WA. November 2016.
[108] Alkureishi ML, Lee WW, Lyons M, Wroblewski K, Farnan J, Arora V. Development and Validation of e-Clinical Evaluation Exercise (e-CEX) Tool to Assess Patient-Centered Electronic Medical Record Use. MESRE Oral Presentation. AAMC Central Group of Education Affairs (CGEA). Ann Arbor, MI. April 2016.
[109] Norcini JJ, Blank LL, Duffy FD, et al. The mini-CEX: a method for assessing clinical skills. Ann Intern Med 2003;138(6):476–81.
[110] Accreditation Council for Graduate Medical Education (ACGME) Milestones. Available from: <http://www.acgme.org/acgmeweb/tabid/430/ProgramandInstitutionalAccreditation/NextAccreditationSystem/Milestones.aspx> [accessed 19.7.16].
[111] Makoul G. The SEGUE Framework for teaching and assessing communication skills. Patient Educ Couns 2001;45(1):23–34.
[112] Kolb DA. Experiential learning: experience as the source of learning and development, 1. Englewood Cliffs, NJ: Prentice-Hall; 1984.
[113] Kolb DA., Fry R.E. Toward an applied theory of experiential learning. MIT Alfred P. Sloan School of Management; 1974.
[114] Lee WW, Alkureishi MA, K Wroblewski, K Farnan J, Arora VA. Need to Reboot? Retention of Patient-Centered EMR Use Skills. Plenary Presentation, Clerkship Directors of Internal Medicine (CDIM) National Meeting, held as part of Academic Internal Medicine Week (AIMW). Washington DC. September 2014.
[115] Alkureishi ML, Lee WW, Webb S, Arora V. Novel Champions for Professionalism in Electronic Medical Record (EMR) Use: Integrating Patient-Centered EMR Use Skills and Documentation Expectations into Required EMR Training. Innovations in MedicalEducation (IME) Oral Presentation. AAMC Central Group of Education Affairs (CGEA) Ann Arbor, MI. April 2016.
[116] Guidelines for Teaching Physicians, Interns, and Residents. Washington, DC: Medicare Learning Network, Center for Medicare and Medicaid Services, Department of Health and Human Services, 2008. Available from: <https://www.cms.gov/Outreach-and-Education/Medicare-Learning-Network-MLN/MLNProducts/Downloads/Teaching-Physicians-Fact-Sheet-ICN006437.pdf> [accessed 19.7.16].
[117] Joe RS, Otto A, Borycki E. Designing an electronic medical case simulator for health professional education. KM&EL 2011;3(1):63–71.

[118] Kushniruk A, Borycki E, Joe R, Otto T, Armstrong B, Ho K. Integrating electronic health records into medical education: considerations, challenges, and future directions. In: Ho K, Jarvis-Selinger S, Novak Lauscher H, Cordeiro J, Scott R, editors. Technology Enabled Knowledge Translation for eHealth. New York: Springer; 2012. p. 21–32.

[119] Kushniruk A, Borycki E, Kuo MH, Parapini E, Wang SL, Ho K. Requirements for prototyping an educational electronic health record: experiences and future directions. Stud Health Technol Inform 2014;205:833–7.

[120] Prenksy M. Digital natives, digital immigrants. Horizon MCB Univ Press 2001;9(5):1–6.

CHAPTER

11

Incorporating Patient's Perspectives in Educational Interventions: A Path to Enhance Family Medicine Communication in the Age of Clinical Information Systems

Margarida Figueiredo-Braga[1], Dilermando Sobral[1] and Marcy Rosenbaum[2]

[1]University of Porto, Porto, Portugal [2]University of Iowa Carver College of Medicine, Iowa City, IA, United States

INTRODUCTION

Computer use has been consistently associated with better clinical performance, preventive care, and access to clinical information. Evidence is less convincing for diagnosis, and demonstration of a positive effect on patient outcomes is even more scarce. Digital communication within the consultation augments the ability of the physician to explain health care issues, to gather biomedical information and to engage in shared decision making (if desired) [1,2]. From the health professional's point of view, a recent review [3] concluded that the use of computers may improve their "information seeking, adherence to guidelines and clinical decision making". However, with the widespread use of computers and electronic health records in health care [4], it becomes critical to explore the communication between health professionals, computers, and patients. Communication problems between health professionals and patients are still reported including inability to share information, to address emotions and concerns, inability to meet individual needs and expectations, and the lack of respect and involvement [5–9]. Clinicians often report a lack of self-confidence in adequately communicating with patients, a problem that may lead to

DOI: http://dx.doi.org/10.1016/B978-0-12-805362-1.00011-5

avoidance of communication, using the intrusion of the computer as justification. Despite the advantages brought by the growing presence of technology [10], its impact on the health professional-patient interaction is still controversial. Some authors report that the use of the computer in the consultation interferes with physician's attentive behavior, psychosocial information gathering, and emotional disclosure by the patient [4,11–14].

Recent literature reviews have highlighted the need for increased computer-specific communication training to mitigate adverse effects and for continued acknowledgment of patient perspectives, reporting strong evidence that technological training could reduce adverse behavioral effects of technology use [2,13,15]. The authors have published a study [16], where the impact of computer use on both physicians and patients was evaluated regarding consultation length, confidentiality, and ability to maintain eye contact, listen, collect information, provide information, and understand the patient. Physicians reported a negative impact of the computer on patient-physician communication, but this was not shared by patients, who considered that the computer had an overall positive impact on patient-physician communication. Previous studies have also found that patients did not view the impact of the computer in the relationship with their doctor as negative [14,17–23].

EDUCATIONAL INTERVENTIONS

Literature Review

In medical education, students require training to become skilled users of examination room computing and computer use in other healthcare settings, but formal pedagogy at the undergraduate level is scarce [24]. Curriculum development for integrating digital resources using a patient-centered perspective needs to be implemented within a longitudinal framework [25] that will extend previously learned strategies "to the new contexts and specialties in which they are learning, while at the same time developing new and advanced skills" [26]. Communication skills training interventions have been described and implemented during internships [27–31]. A positive effect on physician-computer-patient communication patterns of residents and trainees was also found following brief educational interventions [32,33].

Educational interventions to improve computer integration in clinical communication are desirable and needed at multiple time points in the education and practice of medicine. To train senior physicians may be even more important since they act as role models for trainees with less professional experience.

Educational Interventions: Length, Content, and Format of the Training Sessions

We developed a half day training program for small groups [10–15] of physicians and other health professionals working in General Practice Units, aiming to enhance their communication ability and to improve the integration of the computer into patient-professional interactions, safeguarding the clinical relationship. Accumulated evidence shows the efficacy of skills-focused and practice-oriented training programs on communication skills learning [27–30] and patient-physician-computer communication ability [32,33].

Didactic material was made available online before each educational intervention and a list of selected references was provided to the health professionals after each session. The training module was based on a literature review [33] and incorporated theoretical and practical content targeting the professional's knowledge of patient-computer-physician communication skills and their ability to identify resources and solutions. After a brief theoretical introduction addressing basic communication skills (i.e., attentive listening, verbal and nonverbal communication, empathy and patient centeredness), specific strategies to integrate the computer in interactions with patients were explained.

Strategic division of the interaction between patient and computer-centered focus, prioritizing data entry before or after the patient's presence, and keeping patients engaged by sharing the screen or keeping the conversation active while typing, were suggested and discussed [13,34].

The relevance of the consultation room configuration was underscored to avoid the computer coming between the health professional and the patient. For example, the physician was suggested to sit at a 45 degrees to the patient when looking at the computer monitor, which allowed for the possibility of sharing the information that is being entered—an advantageous way to maintain patients' involvement and to confirm the accuracy of the information going into the electronic record [16,35–37] (Fig. 11.1).

Suggestions such as using a tablet computer to type notes, keeping the computer out of the office or more feasibly using mobile computer monitors and keyboards, were delivered, in order to convert digital resources into efficient and nonintrusive tools. Participants were discouraged from keeping the computer behind a desk, which requires the health professional to turn his or her back to the patient while typing, thus interrupting eye contact. Other problem-focused techniques such as enhancing typing and surfing ability, templates for documentation, voice to text software, and optical scanners were also discussed.

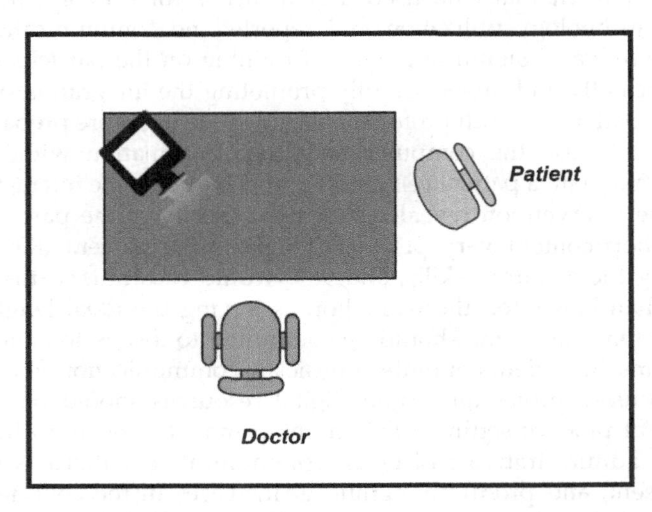

FIGURE 11.1 Location of computer screen on the desk.

This module follows the steps for teaching patient-centered interviewing stages and skills [38], embedded with computer-related skills, a valuable tool for teaching or evaluating these skills, described by Shachak and Reis and Ventres et al. [13,34].

Practical exercises with clinical *vignettes* were preceded by video-recorded demonstrations of effective and ineffective patient-professional-computer communication skills. Experiential learning was achieved through role-playing interviewing in which participants acted as a patient allowing health professionals to observe and reflect upon their own performance concerning the integration of the computer in a simulated clinical setting. This format was reported previously to increase confidence and patient-centeredness in communication skills training courses [39]. The health professional's sociodemographic and professional characteristics, previous communication skills training and confidence in their capacity to adequately integrate the computer in the consultation were assessed before and after the intervention.

Fifty-four health professionals have attended the described short training interventions. They had extensive professional experience, working in clinical settings for a mean period of 20.5 (SD = 13.2) years and using the computer for 10.5 (SD = 5.8) years. Physicians spent about 73% of consultation time interacting with the computer, which they felt to have had a negative impact on confidentiality, ability to look at the patient, and ability to listen to the patient. Most of the participants in the training sessions were family physicians, and approximately half have had previous computer skills and communication skills training. Participants were asked to describe their confidence in appropriately using the computer while preserving the health professional–patient relationship before and after the sessions. Compared to ratings at the end of the program, as a first assessment they were less confident in their ability to avoid the computer diminishing their attention toward the patient, becoming an obstacle to look at the patient, and compromising time management. After the training, participants similarly rated significantly higher capacity to maintain a patient-centered communication while using the computer, similar to other reports [32,33]. The effect of the training was evident in participants who used digital means for a long time, had less formal training in digital technology utilization and reported no communication skills teaching before training. There was a significant effect of training on the participants' confidence in communicating efficiently and in successfully promoting the integration of the computer in the interaction with patients. Health professionals noted feeling more prepared to avoid interrupting the patient, to use the computer to gather information without disturbing the relationship and to maintain a patient centered approach during the interaction.

Evaluation of the intervention revealed that the majority of the participants (83%) considered the workshop content very pertinent. Eighty-nine percent (89%) reported being capable of applying the acquired skills, and 98% would recommend the training to other colleagues. More than half rated the workshop as having the ideal length and methodology, but suggested that more time should be committed to discussion and feedback.

Training programs in patient-computer-physician communication skills that emphasize specific solution-focused, professional, and digital resources should be developed further and tested in clinical practice settings. This training model appears to help change physicians, nurses, and administrative assistants' communicative pattern with patients when computers are present, and produced significant increases in the confidence of the health professionals. Nevertheless, long-term efficacy evaluation will show whether it is possible to transfer the results of short training programs into persistent behavioral changes. Monitoring of the acquired skills and course outcomes are ongoing through the development of local intervention programs. Participants from one of the groups attending the training program are presently developing local projects to improve communication between patients and health professionals, to be evaluated during 2017. Our experience of including different health professions appears to reinforce the advantages to extending training to all the health professionals involved in patients' ambulatory care units. The small groups model favored peer and team based learning, and sharing different experiences and challenges enriching the educational quality of the sessions as well as enhancing the group problem solving achievement. Training programs addressing strategies to improve doctors' communication skills while using the computer are still scarce. Feasible and economic interventions are needed, complemented by the study of which variables are associated with the desired learning outcomes. Health professionals' characteristics, clinical context, and previous skills level assessment may be helpful to increase training efficacy [40].

Development and Inclusion of a Training Module of Computer Communication Skills in the Undergraduate CC Course (Medical Students) and in the Postgraduate and Master Courses

Medical, pharmacy, psychology, nutrition, and nurse students and professionals are attending undergraduate and postgraduate communication skills courses at our Medical School [41–43]. Inclusion of training modules in patient-computer-physician/student communication is ongoing to enhance students and professionals' communication efficacy to deal with the computer within the health professional-patient relationship. Communication skills training programs applied during medical courses have been shown to improve students' knowledge [44,45], attitudes [46,47], confidence [47], empathy, and patient-centeredness [48–52]. Medical and other health professions students, although skilled in personal use of digital resources, require training to become skilled users at a professional level. Theoretical approaches during early curricular years have to be supplemented by the integration of role-modeling by clinical teachers during clerkships [53] Regarding computerized settings, undergraduate curriculum needs to foster students' effective communication and capacity to provide patient centered care [24] in the presence of the computer. Students who were formally taught on integrated computer communication skills demonstrated better performance whilst utilizing a computer during their clerkships [54,55], emphasizing the need for curriculum development throughout all years of training. At the postgraduate level, specific computer use skills should be taught during early stages of internship to enhance their efficacy [56].

In our experience, senior physicians in practice for more than 15 years reported significantly lower undergraduate ($p = 0.001$) and postgraduate ($p = 0.033$) formal communication skills training than their trainees and have acquired computer skills predominantly from personal experience and advice from colleagues [16]. Experienced health professionals attending educational programs to improve the integration of the computer in patient–professional interactions have stated the need for training in late professional career stages (Sobral D and Figueiredo-Braga M, in press).

ASSESSMENT TOOLS FOR PHYSICIAN-PATIENT-COMPUTER COMMUNICATION

Brief Literature Review

The importance of physician-patient communication for a successful and effective medical encounter is well recognized as the "heart of medicine" [57]. Several guidelines have been created to define physicians' communication tasks, strategies, and skills that should be carried out during the medical encounter: the task approach [58], the three function model [59], the four habits model [60], the Smith model [61], the Kalamazoo consensus statements [62], the Calgary-Cambridge guides [63], the SEGUE framework [64], the Macy Model Checklist [65], the MAAS-Global [66] and the Roter Interaction Analysis System [67]. Medical education organizations require medical students to demonstrate communication competencies in order to receive their certifications [68,69], as described in the study of Assis-Hassid et al [70].

In the last decades, computers and more specifically, Electronic Medical Records (EMRs) are being increasingly used in healthcare organizations, and in primary care settings in particular, and the presence of this third actor in the consulting room has potentially threatened the therapeutic relationship between patient and physician. However, all the guidelines listed above were developed prior to the large-scale introduction of EMRs into healthcare and do not take into account the effects of EMR use in physician's interaction with patients.

Presently specific EMR communication skills assessment tools are limited to checklists of recommended communication tasks to be carried out by physicians in the computerized exam room: the EHR-specific communications skills checklist [71], and the e-SEGUE [70]. The next sections detail this last instrument and its validation process in a Portuguese population.

The e-SEGUE

The e-SEGUE is a communication skills assessment tool that integrates computer-related communication skills in the SEGUE communication assessment framework [64]. The scale follows the SEGUE structure, which divides the medical encounter into five stages: **S**et the stage, **E**licit information, **G**ive information, **U**nderstand the patient's perspective and **E**nd the encounter, and is based on the evidence that the ancient dyadic relationship between doctor and patient has become a triadic relationship with the introduction of the computer into the consultation room [12,72]. This fact has given rise to a new construct, Patient-Doctor-Computer Communication (PDCC), defined as the physician's ability to provide patient-centered care while using the computer during the medical encounter.

From the original 27 skills generated from the literature review, the e-SEGUE actually includes 23 PDCC-related behaviors, and is extremely useful for assessing, training and enhancing patient-doctor communication in a computerized setting.

Portuguese Validation—Medical Students, Family Doctors, and Experts

Using the original validation methodology of this instrument as a benchmarking tool, a similar study was performed in Portugal, after the e-SEGUE translation and adaptation process. Data were collected by distributing paper-based questionnaires to medical students and by a Google form sent via email to all other respondents (family physicians and trainees, nurses, physicians attending a clinical communication postgraduation, and faculty members teaching communication skills).

More than 500 respondents rated the relevance of each of the behaviors regarding its importance for establishing an effective physician-patient communication while using the computer. Preliminary results showed interesting differences in ratings between the different groups of participants reflecting different levels of agreement in discrepancy with the original study. Curiously, one item,—*"Verify patient's literacy, primary language, and visual acuity to optimize computer use,"* remained under 50% of agreement among all groups. Medical students reported a higher level of agreement with the item *"Read back what you have written"* than the total sample. Regarding the physicians sample, the faculty staff ($n = 15$) and the physicians completing the Communication Skills course ($n = 12$) reported similar concordance rates to that of family physicians without formal clinical communication training.

Results analyses and discussion are continuing.

PATIENTS PERSPECTIVES AND EXPERIENCE REGARDING COMPUTER USE IN THE CONSULTATION

Brief Review of Patients' Perspectives Regarding Computer Use

We previously referred to divergent perspectives of patients and providers regarding the interference of digital resources with clinical practice, quality of care, and the health professional-patient relationship. Studies focused on patients' perspectives regarding the computer use in the clinical encounter examined various aspects of patient-physician communication. As pointed out in a recent review [37], in most studies, patients reported a positive impact or no significant impact of consulting room technology on patient-physician communication. However, others have stated that communicational aspects of patient-physician relationship significantly declined after computer implementation, namely, eye contact and mutual gaze [12] and attentive behavior [73].

Physician styles of interacting with the computer and the patient as well of patient styles of interacting with the computer and their physicians have also been analyzed in a number of studies. Pearce et al. [72,74] have classified patient styles of interaction with the computer and the physician as either dyadic or triadic, as they focused mainly on the clinician, or deal with the computer as part of the consultation. Dyadic patients tended to ignore the computer while the clinician works on the computer and to draw the clinician's attention back to them; triadic patients tended to actively involve the computer in the consultation by pointing at it, leaning over, or moving their seat so that they could see the computer screen.

Patients' Perspectives and Experience

As quoted above, physicians' perception of the negative impact of the computer use was not shared by patients who even considered that computer had a positive impact on patient-physician communication [1]. In the same study, physicians reported, on average, a higher patient-centered score, assessed with the Patient-Practitioner Orientation Scale (PPOS) [75] when compared to patients. Patients seem to prefer a paternalistic approach, which may be culturally driven [76]. From patients' perspective, specific patient centered strategies were negatively associated with consultation length, confidentiality, and ability to look, listen, talk, give information, and understand the patient while using the computer.

In a recent study [77], a convenience sample of 392 adult patients was surveyed for the perceived impact of the presence and utilization of digital resources in the interaction with their family physician. More than half were female, professionally active, 253 (63.8%) lived with a spouse, and 30.6% had completed 12 years of school. They had seen the current doctor for more than ten years and had on average 3.5 consultations per year. Their experience in communicating with their family physician was analyzed using the QUality Of care Through the patients' Eyes COMMunication (QUOTE-COMM) questionnaire, a 4-point Likert-type scale with 13 items. The items can be divided into an affect-oriented scale, which measures attentive and empathic behavior by the doctor, and a task-oriented scale including exchanging information and advice, diagnosing and problem solving [78].

Overall, patients rated high the family practitioners communication, both regarding affect-oriented (mean score = 3.61, SD = 0.52) and task-oriented (mean score = 3.57; SD = 0.55) aspects. The number of years with their family physician and QUOTE-COMM score were not related, but a positive correlation between number of consultations in the previous year and QUOTE-COMM score ($r = 0.11$; $p < 0.05$) was found. Higher family physician communication quality was associated with a positive impact of computer use by the physician ($r = 0.37$; $p < 0.001$), although a negative correlation between QUOTE-COMM score and time spent interacting on the computer ($r = -0.19$; $p < 0.001$) was detected in the patients' evaluation. These findings suggest that better communication skills are related to a more positive impact of computer use reported by physicians.

In summary, patients evaluated health professionals' communication capacity as negatively associated with time spent interacting with the computer during the consultation, although they appreciated the computer as a benefit. No differences were found regarding patient gender and only the affect-oriented scale showed to be positively correlated with patients' age.

FUTURE TRENDS: DESIGN OF EDUCATIONAL INTERVENTIONS, INCORPORATING PATIENT'S PERSPECTIVES

Scrutiny of patient perspectives regarding the use of the computer in the consultation is essential [74], taking into account the reported divergent standpoints between physicians and patients [16,79]. Their valuable contribution may inform practical remarks on patient-physician-computer communication strategies to integrate in future educational interventions, and tailor assessment of acquired and trained skills. The adaptation of communication skills assessment tools to be applied to patients in different contexts will facilitate the evaluation of their experience and preferences regarding an integrated computer communication ability.

The development and evaluation of training programs should integrate patient's perspective, grounded in the "need for clinicians to be aware of how patients—the most important stakeholder in healthcare provision—perceive the new technologies implemented in the consultation room." [2]

References

[1] Beaulieu MD, Haggerty JL, Beaulieu C, Bouharaoui F, Lévesque JF, Pineault R, et al. Interpersonal communication from the patient perspective: comparison of primary healthcare evaluation instruments. Healthc Policy 2011;7:108–23 (Special Issue).
[2] Kazmi Z. Effects of exam room EHR use on doctor–patient communication: a systematic literature review. Inform Prim Care 2013;21(1):30–9.
[3] Mickan S, Atherton H, Roberts NW, Heneghan C, Tilson JK. Use of handheld computers in clinical practice: a systematic review. BMC Med Inform Decis Mak 2014;14(56).
[4] Asan O, Xu J, Montague E. Dynamic comparison of physicians' interaction style with electronic health records in primary care settings. J Gen Pract (Los Angel) 2013;2 Epub 2013 Dec 10.
[5] Beckman HB, Markakis KM, Suchman AL, Frankel RM. The doctor–patient relationship and malpractice. Arch Intern Med 1994;154(12):1365–70.
[6] Epstein RM, Alper BS, Quill TE. Communicating evidence for participatory decision making. JAMA 2004;291 (19):2359–66.
[7] Grol R, Wensing M, Mainz J, Jung HP, Ferreira P, Hearnshaw H, et al. Patients in Europe evaluate general practice care: an international comparison. Brit J Gen Pract 2000;50(460):882–7.
[8] Dosanjh S, Barnes J, Bhandari M. Barriers to breaking bad news among medical and surgical residents. Med Educ 2001;35(3):197–205.
[9] Stewart M, Brown JBDA, McWhinney IR, Oates J, Weston WW, Jordan J. The impact of patient-centered care on outcomes. J Fam Pract 2000;49(9):796–804.
[10] Chaudhry B, Wang J, Wu S, Maglione M, Mojica W, Roth E, et al. Systematic review: impact of health information technology on quality, efficiency, and costs of medical care. Ann Intern Med 2006;144(10):742–52.
[11] Makoul G, Curry RH, Tang PC. The use of electronic medical records: communication patterns in outpatient encounters. J Am Med Inform Assoc 2001;8:610–15.
[12] Margalit RS, Roter D, Dunevant MALS, Reis S. Electronic medical record use and physician-patient communication: an observational study of Israeli primary care encounters. Patient Educ Couns 2006;61:134–41.
[13] Shachak A, Reis S. The impact of electronic medical records on patient–doctor communication during consultation: a narrative literature review. J Evaluat Clin Pract 2009;15:641–9.
[14] Rethans J-J, Höppener P, Wolfs G, Diederiks J. Do personal computers make doctors less personal? Br Med J 1988;296(6634):1446–8.
[15] Rozenblum R, Donzé J, Hockey PM, Guzdar E, Labuzetta MA, Zimlichman E, et al. The impact of medical informatics on patient satisfaction: a USA-based literature review. Int J Med Inform 2013;82(3):141–58.
[16] Sobral D, Rosenbaum M, Figueiredo-Braga M. Computer use in primary care and patient-physician communication. Patient Educ Couns 2015;98:1568–76.
[17] Garrison GM, Bernard ME, Rasmussen NH. 21st-century health care: the effect of computer use by physicians on patient satisfaction at a family medicine clinic. Fam Med 2002;34(5):362–8.
[18] Street Jr RL, Liu L, Farber NJ, Chen Y, Calvitti A, Zuest D, et al. Provider interaction with the electronic health record: The effects on patient-centered communication in medical encounters. Patient Educ Couns 2014;96(3):315–19. Available from: <http://dx.doi.org/10.1016/j.pec.2014.05.004> [Epub ahead of print]
[19] Buscató CR, Yuste NE, Toirán AS, Díaz SB, Font J. Opinión de profesionales y pacientes sobre la introducción de la informática en la consulta. Aten Primaria 2005;36(4):194–7.
[20] Callen J, Bevis M, McIntosh J. Patients' perceptions of general practitioners using computers during the patient-doctor consultation. Health Inform Manag 2005 2005;34(1):8–12.
[21] Stewart RF, Kroth PJ, Schuyler M, Bailey R. Do electronic health records affect the patient–psychiatrist relationship? A before & after study of psychiatric outpatients. BMC Psychiatr 2010;10:3.

[22] Rose D, Richter LT, Kapustin J. Patient experiences with electronic medical records: lessons learned. J Am Assoc Nurse Pract 2014;26:674–80.
[23] Irani JS, Middleton JL, Marfatia R, Omana ET, D'Amico F. The use of electronic health records in the exam room and patient satisfaction: a systematic review. J Am Board Fam Med 2009;22(5):553–62.
[24] Wald HS, George P, Reis SP, Taylor JS. Electronic health record training in undergraduate medical education: bridging theory to practice with curricula for empowering patient- and relationship-centered care in the computerized setting. Acad Med 2014;89(3):380–6.
[25] Kern DE, Thomas PA, Howard DM, Bass EB. Curriculum development for medical education: a six-step approach. Baltimore, USA: John Hopkins University Press; 1998.
[26] Martin LR, Dimatteo MR. The Oxford handbook of health communication, behavior change, and treatment adherence. Oxford, UK: Oxford University Press; 2013.
[27] Roth CS, Watson KV, Harris IB. A communication assessment and skill-building exercise (CASE) for first-year residents. Acad Med 2002;77(7):746–7.
[28] Rajavel Murugan P, Padmavathi T. Effectiveness of educational intervention to enhance communication skills among interns panacea. J Med Sci 2016;6(1):20–5.
[29] Yuen JK, Mehta SS, Roberts JE, Cooke JT, Reid MC. A brief educational intervention to teach residents shared decision making in the intensive care unit. J Palliat Med 2013;16(5):531–6.
[30] Markin A, Cabrera-Fernandez DF, Bajoka RM, Noll SM, Drake SM, Awdish RL, et al. Impact of a simulation-based communication workshop on resident preparedness for end-of-life communication in the intensive care unit. Crit Care Res Pract 2015;2015:534879 Epub 2015 Jun 25.
[31] Kelm Z, Womer J, Walter JK, Feudtner C. Interventions to cultivate physician empathy: a systematic review. BMC Med Educ 2014;14 Epub 2014 Oct 14.
[32] Reis S, Cohen-Tamir H, Eger-Dreyfuss LL, Eisenburg O, Shachak A, Hasson-Gilad DR, et al. The Israeli patient-doctor-computer communication study: an educational intervention pilot report and its implications for person-centered medicine. Int J Pers Cent Med 2011;1(4):776–81.
[33] Duke P, Frankel RM, Reis S. How to integrate the electronic health record and patient-centered communication into the medical visit: a skills-based approach. Teach Learn Med 2013;25(4):358–65.
[34] Ventres W, Kooienga S, Marlin R. EHR's in the exam room: tips on patient-centered care. Fam Pract Manag 2006;13(3):45–7.
[35] Kumarapeli P, de Lusignan S. Using the computer in the clinical consultation; setting the stage, reviewing, recording, and taking actions: multi-channel video study. J Am Med Inform Assoc 2013;20:e67–75.
[36] Shachak A, Hadas-Dayagi M, Ziv A, Reis S. Primary care physicians' use of an electronic medical record system: a cognitive task analysis. J Gen Intern Med 2009 Mar;24(3):341–8.
[37] Crampton NH, Reis S, Shachak A. Computers in the clinical encounter: a scoping review and thematic analysis. J Am Med Inform Assoc 2016 May;23(3):654–65.
[38] Fortin VI, Auguste H, Dwamena FC, Frankel RM, Smith RC. . Smith's patient centered interviewing: an evidence-based method. 3rd ed. New York, USA: McGraw-Hill; 2012.
[39] Nørgaard B, Ammentorp J, Ohm Kyvik K, Kofoed PE. Communication skills training increases self-efficacy of health care professionals. J Contin Educ Health Prof 2012;32(2):90–7.
[40] Bragard I. Communication skills training for residents: which variables predict learning of skills? Open J Med Psychol 2012;1:68–75.
[41] Carvalho IP, Pais VG, Almeida SS, Ribeiro-Silva R, Figueiredo-Braga M, Teles A, et al. Learning clinical communication skills: outcomes of a program for professional practitioners. Patient Educ Couns 2011;84:84–9.
[42] Carvalho IP, Pais VG, Silva FR, Martins R, Figueiredo-Braga M, Pedrosa R, et al. Teaching communication skills in clinical settings: comparing two applications of a comprehensive program with standardized and real patients. BMC Med Educ 2014;14:91 Epub 2014 May 9.
[43] Carvalho IP, Ribeiro-Silva R, Pais VG, Figueiredo-Braga M, Castro-Vale I, Teles AAS, et al. Teaching doctor-patient communication-a proposal in practice. Acta Med Port 2010;23(3):527–32.
[44] Von Lengerke T, Kursch A, Lange K. The communication skills course for second year medical students at Hannover Medical School: an evaluation study based on students' self-assessments. GMS Z Med Ausbild 2011;28 Doc54.
[45] Tiuraniemi J, Läärä R, Kyrö T, Lindeman S. Medical and psychology students' self-assessed communication skills: a pilot study. Patient Educ Couns 2011;83:152–7.

[46] Aper L, Reniers J, Koole S, Valcke M, Derese A. Impact of three alternative consultation training formats on self-efficacy and consultation skills of medical students. Med Teach 2012;34:e500–7.
[47] Hausberg MC, Hergert A, Kröger C, Bullinger M, Rose M, Andreas S. Enhancing medical students' communication skills: development and evaluation of an undergraduate training program. BMC Med Educ 2012;12:16 Epub 2012 Mar 24.
[48] Noble LM, Kubacki A, Martin J, Lloyd M. The effect of professional skills training on patient-centredness and confidence in communicating with patients. Med Educ 2007;41:432–40.
[49] Joekes K, Noble LM, Kubacki AM, Potts HWW, Lloyd M. Does the inclusion of "professional development" teaching improve medical students' communication skills? BMC Med Educ 2011;11:41.
[50] Fernandez-Olano C, Montoya-Fernandez J, Salinas-Sanchez AS, Fernndez-Olano C, Montoya-Fernndez J, Salinas-Snchez AS. Impact of clinical interview training on the empathy level of medical students and medical residents. Med Teach 2008;30:322–4.
[51] Winefield HR, Chur-Hansen A. Evaluating the outcome of communication skill teaching for entry-level medical students: does knowledge of empathy increase?. Med Educ 2000;34:90–4.
[52] Ozcan CT, Oflaz F, Bakir B. The effect of a structured empathy course on the students of a medical and a nursing school. Int Nurs Rev 2012;59:532–8.
[53] Taveira-Gomes I, Mota-Cardoso R, Figueiredo-Braga M. Communication skills in medical students—an exploratory study before and after clerkships. Porto Biomed J 2016;1(5-5):173–80. Available from: http://dx.doi.org/10.1016/j.pbj.2016.08.002.
[54] Borycki E, Joe RS, Armstrong B, Bellwood P, Campbell R. Educating health professionals about the electronic health record (EHR): removing the barriers to adoption. Knowl Manag E-Learning 2011;3:51–62.
[55] Stellefson M, Hanik B, Chaney B, Chaney D, Tennant B, Chavarria EA. eHealth literacy among college students: a systematic review with implications for eHealth education. J Med Internet Res 2011;13(4) 2011;13(4).
[56] Back AL, Arnold RM, Baile WF, et al. Efficacy of communication skills training for giving bad news and discussing transitions to palliative care. Arch Intern Med 2007;167(5):453–60.
[57] Nasca TJ, Philibert I, Brigham T, Flynn TC. The next GME accreditation system–rationale and benefits. N Engl J Med 2012;366(11):1051–6.
[58] Pendleton D, Schofield T, Tate P, Havelock P. The consultation: an approach to learning and teaching. Oxford: Oxford University Press; 1984.
[59] Bird J, Cohen-Cole SA. The three-function model of the medical interview: an educational device. Adv Psychosom Med 1990;20:65–88.
[60] Frankel RM, Stein T. Getting the most out of the clinical encounter: the four habits model. Perm J 1999;3:79–88.
[61] Smith RC, Marshall-Dorsey AA, Osborn GG, Shebroe V, Lyles JS, Stoffelmayr BE, et al. Evidence-based guidelines for teaching patient-centered interviewing. Patient Educ Couns 2000;39:27–36.
[62] Makoul G. Essential elements of communication in medical encounters: the Kalamazoo consensus statement. Acad Med 2001;76:390–3.
[63] Kurtz SM, Silverman JD. The Calgary–Cambridge referenced observation guide: an aid to defining the curriculum and organising the teaching in communication training programmes. Med Educ 1996;30:83–9.
[64] Makoul G. The SEGUE Framework for teaching and assessing communication skills. Patient Educ Couns 2001;45:23.
[65] Kalet A, Pugnaire MP, Cole-Kelly K, Janicik R, Ferrara E, Schwartz MD, et al. Teaching communication in clinical clerkships: models from the macy initiative in health communications. Acad Med 2004;79:511–20.
[66] van Thiel J, Ram P, van Dalen J. MAAS-global manual. Maastricht. Maastricht University; 2000.
[67] Roter D, Larson S. The Roter interaction analysis system (RIAS): utility and flexibility for analysis of medical interactions. Patient Educ Couns 2002;46:243–51.
[68] Frankel R, Alfschuler A, George S, Kinsman J, Jimison H, Robertson NR, et al. Effects of exam-room computing in clinician-patient communication. J Gen Intern Med 2005;20:677–82.
[69] Stein T, Frankel RM, Krupat E. Enhancing clinician communication skills in a large healthcare organization: a longitudinal case study. Patient Educ Couns 2005;58:4–12.
[70] Assis-Hassid S, Reychav IHT, Pliskin JS, Reis S. Enhancing patient-doctor-computer communication in primary care: towards measurement construction. Israel J Health Policy Res 2015;4:4–17.

[71] Morrow JB, Dobbie AE, Jenkins C, Long R, Mihalic A, Wagner J. First-year medical students can demonstrate EHR-specific communication skills: a control-group study. Fam Med 2009;41:28–33.
[72] Pearce C, Dwan K, Arnold M, Phillips C, Trumble S. Doctor, patient and computer-a framework for the new consultation. Int J Med Inform 2009;78(1):32–8.
[73] Onur Asan Henry N, Young Betty. Chewning, Enid Montague. How physician electronic health record screen sharing affects patient and doctor non-verbal communication in primary care. Patient Educ Couns 2015;98(3):310–31.
[74] Pearce C, Arnold M, Phillips C, Trumble S, Dwan K. The patient and the computer in the primary care consultation. J Am Med Inform Assoc 2011;18:138–42.
[75] Krupat E, Rosenkranz SL, Yeager CM, Barnard K, Putnam SM, Inui TS. The practice orientations of physicians and patients: the effect of doctor-patient congruence on satisfaction. Patient Educ Couns 2000;39:49–59.
[76] Swenson SL, Buell S, Zettler P, White M, Ruston DC, Lo B. Patient-centered communication: do patients really prefer it? J Gen Intern Med 2004;19:1069–79 2004.
[77] Sobral D, Figueiredo-Braga M. Family physician communication, quality of care and the use of computer in the consultation—the patient's perspective. Original research article. Proc Comput Sci 2016;100:594–601.
[78] van den Brink-Muinen A, van Dulmen AM, Jung HP, Bensing JM. Do our talks with patients meet their expectations? J Fam Pract 2007;56(7):559–68.
[79] Garcia-Sanchez R. The patient's perspective of computerized records: a questionnaire survey in primary care. Inform Prim Care 2008;16:93–9.

CHAPTER

12

Strategies Through Clinical Simulation to Support Nursing Students and Their Learning of Barcode Medication Administration (BCMA) and Electronic Medication Administration Record (eMAR) Technologies

Richard G. Booth[1], Barbara Sinclair[1], Gillian Strudwick[2], Jodi Hall[3], James Tong[1], Brittany Loggie[1] and Ryan Chan[1]

[1]Western University, London, ON, Canada [2]Centre for Addiction and Mental Health (CAMH), Toronto, ON, Canada [3]Fanshawe College, London, ON, Canada

INTRODUCTION

Globally, information and communication technologies have been widely adopted in the provision of healthcare services. With increased utilization of these forms of healthcare innovation, many schools of nursing have undertaken significant revisions to curricula and educational delivery practices in an effort to respond to the changes brought about by the increased prevalence of health technologies. The recent addition of informatics and other technological competencies by nursing education and their accreditation bodies [1,2] has signaled a new collective appreciation for the importance of health technology use in clinical practice.

Health Professionals' Education.
DOI: http://dx.doi.org/10.1016/B978-0-12-805362-1.00012-7

Although there has been significant change and evolution within nursing education toward the receptivity of informatics, some educational practices and teaching-learning approaches still remain aligned to processes of a predigital healthcare environment. For instance, a range of nursing-sensitive topics like documentation practices, medication administration, and the development of nurse–patient therapeutic relationships in digital contexts have not been comprehensively researched within the profession. As well, it remains common practice to find schools of nursing who continue to teach medication administration procedures aligned to paper-based record keeping processes, even in light of the growing prevalence of barcode medication administration (BCMA) and electronic medication administration record (eMAR) technologies within clinical practice areas. With increasing use of health technology in clinical practice [3], it is essential that nurse educators develop deeper understandings of how technologies like BCMA and eMAR can be integrated meaningfully in nursing education. To assist in this endeavor, the purpose of this chapter is twofold: (1) to provide insights related to the importance of teaching medication administration practices using BCMA and eMAR within nursing education; and, (2) based on insights and lessons-learned, to outline a range of suggestions toward developing, implementing, testing, and designing curricula for simulation incorporating BCMA and eMAR.

BCMA/EMAR AND THE IMPORTANCE TO NURSING EDUCATION

Barcode Medication Administration (BCMA) and Electronic Medication Administration (eMAR) Technologies

According to the Mosby Medical Dictionary [4], *medication administration* is defined as the act of "preparing, giving and evaluating the effectiveness of prescription and nonprescription drugs" (p. 1929). As outlined by Hughes and Biegen [5], nurses are heavily involved in many aspects of medication administration within healthcare. Until recently, most medication administration practices and processes undertaken by nurses have utilized various paper or analog methodologies for record keeping, reconciliation, and administration. With the wider-scale diffusion of barcode medication administration (BCMA) and electronic medication administration (eMAR) technologies across various elements of the healthcare system, developing evidence informed teaching approaches toward this new form of health technology has become an immediate concern for nurse educators [6]. Unfortunately, there is currently little research or guidance for nursing educators toward implementing BCMA and eMAR education within nursing curricula. This lack of knowledge base to inform curriculum development and delivery is acutely noticeable within the nursing simulation literature, where there are only a few articles [6–9] outlining how best to develop and evaluate BCMA and eMAR sensitive education for students.

BCMA and eMAR technologies are defined by the United States' Agency for Healthcare Research and Quality (AHRQ) [10] as interrelated systems that are used to ensure various elements of a medication administration process are in compliance with various safety *medication rights* procedures commonly exercised across many jurisdictions (i.e., that the medication is being given to the correct patient, at the correct time, route, and dose).

BCMA and eMAR automate some of these *medication rights* procedures [11], which have been traditionally accomplished through a combination of paper record keeping and safety checks manually conducted by the administering nurse (or related clinician). For instance, BCMA systems use a series of barcodes that are affixed to medication packages and identification wristbands for patients (Fig. 12.1).

Through barcode validation of both the medication and patient, the BCMA system validates that the medication about to be administered to the patient is the correct type, time, dosage, and route. Commonly operating alongside a BCMA is an eMAR system. An eMAR is a related platform that documents and records the administration of medications

FIGURE 12.1 Barcoded medications used for simulated nursing practice.

Patient Name: Smith, Mabel Age: 80 Allergy: No known drug allergies Date: 2017-02-09
PIN: 000 001 010 Physician: Dr. Kneecap
Patient Scan:

Order	Time	Scanned Item	Check	Time Stamp	Signature	Comments
Calcium Carbonate 1 250 mg orally twice daily	0800					
	1700					
Dalteparin 5 000 units subcutaneously daily	0800					
Docusate Sodium 100 mg orally twice daily	0800					
	1700					
Vitamin D 1,000 international units orally daily	0800					
PRN Medications						
Acetaminophen 325 mg with Codeine 30 mg 1 tablet orally every 4 hours PRN	PRN					
Morphine 5 mg subcutaneously every 4 hours PRN	PRN					

Medication Lookup

FIGURE 12.2 Interface of a simulated Electronic Medication Administration Record (eMAR) used for nursing education.

(Fig. 12.2). Much like a paper-based medication administration record, an eMAR functions in the same manner.

For the purposes of the remainder of the chapter, BCMA and eMAR will be discussed in conjunction with each other (i.e., BCMA/eMAR), due to their close and mutually dependent nature. The BCMA/eMAR system is commonly accessible by a nurse through the use of a wireless computer terminal (and the related hardware like a monitor, keyboard, mouse, and barcode scanner), sometimes mounted on a mobile workstation, to afford clinicians a portable unit that the patient's bedside (Fig. 12.3).

Medication Rights Framework and Nursing Education

Commonly, medication administration education has utilized a range of strategies to ensure students prepare and administer medications in a safe manner. One mechanism that has been used within nursing education extensively is the *medication rights* framework (i.e., right patient, medication, time, route, and dosage) [12]. This *medication rights* framework has been expanded over the years by various groups, and is currently suggested to contain anywhere from five to ten individual *rights* [12–15]. Unfortunately, there has been little consensus or global collaboration regarding the medication *rights* expansion, and the published work often lacks agreement on recommendations beyond the classical interpretation of the five original rights. Although *medication rights* are still important in the procedural elements of medication administration, the advent of BCMA/eMAR technology into practice has fundamentally changed how nurses administer medication in practice—including the explicit operationalization of *medication rights*. As outlined by Novak et al. [16], technology like BCMA/eMAR often enforce "rigid interpretations" (p. e333) of the

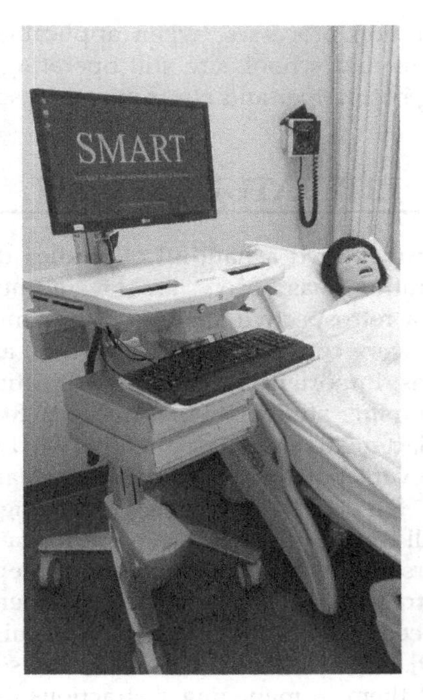

FIGURE 12.3 BCMA/eMAR mobile workstation used for simulation.

medication rights ontology, and subsequently require end-users like nurses to modify their practice, workflows, and behaviors to conform to the technology's stepwise requirements. Although a strict interpretation of *medication rights* is beneficial in many instances in terms of improved accuracy and safety, considerable changes to previously established patterns of nursing workflow can occur [17]. Along with changes to clinician workflow, BCMA/eMAR technologies can also facilitate new process challenges [11] and introduce latent medication errors [18] to the administration process. As outlined by Koppel et al. [19,20], a number of human-machine interface and interaction processes that are active in BCMA/eMAR administration can give rise to new types of medication errors that are highly unlikely within a paper-based administration process. Although there is significant research literature outlining the safety benefits of BCMA/eMAR technology [21–23], nursing educators must be proactive and aware of the unexpected consequences that can occur when historically established medication administration processes are supplanted by new methods brought by BCMA/eMAR (i.e., workarounds, generation of latent medication errors, etc.). Therefore, given the increasing use of BCMA/eMAR technology within clinical practice, it is essential that nursing educators reconceptualize and develop evidence-based approaches to medication administration instruction and learning in their curricula. As outlined above, merely transposing the classical *medication rights* framework into a digital environment (and its subsequent instruction to students) is not appropriate due to the significant differences in processes and workflows required with BCMA/eMAR. For the remainder of the chapter, the authors outline the differences between paper-based and digital medication administration processes (when applicable), as it is appreciated that many healthcare organizations and schools are still operating in a hybrid model fashion (i.e., simultaneously utilizing both paper and electronic processes).

LITERATURE REVIEW

A review of the literature identified a handful of studies that investigated or detailed how medication administration has been taught in nursing education to date [8,9,12,24–26]. For instance, a retrospective analysis of nursing student medication errors reported that system factors were common causes of errors, and these factors were often not taught or highlighted as important in nursing education [26]. As well, previous research has reported that nursing students are commonly fearful of making medication errors during their clinical placement, and subsequently relied in some ways on their clinical instructor to catch or prevent potential errors [24]. Similarly, Orbaek et al. [8] found that even when students felt confident using technology to support medication administration, they were still fearful of committing a serious error.

In a study aimed at understanding nursing students' perceptions of teaching strategies for safe medication administration, students reported that demonstrations, learning from their peers, repetition and feedback were important mechanisms to gain competence in medication administration [9]. As well, these students expressed an interest in developing strategies that could support them in managing distractions, interruptions, and computer generated alerts that they anticipated experiencing in clinical environments. The authors of the study suggested that simulation activities should be created to mimic real-world situations and the related contextual conditions of clinical practice.

To date, there has been limited direct research exploring educational strategies to best support nursing students' use of BCMA or eMAR technology in education [27–29]. For example, Nickitas et al. [30] discussed the process by which a school of nursing acquired and implemented an electronic health record inclusive of BCMA/eMAR technology into the curriculum. Teaching strategies included students demonstrating their ability to navigate through an electronic health record, documentation within the record, identification of standardized terminology, and comprehension of privacy implications related to technology. Training lab sessions were used to introduce and use the technology with nursing students, however the duration of these sessions were limited, and as such the instructors involved in the curriculum delivery did not feel it was the best use of student laboratory time.

Cherry [25] conducted a study examining nursing faculty knowledge and confidence after a *Safe Medication Practices* seminar, which emphasized the importance of using simulation in teaching and learning medication administration. In this study, nursing faculty were taught how to conduct high-fidelity medication scenario exercises with their students. At the end of the seminar, nursing faculty reported improved knowledge and confidence in teaching appropriate and relevant medication administration skills to their students.

In a paper by Angel et al. [7], the authors discussed how the BCMA/eMAR system was integrated into an associate degree nursing program. BCMA/eMAR technology was obtained and incorporated into the nursing skills lab where high-fidelity interprofessional simulation scenarios were developed, and delivered to students. Students utilized barcode scanners and the BCMA/eMAR system to administer simple medications in their first semester as students. As they continued through their nursing program, the complexity of the BCMA/eMAR-related scenarios also evolved. In this study, the majority of the students indicated that the use of the system in a clinical simulation environment enhanced the realism of using and learning about the technology. As such, using simulation to teach BCMA/eMAR skills to nursing students is reported by Angel et al. to be a potentially effective mechanism to integrate theoretical knowledge with other technologically driven elements of the medication administration process.

In summary, research that has investigated best practices associated with teaching and learning medication administration broadly have highlighted the need for learning to take place in environments that attempt to replicate real life clinical contexts and processes. As such, simulation scenarios using relevant BCMA/eMAR has been suggested to be an effective mechanism from which to promote the integration of theoretical and practical medication administration learning.

DEVELOPING OR PROCURING A BCMA/EMAR PLATFORM FOR SIMULATED EDUCATION

Although increasing in popularity, BCMA/eMAR platforms that are customized and developed for simulated practice are far from ubiquitous within nursing education. Due to a number of likely technical, financial, and pedagogical reasons, the diffusion of BCMA/eMAR technology into nursing education appears only now to be gaining traction. Beyond a few published accounts and dissertations, there is a discernable lack of knowledge regarding how to develop or implement BCMA/eMAR systems into nursing education. In order to generate usable recommendations, a range of suggestions and insights will be presented regarding the current repertoire of options available to educators wishing to implement BCMA/eMAR into their simulated clinical education.

BCMA/eMAR Technology

To date, there are three main methods from which to procure an educational BCMA/eMAR system that can be used for simulated nursing education: (1) use a real BCMA/eMAR system, sold by a technology vendor; (2) use a simulated BCMA/eMAR simulator which has been designed for nursing or clinical education; and (3) use a BCMA/eMAR simulator that has been developed in-house or retrofitted from open access software. All three options possess benefits and limitations, and are highly dependent on the context in which the school of nursing operates as well as available technical and financial resources.

Using a Real BCMA/eMAR System

Within many clinical and health system environments, a range of vendor supplied BCMA/eMAR technology exists. These *real* health information systems are commonly connected to the larger information system of the healthcare organization and are certified technology, conforming to a range of relevant federal and international standards. Commonly, vendors who design and develop BCMA/eMAR also develop training domains or platforms, based on their original product. These training domains typically have dummy data inserted into the BCMA/eMAR or allow educators to enter their own patient information scenarios into the system. Similarly, training domains are typically kept separate from the actual clinical BCMA/eMAR system and its related data, in order to avoid any potential for a privacy breach of personal health information.

A benefit of obtaining access to the training domain of a real BCMA/eMAR platform is that students will be afforded the opportunity to learn and train with a system that is actually used within practice settings. From a fidelity perspective, training domains of BCMA/eMAR are commonly the same platform with identical features of the real system. This can allow students the opportunity to use a system they will likely experience later within their clinical placements. Unfortunately, one of the major drawbacks of using the training domain of a real BCMA/eMAR is cost. Due to the proprietary nature of vendor supplied BCMA/eMAR systems, significant licensing and maintenance fees are commonly required. These potential licensing fees are also amplified when the cost of the hardware to support the BCMA/eMAR software is factored into the overall pricing footprint of the system.

Using a Simulated BCMA/eMAR Platform

Over the last few years, a range of products has entered the market that allows educators to obtain a simulated BCMA/eMAR platform for students. Commonly generated by academic publishing companies or videogame producers, these sorts of simulated BCMA/eMAR platforms replicate aspects of the functionality and fidelity of a real system, but at a reduced cost of a real system. A benefit of using a simulated BCMA/eMAR platform is the predeveloped specificity of the software to act as training and learning tool. Unlike real BCMA/eMARs, these sorts of simulated BCMA/eMAR platforms tend to be more learner centric and purposefully scaffolded for consistency in the learning experience. Similarly, since these systems are designed for learning, they typically do not possess the same levels of complex functionality of a real BCMA/eMAR, which may be beneficial for a novice learner who is still learning the basics of medication administration processes, workflow, and clinical judgment. These types of simulated BCMA/eMAR platforms are often accessible via the Internet and a computer web browser, removing the need for a school of nursing to host the platform internally. Regardless, simulated BCMA/eMAR platforms usually operate through a purchased service model, with ongoing licensing fees required to access the system. Although fees vary, it is not uncommon to find pricing models of per student, per semester. Similarly, like real BCMA/eMAR, the cost of the hardware to complement the simulated BCMA/eMAR also needs to be factored into the pricing and sustainability equation.

Using a Homegrown or Open Source BCMA/eMAR Platform

An approach undertaken by some schools of nursing has been to develop their own simulated BCMA/eMAR platform for educational purposes. To date, there are a few examples of homegrown or open source implementations of electronic health technology like BCMA/eMAR published in the nursing literature [7,31–33]. Many of these simulated systems were developed by faculty or interprofessional teams with expertise in various elements of software design and clinical practice. Other systems of this complexion have been obtained through open source code of BCMA/eMAR technology, which has been made freely available. Although developing a homegrown or implementing an open source BCMA/eMAR system can be cost effective, it is a task that requires significant knowledge in terms of programing, development, and upkeep. Unlike the two other options provided above, all upkeep and maintenance of these homegrown or open access BCMA/eMAR systems are typically dependent on faculty and staff, and their related technical resources to support operations.

Key Principles for an Educational BCMA/eMAR

Regardless of the BCMA/eMAR procurement approach, a few key principles should be kept in mind. First, the amount of system functionality should be purposefully and tactfully attenuated for new learners. Although it may be tempting to seek a platform that offers the most varied and complex array of features (e.g., decision support, seamless integration with electronic health records, etc.), systems with a large number of features may detract from the actual learning of new learners. Therefore, it is recommended that a balance be sought, weighing the current and future learning needs of students, the types of functionality that are essential to embed into BCMA/eMAR practice, and obtaining a system that is reasonably customizable by educators (i.e., allowing faculty to turn on/off specific functionality to match the current knowledge of students and prevent cognitive overload). Second, a system that is generic enough to build student competency in the process elements, workflow, and clinical reasoning required to conduct BCMA/eMAR is likely of more importance to new learners than the actual technical fidelity of the system. Since there are numerous types of BCMA/eMAR in operation within healthcare environments, using a platform agnostic approach to teaching-learning may be beneficial for students and their future transferability of knowledge between simulated and practice settings. Finally, the cost and sustainability of BCMA/eMAR needs to be heavily vetted and planned for in the process of procurement and implementation. Since use of BCMA/eMAR is likely to increase in prevalence over the coming decade, educators need to ensure that investments are made immediately, yet wisely, to support the learning of future nursing students.

BEST PRACTICES RELATED TO SIMULATED BCMA/EMAR MEDICATION ADMINISTRATION

In this section, a range of topics will be explored in regards to the teaching aspects of simulated BCMA/eMAR processes within nursing education. First, best-practice approaches to teaching-learning medication administration in electronic contexts will be highlighted. Second, insights related to meaningful approaches from which to embed medication administration education into simulation and theory will be discussed.

Best Practices to Teaching-Learning Medication Administration

To date, the nursing literature exploring the use of BCMA/eMAR technology in education has not been systematically developed or synthesized. Therefore, the majority of the insights provided in these following sections are derived from published research arising from the health informatics literature (namely, exploring practicing clinicians' usage of these forms of clinical technology) and through experiential knowledge of the chapter authors.

Like all educational practices, it is important for an educator to understand and appreciate the contextual environment in which teaching-learning activities occur. Currently, for many schools of nursing, BCMA/eMAR education present as a significant departure from traditional approaches to teaching medication administration. Given the established processes related to medication administration and its education, the introduction of a new process and related technology like BCMA/eMAR should be conceptualized both as a curriculum evolution and also a culture change for simulated practice education. Therefore, it is recommended that educators who attempt to implement BCMA/eMAR into clinical simulation do so in a collaborative approach, which appreciates the subtle yet significant effects that this type of curricular change can have on other various dependencies. These include but are not limited to: faculty development and clinical competency; predeveloped simulation content; student assessment methods; interprofessional education; and other financial considerations related to procurement, implementation, and sustaining BCMA/eMAR platforms.

Reinterpretation of the Medication Administration Process Underpin by BCMA/eMAR

Approaches in nursing education have commonly used the structured *medication rights* ontology from which to frame the processes or administration practices education [12]. Although the use of a rigid rights based approach is still acceptable from a practice perspective, this type of stepwise approach to medication administration education may not always translate effectively to certain elements of the administration process underpinned by BCMA/eMAR. As outlined in the background of this chapter, *rights* based approaches to medication administration sometimes force an overly structured interpretation of process, which can be amplified when enforced by BCMA/eMAR platforms. Since many BCMA/eMAR platforms have been developed to ensure compliance to various medication rights rather than the actual organization of work [19], conflict can occur when a learner attempts to operationalize their interpretation of administration processes, against a digitized system that also enforces its own specific (commonly stepwise) procedure. Due to the complex and dynamic nature of nursing practice and medication administration, a *one size fits all* approach to performing complex tasks like BCMA/eMAR medication administration can lead to potential process collisions [16] between the user and the technology.

These sorts of *collisions* are commonly seen when a user, like a student or nurse, attempts to execute a mental schema of a process in a specific and clinically logical order, but is impeded or cognitively persuaded by the BCMA/eMAR system to perform the action in a different manner. For instance, the literal operationalization of the *medication rights* ontology (right patient, medication, dose, route, and time) learned in a paper-based context may differ substantively within an electronic environment, whereby various login, barcode scanning, and interface navigation requirements now complement or replace various previous manual tasks. The authors of this chapter have found (manuscript forthcoming) that previous knowledge and competency with paper-based medication administration processes have a lingering tendency to cause students to execute mental process schemas that are different from the mandated processes required in, and by, BCMA/eMAR systems. Thus, the inability for a learner to execute their previously conceptualized mental schema as related to medication administration in a paper-based environment sometimes leads to cognitive dissonance and the potential for non-clarity in the newly learned BCMA/eMAR process.

Similarly, although BCMA/eMAR platforms typically mandate a structured and stepwise approach to medication administration, other research has uncovered that workarounds and procedural deviations using this type of technology are common [11,20,34,35]. For instance, Carayon et al. [34] found that nurses generated 18 different sequences for medication administration in 59 work process observations, with 16% of these observations possessing actions that were potentially unsafe. This insight provided by Carayon et al. is extremely important for educators in terms of simulation best practices, given the fact that BCMA/eMAR technologies are commonly designed to improve safety and standardize various medication administration processes. The existence of various workarounds and deviations from administration policies is further evidence that medication administration, within a digitized environment, is far from being easily standardized to one discrete process or workflow. Therefore, as previously mentioned, enforcing a classical *medication rights* framework upon BCMA/eMAR as a mechanism from which to educate students to procedural elements of medication administration is likely to foster a continuum of latent issues or procedural risks. Instead, returning to fundamentals of medication administration that explore clinical insight, judgment, and decision making [12,36] is recommended as the primary point of insertion of BCMA/eMAR education into simulation. Generating simulations where students can comment, reflect, and respond to situations that enact clinical decision-making (prior to the addition of psychomotor or human-technical interface components of BCMA/eMAR) is also suggested to precede all interaction with clinical technology. Second, reducing the foci or expectation that students enact a *medication rights* mnemonic or ontology in a prescriptive fashion may also assist in the transference of knowledge to BCMA/eMAR methodologies. Although there is value in the mnemonic offered by the *medication rights* framework, encouraging students to remain flexible in the application of their learning and openness to evolve processes guided by both clinical judgment and the structure of the BCMA/eMAR platform is required. Medication administration cannot be viewed as a static learning artifact; rather, it should be conceptualized as a dynamic, human-technical interaction between the human user and the technology. Much like driving an automobile is a symbiotic relationship between a human and vehicle (i.e., the vehicle cannot operate without input from the human driver), medication administration using technology should be viewed and taught in a similar fashion.

To maximize this learning dynamic between human user and technology, it is further recommended that the entire medication administration process not be taught in one discrete learning experience; instead, the process of administering medication should be purposefully divided into meaningful sections presented over time to allow students to internalize their learning. Also, by subdividing the medication process into preplanned sections, ongoing reflection by the educator can be inserted, reminding learners to be cognizant that clinical judgment and insight are equally as important as the decision support and patient/medication validation features afforded by BCMA/eMAR technology. By making learners aware that BCMA/eMAR technology are not infallible (and can *nudge* a user toward a different decision pathway, which may not be congruent with clinical best practices) is a discussion that should be offered early on in the education of BCMA/eMAR. Insomuch, decision support and the automation of tasks within BCMA/eMAR environments should be conceptualized as a supportive tool to inform and assist clinical practice; not a replacement for human judgment, insight, and clinical decision-making.

Finally, learners should be afforded the opportunity to interact with the system in a non-structured fashion. Drawing from other research [37], it was found that nurses best learned the use of technology when provided with informal learning opportunities from which to experience and use health technology. Commonly, formalized education sessions (e.g., classroom and lecture) are used to train students and nurses in the use of health technology like BCMA/eMAR. Although this approach is cost and time effective, it demotes the potential for learners to explore the system in a more non-restricted fashion and to learn informally through peers, instructors, and experimentation. Using a combination of both formal and other informal learning opportunities (e.g., *in situ* opportunities to examine and interact with the system without dedicated curricula requirements or assessment) has been suggested to be a valuable approach to help users become better acquainted with health technology, develop new patterns of usage, and deeper learning of systems [37].

Embedding Medication Administration Content Into Nursing Curricula

As outlined above, given the changes to the medication administration processes afforded by BCMA/eMAR, embedding this type of educational content into nursing education needs to be carefully planned and implemented. It is common for nursing students to receive labs and other simulated sessions related to medication administration. This type of experiential knowledge is eventually combined with practice experience gained within real clinical environments, where skills learned in simulation can be operationalized. Therefore, as an educator, it is vitally important that medication administration and its education be viewed as a longitudinal process that scales appropriately with the student's understanding, maturity, and clinical judgment abilities. Similarly, the authors of this chapter view the learning and education of medication administration to be more than a clinical skill delivered within practice settings; it is an approach to knowledge/knowing and a skill set, which should be incorporated into relevant theory courses, where experiential learning opportunities are possible. The following section will outline and describe some of the recommendations toward the functional integration of BCMA/eMAR medication administration knowledge into nursing education, and specific recommendations for simulated and theory-based educational contexts.

Simulated Practice Setting

Embedding BCMA/eMAR medication administration opportunities within simulated practice settings is both an important and necessary action for nursing educators in order to prepare the next generation of nurses to operate in digitally intense clinical environments.

As reported in the literature, nursing students typically progress in their skills, competency, and comfort related to medication administration over the duration of their nursing education [12]. Similarly, medication administration is not conceptualized by the chapter authors to be a cross-sectional task or skill taught within curricula; rather, as a longitudinal experience that is built upon and refined over time, through repeated experiences and assessment. Therefore, we recommend that *how* medication administration is taught or delivered in simulated settings should be evolved from an episodic experience, and embedded as an ongoing longitudinal experience that appropriately scaffolds in conjunction to the learner's abilities and progression in the nursing program. Within simulated practice, a unique opportunity is provided to educators to maximize this scaffolding process, through simulation scenarios that can be preplanned to be longitudinal in nature, yet stable enough to offer a consistent and evolving experience for the learner over time. The use of longitudinal or unfolding situations that both mimic a natural progression of a patient's illness journey and increasing complexity of medication administration activities, are proactive approaches to scaffolding learning opportunities for students [38,39]. In a recent study conducted by the authors, it was found that junior nursing students struggled with the complexity of medication administration when confounded with elements of increasing patient acuity, and medication preparation or order complexity. The further addition of a BCMA/eMAR system into the process complicated the situation to the point where students had difficulty internalizing the features of the technology, and what role they played in the actual administration process. Similarly, it was also found that the preplanned progression through medication administration complexity (i.e., different types of dosages, order types, etc.) over the semester long course was at times still too complex for students, as the presence of the BCMA/eMAR in the process exacerbated the learning curve of students in unintended ways. For instance, the cognitive and clinical judgment leap between administrations of a single oral medication for a stable adult patient, versus a repeated sublingual PRN (as needed) medication (i.e., nitrospray) was more significant than first realized. Given the new learners' lack of developed understanding and appreciation of the procedural elements of the BCMA/eMAR and insight toward PRN medication orders, students became confused when confronted with these added clinical variables. In this case, the BCMA/eMAR system became one of the major variables that disrupted student clinical insight toward the completion of the complex PRN administration. Students became overly confounded by the technical requirements of the BCMA/eMAR (i.e., barcode scanning and procedural requirements mandated by the system), and subsequently, had difficulty interpreting other elements of the administration task and clinical presentation of the simulated patient. Due to this confusion, it was deemed important to normalize the progression of medication administration to avoid insertion of more complex administrative processes (like PRN) medications until educators were assured that a solid, yet flexible understanding of basic oral medication administration was achieved by students. It is hoped that through this prolonged *in situ* simulation experience, students can be provided appropriately scaled clinical situations that evolve purposefully over time to allow mastery of clinical judgment and the related psychomotor tasks involved in BCMA/eMAR medication administration.

Outside of Simulation Settings

Although the majority of experience and learning BCMA/eMAR will likely occur within simulated environments, a large repertoire of skills and insights related to medication administration can be offered to students during relevant theoretical courses. For instance, there is a growing awareness and insight within nursing education that informatics skills and competencies are required by undergraduate nursing students [1,2], in order to ensure they are competent to practice in digital health environments. Given the need to increase informatics education within undergraduate nursing education, we suggest that BCMA/eMAR is a robust and relevant topic that could be embedded into a host of nursing professional practice and other theoretical courses. Embedding BCMA/eMAR content into these types of courses will result in a different approach to understanding various elements of BCMA/eMAR, and in conjunction with simulation opportunities, provide a larger and more rounded approach to digital medication administration education. For instance, developing case situations related to medication administration that juxtapose the processes conducted in paper versus electronic can generate meaningful opportunities for students to deconstruct processes contained within each administrative practice. Positioned under the lens of Nursing Informatics, the interaction of nurses and technology in the use of BCMA/eMAR could be developed as a case study for deconstruction in a nursing theory course, whereby such technology is considered within the context of fostering or disrupting nurse–patient relationships. Given the long lineage of nursing scholarship exploring the role of technology in, for, and against elements of the nurse–patient relationship [40–42], operationalizing the delivery of BCMA/eMAR education in this fashion may be a pedagogically advantageous opportunity to reinforce how other elements of the nursing role are changed and modified by health technology.

Finally, if the school of nursing has a dedicated informatics course or related section of the program where informatics issues are discussed, it is suggested that students be tasked with developing workflow and process mappings of the medication administration process underpinned by BCMA/eMAR. By having students map and discuss the work processes involved in administering a medication using BCMA/eMAR, a deeper level of understanding and conceptualization of processes can be obtained. Similarly, this type of constructivist and exploratory learning approach can also provide students an opportunity to realize various elements where BCMA/eMAR systems sway decision making of humans. Junior students who are learning medication administration through BCMA/eMAR may not have the opportunity to reflect, discuss, or critique the medication administration processes, as a primary learning activity within simulated settings. Therefore, generating opportunities for these students to conduct critical analyses of medication administration underpinned by BCMA/eMAR, with the sole purpose of generating deeper insights and recommendations for nursing practice may be a robust way to deepen the knowledge of this form of medication administration process that is difficult to obtain during a simulated clinical situation.

EVALUATION, QUALITY ASSURANCE, AND RESEARCH

Ongoing quality assurance and improvement within the simulation experience should be sought, for the benefit of both nursing students and also curriculum development. Although there is a long lineage of research exploring simulation published in the literature [43], there is a shortage of informative and methodological works describing how best to evaluate health technology used in, and for, simulated practice. Given aforementioned changes the introduction of BCMA/eMAR administration practices bring to clinical practice, developing a sensitive methodology to assess this type of technology used in simulated practice is essential if nursing educators are to truly understand the impacts of this type of technology. Therefore, in this section the authors of this chapter will provide a range of insights in relation to the evaluation of BCMA/eMAR within a simulated setting. Traditional research methodologies that quantitatively attempt to examine individual student or faculty perceptions and attitudes will be minimized, given the limitations of these approaches toward ascertaining a broader understanding of the phenomenon of interest. Instead, a different evaluation technique will be introduced that is sensitive to the dynamic and nuanced nature of how BCMA/eMAR influence both students and the simulation experience in general.

Evaluating BCMA/eMAR Used in Simulation

As with any research project, first understanding the purpose, requirements, and question is pertinent—this is especially true within research exploring BCMA/eMAR, due to the relative newness of this form of technology within many nursing education programs. Second, developing a team with knowledge and skills toward evaluation is also essential. Given the multi-layered requirements of traditional research (e.g., funding, ethical approval, and methodological rigor), in-depth discussions with the research team regarding the development of a robust plan and protocol should be developed prior to commencing any level of formative evaluation.

For the purposes of this chapter, the type of evaluation methodology that will be discussed in depth could be considered a form of quality improvement or assurance research. This type of research is typically conducted to improve or refine a process or innovation, using a pragmatic evaluation approach to inform future actions [44]. For the purposes of space, the intricacies of a formal research process (e.g., ethics approval, analysis of finding, etc.) will not be provided here; rather, one observational methodology for deconstructing processes of BCMA/eMAR administration will be described in depth, with the hope that it can be used as a blueprint for other educators to follow or evolve.

Observational Approaches

Observational research approaches commonly use a range of naturalistic data collection methods, with data collection occurring in the context of the phenomena of interest. In using this sort of approach, a researcher is able to gain a realistic and descriptive interpretation of process and actions related to the phenomena under examination. Due to the interpretive nature of many observational methods, this type of research approach is commonly qualitative in nature, or mixed methods, with numerical or statistical data being used to triangulate and extend qualitative interpretations.

Within clinical simulation research, observational approaches have been used extensively to examine a range of topics, including: student assessments, mock codes, and clinical return demonstrations [45,46]. Technology assessment and evaluation within simulation is also highly related to this type of research method, given the dynamic and interactive relationships that can be shaped between students/faculty and the system under use. Since using a BCMA/eMAR system is a human-technical interaction, observational methods are valuable approaches to capture the process of work completed by students (e.g., workflow, decision making actions, etc.), their interactions with the technological system in question (e.g., usability, location of device, and influence of the technology's decision support), and the larger environment where action occurs (e.g., context, setup of the clinical simulation lab, curricula and other forces that may represent as important). A trained observer may be able to identify deficiencies in the medication administration process, which may be related to the students' knowledge, functionality of the technology, environmental context or other element. Ideally, the observer would understand the context of the clinical environment, student expectations with regards to medication administration as well as the functionality of the BCMA/eMAR system being used.

Workflow Analysis

For the purposes of BCMA/eMAR evaluation, conducting an observational *workflow analysis* [47] is an informative method from which to begin a quality improvement assessment. A workflow analysis helps to generate a detailed mapping of various processes, decision-making, and activities that occur within a specific set of actions. Conducting a workflow analysis to examine a learning process, or, build better simulation scenarios is recommended as pragmatic uses of this method. In its simplest form, a workflow analysis should attempt to capture basic elements of a process: *tasks* and *decisions* [47]. When a researcher is observing a process, *tasks* take the form of actions or behaviors that are conducted by a specific person or persons [47]. For instance, a task in a BCMA/eMAR medication administration could be "nursing student barcode scans a medication." Subsumed in this task element is both the action (i.e., barcode scanning a medication) and the individual or object that is conducting the action (i.e., nursing student). When observing a process, the identification of tasks in a process is easy to do, given that every action conducted by a student or faculty member could be further divided and subdivided into a hierarchy of subtasks. As an example, barcode scanning a medication by a nursing student could be conceivably further divided into subtasks, including, "nursing student picks up barcode scanner." Although there are no firm rules on how granular the development of tasks should be for a given project, the authors of this chapter encourage educators to return to the purpose and objectives of their study to help gauge the level of analysis that will be required when developing the specificity of tasks related to a given process.

Much like *tasks*, workflow analyses also seek to capture and record all actions that resemble *decisions* made by various people within the given process. Decisions in a workflow analysis identify points in a process where a person is required to make a selection of choices, based on other potential options [47]. For instance, in BCMA/eMAR there are a range of important decisions that are required during the process of administration, which require some level of human involvement and decision-making. The determination of whether a medication is prescribed for a patient; whether a medication is contraindicated with other medications; or, if a medication is even available locally to be administered, could all be conceivably labeled as distinct *decisions* that require a person to apply some level of judgment in order to inform subsequent tasks and decisions. Much like *tasks*, the granularity of what is conceived to be a *decision* should be vetted carefully to ensure the selected *decisions* are informative and representative of the level of analysis of the overall study. Therefore, although it is possible to generate a massive host of *decisions* related to a given process, efforts should be taken to refine and collapse these decisions into the most meaningful and informative decisions that are representative of the larger process in general.

Decisions in workflow analyses also include other dependencies that need to be conceptualized and rationalized during the collection of data. All decisions undercovered from a workflow analysis should be answerable with "yes" or "no" responses [47]. Although it is appreciated that some decisions may not be addressable with a binary response like yes/no, for the purposes of simplicity, it is advocated all outlined decisions have the ability to be sufficiently addressed with a yes or no response. By using a binary "yes" or "no" contingency, decisions can be made actionable, and lead to further tasks/decisions as mapped in the workflow analysis.

For clarity purposes, a workflow diagram of a medication administration process involving the delivery of one oral medication to a simulated patient is found below (Fig. 12.4). Removed from the workflow diagram (Fig. 12.4) are other various nursing-sensitive tasks, including hand washing, nurse–client interaction opportunities, and other psychosocial tasks related to client care. This has been purposefully done to demonstrate the skeleton elements of the medication administration process underpinned by BCMA/eMAR. This workflow analysis was generated through observation of nursing students and faculty, in the simulated medication administration process, with BCMA/eMAR. Similarly, this workflow analysis contains the recorded and documented observations of a number of nursing students. Through qualitative reduction, the most representative processes observed between the different students have been combined and synthesized to provide a pragmatic understanding of common BCMA/eMAR process.

Along with observing and recording salient tasks and decisions of a given process to develop a workflow analysis, the authors of this chapter also advocate keeping a running roster of clinician or nursing-sensitive tasks that may not be essential within the larger medication administration process, but be extremely important to the nursing or client care role. For instance, hand hygiene, cleaning of the BCMA/eMAR devices, positioning of the workstation cart within the client room, opportunities for meaningful nurse–client interaction, and other care delivery factors should be collected when conducting observational workflow analyses. Although these variables and decision points are typically more subtle than the larger workflow decisions and tasks (as described above), maintaining awareness of these more recluse actions is invaluable to ensuring the findings of the workflow are grounded and pragmatic for practice purposes. Therefore, it is suggested that when conducting workflow analyses, two iterative, yet separate data collection processes occur simultaneously. The first set of data collected is focused exclusively upon the traditional tasks and decisions of a given process. The second type of data collected is sensitive toward other tasks and processes, which are nurse or client focused in nature, and do not directly influence the process of medication administration in an overt or noticeable fashion (Fig. 12.5). By combining the two iterative workflow analyses' data, both process

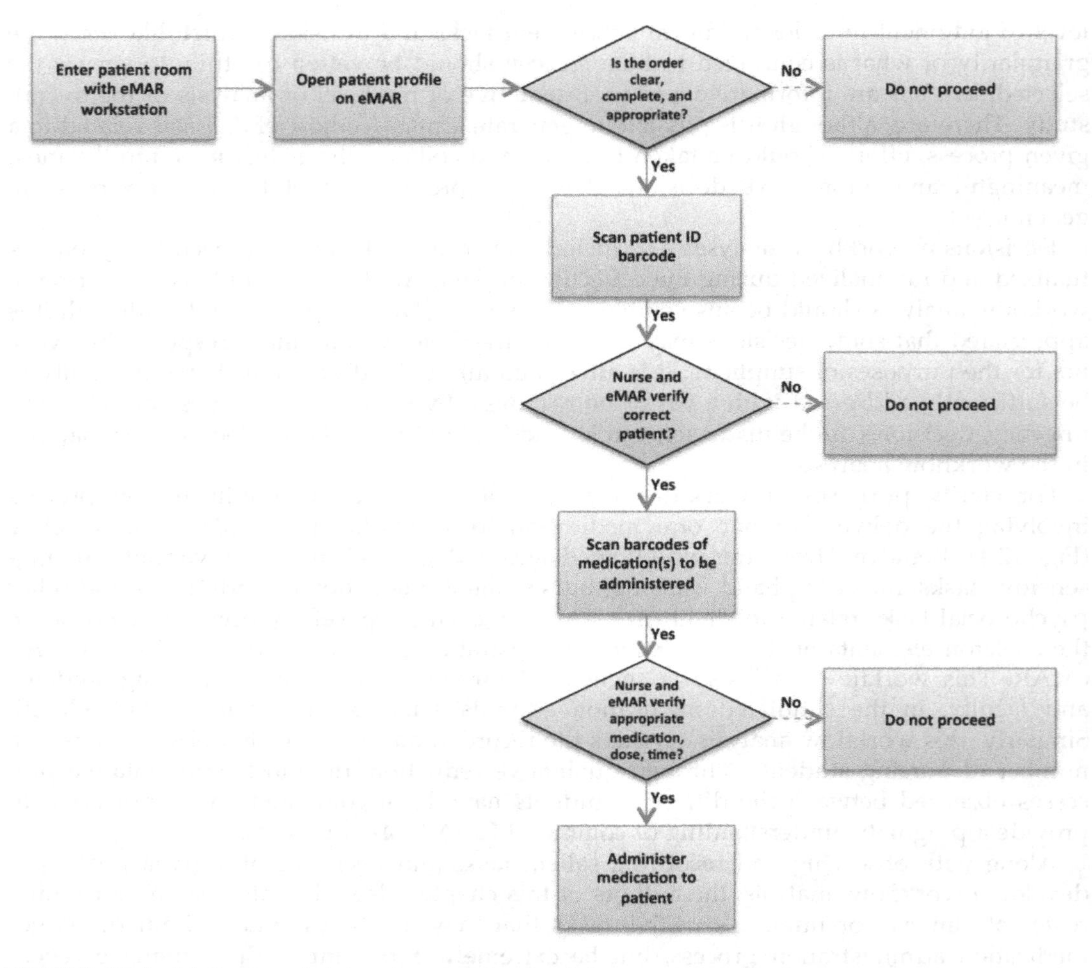

FIGURE 12.4 Medication administration workflow involving one oral medication.

elements related to medication administration and nursing and patient-sensitive considerations can be delineated.

Overall, workflow analysis can be used as a mechanism to allow an educator to unpack the complex processes, actions, and decisions contained with medication administration. Similarly, this type of qualitative evaluation process can serve a range of evaluation requirements, and be used in any phase of the development, refinement, or implementation of simulated medication administration practices. As an approach that seeks logically and concisely recorded real-world actions of users manipulating a technological system within a clinical context, workflow analyses may be helpful for educators who are wishing to gain better understandings regarding the dynamic presence generated by BCMA/eMAR in education settings (Box 12.1).

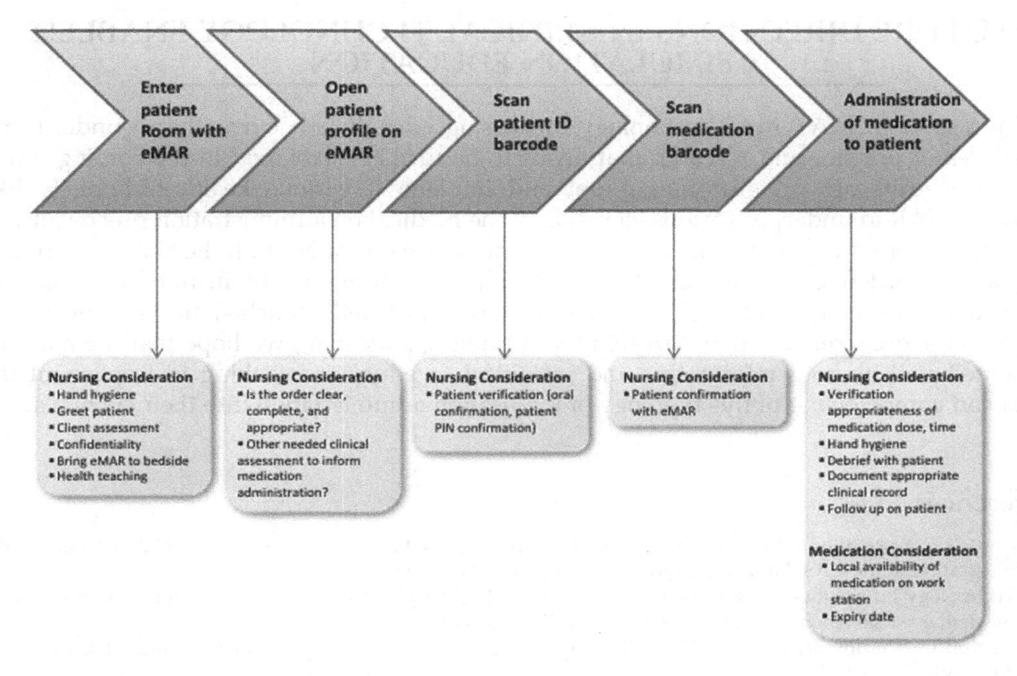

FIGURE 12.5 Other nursing and medication sensitive data collected during a workflow analysis of BCMA/eMAR medication administration.

BOX 12.1

EXAMPLES OF VARIOUS PUBLISHED WORKFLOW ANALYSES RELATED TO ELEMENTS OF BCMA/EMAR

The four example papers provided discuss both the process and outcomes of better understanding how clinicians' use BCMA/eMAR technology. Process flows associated with medication administration are detailed in the 2011 study by Grigg et al. [48]. These tools can be used to understand the workflow of conducting medication administration among nurses and nursing students in various clinical contexts. As well, the importance of understanding workflow through observational methods are clearly highlighted in the Cheung et al. [49] study reviewing the implementation of a computerized provider order entry system in an intensive care unit setting. Variables that are related to BCMA/eMAR technology workflow are further detailed in the Dasgupta et al. [50] article. Lastly, in the Kelly et al. [11] article, the authors describe a 4-phased BCMA/eMAR evaluation program to evaluate how current practices align with best practices associated with technology use. Organizations can adopt this program in an effort to improve their use of BCMA/eMAR among frontline clinicians, namely nurses.

FUTURE DIRECTIONS IN CLINICAL TECHNOLOGY ENABLED SIMULATION EDUCATION

The use of BCMA/eMAR technology in nursing education is currently an underdeveloped area of practice and research that needs to be further developed in the coming years. As more healthcare organizations adopt and implement various health technology like BCMA/eMAR to underpin various elements of the medication administration process, it will be essential for educators to ensure nursing students are educated with the skills, competencies, and knowledge to use these forms of clinical medication administration technology safely and effectively. Although this chapter has only briefly touched upon some of the important aspects related to BCMA/eMAR in nursing education, we hope that the material presented will be both informative and insightful to educators wishing to implement the skills and competencies of this typology of medication administration into their curricula.

References

[1] Canadian Association of Schools of Nursing (CASN). Nursing Informatics Teaching Toolkit: Supporting the integration of the CASN nursing informatics competencies into nursing curricula; 2013.

[2] Technology Informatics Guiding Educational Reform (TIGER). Overview informatics competencies for every practicing nurse: recommendations from the TIGER Collaborative; 2007.

[3] Huston C. The impact of emerging technology on nursing care: warp speed ahead. Online J Issues Nurs 2013;18:1.

[4] Mosby. Mosby's Medical Dictionary. 10th ed. St. Louis: Elsevier; 2017.

[5] Hughes RG, Blegen M. Chapter 37. Medication administration safety. Patient Saf Qual An Evidence-Based Handb Nurses. 2008.

[6] Anest R. Teaching patient safety with a functional electronic medication record. J Nurs Educ 2013;52:303. Available from: http://dx.doi.org/10.1016/j.ecns.2009.07.006.

[7] Angel VM, Friedman MH, Friedman AL. Integrating bar-code medication administration competencies in the curriculum: implications for nursing education and interprofessional collaboration. Nurse Educ Perspect 2016;37:239–41. Available from: http://dx.doi.org/10.1097/01.NEP.0000000000000038.

[8] Orbaek J, Gaard M, Fabricius P, Lefevre RS, Møller T. Patient safety and technology-driven medication—a qualitative study on how graduate nursing students navigate through complex medication administration. Nurse Educ Pract 2015;15:203–11. Available from: http://dx.doi.org/10.1016/j.nepr.2014.11.015.

[9] Krautscheid LC, Orton VJ, Chorpenning L, Ryerson R. Student nurse perceptions of effective medication administration education. Int J Nurs Educ Scholarsh 2011;8:1–15. Available from: http://dx.doi.org/10.2202/1548-923x.2178.

[10] Agency for Healthcare Research and Quality (AHRQ). Bar-coded Medication Administration | AHRQ National Resource Center; Health Information Technology: Best Practices Transforming Quality, Safety, and Efficiency. <https://healthit.ahrq.gov/ahrq-funded-projects/emerging-lessons/bar-coded-medication-administration>.

[11] Kelly K, Harrington L, Matos P, Turner B, Johnson C. Creating a culture of safety around bar-code medication administration: an evidence-based evaluation framework. J Nurs Adm 2016;46:30–7. Available from: http://dx.doi.org/10.1097/NNA.0000000000000290.

[12] Bourbonnais FF, Caswell W. Teaching successful medication administration today: more than just knowing your "rights". Nurse Educ Pract 2014;14:391–5. Available from: http://dx.doi.org/10.1016/j.nepr.2014.03.003.

[13] Elliott M, Liu Y. The nine rights of medication administration: an overview. Br J Nurs 2010;19:300–5. Available from: http://dx.doi.org/10.12968/bjon.2010.19.5.47064.

[14] College of Registered Nurses of Nova Scotia. Medication guidelines for registered nurses. Halifax, Canada; 2014.

[15] Bonsall L. 8 rights of medication administration. Riverwoods, IL: Lippincott Nursing Center; 2011.
[16] Novak LL, Holden RJ, Anders SH, Hong JY, Karsh BT. Using a sociotechnical framework to understand adaptations in health IT implementation. Int J Med Inform 2013;82:e331–44. Available from: http://dx.doi.org/10.1016/j.ijmedinf.2013.01.009.
[17] Staggers N, Wier C, Phansalkar S. Patient safety and health information technology: role of the electronic health record. In: Hughes R, editor. Patient safety and quality: an evidence-based handbook for nurses. Rockville, MD: Agency for Healthcare Research and Quality; 2008.
[18] Samaranayake NR, Cheung STD, Chui WCM, Cheung BMY. Technology-related medication errors in a tertiary hospital: a 5-year analysis of reported medication incidents. Int J Med Inform 2012;81:828–33. Available from: http://dx.doi.org/10.1016/j.ijmedinf.2012.09.002.
[19] Koppel R, Metlay JP, Cohen A, Abaluck B, Localio AR, Kimmel SE, et al. Role of computerized physician order entry systems in facilitating medication errors. JAMA 2005;293:1197–203. Available from: http://dx.doi.org/10.1001/jama.293.10.1197.
[20] Koppel R, Wetterneck T, Telles JL, Karsh B-T. Workarounds to barcode medication administration systems: their occurrences, causes, and threats to patient safety. J Am Med Inform Assoc 2008;15:408–23. Available from: http://dx.doi.org/10.1197/jamia.M2616.
[21] Poon EG, Keohane CA, Bane A, Featherstone E, Hays BS, Dervan A, et al. Impact of barcode medication administration technology on how nurses spend their time providing patient care. JONA J Nurs Adm 2008;38:541–9. Available from: http://dx.doi.org/10.1097/NNA.0b013e31818ebf1c.
[22] Poon EG, Keohane CA, Yoon CS, Ditmore M, Bane A, Levtzion-Korach O, et al. Effect of bar-code technology on the safety of medication administration. N Engl J Med 2010;362:1698–707. Available from: http://dx.doi.org/10.1056/NEJMsa0907115.
[23] Morriss FH, Abramowitz PW, Nelson SP, Milavetz G, Michael SL, Gordon SN, et al. Effectiveness of a barcode medication administration system in reducing preventable adverse drug events in a neonatal intensive care unit: a prospective cohort study. J Pediatr 2009;154. Available from: http://dx.doi.org/10.1016/j.jpeds.2008.08.025.
[24] Betts K. Nursing students' knowledge and training during the medication administration process [Doctoral dissertation]. Walden University; 2014.
[25] Cherry SS. Medication safety: improving faculty knowledge and confidence by Sharon S. Doctorate of Nursing. Gardner-Webb University; 2013.
[26] Harding L, Petrick T. Nursing student medication errors: a retrospective review. J Nurs Educ 2008;47:43–7.
[27] Fauchald SK. An academic-industry partnership for advancing technology in health science education. Comput Inform Nurs 2008;26:4–8. Available from: http://dx.doi.org/10.1097/01.NCN.0000304763.94789.02.
[28] Melo D, Carlton KH. A collaborative model to ensure graduating nurses are ready to use electronic health records. Comput Inform Nurs 2008;26:8–12. Available from: http://dx.doi.org/10.1097/01.NCN.0000304764.02414.01.
[29] Vestal VR, Krautwurst N, Hack RR. A model for incorporating technology into student nurse clinical. Comput Inform Nurs 2008;26:2–4. Available from: http://dx.doi.org/10.1097/01.NCN.0000304762.94789.4b.
[30] Nickitas DM, Nokes KM, Caroselli C, Mahon PY, Colucci DE, Lester RD. Increasing nursing student communication skills through electronic health record system documentation. Comput Inform Nurs 2010;28:7–11. Available from: http://dx.doi.org/10.1097/01.NCN.0000336491.11726.4e.
[31] Wyatt TH, Li X, Indranoi C, Bell M. Developing iCare v.1.0: an academic electronic health record. Comput Inform Nurs 2012;30:321–9. Available from: http://dx.doi.org/10.1097/NXN.0b013e31824af81f.
[32] Rubbelke CS, Keenan SC, Haycraft LL. An interactive simulated electronic health record using google drive. Comput Inform Nurs 2014;32:1–6. Available from: http://dx.doi.org/10.1097/CIN.0000000000000043.
[33] Mannino EJ, Cornell G. Teaching electronic charting with simulation and debriefing in early fundamentals. Dean's Notes 2014;35:1–4.
[34] Carayon P, Wetterneck TB, Hundt AS, Ozkaynak M, DeSilvey J, Ludwig B, et al. Evaluation of nurse interaction with bar code medication administration technology in the work environment. J Patient Saf 2007;3:34–42. Available from: http://dx.doi.org/10.1097/PTS.0b013e3180319de7.
[35] Van Onzenoort HA, Van De Plas A, Kessels AG, Veldhorst-Janssen NM, Van Der Kuy PHM, Neef C. Factors influencing bar-code verification by nurses during medication administration in a Dutch hospital. Am J Heal Pharm 2008;65:644–8.
[36] Pauly-O'Neill S. Beyond the five rights: improving patient safety in pediatric medication administration through simulation. Clin Simul Nurs 2009;5:e181–6. Available from: http://dx.doi.org/10.1016/j.ecns.2009.05.059.

[37] Booth RG. Nurses' learning and conceptualization of technology used in practice [Doctoral thesis]. Western University; 2013.
[38] Oudshoorn A, Sinclair B. Using unfolding simulations to teach mental health concepts in undergraduate nursing education. Clin Simul Nurs 2015;11:396–401. Available from: http://dx.doi.org/10.1016/j.ecns.2015.05.011.
[39] Frey-Vogel AS, Scott-Vernaglia SE, Carter LP, Huang GC. Simulation for milestone assessment: use of a longitudinal curriculum for pediatric residents. Simul Healthc 2016;11:286–92. Available from: http://dx.doi.org/10.1097/SIH.0000000000000162.
[40] O'Keefe-McCarthy S. Technologically-mediated nursing care: the impact on moral agency. Nurs Ethics 2009;16:786–96. Available from: http://dx.doi.org/10.1177/0969733009343249.
[41] Yeo MT. Implications of 21st century science for nursing care: interpretations and issues. Nurs Philos 2014;15:238–49. Available from: http://dx.doi.org/10.1111/nup.12066.
[42] Barnard A. Philosophy of technology and nursing. Nurs Philos 2002;3:15–26.
[43] Cant RP, Cooper SJ. Simulation-based learning in nurse education: systematic review. J Adv Nurs 2010;66: 3–15. Available from: http://dx.doi.org/10.1111/j.1365-2648.2009.05240.x.
[44] Boaden R, Harvey G, Moxham C, Proudlove N. Quality improvement: theory and practice in healthcare. Coventry; 2008.
[45] Leighton K, Scholl K. Simulated Codes: understanding the response of undergraduate nursing students. Clin Simul Nurs 2009;5:e187–94. Available from: http://dx.doi.org/10.1016/j.ecns.2009.05.058.
[46] Selim AA, Ramadan FH, El-Gueneidy MM, Gaafer MM. Using Objective Structured Clinical Examination (OSCE) in undergraduate psychiatric nursing education: is it reliable and valid? Nurse Educ Today 2012;32: 283–8. Available from: http://dx.doi.org/10.1016/j.nedt.2011.04.006.
[47] Canada Health Infoway. A framework and toolkit for managing ehealth change: people and processes. Toronto; 2013.
[48] Grigg SJ, Garrett SK, Craig JB. A process centered analysis of medication administration: identifying current methods and potential for improvement. Int J Ind Ergon 2011;41:380–8. Available from: http://dx.doi.org/10.1016/j.ergon.2011.01.014.
[49] Cheung C, Goldstein M, Geller E, Levitt R. The effects of CPOE on ICU workflow: an observational study. AMIA Annu Symp Proc 2003;150–4.
[50] Dasgupta A, Sansgiry SS, Jacob SM, Frost CP, Dwibedi N, Tipton J. Descriptive analysis of workflow variables associated with barcode-based approach to medication administration. J Nurs Care Qual 2011;26: 377–84. Available from: http://dx.doi.org/10.1097/NCQ.0b013e318215b770.

SECTION VI

LOCAL AND REGIONAL INTERVENTIONS

CHAPTER

13

From Competencies to Competence: Model, Approach, and Lessons Learned From Implementing a Clinical Informatics Curriculum for Medical Students

William Hersh, Fran Biagioli, Gretchen Scholl, Jeffery Gold, Vishnu Mohan, Steven Kassakian, Stephanie Kerns and Paul Gorman

Oregon Health & Science University, Portland, OR, United States

INTRODUCTION

Physicians of the 21st century face a very different information landscape than their predecessors of prior centuries. Not only has the rate of growth of scientific knowledge of medicine increased, but new applications of data and information are required to deliver care that is effective, safe, and affordable. Medical education has historically mainly focused on the biomedical model of health and disease. Yet modern physicians will need to manage populations, costs, safety, and other aspects of care beyond conventional health and disease [1].

One of the essential areas of competence for the new requirements of 21st century medical practice is *clinical informatics*, which involves a range of skills that have the potential to improve patient documentation, access to knowledge, improved quality and safety of care, and ability to function in new models of care delivery [2]. If clinical informatics is an

Health Professionals' Education.
DOI: http://dx.doi.org/10.1016/B978-0-12-805362-1.00013-9

essential competency for 21st century medical practice, then it must be introduced along with the rest of the curriculum in undergraduate medical education (UME). A confluence of events and activities enabled us to implement clinical informatics in a major way in our UME curriculum at Oregon Health & Science University (OHSU). These included:

- A recent overhaul of our UME curriculum [3]
- Grant funding from the American Medical Association [4]
- Establishment of the subspecialty of clinical informatics and establishment of fellowships accredited by the Accreditation Council for Graduate Medical Education (ACGME) [5]
- Development of a simulation facility and funding for research in electronic health record (EHR) simulation [6]
- Maturation of our commercial EHR system, Epic Systems Corp. (Verona, WI)

OPPORTUNITY

The OHSU medical school curriculum transformation provided a fresh opportunity to incorporate clinical informatics into the UME curriculum. The Dean (the late Mark Richardson) recognized and upheld informatics as an essential element of a UME curriculum for the 21st century, and provided vocal top-down support for its incorporation from the outset. Aligned with this leadership, the transformation steering committee convened a Working Group on Integration of Biomedical Informatics and Technology as one of the key planning groups for the new curriculum.

Once detailed planning was underway, we formed an informatics curriculum group composed of key faculty from our Department of Medical Informatics & Clinical Epidemiology (DMICE) and other clinical departments. This group met weekly to develop the initial architecture of the curriculum by (1) defining competencies and learning objectives in clinical informatics; (2) mapping these competencies to the ACGME competency domains, (3) proposing a timeframe for staging the introduction of these competencies into the UME curriculum appropriate to the learners' stage of development, and (4) devising an overall strategy for integrating informatics into the evolving new curriculum. This process and the results are described in a previous publication [7]. We aimed for these competencies to represent how practicing clinicians would use, apply, and evaluate informatics in their care of patients. These included not only the two main areas one might think of in terms of competence, that is, using the EHR and searching for knowledge, but also looked at aspects of practice likely to increase in the future, such as use of clinical decision support (CDS), quality measurement and improvement, patient safety, and patient engagement. Our process yielded 13 broad areas of competency, each with specific subareas, which are shown in Table 13.1.

In the rest of this chapter, we describe details of our curriculum, starting with general principles, describing major areas of implementation, and discussing challenges and lessons learned. As with all curricula, our efforts are a work in progress, and this chapter represents a snapshot of our progress as the third year of students enters our UME program.

TABLE 13.1 Competencies and Learning Objectives/Milestones in Clinical Informatics Guiding Curriculum, With Additional Mapping to the Timing of the Learning Objective/Milestone Introduction in the Curriculum and the One or MACGME Core Competencies of the Clinical Informatics Competency

Clinical Informatics Competency	Learning Objectives/Milestones	When Start?	PC	MK	PBLI	ICS	Prof	SBP
Find, search, and apply knowledge-based information to patient care and other clinical tasks	Information retrieval/search-choose correct source for specific task, search using advanced features, apply results	Early	x	x	x			x
	Evaluate information resources (literature, databases, etc.) for their quality, funding sources, biases	Early						
	Identify tools to assess patient safety (e.g., medication interactions)	Early						
	Utilize knowledge-based tools to answer clinical questions at the point of care	Mid						
	Formulate an answerable clinical question	Mid						
	Determine the costs/charges of medications and tests	Mid						
	Identify deviations from normal (labs/x-ray/results) and develop a list of causes of the deviation	Mid						
Effectively read and write from the electronic health record for patient care and other clinical activities	Graph, display, and trend vital signs and lab values over time	Early	x		x			x
	Adopt a uniform method of reviewing a patient record	Early						
	Create and maintain an accurate problem list	Early						
	Recognize medical safety issues related to poor chart maintenance	Early						
	Identify a normal range of results for a specific patient	Mid						
	Access and compare radiographs over time	Mid						
	Identify inaccuracies in the problem list/history/med list/allergies	Mid						
	Create useable notes	Mid						
	Write orders and prescriptions	Mid						
	List common errors with data entry (drop down lists, copy and paste, etc.)	Mid						

(Continued)

TABLE 13.1 (Continued)

Clinical Informatics Competency	Learning Objectives/Milestones	When Start?	PC	MK	PBLI	ICS	Prof	SBP
Use and guide implementation of clinical decision support (CDS)	Recognize different types of CDS	Early	x	x	x			x
	Be able to use different types of CDS	Late						
	Work with clinical and informatics colleagues to guide clinical decision support use in clinical settings	Late						
Provide care using population health management approaches	Utilize patient record (data collection and data entry) to assist with disease management	Early	x		x			x
	Create reports for populations in different healthcare delivery systems	Late						
	Use and apply data in accountable care, care coordination, and the primary care medical home settings	Late						
Protect patient privacy and security	Use security features of information systems	Early	x				x	
	Adhere to HIPAA privacy and security regulation	Early						
	Describe and manage ethical issues in privacy and security	Early						
Use information technology to improve patient safety	Perform a root-cause analysis to uncover patient safety problems	Late	x			x		x
	Familiarity with safety issues	Early						
	Use resources to solve safety issues	Late						
Engage in quality measurement selection and improvement	Recognize the types and limitations of different types of quality measures	Early	x		x			x
	Determine the pros and cons of a quality measure, how to measure it, and how to use it to change care	Mid						
Use health information exchange (HIE) to identify and access patient information across clinical settings	Recognize issues of dispersed patient information across clinical locations	Early	x		x	x		x
	Participate in the use of HIE to improve clinical care	Late						
Engage patients to improve their health and care delivery though	Instruct patients in proper use of a personal health record (PHR)	Early	x			x	x	x

(Continued)

TABLE 13.1 (Continued)

Clinical Informatics Competency	Learning Objectives/Milestones	When Start?	PC	MK	PBLI	ICS	Prof	SBP
personal health records and patient portals	Write an e-message to a patient using a patient portal	Early						
	Demonstrate appropriate written communication with all members of the healthcare team	Mid						
	Integrate technology into patient education (e.g., decision making tools, diagrams, patient education)	Late						
	Educate patients to discern quality of online medical resources (Web sites, apps, patient support groups, social media, etc.)	Early						
	Maintain patient engagement while using an EHR (eye contact, body language, etc.)	Mid						
Maintain professionalism through use of information technology tools	Describe and manage ethics of media use (cloud storage issues, texting, cell phones, social media professionalism)	Early					x	
Provide clinical care via telemedicine and refer those for whom it is necessary	Be able to function clinically in telemedicine/telehealth environments	Late	x			x		x
Apply personalized/precision medicine	Recognize growing role of genomics and personalized medicine in care	Early	x	x				
	Identify resources enabling access to actionable information related to precision medicine	Early						
Participate in practice-based clinical and translational research	Use EHR alerts and other tools to identify patients and populations for offering clinical trial participation	Mid			x			x
	Participate in practice-based research to advance medical knowledge	Late						

The ACGME core competencies and their abbreviations include [8]:

1. Patient Care and Procedure Skills (PC)
2. Medical Knowledge (MK)
3. Problem-Based Learning and Improvement (PBLI)
4. Interpersonal and Communication Skills (ICS)
5. Professionalism (Prof)
6. System-Based Practice (SBP)

OVERVIEW OF UME CURRICULUM

The new curriculum at OHSU represented a complete transformation of UME to an innovative program that integrated basic sciences, clinical sciences, and health systems sciences. During the 18-month *Foundations* phase, learning is predominantly classroom based, complemented by *Preceptorship* experiences in clinical settings. This is followed by the *Clinical Experiences* phase during which learning is predominantly through guided practice in clinical settings, complemented by periodic classroom-based *Intersessions* to allow time for "deep dives" that reinforce and reintegrate foundational concepts. In addition, every student must complete a *Scholarly Project* under the mentorship of OHSU faculty in an individual area of focus. The Foundations phase is divided into seven blocks organized around organ systems: Fundamentals; Blood and Host Defense; Skin, Bone, and Muscled; Cardiac, Pulmonary, and Renal; Hormones and Digestion; Nervous System and Function; and Developing Human. Each block combines basic and clinical sciences, and integration is achieved in part by interweaving "threads" of traditional disciplines such as anatomy, pharmacology, or microbiology throughout these curricular components. An overview of the curriculum is shown in Fig. 13.1.

One of the threads in the new curriculum is *Epidemiology, Evidence-Based Medicine, and Informatics*. Within this thread, we have implemented an extensive curriculum in clinical informatics. Even prior to the curriculum, our students had exposure to various aspects of clinical informatics. This was mainly via faculty from our Department of Family Medicine,

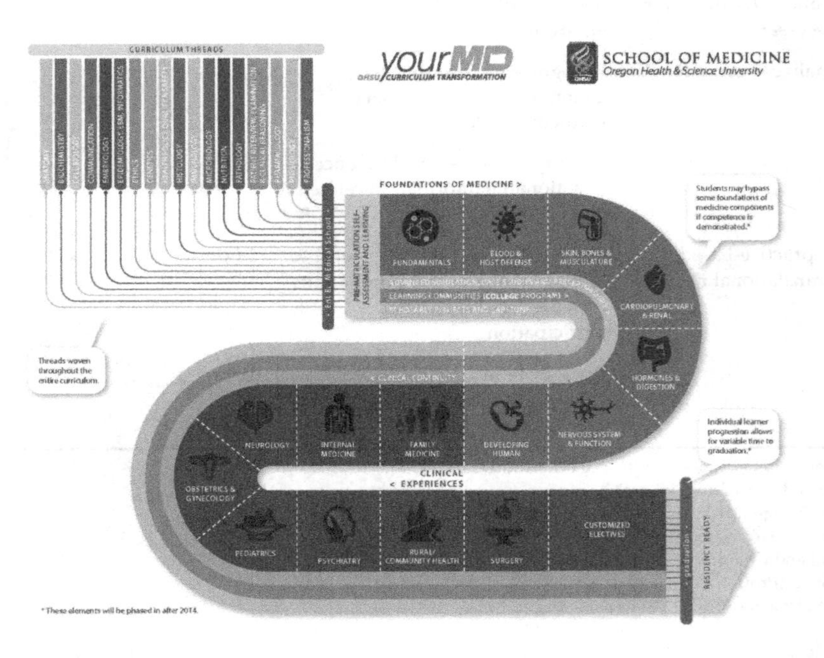

FIGURE 13.1 Overview of OHSU UME curriculum. Clinical informatics is part of the Epidemiology, EBM, and Informatics curriculum thread.

which developed a number of activities around practical use of the EHR. With the revision of the curriculum, we have been able to expand the clinical informatics curriculum substantially from the foundation of the competencies listed in Table 13.1. We have been able to engage faculty from an academic department (DMICE), which has a mission to advance research and education in all aspects of biomedical informatics. The authors of this chapter formed a working group of faculty who have implemented the curriculum based on the competencies.

PRINCIPLES

Our first tasks in integrating informatics into the new curriculum were to recognize constraints or limitations as well as resources and facilitators of implementation, then to set priorities and devise strategies for a staged implementation.

Constraints, Limitations, Assumptions

Many potential constraints and limitations had to be taken into account to implement this clinical informatics curriculum, most of which are likely not unique to our institution. First, the ever-present competition among the disciplines for time in the curriculum, in this case complicated by compression of the 2 year basic sciences phase to an 18-month Foundations phase. To address this has required persistence and an opportunistic approach to identify topics and sessions that lend themselves to integration with our defined informatics competencies and objectives. Examples include information retrieval (IR) and critical appraisal sessions to provide closure to weekly cases, and precision medicine in the discussion of treatment of cystic fibrosis.

A second challenge has been a degree of resistance among faculty, educators, and even students, who may not yet recognize the importance of informatics and other "third science" topics in a twenty first century UME curriculum. Faculty and educators may be accustomed to thinking informatics means computers and printers, not legitimate medical curriculum. Students may be difficult to refocus on skills they do not perceive are important to US Medical Licensure Exam (USMLE) Step 1 performance. To legitimize the place of informatics in the UME curriculum has required persistence, buttressed by top-down support from the Dean and School of Medicine leadership. As a result, each year we have been able to make progress in expanding the clinical informatics content in the curriculum.

A third significant challenge, which may be somewhat unique to our curriculum structure, has been the push for integration. While seamless integration of informatics into a case-based curriculum may seem the ideal, so that informatics becomes as routine as the stethoscope, in practice this requires a very substantial amount of planning and faculty collaboration across disciplines. Furthermore, instructional material that is tightly integrated with a specific case requires updates whenever cases are changed or repositioned in the curriculum. As a result, we have had to revise our approach to make the informatics curriculum less tightly integrated so that it is more amenable to revisions in subsequent years.

Technology considerations comprise the final major challenge. Much institutional collaboration is required to address issues such as protecting protected health information (PHI) in training sessions, use of production EHR environments or creation of simulation EHR environments, integration of robust functionality that may not be available or affordable in nonproduction environments, such as Picture Archiving and Communications Systems (PACS), health information exchange (HIE) capability, patient portals, etc. Beyond this, for the most part EHR systems have not been designed with students in the workflow, and institutional policies may be interpreted to preclude student participation in some EHR tasks, even as residency programs expect that students will be competent at these tasks at graduation. Local institutional policies may need to be revised to fully integrate students into EHR workflows and ensure competence by the time of graduation.

Facilitators and Resources

Fortunately, our curriculum benefitted from significant existing resources and facilitators at our institution. Most important among these has been the expressed and tangible support from the Dean and School of Medicine leadership. From the outset of the transformation process, and at key stages in implementation, support from the Dean's Office has enabled us to make progress in gradually expanding clinical informatics in the curriculum, a process augmented by deep involvement of informatics faculty in the curriculum development process from its inception (PG). Equally important has been a culture of continuous improvement encouraged by leadership and built in to the curriculum transformation process, enabling us to try, to fail, and to improve our approaches.

A second essential resource and facilitator of informatics curriculum implementation has been the addition of an educational informatician to support these efforts. We were fortunate to identify someone (GS) who is an "informed insider," familiar with the institutional IT department, personnel, and procedures and possessed of extensive experience in EHR training and creation and use of simulation environments for research and education. These capabilities have been essential in meeting the technical and operational challenges of a simulated EHR environment, and in developing and delivering training activities.

A third key ingredient for successful implementation has been prior work and already existing curricular elements created by our faculty, including a robust set of clinical EHR-related exercises in the Family Medicine clerkship (FB); ongoing research and training with simulated EHR environments (JG, VM), existing curricular elements in topics such as evidence-based medicine and systems safety (PG), extensive departmental experience in informatics curriculum development from the American Medical Informatics Association (AMIA) 10x10 program to certificate, masters, doctoral, and postdoctoral programs in biomedical informatics (WH) [9].

Finally, as we have implemented the informatics and health systems science curriculum at OHSU, there has been a growing national consensus on the importance of these topics in UME and GME curriculum, evidenced by the forthcoming American Medical Association (AMA) book on health systems science [10], the recent expansion of this content in the USMLE Content Outline, and the Association of American Medical Colleges (AAMC) collaboration to develop a set of Core Entrustable Professional Activities (EPAs) for Entering Residency, two of which are focused on informatics related skills: EPA 4,

"Enter and Discuss Orders and Prescriptions" and EPA 5, "Document a Clinical Encounter in the Patient Record." This convergence of efforts in medical education at the national level has provided the necessary additional momentum to help us overcome some residual curricular inertia.

Integration Strategies

With these resources and constraints in mind, we developed principles and strategies to guide our implementation efforts. First among these was to develop priorities and milestones that would be compelling to our faculty colleagues in arguing for expanding the informatics footprint in the curriculum. For the preclinical or Foundations phase, where so much basic and clinical science is competing for curriculum time and students' attention, we used the rule of thumb: "boards or wards"—that is, if students need it to do well on their USMLE Step 1 exam, or if it will help them perform well and learn on their clinical experiences, then it needs to be included in the curriculum. This rule of thumb has been helpful in decisions about curriculum time allocation, and ideal placement of elements of our informatics curriculum. Beyond this, we developed two basic milestones to guide our informatics curriculum implementation:

Milestone 1. At end of preclinical time, learners can
- Access and appraise latest medical knowledge
- Protect PHI
- Access and enter data in EHR
- Engage patients with health IT

Milestone 2. By graduation, learners are competent users of health IT and data to improve patient and population health and improve health systems

With these in mind, we developed a set of principles and strategies for implementation:

1. Matriculation to graduation. We set out to develop a four-year curriculum, including all components of the UME program as potential contexts for informatics incorporation. This principle enables to look across the entire 4 year continuum of UME to identify the ideal contexts for introducing informatics topics, including Foundations preclinical classrooms, Clinical Experiences in clinics or on the wards, Intersessions where students return to the classroom for deep dives, and their Scholarly Project, where they will use many informatics related skills and may in some cases choose an informatics focused project.
2. Day One integration of EHR. In 21st century practice, the EHR has become ubiquitous. Our curriculum is designed to make the EHR a routine part of medicine from the moment the students begin learning—using the EHR to introduce the cases that are the basis for our case-based curriculum, and adding new EHR and informatics skills every week during their Foundations curriculum.
3. Staged implementation and Continuous Improvement. The full set of competencies we developed [7] is an ambitious agenda. Recognizing this fact and the constraints mentioned above, we set priorities for early inclusion, and implemented our curriculum

in stages, with an expectation of gradually introducing the competencies according to identified priorities, and of learning from error in the process.

4. Integration with weekly curriculum content. To make our informatics relevant and necessary, we set out to integrate it tightly with the case material and basic and clinical science content. This has been accomplished largely through the use of weekly Clinical Informatics Pearls. Pearls are very brief pre-recorded videos of informatics faculty explaining a single, simple skill that is useful in practice. Wherever possible, performing these skills connects the student to information that is related to the weekly case. For example, a Clinical Informatics Pearl explaining how to display trends in data might use the complete blood count (CBC) as an example, during a week when they will be learning about anemia or cytopenias.
5. Cotton ball in the water glass. This principle, suggested by Howard Silverman of the University of Arizona, is meant to make the informatics content or skill so tightly integrated it is almost not noticeable—a glass with a cotton ball will hold almost as much water as the glass without the cotton ball. An example of this sort of integration might be a session where the informatics skills are taught and used as part of a session teaching some clinical or basic science content. One example is where students learn about infectious disease and outbreak investigation and control, in which they use the EHR and simulated patients to order lab tests, obtain results a few days later, and "discover" an infectious disease outbreak that they must then investigate and use systems based practice principles to control.
6. Opportunistic, available, and flexible. It has been essential to our approach that we be opportunistic in identifying places in the curriculum where informatics competencies may be introduced, and that we be available and responsive when these opportunities present themselves, and that we be flexible sometimes modifying instructional methods (lectures, skills labs, enrichment week sessions, clinical informatics pearls) to fit existing curriculum structure and constraints.

IMPLEMENTATION

After establishing the competencies and principles of the curriculum, we then set out to implement it. We realized that we would require a wide range of modalities used to deliver the curriculum. These have included:

1. Lectures—large-group traditional lectures, aiming for some active learning
2. Large-group interactive—sessions that involve interaction with students, such as breaking down into small groups and reporting out to larger group
3. Pearls—short (7–20 minute) asynchronous recorded lectures and demonstrations of a variety of informatics topics and skills
4. Clinical skills labs—small group learning, practice, and integration
5. Enrichment sessions on specialized topics
6. Clinical skills assessments
7. Intersessions—students return after Foundations for four two-week deeper dive courses

TABLE 13.2 Curricular Activities for Various Learning Objectives/Milestones

PRECLINICAL FOUNDATIONS PHASE	
Clinical Informatics Pearls	About 50 weekly short online videos of skills—protecting PHI, EHR skills, and information retrieval skills are included in each block
Clinical Informatics Assessments	About monthly assessment of EHR skills from Pearls
Orientation to EHR: Epic Workshop	Orientation week intro to Epic
Orientation to Knowledge Sources	Orientation week intro to online resources
Introduction to Third Science	Overview of health systems sciences threads
Information Is Different Now That You're a Doctor	Overview of role of information in medicine
Connecting Population Health and Value-Based Care	Value-based and population-based care principles
Informatics and Evidence-Based Medicine Series	Weekly closure sessions wrap up case issues with EBM approach
Informatics and Evidence-Based Medicine Skills Lab	Skills lab to introduce EBM and informatics skills
Clinical Informatics Skills Assessment (EHR)	Assessment to demonstrate skills
Epidemiology and Evidence-Based Medicine Series-10 weekly Skills Labs	Ten weekly skills labs cover epidemiology and EBM
Precision medicine: CF genes and therapies	Integrated into pulmonary block, precision medicine tools and examples
Improving Quality with Population Data	Local examples of successful improvement science application
Improving Quality with Population Data Skills Lab	Skills lab on quality measures and data interpretation
Infectious Disease Outbreak	EHR skills and systems based practice principles integrated into GI case
Evidence-Based Medicine Literature Appraisal-3 skills labs	Weekly sessions based on Users Guides to literature
Inpatient EHR Skills	Skills lab combining EHR skills needed for inpatient rotations
Assessment: EHR skills diabetic meds and labs	Assessment of order entry and documentation
Outpatient EHR Skills	Skills lab combining EHR skills needed for ambulatory rotations
Assessment: EHR Skills	Assessment of outpatient EHR skills
Advanced EHR Skills	Chart hygiene and record management
Assessment: interpret medical literature	Assessment of EBM appraisal skills

(*Continued*)

TABLE 13.2 (Continued)

CLINICAL EXPERIENCES–FAMILY MEDICINE ROTATION	
EHR Objective Structured Clinical Examination	Assess use of EHR in standardized patient encounter
EHR Chronic Disease Management Workshops	Use EHR data in population-based care, chronic disease
Telemedicine OSCE (TeleOSCE)	Telemedicine encounter on rural rotation
Population data seminar	FM department data from EHR for population care
Order-writing EHR workshop	Use EHR to pend orders in FM clerkship encounter
EHR-related educational activities	Information retrieval and information literacy
INTERSESSIONS	
Honing EHR Skills Workshops	Between clerkships, building on EHR skills—4 required
Formulating the question workshop	Between clerkships, building EBM skills—4 required
Literature search workshop	Between clerkships, building lit search skills—4 required
Decision Aids	Decision aids session, integrated in cognitive impairment intersession
Evidence-Based Policy for Cancer Diagnosis and Treatment	State health policy group using EBM approach to set policies
Population-based care and quality improvement	Quality improvement in dementia care of veterans
Public Health and System Management of Outbreaks	Informatics, evidence, and systems based practice for outbreak response
Prescription Opioid Overdose: Public Health Perspective	Public health, systems based practice related to opioid epidemic

Table 13.2 lists the activities we have implemented in the curriculum. Improvements based on student feedback and faculty observation were made for the new classes, and it is likely that the curriculum will continue to evolve not only due to feedback but also changes in technology and its use. The details are listed in this section.

Electronic Health Records

OHSU medical students have had training in the EHR that predates our recent curriculum revision in 2014. These established activities have been incorporated into our new overall approach. Since 2009, students have been trained in the logistics of OHSU's Epic EHR and provided a login and password at the beginning of medical school. Epic has allowed us (as it does for all of its customers) to use a training environment based on our operational implementation for use by all clinical students at our university. In order to facilitate systemic integration of the EHR into the revised curriculum, we established a dedicated training (simulation) environment solely for medical student education in 2014. Finally, OHSU has also established institutional policy defining specifics of what students are allowed and encouraged to do within patient EHR Records. Pocket cards are distributed to students to further the communication with hospital and ambulatory clinical teachers [11].

Of course, our EHR curriculum is not just focused on Epic. Our students also gain some exposure to the Veterans Information Systems and Technology Architecture (VistA) EHR, which they use during clinical experiences at the nearby Portland Veterans Affairs Medical Center. In addition, our instruction aims to focus on generalizable EHR skills that happen to be implemented within the Epic system.

In the new curriculum, students receive EHR training during orientation that enables them to access the EHR from the very onset of their medical school education. They also receive initial information regarding how they will receive exemplar cases, and where to expect short videos and weekly assignments. The training includes topics pertaining to navigation and basic documentation within the EHR. Consequently, weekly use of the EHR is expected, so the students can become proficient by the time they participate in rotations.

Each week during all of the blocks, the students are assigned a patient. This patient includes data that is built into the EHR, incorporating exemplar case elements that will be covered during the week. The case can be built to be released incrementally as well—history and physical on Monday, labs on Tuesday, and assessment and plan on Friday. In subsequent blocks, some exemplar cases spiral back to earlier blocks. A patient seen in the Fundamentals block, for example, might be seen again in Cardiopulmonary and Renal. These virtual patients are housed in a separate simulation instance of Epic dedicated for medical students' use; cases are specifically created to meet the educational goals of students and loaded into the simulation environment for access during block weeks.

In addition to the exemplar case, the students are given a weekly EHR/Informatics assignment. The EHR assignments include such tasks as finding data, placing orders, and writing notes as well as short answer questions relating to IR or other informatics topics. These topics are delivered in a series of short (<8 minutes), online, "pearls" that include information and video demonstrations that foster optimal EHR use and promote best practices of EHR interface navigation.

Students are also taught EHR skills during the preclinical curriculum in large-group sessions followed by small group discussion. EHR communication skills and ways to prevent common EHR-related errors are discussed. The class reviews alterations in body language and room set-up, and methods of utilizing the EHR to augment patient education as well as how to establish and maintain patient rapport when utilizing an EHR during a patient encounter. Students complete a prehomework assignment in which they are instructed to craft the content of a written message sharing an abnormal glucose result with a patient. Small groups facilitated by faculty discuss the student writing samples and then discuss in which scenarios patient portal written communication versus telephone versus in person visits would be best to share information with patients. Students review and critique video clips with patient scenarios and discuss how to alter the use of the EHR in certain patient scenarios.

Other preclinical activities include EHR inpatient and outpatient chart preview workshops. Faculty and chief residents are invited to demonstrate to students how they efficiently prepare for a new admission from the Emergency Department, rounding on a patient, and preparation for an outpatient visit. Students are able to follow along synchronously. We also provide a Chart Hygiene Workshop, where faculty present a complex chart that includes errors. Students are asked to review the chart as well as correct and add documentation. Activities like these help to build EHR related proficiencies in the context of case-based learning.

Students also get an EHR Skills Honing Workshop during their Intersession activities. These consist of focused curricula on the topics of cancer, cognitive impairment, infectious disease, and pain. Students are also asked to identify skills they would like to refine as well as instructor-led activities.

Another part of the curriculum is an EHR Objective Structured Clinical Examination (OSCE) [12]. Utilizing a standardized patient actor and a simulated EHR, the EHR OSCE case teaches the skill of conducting an ambulatory visit utilizing an EHR. EHR OSCE checklist and teaching items include establishing and maintaining rapport with the patient despite computer use, utilizing the computer to augment care and explanations, data reconciliation with patients to ensure EHR record accuracy, allergy, and medication review and reconciliation before prescribing, and electronic medication ordering. The EHR OSCE was implemented in the midst of the required third year clerkship OSCE during a formative feedback Session in 2010–2016 and now will be utilized within the preclinical years Clinical Skills Course.

Since 2011, we have also offered simulated EHR Chronic Disease Management Workshops [13,14]. During these workshops, students are taught to thoroughly review a patient record, maintain an accurate EHR patient record, and to apply evidence-based guidelines in the care of a complex simulated patient. Students review the EHR notes and data and update the medically complex simulated patient chart. The student updates the problem list, medical history, and allergy list sections of the chart in a manner such that the chart accurately summarizes the contents of the patient data contained in the chart. Students then interpret the EHR data and apply evidenced guidelines and create and place order sets for their simulated patients. Initially these workshops were part of one required clerkship; however, students reported the information as most useful before the start of the clinical portion of their education. As of 2014, one of the two workshops was moved to the transition to clerkship seminars and in 2015, both workshops were moved to the end of the 18-month preclinical curriculum. Since 2011, Internal Medicine interns also all receive these workshops as part of the orientation.

In 2013, we added a telemedicine OSCE (TeleOSCE) [15,16]. Students are tasked with caring for a patient utilizing a telemedicine. Standardized patients and a simulated telemedicine interface are used to mimic caring for a patient via a video visit. Students learn to utilize this technology as a mechanism to improve access to care for underserved and remote/rural populations. Use of telemedicine is a required aspect of the student curriculum and the TeleOSCE is used to measure the student skill level in the application of their medical knowledge in various patient scenarios.

Starting in 2014, we added a population data seminar. During a required clinical clerkship medical directors share clinic population data with students and task students with sharing ideas regarding system methods of improving clinical performance measures such as pneumococcal vaccine or mammogram rates or determining optimum provider panel size.

In 2015, we established an order-writing EHR workshop [17]. Utilizing a simulated EHR, clinical year students review EHR data, apply evidence guidelines, and write orders and notes for a simulated patient. Clinical scenarios and simulated charts represent various simulated patient scenarios in which it would be appropriate to initiate diabetic pharmaceutical management, disease, and comorbidity monitoring laboratory work, and prevention with vaccinations and medications. EHR competency is a required part of the curriculum and students chart work and notes are evaluated by faculty.

Most recently, we added additional required EHR-related educational activities [18]. As part of required clinical year curriculum, students must complete and log educational activities involving the clinical use of the EHR. If the clinical site does not permit order writing, the student must log the completion in a simulated environment. Educational activities specific to EHR and informatics skills include: Write orders for the best evidenced based prevention for a child, and adult and an elder patient; Participate in the transitions of care by locating reviewing and summarizing data available in an HIE; Review and reconcile EHR problem list and write orders for needed disease prevention and management based on application of evidenced guidelines; and communicate in writing with a patient using a secure patient portal.

Information Retrieval and Evidence-Based Medicine

Our previous curriculum had a number of activities in IR and evidence-based medicine (EBM), but was mostly focused on basic skills. Students were oriented to the library and basic resources available. They also received some instruction in EBM, both in preclinical as well as clinical years. In the new curriculum, we significantly expanded offerings in these areas.

In IR, a number of our pearls are focused on basic principles and skills. Students are instructed on the type of resources available and basic concepts of indexing and advanced techniques used in retrieval. Other activities involve the application of these skills. For example, in the Intersession, students are required to search for literature to answer questions pertinent to the clinical theme of the section.

Students are also exposed to EBM from very early in the first year. They are taught skills based on the ask-access-appraise-apply-evaluate model. EBM and IR are intertwined, for example, finding best evidence is used as a context for searching resources like PubMed. Likewise, features of PubMed relevant to EBM, such as publication type limits, are emphasized in the teaching of IR.

We have also provided a Clinical Inquiry Small Group Discussion during a required clinical year clerkship since 2012. Students are tasked weekly with identifying a common clinical question in the course of the care of ambulatory patients. Using point of care information resources, students identify a discreet clinical question and research evidenced answers. Students document their findings in a short half-page summary with three quality references. Faculty facilitators evaluate students in their skill of defining a question, utilizing and interpreting evidenced primary resources, and interpreting and sharing their findings with other students.

Other Competencies

As noted in our competencies paper, we recognized that there were a number of other areas of medical student competency in clinical informatics beyond the EHR and IR. This section highlights some activities from the list in Table 13.2.

In addition to competencies, we also provide a number of sessions covering informatics generally. For example, in the first Fundamentals block, students receive a lecture entitled, *Information is Different Now That You're a Doctor*. This lecture introduces the notion that being a physician establishes a special relationship with information that embodies trust, professionalism, and expertise. Shortliffe has noted that this specialness of information is as important that rivals that of seeing patients [19]. Early on, students also receive exposure to applying informatics in practice as well as pursuing it as a subspecialty.

Students also receive instruction focusing on the other competencies. They are introduced to the practical application of CDS and population health management. A series of sessions focused on the critical aspects of privacy, security, and confidentiality. Recognizing that these future physicians are likely to have the quality of the care they deliver measured and provided as feedback, we carry out a large-group interactive session where the principles of quality measurement and improvement are introduced, followed by providing by searching the National Quality Forum (NQF; Washington, DC) Web site and applying appropriate measures to a small synthetic data set consisting of diabetes-related ICD-9 codes and HgbA1C measurements (or lack of them) tied to a practice of 10 physicians. Completing the exercises correctly identifies physician outliers in terms of obtaining HgbA1C and its results across the physicians.

Students are also exposed to various aspects of patient engagement and professionalism. In addition, they are provided curricular activities on the topics of personalized (now called precision) medicine, and clinical and translational research.

Challenges and Lessons Learned

The above activities have enabled us to learn a great deal about teaching clinical informatics in UME and beyond. We will likely continue to revise the curriculum based on these lessons as well as feedback from students and the experiences of other medical schools. In this section, we describe facilitators, challenges, and lessons learned.

Some of the facilitators in helping our work include a supportive culture, both in the Dean's Office and in two major departments involved, DMICE and Family Medicine. Our group also functions well as a team, with a commitment to providing clinical informatics instruction that focuses on the practical aspects of what 21st century physicians will need to know as they enter practice in the next 5–10 years.

Nonetheless, we have faced a number of challenges. Probably the major challenge we have faced is competing for curriculum time with all of the other subjects. Since informatics is part of a thread, and not a block, thread directors are likely to respond that while informatics is important, it should be in a different block that has more time availability (none of which actually do). This attitude is somewhat mitigated by support from leads of the overall curriculum.

One way we have (partially) overcome the time limitation is to develop asynchronous content, namely our informatics pearls. These include EHR-related tasks such as efficiently finding data, placing orders, writing notes, security features. They also include topics such as IR (PubMed and other resources), patient privacy, EBM, and HIE. The pearls are recorded by faculty and other staff proficient in EHR use or other areas of informatics. The pearls are intended to be generic enough that if the exemplar cases within a block change, they can be rearranged easily. One challenge with the asynchronous content is that they do not seem part of the core curriculum for some students.

Success in facilitating this process has included having an "informed insider." This person must have a thorough knowledge of EHR training, the EHR systems and processes and adult education. Working knowledge of how the EHR team is structured, and how it fits into IT is also important. Another aspect is to have a connection to the analysts on the EHR team. We have also had support of our IT department in obtaining an instance of Epic and maintaining it.

Another challenge is that while we have effectively dealt with licensing and hardware needs to support use of our Epic instance, there is also the challenge of the inability to access other systems associated with EHRs. We do not have a learning instance of several modules, such as the PACS, HIE (Epic CareEverywhere), and the patient portal (Epic MyChart). Likewise, we do not have access to the other EHR systems beyond Epic, with the exception of VistA.

An additional challenge we have learned is the need for assessments, which we are still trying to develop for the many areas covered. Students tend to triage their study time based on what is likely to be examined, and will be more likely to study our material if they know they will be assessed about it.

LESSONS LEARNED AND FUTURE DIRECTIONS

Consistent with our strategy of continuous improvement, we began early to make changes and have continued to do so, including the following.

1. Assessments. We initially intended that informatics skills introduced in Pearls would be essential and useful by virtue of their integration into weekly cases—you would need the skill to get the data about the case. This has been slow in developing, and as a result, students' engagement with clinical informatics skills was disappointing. An effective strategy to remedy this has been periodic Clinical Skills Assessments that assess skills covered in the online pearls. Scores are incorporated into their block grades and count toward passing each block.
2. Required Assignments. For the same reason, we increasingly require students to complete simple assignments based on the skills introduced in informatics pearls. For example, they may be required to review a chart and locate a particular piece of information, then "send it to their attending" using secure messaging.
3. Expanded EHR skills in the second year. Based on feedback from clerkship directors and faculty, we expanded the EHR skills program that consisted mainly of weekly pearls, plus a single workshop. In the second year of the curriculum, we added two EHR workshops

and an EHR Assessment. Titled, "How to be a star on the wards," these sessions put together individual skills to prepare students for prerounding and previsit chart review, order entry, and initial documentation in the inpatient and ambulatory settings.

4. Expanded didactic sessions: Consistent with our staged roll-out plan, we have expanded didactic sessions in version 2 of the curriculum, adding (1) Informatics and Evidence-Based Medicine sessions that role model these skills by answering questions arising out of weekly cases; (2) an overview of information in medicine ("Information is different now that you are a doctor"); (3) an overview of modern population-based care and value-based purchasing; and (4) a new session demonstrating effective use of EHR data in quality improvement.

For many years, advocacy of clinical informatics in the medical student curriculum fell of deaf ears. But with support of the school leadership, along with an expert and committed faculty working group, we have made significant headway in providing instruction to develop competencies that we believe are needed for the practice of medicine in the 21st century. As with all curricula, our effort will always be a work in progress, but we will continue to pursue improvement and increased competence in our learners. We also hope that a small number of those students will become board-certified and pursue careers in the new subspecialty.

Acknowledgments

The work in this chapter was funded in part by a grant from the American Medical Association Accelerating Change in Medical Education initiative. We also acknowledge the support of the late Mark Richardson, MD, MBA, Dean of the OHSU School Medicine as well as George Mejicano, MD, Senior Associate Dean for Education.

References

[1] Gonzalo JD, Haidet P, Papp KK, Wolpaw DR, Moser E, Wittenstein RD, et al. Educating for the 21st-century health care system: an interdependent framework of basic, clinical, and systems sciences. Acad Med 2017;92:35–9.

[2] Hersh W, Ehrenfeld J. Clinical informatics. In: Skochelak SE, Hawkins RE, Lawson LE, Starr SR, Borkan JM, Gonzalo JD, editors. In Health systems science. New York, NY: Elsevier; 2017. p. 105–16.

[3] MD Curriculum Transformation. Oregon Health & Science University. Available from: <https://www.ohsu.edu/xd/education/schools/school-of-medicine/about/strategic-initiatives/md-curriculum-transformation.cfm>; [accessed 11.06.17].

[4] Creating the Medical School of the Future. Available from: <https://www.ama-assn.org/education/creating-medical-school-future>; [accessed 11.06.17].

[5] Detmer DE, Shortliffe EH. Clinical informatics: prospects for a new medical subspecialty. J Am Med Assoc. 2014;311:2067–8.

[6] Simulation at OHSU. Available from: <http://www.ohsu.edu/xd/education/simulation-at-ohsu/>; [accessed 11.06.17].

[7] Hersh WR, Gorman PN, Biagioli FE, Mohan V, Gold JA, Mejicano GC. Beyond information retrieval and EHR use: competencies in clinical informatics for medical education. Adv Med Educ Pract. 2014;5:205–12. Available from: http://www.dovepress.com/beyond-information-retrieval-and-electronic-health-record-use-competen-peer-reviewed-article-AMEP.

[8] Swing S. The ACGME outcome project: retrospective and prospective. Med Teach 2007;29:648–54.

[9] Hersh WR. The full spectrum of biomedical informatics education at Oregon Health & Science University. Methods Inform Med 2007;46:80–3.

[10] Skochelak SE, Hawkins RE, Lawson LE, Starr SR, Borkan JM, Gonzalo JD, editors. Health systems science. New York, NY: Elsevier; 2017.

[11] Clerkship Cards Internal Medicine. Available from: <http://www.ohsu.edu/xd/education/schools/school-of-medicine/academic-programs/md-program/curriculum/upload/Clerkship-Cards-Internal-Medicine.pdf>; [accessed 11.06.17].

[12] Biagioli FE, Elliot DL, Palmer RT, Graichen CC, Rdesinski RE, Kumar KA, et al. Electronic health record objective structured clinical examination: assessing student competency in patient interactions while using the electronic health record. Acad Med 2017;92:87–91.

[13] Milano CE, Hardman JA, Plesiu A, Rdesinski RE, Biagioli FE. Simulated electronic health record (Sim-EHR) curriculum: teaching EHR skills and use of the EHR for disease management and prevention. Acad Med 2014;89:399–403.

[14] Biagioli FE, Milano CE, Hardman JA, Scholl GR. Teaching chronic disease management surveillance and prevention using simulated electronic health records-resource 561. MedEdPortal; 2012. Available from: https://www.mededportal.org/icollaborative/resource/561.

[15] Palmer RT, Biagioli FE, Mujcic J, Schneider BN, Spires L, Dodson LG. The feasibility and acceptability of administering a telemedicine objective structured clinical exam as a solution for providing equivalent education to remote and rural learners. Rural Remote Health 2015;15:3399.

[16] Palmer RT, Robinson R, et al. Measuring the impact of a telemedicine simulation on medical students. STFM Conference on Medical Student Education, Anaheim, CA. <http://www.stfm.org/Conferences/ConferenceonMedicalStudentEducation/PastConferences/PastAbstracts,Brochures,andHandouts/2017STFMConferenceonMedicalStudentEducation?m = 6&s = 18541>; 2017.

[17] Lahlou R, Wiser E, et al. Teaching diabetes management and EHR competencies using an EHR workshop. STFM Conference on Medical Student Education, Anaheim, CA. <http://www.stfm.org/Conferences/ConferenceonMedicalStudentEducation/PastConferences/PastAbstracts,Brochures,andHandouts/2017STFMConferenceonMedicalStudentEducation?m = 6&s = 18511>; 2017.

[18] Palmer R., Biagioli F.E., Cawse-Lucas J., Keen M., Kost A.R., O'Neill P., et al. EPAs and competencies: placing UME family medicine in a competency-based framework. In: Society of teachers of family medicine medical student education conference, Phoenix, AZ; 2016.

[19] Shortliffe EH. Biomedical informatics in the education of physicians. J Am Med Assoc 2010;304:1227–8.

CHAPTER

14

Nurse Education in the Digital Age—A Perspective From the United Kingdom

Sharon Levy
University of Edinburgh, Edinburgh, United Kingdom

INTRODUCTION

This chapter is set to give readers an overview of the development of nursing across the United Kingdom (UK), their education, and the way they fit with what is now often referred to as "digital health." It will elaborate on the progress made in engaging the workforce in the eHealth agenda and articulate the drivers for needed changes to the education of professional care providers. A case study, developing two new courses for undergraduate and postgraduate programmes, will be used to demonstrate the approach taken by one leading academic institution in Scotland. Such a development reaffirms the University's position at the forefront of developments in nursing education in the United Kingdom 60 years on.

Nursing Studies at the University of Edinburgh (Scotland) became a trailblazer, some 60 years ago, by being the first higher education institution in Europe to offer a nursing programme to degree level. We recently celebrated the achievement by noting the impact we had through our graduates, on the development of the caring profession at national and international levels. However, the education landscape has changed significantly since those early pioneering days and nursing in the United Kingdom is now an all-graduate profession. Higher Education Institutions (HEI) offer programmes lasting 3 or 4 years (Honors studies) to undergraduate students leading to professional registration. The curricula enables students to have 50% of their time in clinical practice and they can choose the nursing branch (or field of practice) they wish to specialize in from the first day of their studies. Those branches include Adults, Children, Mental Health, Learning Disabilities, and Midwifery (The term nursing in this paper includes midwives.).

Health Professionals' Education.
DOI: http://dx.doi.org/10.1016/B978-0-12-805362-1.00014-0

Once qualified and in order to maintain their registration, nurses complete, on an annual basis, mandatory days that serve to fulfill their need for Continuous Professional Development (CPD). To advance their practice and continue to offer safe care, clinicians may also engage in Post Graduate (PG) studies and complete a Master's program in a nursing related subject. A few opt to undertake Doctoral studies, which is advocated for nurse consultants in clinical posts within the National Health Service (NHS).

INFORMATION FLOWS IN CLINICAL SETTINGS

The clinical environment has also seen tremendous changes in the last 60 years, capitalizing on advances in technology and biomedical sciences. Clinical information technology (IT) systems were introduced in the United Kingdom in an effort to modernize the healthcare arena and ensure it remains effective, efficient, and able to sustain the ethos of being free at the point of need. Managing information is seen as a key factor in providing a seamless and smooth patient journey, between primary and acute care settings. In fact, the desire to manage information flows led to a bold attempt to establish a national (English) "spine-like" infrastructure to aid the link between various regional clinical IT systems. However, despite billions of pounds being spent on the National Programme for IT (NPfIT) this project, like many other large scale IT developments in the UK public sector, essentially failed [1]. At that time, the "big bang" approach to clinical IT implementation was not adopted in Scotland, Northern Ireland or Wales. Instead, clinicians in different geographical "Health Boards" or "Trusts" continued to use a myriad of legacy systems along with a few nationally procured systems such as the Picture Archive and Communication System (PACS). The current focus in Scotland, for example, is on exploring a mechanism to amalgamate various data sources and render information in a way that fits the needs of clinicians and patients alike.

Having access to the right information, when and where the patient is being seen, remains a critical factor to delivering safe and effective care. Dedicating needed financial resources to update and upgrade clinical IT systems also remain a challenge in a cash strapped UK wide NHS. Training and supporting those who use the systems most often—nursing staff—is an extensive and expensive undertaking. What is needed, I argue here, is a way of embedding informatics competencies or digital capabilities to education programme for the future nurse.

NURSES AND IT—READY FOR THE REVOLUTION?

The term revolution is often associated with images of violence and bloodshed and may not fit well with the common public perception of the caring professional. Yet, as far back as 1983 Berg argued that: "[T]he choice is there and the time to make the choice is now. The decision must be whether to act traditionally and have change thrust upon the profession [nursing] from the outside or to anticipate this revolution in nursing practice, familiarize nurses with it, and prepare them to take an active part in the introduction of computers into the nursing community" [2].

The commitment to spending billions of pounds on clinical IT systems, in the NHS in England back in 2003, led to a flurry of activity concerning the development of systems and implementation plans. During the early days of the NPfIT, the issues concerning workforce readiness and the impact the programme may have on clinicians were not set as a priority. The medical profession and their trade union were very vocal in expressing concerns about the approach taken but the voice of nursing was noticeably limited.

The first survey concerning nurses' attitudes to IT development in the United Kingdom was initiated by the Royal College of Nursing (RCN) in 2004 [3]. Engaging with an online survey participants ($n = 2020$) asserted they were ill informed about planned developments and qualitative data analysis affirmed many felt excluded from decisions concerning the purchase, design and implementation of clinical IT systems [4]. Similar findings were noted in the following year when quantitative data from the online survey were analyzed ($n = 1776$). As the largest professional nursing organization in the United Kingdom, with 380,000 members in 2005, the RCN developed an eHealth strategy and associated work streams. Efforts resulted in the production of a few educational resources and key lobbying messages for government to promote better engagement with the largest group of professionals working in the NHS.

WORKFORCE ENGAGEMENT

Staff engagement is an established critical factor in successful implementation of eHealth in the work place [5]. Giodarno et al. also argued [6] that the most effective strategy for engaging the workforce is to include front line staff in the earliest stages of project design and planning. The work of the RCN resulted in significant progress in staff awareness about UK wide IT national programmes, as captured in subsequent eHealth surveys in 2006 and 2007. Yet, concerns remained about the level of informatics knowledge and skills amongst those nurses who were now asked to be involved in shaping the future of clinical IT systems across the United Kingdom.

The general lack of professional training to aid development of core informatics skills, within the UK nursing workforce, was further highlighted in the final eHealth survey by the RCN in 2012 [7]. Participants ($n = 1158$) were asked to note the training they had received regarding topics such as electronic patient record keeping, standardized clinical terminology, and remote communications with patients and Telehealth/Telecare. The results demonstrated that, whilst general IT and information governance training were offered by employers, there was no provision of formal education for the use of patient data to improve practice.

Considering the link between staff training, usage and attitudes to IT, Ward et al. [8] argued that experience and confidence in IT use influence positively staff attitudes. Training aimed at achieving general competencies in IT and skills in using digital equipment was found to aid effective implementation of clinical computerized systems. However, the review identified enhancing professional practice to be a core issue in the sustained use of such systems, suggesting education as key to demonstrating the value of technology use. Such technology had to be seen as anchored in current professional practice and values, including person centered care (PCC). The review concluded by calling for healthcare professionals to be trained in using IT in clinical practice as part of their academic journey, in both undergraduate and postgraduate education.

NURSING INFORMATICS AND LEADERSHIP

Unlike the United States, where the TIGER (Technology Informatics Guiding Education Reform) initiative was developed, there is no agreed national core informatics (The terms informatics, digital health and eHealth are used here to note similar spheres.) competency framework to guide UK-based nursing education providers. In contrast, the Academy of Medical Royal Colleges [9] produced a tool to assess trainees' eHealth competence acquisition for their members and ensure young doctors are ready to work effectively with clinical IT systems. It must be noted, however, that unlike Medical Royal Colleges, the RCN does not set, provide or monitor the education of the profession. This task is reserved for the regulatory body, the Nursing and Midwifery Council in the United Kingdom. In the few academic nursing departments, where staff have an interest and knowledge of nursing informatics, specific courses are being offered. In other institutions where both the faculty and the curricula are "informatics free," the teaching of such knowledge is currently missing.

The tension between faculty members who may be "digital immigrants" [10] struggling with technology use, and students who are digital natives, resembles the friction between professional nurses who have just graduated and their senior managers in clinical practice. New nurses may not be "encouraged" to use digital tools that could be perceived by their senior charge nurses as hindering rather than supporting person centered care. Yet, it is argued [11] that managers have a key role to play in facilitating successful implementation of clinical IT systems. Their attitudes, knowledge and skills in using technology are critical. Indeed, senior "buy in," manifested through leadership and explicit commitment from the "top of the organization," was found to be the most important factor in successful implementation of clinical information systems (ibid).

eLEARNING AS A MODE OF NURSING EDUCATION

Effective clinical leadership rather than efficient management is one of the courses offered at the University of Edinburgh to postgraduate nursing students. This course was transformed from a face to face mode of delivery to being offered online through "flexible learning and pedagogy" [12]. Flexibility is offered by maximizing three separate variables along the learning pathway: pace, place, and mode. Online learning enables students to progress at a pace that suits their individual circumstances and enables healthcare professionals who work full time to fit learning around their available spare time. The place of learning is not limited to the classroom and students may do the learning at home, whilst commuting or even at work. The mode encapsulates the use of a range of learning technologies and social media channels that augment and supplement the learning of PG students. Creating a sense of "closeness at a distance," through digital media, brings learners together and enables the sharing of stories and experiences in a safe and welcoming virtual environment.

Moreover, the adoption of distance and self-directed learning is often used to support active engagement with (learning) technology and as a gateway to personal and professional development. Evidence suggests [13] that some learners are more satisfied with online learning, compared with face to face learning, and the use of multimedia channels supports engagement with the subject matter and enhances the overall learning experience. In fact, Ferrari argues that graduates who hone their digital competencies are set to be "future proofed" and gain from the opportunity to progress their practice beyond acting as technology operators onto becoming knowledge workers[14].

Most young students at the University of Edinburgh have sufficient IT skills to embark on learning that capitalizes on digital and innovative pedagogies. This is due to the fact that digital capabilities are being honed at school and undergraduate student nurses are no exception. They have strong foundations on which to expand their digital skill set. Such skills include elements such as:

- Information, media and data literacy
- Digital creation scholarship and innovation
- Digital communication, collaboration, and participation
- Digital learning and personal/professional development
- Digital identity and well-being (see Jisc digital capability framework (https://digitalcapability.jiscinvolve.org/wp/files/2015/06/1.-Digital-capabilities-6-elements.pdf))

The PG student cohort, at Nursing Studies, is somewhat different and includes professionals who may have many years of practice, but limited knowledge of using digital tools. To capitalize on the experience and expertise, we developed in online learning and teaching, a new programme entitled Advanced Clinical Skills. The eHealth Competency framework [9] was a useful tool to craft learning that fits the needs of clinicians who wish to be at the forefront of advanced clinical practice. The framework offered detailed competencies in relation to knowledge, skills, and behavior in areas such as:

- Clinical leadership and management
- Working with information and clinical care records
- Knowledge management

Yet, key nursing concepts were not included:

- Using technology to support shared decision-making and self-care
- Using technology to share information for integrated PCC

ADVANCED NURSING PRACTICE IN THE UNITED KINGDOM—EMBEDDING INFORMATICS INTO THE PG CURRICULA (CASE STUDY)

Recent years have seen a strong political push to extend the scope of nursing practice and embed advanced technical skills in PG education. This was partly driven by a marked reduction in the hours junior doctors were permitted to work to fit in standardization of medical work/education practices across Europe. The professional ethos, nevertheless, was not to create "mini doctors" but rather introduce the concept of "maxi nurses." Such professionals continue to be able to act as a core member of the multidisciplinary team that care for patients and their relatives.

However, the clinical reality is that registered nurses spend less time with patients and more time focused on administration tasks. Very often these tasks involve the use of IT which leads to potential resentment and association of technology with tasks away from the bedside. Nursing at a distance, culminating in the emergence of eNursing or cyber nursing practice, disrupts the physical, narrative, and moral proximity to patients [15]. Yet, many nurses come into the profession with a desire to be with the patient and care for them in a holistic way. Moreover, in the developed world patients spend less time being cared for in hospitals, where care around the clock is provided, leading to fewer opportunities for patients to interact with clinical staff. It seems that the essence of nursing practice in the United Kingdom is increasingly characterized not by specific clinical skills but by values, attitudes, and nontechnical skills such as compassion and person centeredness [16].

DEVELOPING AN ONLINE COURSE

To address the prevailing perception of IT in the clinical environment as an administration driven directive, rather than an essential clinical tool, it was decided to develop and name the first core course in the new online programme as "Person Centred Care in practice." This course in one of three core segments that the students must complete. The others include the "Evidence Based Practice" course and "Assessment, Clinical decision making and diagnostic skills." These core modules account for a third of the needed credits to get an MSc award (180 credits in total) and students also need to complete a dissertation, which counts for another third of the total sum. This means students can choose to complete other online courses that fit with the overall theme of the program, including courses such as Telecare, or mHealth. The program was set to "produce" graduates who have effective skills in the use of digital tools for learning, practical skills to design a dynamic package of care that includes assistive technologies as well as excellent team working skills in an eLearning environment.

The platform for delivering the course is BlackBoard "Learn" and students are asked to engage by using the asynchronous discussion boards and interact remotely using the virtual synchronous classroom setup. The Person Centred Care course, which was developed first, was delivered initially in a blended way—requiring students to attend a number of face-to-face sessions, which were complemented by content and activities online. These sessions were recorded and edited to fit an online environment—following guidance on good pedagogy for distance education. The second cohort of students had no physical interactions with teaching staff and had to come armed with core digital competencies—offered as a preenrolment course. The first task all online students had to complete was introducing themselves and their clinical background. Like any introduction—there are those who were shy or reserved and those students who were eager to explore the boundaries of a new learning platform and exploit the possibilities. The course ran with five cohorts of students and students' evaluation was very positive. Initially, a two week period (out of a 12 week course) was dedicated to exploring a Person Centred Record, where the focus was on managing data, information sharing practices and the practice impact of the emergent ePatients.

The advisory board that mentored and monitored the progress of our new online program included a few service users (patients/carers). They questioned the decision to include the PCC course as part of the program, arguing that nurses should already possess the essence of person centeredness. They suggested that by developing the course and making it compulsory, we indicate that attributes of caring and compassion are reserved to those practicing at an advanced level only. Students, in contrast, had no problems with articulating why PCC is a core part of an advancing clinical skills programme, claiming that the in-depth knowledge of the theory and rationale for practice enable them to be role models to less advanced colleagues in practice. The same students also asked that the two dedicated eHealth weeks (Person Centred Record) were removed. The Telecare optional course had limited uptake and the mHealth course that is run by the Medical Schools remains active despite small numbers of learners, none of whom are nurses.

SHOULD A DIFFERENT APPROACH BE USED?

If there is such resistance to learning about or seeing eHealth as part of a nurse's clinical role or indeed part of an advanced practitioner's remit—should the focus be shifted to undergraduate education? This is the essence of what Bond and Procter [17] suggest in their United Kingdom focused "prescription for nursing informatics in pre-registration nurse education". However, rather than looking to map nursing education, delivered over a 4 year period, onto an informatics competency framework, the focus taken was on developing a standalone course. The view was that students who completed the course would become change agents within clinical settings and amongst peers, to fire up the needed revolution.

eHEALTH AND DIGITAL NURSING—DEVELOPING AN HONORS OPTION COURSE FOR UNDERGRADUATE NURSES

The rationale for creating the new course was to entice students to take this option. The course description offered a general background noting that the provision of healthcare in Scotland and across the globe is changing to fit the demands of those who require professional services and those who pay for them. Moreover, students were alerted to the fact that demographic changes, new disease patterns, and advances in technology all impact on the way care is planned and delivered. The focus on wellness, ill health prevention and care closer to home also affects the "skill mix" composition and the competencies needed by healthcare staff. It was suggested that adaptability, innovative practice, and creative thinking are amongst the vital skills needed by those who face the evolving nature and rapid pace of change in 21st century health care systems. The course was set to give students the opportunity to acquire knowledge and skills with which to examine core concepts, theories, and policies concerning eHealth. It offered an insight into current trends and future developments of digital nursing, whilst highlighting the benefits, as well as the risks, to users, healthcare providers and society as a whole.

The delivery of the course consisted of face-to-face lectures, online virtual sessions and asynchronous interactions. Students had an opportunity to both get involved in a mock clinical examination using video conferencing and to visit a clinical nurse informatician in the local NHS hospital. In the first lecture, students were asked to switch off all their gadgets and we reverted to a session with paper handouts and a printed workbook for them to complete. Core informatics theories were then introduced and discussed in the classroom. The second session was "paper-light" where a smartboard was used as teaching aid and digital content was shown (TED talk). Students then had to work in groups and present the summary of their discussion on a flipchart. For the third session all the content was available in advance on the "Learn" platform and the face to face session was focused on exploring some of the challenges student had with the content they worked on as independent learners. In the online-only week, students had to engage on the discussion boards and gave a short presentation on the virtual classroom to peers. The final summative assessment focused on an information prescription concept where students had to appraise digital content and guide patients in using online health resources.

The progression from paper-based teaching to paperless interaction was centered on knowledge and skills acquisition using experiential learning. Students were encouraged to explore and experiment and be creative, whilst cognizant of the possible challenges to their own professional practice and delivery of PCC. An attempt was made to move away from introducing technologies (i.e., social media) merely as a potential risk that should be carefully managed to a powerful tool that could enhance innovative practice. Students were also asked to showcase their learning to those who did not take the course and lead group work with peers on the benefits and risks of managing care using Electronic Patient Records.

Lessons learned from the first cohort of students will inform future delivery of this course if sufficient numbers opt to take it. The small number of students ($n = 35$) in a typical yearly UG intake makes it difficult to ensure such a course remains a viable option. Luckily, work is underway to offer a national solution in both England and Scotland for nurses, midwives and allied health professionals. The vision is that appropriate content will be authored and offered to HEIs on a digital platform, free of charge. These learning "chunks" could be bolted on to current curricula to enhance attainment of informatics knowledge and skills—couched in topics such as record keeping and ethics. The RCN is also planning the delivery of a range of new products to support needed work—as affirmed in the 2016 annual Congress (https://www2.rcn.org.uk/newsevents/congress/2016/agenda/debates/10-e-nursing).

CONCLUSION

This chapter offered a selective review of progress concerning the UK eHealth agenda as it applies to professional nursing. It elaborated on efforts made to promote more and better clinical engagement and described a few drivers to needed educational reform. The chapter grounded the discussion about future education of healthcare professionals to advancement of practice in a technology enabled care environment. The discussion then shifted to considering ways to embed informatics principles and theories in postgraduate nurse education whilst discussing a case study of an online programme. Content development for an undergraduate course was also described and discussed, to reflect on an attempt to engage with students before they qualified as professionals. It seems that whilst a small and successful local initiative may support a few bright learners, a different scale is needed to make a big impact on the nursing profession across the United Kingdom. A national agenda with coproduction of content with domain experts, aimed at those who aspire to become professional healthcare providers, is suggested as the best way forward.

References

[1] Hendy J, Reeves BC, Fulop N, Hutchings A, Masseria C. Challenges to implementing the national programme for information technology (NPfIT): a qualitative study. BMJ 2005;331(7512):331–6.

[2] Berg C. The importance of nurses' input for the selection of computerized systems. In: Scholes M, Bryant Y, Barber B, editors. The impact of computers on nursing: an international review. Amsterdam: Elsevier; 1983. p. 42–58.

[3] Royal College of Nursing. Nurses and NHS IT developments results of an online survey London: Royal College of Nursing; 2004. Available from: <https://www.rcn.org.uk/__data/assets/pdf_file/0004/627250/nurses-it-devs-survey2004.pdf> [cited 05.11.16].

[4] Royal College of Nursing. Speaking up: nurses and NHS IT developments 2004. Available from: <https://www.rcn.org.uk/professional-development/publications/pub-002477> [cited 02.09.16].

[5] Mair FS, May C, O'Donnell C, Finch T, Sullivan F, Murray E. Factors that promote or inhibit the implementation of e-health systems: an explanatory systematic review. Bull World Health Org 2012;90(5):357–64.

[6] Giordano R, Clark M, Goodwin N. Perspectives on telehealth and telecare: learning from the 12 whole system demonstrator action network (WSDAN) sites. London: King's Fund; 2011.

[7] Royal College of Nursing. Positioning nursing in a digital world London: Royal College of Nursing; 2012. Available from: <https://www.rcn.org.uk/__data/assets/pdf_file/0020/530390/004_440.pdf> [cited 25.11.15].

[8] Ward R, Stevens C, Brentnall P, Briddon J. The attitudes of health care staff to information technology: a comprehensive review of the research literature. Health Info Libr J 2008;25(2):81–97.

[9] Academy of Royal Colleges. eHealth Competency framework 2011. Available from: <http://www.aomrc.org.uk/wp-content/uploads/2016/05/EHealth_Competency_Framework_0611.pdf> [cited 02.09.16].

[10] Prensky M. Digital natives, digital immigrants part 1. On the Horizon 2001;9(5):1–6.

[11] Doolan DF, Bates DW, James BC. The use of computers for clinical care: a case series of advanced US sites. J Am Med Info Assoc 2003;10(1):94–107.

[12] Gordon N. Flexible Pedagogies: technology-enhanced learning. University of Hull, January. 2014. Available from: <http://www.heacademy.ac.uk/resources/detail/flexiblelearning/flexiblepedagogies/tech_enhanced_learning/main_report>.

[13] Petty J. Interactive, technology-enhanced self-regulated learning tools in healthcare education: a literature review. Nurse Educ Today 2013;33(1):53–9.

[14] Ferrari A. Digital competence in practice: an analysis of frameworks. Sevilla: JRC IPTS; 2012. Available from: <http://dx.doi.org/10.2791/82116>.

[15] Malone RE. Distal nursing. Social Sci Med 2003;56(11):2317–26.

[16] Rolfe G. Advanced nursing practice 1: understanding advanced nursing practice. Nurs Times 2014;11(27):20–3.

[17] Bond CS, Procter PM. Prescription for nursing informatics in pre-registration nurse education. Health Info J 2009;15(1):55–64.

CHAPTER

15

Effectiveness of Training Strategies That Support Informatics Competency Development in Healthcare Professionals

Bev Rhodes[1], *Anne Short*[2] *and Tracy Shaben*[3]

[1]**Alberta Health Services, Medicine Hat, AB, Canada** [2]**Alberta Health Services, Brooks, AB, Canada** [3]**Alberta Health Services, Edmonton, AB, Canada**

INTRODUCTION

With the introduction of new information and communication technologies (ICT) into healthcare environments, health care professionals must receive education in a meaningful manner that enhances learning, which includes understanding and behavior change. The group of health care professionals (HCPs) would include clinical staff such as nurses, physicians, allied health disciplines, and students.

Many implementations of ICT have the "potential to improve healthcare quality, increase patient safety, and reduce costs" [1]. But most ICT education curricula focus on training HCPs on the steps involved in how to use the system. Unfortunately, this approach does not guarantee that learning with understanding has occurred. Training and learning are often used interchangeably but over the past 10 years, the authors recognized that there was a distinct difference between "training" and "learning" during the education of HCPs in ICT [2].

Training is defined as "organized activity aimed at imparting information and/or instructions to improve the recipient's performance or to help him or her attain a required level of knowledge or skill" [3]. Learning is defined as "Measurable and relatively permanent change in behavior through experience, instruction, or study. Individual learning is selective, group learning... depend largely on power playing in the group" [3].

As organizations introduce ICT into clinical care workflows, the United States' Office of the National Coordinator for Health Information Technology (ONC) states that training

Health Professionals' Education.
DOI: http://dx.doi.org/10.1016/B978-0-12-805362-1.00015-2

and support must be available and sufficient to ensure HCPs are using the technology safely [4]. In Canada and throughout the world, similar digital health funding agencies have also stated the same patient quality and safety outcomes, which result from the maturity of ICT use [5]. For ICT to be successful, staff must be willing to learn and use the new technology [2,6]. As ICT expands within the various healthcare settings, many of the clinical workflows are also impacted. For the ICT introduction to be successful, we must identify training strategies that promote learning to support the implementation of the ICT into workflow processes.

Training strategies will include the various modes of delivery such as traditional classroom as well as electronic delivery methods. Strategies will include learning objectives to support specific curriculum including software functionality and HCPs' safe usage of the ICT [7]. These training strategies will need to address the unique learning needs for the HCPs and their clinical workflows. ICT will also have differences in functionalities based on the HCPs' roles [8]. When looking at factors that support learning and acceptance of ICT, areas such as benefits of use, ease of use by design, and build to support workflows are key fundamentals [9].

The learning experience and mastery of ICT is not a single event. As HCPs interact with the technology, they may identify areas of their clinical practices that require the use of additional and/or advanced functionality. In addition, HCPs will need to keep up to date with the latest software releases and practice standard changes that affect system use. Monitoring for long-term acceptance and continuing competency with the use of ICT is an important aspect of any ICT implementation. Determining a competency framework provides a theoretical and practical focus for evaluation leading to an effective program that supports informatics competency development for HCPs.

In this chapter we will explore our lived experiences discussing the Kirkpatrick Training model in the use of various modes of training delivery that enabled our HCPs to learn and apply knowledge of why and how to safely use the ICT. We will also discuss our discovery of the importance of including how to incorporate the ICT into clinical workflows within the training activities versus focusing on ICT functionality training alone. We will reflect on the impact on learning when utilizing super users, champions, and mentors during and post training. We will discuss the need for evaluation and evaluation tools to measure the effectiveness of training strategies, the success of learning, and the overall progress toward continuing competency through the use of Bloom's taxonomy and the related competency frameworks [10,11].

COMPETENCY FRAMEWORK

Curriculum development, instructional design with performance assessment, and process implementation tools make up the vital pieces of a training design such as Kirkpatrick's Training Model [12,13] (see Text Box 15.1 and Fig. 15.1) and the six levels of cognitive complexity in Bloom's Learning Taxonomy [14] (Fig. 15.2) provide important elements that greatly impact learning. Learning can be monitored and measured through

TEXT BOX 15.1

KIRKPATRICK'S EVALUATION MODEL (KIRKPATRICK DL. REVISITING KIRKPATRICK'S FOUR-LEVEL MODEL. TRAINING DEV 1996;50(1):54–9.)

Kirkpatrick's Evaluation Model suggests that educational and training interventions may be evaluated at four levels: (1) reaction, which assesses the degree to which participants view the intervention favorably; (2) learning, which refers to the gain in knowledge, skills, and changes in attitudes that occur as a result of the intervention; (3) behavior, referring to the extent to which participants apply what they learned when they return to their job after the training; and (4) results—the extent to which targeted (organizational) outcomes occur following training.

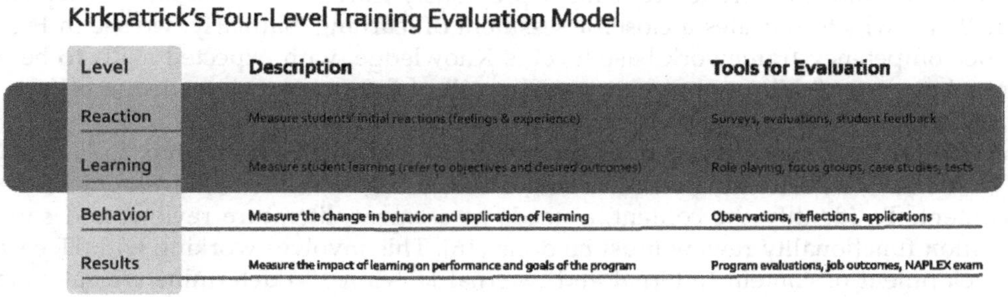

Kirkpatrick's Four-Level Training Evaluation Model

Level	Description	Tools for Evaluation
Reaction	Measure students' initial reactions (feelings & experience)	Surveys, evaluations, student feedback
Learning	Measure student learning (refer to objectives and desired outcomes)	Role playing, focus groups, case studies, tests
Behavior	Measure the change in behavior and application of learning	Observations, reflections, applications
Results	Measure the impact of learning on performance and goals of the program	Program evaluations, job outcomes, NAPLEX exam

FIGURE 15.1 Kirkpatrick's Four-level Training Evaluation Model. *Adapted from Smidt A, Balandin S, Sigafoos J, Reed VA. The Kirkpatrick model: a useful tool for evaluating training outcomes. J Intellect Dev Disabil 2009; 34:266–74.*

the six levels of the competency framework of knowledge, comprehension, application, analysis, synthesis, and evaluation [10] as illustrated by Fig. 15.2 to provide a competency foundation for the overall acceptance of ICT [11].

One of the tools to review training success at Kirkpatrick's base level Reaction as seen in Fig. 15.1 is with an evaluation that reflects the HCPs' immediate reaction to the ICT training event [13,14]. Kirkpatrick considered training to be successful if the learner responded with a positive reaction in the evaluation. In our experience, we have found that this was not the case for training HCPs. HCPs stated they had a positive training event, but when the time came to implement the training at the workplace, it was evident

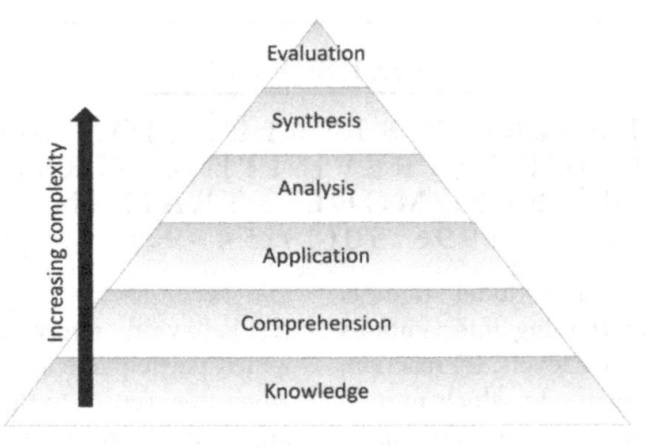

FIGURE 15.2 Bloom's Learning Taxonomy. *From Adams NE. Bloom's taxonomy of cognitive learning objectives. J Med Libr Assoc 2015;103:152–3, used with permission.*

there was limited learning and minimal comprehension. An immediate favorable reaction does not measure learning or acquisition of knowledge [15]. We found that evaluating HCPs training event through Bloom's base level Knowledge was more accurate in assessing HCPs' cognitive ability to "remember previously learned information," as depicted in Fig. 15.2 [10] which indicates a closer assessment of learning. Similarly, we see in Fig. 15.3 that the competency framework base level is Knowledge, with expected skills to be demonstrated such as, "skill in observation and recall of information, knowledge of facts and major ideas" [11] which are also better measurements of learning.

In considering the base levels of Learning and Knowledge (see Text Box 15.1, Figs. 15.2, and 15.3) [10,13,14] a training and learning strategy begins with the development of subject content. To identify the content, a needs assessment, literature review, and software application functionality review must be done [16]. This involves working with IT partners on development of content (internal and external IT teams) to determine crucial concepts that must be included in the ongoing training strategy [17].

Content decisions provide the foundation for the development of scenarios or case studies [18,19]. This becomes "the electronic patient story" that motivates engagement and learning [2,20]. Content decisions have cost implications when developing learning activities including HCPs' time and organizational commitment so understanding the key application functionality and the learning objectives for each element are essential.

Content development is one of the features of a training and learning strategy but not the only component. Englebardt identified additional issues when using technology for training, which included support for the technology and matching the technology options to the content, audience, and purpose [21]. Elements of design to consider in the various modes of delivery of content (classroom, one on one, and e-Learning) include: amount of material (time per session/module), relevance of practice scenarios or case studies with feedback ("making it real"), mixed instructional strategies (simulation, animation, mnemonics, and multimedia), and layering of content to enable critical thinking skills to develop [7,22,23]. Effective instructional design aides in outcomes of knowledge retention, HCPs' satisfaction, and ultimately short- and long-term competency [24].

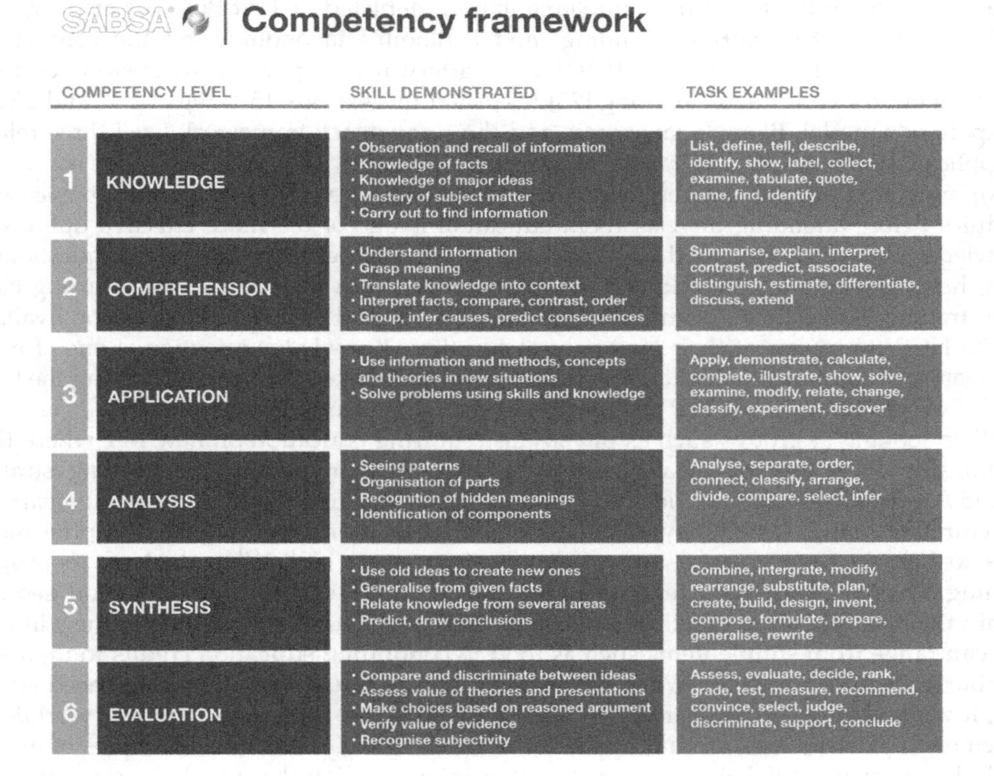

SABSA | Competency framework

COMPETENCY LEVEL		SKILL DEMONSTRATED	TASK EXAMPLES
1	KNOWLEDGE	• Observation and recall of information • Knowledge of facts • Knowledge of major ideas • Mastery of subject matter • Carry out to find information	List, define, tell, describe, identify, show, label, collect, examine, tabulate, quote, name, find, identify
2	COMPREHENSION	• Understand information • Grasp meaning • Translate knowledge into context • Interpret facts, compare, contrast, order • Group, infer causes, predict consequences	Summarise, explain, interpret, contrast, predict, associate, distinguish, estimate, differentiate, discuss, extend
3	APPLICATION	• Use information and methods, concepts and theories in new situations • Solve problems using skills and knowledge	Apply, demonstrate, calculate, complete, illustrate, show, solve, examine, modify, relate, change, classify, experiment, discover
4	ANALYSIS	• Seeing paterns • Organisation of parts • Recognition of hidden meanings • Identification of components	Analyse, separate, order, connect, classify, arrange, divide, compare, select, infer
5	SYNTHESIS	• Use old ideas to create new ones • Generalise from given facts • Relate knowledge from several areas • Predict, draw conclusions	Combine, intergrate, modify, rearrange, substitute, plan, create, build, design, invent, compose, formulate, prepare, generalise, rewrite
6	EVALUATION	• Compare and discriminate between ideas • Assess value of theories and presentations • Make choices based on reasoned argument • Verify value of evidence • Recognise subjectivity	Assess, evaluate, decide, rank, grade, test, measure, recommend, convince, select, judge, discriminate, support, conclude

FIGURE 15.3 SABSA Competency Framework. © *The SABSA Institute CIC. Used with permission.*

MODES OF TRAINING DELIVERY

The first element to contributing to HCPs' learning, competency, and knowledge retention would be consideration of which mode of training delivery would be optimal for the individual or group. The ICT education of HCPs has evolved by implementing informatics tools to encourage learning and knowledge retention. HCPs' learning can be difficult due to a variety of challenges such as individual staff schedules, staff turnover rates within certain care areas, as well as conflicting priorities with clinical group learning [18]. Given these challenges, the use of various learning tools must be applied pragmatically by the educators while being cognizant of teaching and learning styles whether in a classroom setting, through an online electronic learning (e-learning) module, or one-to-one instruction to promote knowledge retention.

An evaluation of three possible training delivery modes can assist in finding the right balance of face-to-face training with e-learning for optimal knowledge retention. Training can occur in a single event, but learning cannot be accomplished through a single training event [2]. Managers must be open to allowing their staff sufficient time within their schedules to learn the ICT during implementation (depicted in Text Box 15.1 and Fig. 15.2. And Fig. 15.3 in Kirkpatrick's training model, Bloom's taxonomy, and the competency framework levels 1 and 2) [10,13,14] as well as afterward supporting the creation of a safe environment for continuous learning [25] (depicted in Text Box 15.1, Figs. 15.2, and 15.3 in Kirkpatrick's model, Bloom's taxonomy, and the competency framework level three related to application of training and transfer of learning) [10,13,14].

An integrated training delivery mode, where the HCPs review the ICT through e-learning modules before attending the classroom education is one of the more effective options for knowledge retention. This method allows learners to preview the system, ask questions in class, have immediate feedback, and be able to collaborate with one another during classroom training [26]. The e-learning module can also be used as a follow up option available for the HCPs to review at their own pace at any time. Knowledge retention is one of many necessary priorities during and after education for HCPs to be successful. Adapting the training delivery to the various HCP groups provides the best learning outcomes [27].

When looking at groups such as physicians requiring education on new ICT where their relationship with the organization is often contractual in nature, the training strategy should address structured learning, supportive learning and individual learning. Structured learning, which may include classroom based or e-learning, should offer incentives to help increase interest and offset physician costs [28]. When looking to provide training to physicians, they tend to have limited time available for education sessions within their daily clinical practice. The use of incentives has shown to bolster their interest and can range from simple items such as food to continuing education credits to monetary reimbursement for time [29]. When looking to schedule structured classroom based education, it works best to group learners into sessions based on common interest or workflows, which enable discussions around impact of ICT on these components. Examples of groups might be internal medicine physicians together in one session that would be separate from orthopedic surgeons.

One-to-One Instruction

Certain groups such as physicians find that one-to-one instruction, commonly referred to as "at the elbow" training provides the best user feedback and integration of learning objectives into clinical workflows [30]. Using this method of training enables the educator to adjust the content for the unique HCP and clinical workflow. It also enables the HCPs to use the new ICT during patient care with someone available to reinforce learning and answer any questions. One to one instruction is also useful for training new HCPs in a specialized discipline that tends to hire one or two people at a time. Having a dedicated software environment for the specific learning content paired with materials focused on the workflow processes enables the HCPs to have as close to the "real" experience as is possible.

One to one instruction can be a method to deliver supportive or continuous refresher content with the availability of mentors and educators present within the clinical department after "go live" when the ICT is operational. The mentor is able to provide updated information on new functionality or changes in the ICT. The mentor can provide timely support to a variety of HCPs within the clinical department as the learners experience questions or concerns during the day-to-day use of ICT.

Traditional Classroom-Based Instruction

One to one education is not always practical and requires copious amounts of human resources to be simultaneously available for several HCPs in their workplace. A practical and common approach is group instruction in a traditional classroom setting. Classroom training has evolved from lecture style training to interactive training. This evolution was inevitable in considering the needs of the adult learner [27]. As educators, we observed that HCPs often presented to the classroom with computer anxiety, placing high expectations of themselves to learn the ICT and to become experts immediately without allowing time to learn [31]. As a result, frustration would cloud motivation and ability to absorb the content required for ICT acceptance. We found that learning and understanding improved by encouraging HCPs to start with the basics and take the opportunity to apply the content during training [31]. It was harder to assess if HCPs were gaining knowledge with comprehension to be able to transfer their learning in the workplace, especially with HCPs who opted not to "play" with the system ahead of time. Some other challenges noted were HCPs who were unable to admit their difficulty in learning how to use the ICT and rejected help or training during the education session. Others who were afraid to "show their weaknesses" would behave confidently indicating they had no issues, but during implementation would lack competency. Finally, there were HCPs who were incapable of following most of the training, which resulted in the need for one to one support in the classroom setting [32].

Class size has a direct influence on the ability of the educator to assess the learning abilities of the learners, thus having the appropriate class size is important [33]. We observed that our optimal classroom-learning environment was a ratio of a maximum of 8–10 learners to one educator. If any of the group of HCPs were information technology (IT) challenged, the educator's focus turned from delivering the educational content to the group to that of a one on one instructor support and this extended the time needed for the training [30]. Utilizing super users or application specialists to assist the educator made the classroom experience more efficient and smoother for delivering the presentation.

Educators must be aware of the varied HCPs' confidence and competency levels when presenting in the classrooms [32]. We found it was best to note the range of HCPs' levels of comprehension while training to gauge timing in delivery between the fastest and slowest learning levels to accommodate HCPs in the classroom. If super users are not available during training, pairing HCPs who have higher abilities with those who have lower abilities is beneficial in supporting and engaging all the HCPs while the educator is presenting [34].

We discovered that during the delivery of classroom content it was important to give HCPs the opportunity to apply the presented information in a sample case study to evaluate whether comprehension had occurred during classroom time, as illustrated in Kirkpatrick's model in Text Box 15.1 [13,35]. Bredfeldt et al. found that when hands-on exercises were used within a classroom delivered training, learners found this time of great use [1].

We recently introduced electronic learning (e-learning) modules as precourse work in hopes that the introduction to the screens and navigation tools of the ICT would be less foreign in the classroom on the first day of in-class training and that these modules could also be used as a follow up reference for HCPs to revisit after classroom education [1]. Some HCPs who went through the e-learning modules reported that they were ineffective as precourse material and felt that the modules would be better as a post education tool.

The following is an example of the ICT training strategy that uses a case study. Our team presented the training and the case study over three days:

Day One: New information was presented, demonstrated, and HCPs had opportunity to apply the information to a case study.

Day Two: Began with a review of the highlights from Day One, followed by additional material presented building on the reviewed information, and the HCPs further developed their care plans based on the case study they began the day before.

Day Three: In the morning we reviewed the past two days, followed by new material building on the previous two days of education. In the afternoon of Day Three, HCPs received a new case study to apply all three days' worth of education independently to ensure they had internalized some of the steps during training, and also, to solidify the need to reference back to the education materials for help when unsure [34]. The educator would review the HCPs' work to identify areas that needed correcting and to determine areas for review.

HCPs reported that the training strategy of using a case study helped enhance learning during training and gave opportunity to apply the training in a test environment. Kirkpatrick's second level of training is defined as "the degree participants acquire the indented knowledge...based on their participation in the learning event" [13]. This level uses tools such as case studies to evaluate training as listed in Fig. 15.1 [13]. The case studies also gave educators the opportunity to evaluate HCPs' comprehension through their ability to "demonstrate an understanding of the facts" (see Fig. 15.2) [14] and at their skill in "understanding information, grasp meaning, and transfer knowledge into context" (see Fig. 15.3) [10], which offered a more accurate measure of learning that showed a change in behavior and a transfer of learning into the workplace.

Evaluation of the mode of training delivery provides important feedback to evaluate the best training strategies that provide the highest level of learning resulting in transfer of learning. Evaluations were handed out at the end of education sessions. Most were circled indicating that the materials used, the presentation, and the educator had significantly met their education needs for training, this type of evaluations collected HCP reactions to training instead of learning as summarized in Table 15.1. There were also some evaluations that were incorrectly marked indicating the need had not been met but, when clarified, the HCPs had meant to circle the opposite. Narrative responses varied, often giving accolades to the presenter; some were valuable in adjusting education content, materials, and presentation. Repetition was a key aspect that HCPs reported helped improve learning. Other areas HCPs reported on were the preference of a whole manual versus smaller modules with handouts, and requesting quick reference cheat sheets. HCPs who used the e-learning modules as precourse work stated it was too hard to follow; however those HCPs returning to work from a leave indicated retaking the course as a refresher was beneficial (Table 15.2).

Evaluations are a useful tool to gauge how the HCPs perceived the education, but it is more difficult to ascertain how much learning transpired during training. An evaluation would need to be very specific in identifying areas of learning from the course. Such as "I am confident in being able to navigate the ICT system" or "I am able to build a comprehensive care plan from this education." It was noted that our evaluations used in 2014 did not have any components to assess if HCPs had learned during their training event. In

TABLE 15.1 Summary of Selected 2014 Evaluation Results

Number of Evaluations Reviewed	Care Manager Update Zone		1. Session Met My Expectations Rating	2. Content Information Useful Rating	3. Presentation Understandable Rating	4. Information presented in an understandable form. Rating	5. Usefulness of the handouts. Rating
	Zone A	Zone B	Rate 1–4	Rate 1–4	Rate 1–4	Rate 1–4	Rate 1–4
			1 = Strongly Agree	1 = Strongly Agree	1 = Strongly Agree	1 = Strongly Agree	1 = Strongly Agree
			2 = Agree	2 = Agree	2 = Agree	2 = Agree	2 = Agree
			3 = Neither Agree or Disagree	3 = Neither Agree or Disagree	3 = Neither Agree or Disagree	3 = Neither Agree or Disagree	3 = Neither Agree or Disagree
			4 = Disagree	4 = Disagree	4 = Disagree	4 = Disagree	4 = Disagree
59	40	19	2.13	1.83	1.98	1.79	1.63

Themes of Post Training Comments: confusing, not enough time, pace is too fast, should be teaching elements when they are live for use. Need a practice client with case study for everyone to try during training.
Training occurred in Jan, Feb, Mar/2014 while system was live, except one element was not ready by the time the training was delivered
Two different types of evaluation forms with slightly different questions were given to staff in the Zone A and B.
Zone A used an evaluation that asked learners to answer questions #1–3 and to make comments.
Zone B used an evaluation that asked learners to answer questions #2–5 and to make comments.
(© the authors and Alberta Health Services. Used with permission).

2016, the evaluations were reworked to include statements of comfort with expected skills as listed in Table 15.2. Performance assessments can be created to assess that the learning objectives have been met. Scenarios can be built using screen shots, and questions reflecting learning content can be answered and scored to determine knowledge at points in time including new user, after significant ICT enhancements, following extended leaves or at any other point in time [11]. Fig. 15.1 shows that most evaluation of learning occurs in relation to Kirkpatrick's Level One-Reaction, and Level Two-Learning. Tools to measure learning used in these levels are the most common follow-through of training, such as immediate post training evaluations, surveys, and application to case studies.

Electronic Learning Tools

The advancement of electronic learning tools has provided another alternative to delivering learning objectives to large groups of HCPs in a time when resources such as classrooms, staff schedules, and qualified trainers are limited. E-learning modules have many development options that can support deeper knowledge retention as well as evaluation of retention and satisfaction [7]. Previous research has looked at the HCPs' satisfaction and retention of knowledge when e-learning modules have been used with mixed results. Higher satisfaction rates have been noted when e-learning modules were customized to the HCPs to meet their unique roles or job profile [18]. For some HCPs using e-learning modules, fear of technology or the lack of personal interaction with other learners or instructors may be barriers for learning [36].

TABLE 15.2 Summary of Selected 2016 Evaluation Results

Number of Evaluations Reviewed	Designation		How Many Years Worked				Skill in Performing Clinical Review Rating		Skill in Understanding Why CAPs triggered Rating		Knowledge of How CAPs impact OMAHA Care Plan Rating		Skill in Adding Suggested Problem Rating		Level of Comfort in Linking Interventions to Problem Rating	
	RN	LPN	< 1	1-2	3-5	> 5	Pre Train	Post Train	Pre Train	Post Train	Pre Train	Post Train	Pre Train	Post Train	Pre Train	Post Train
51	43	8	4	12	14	21	3.37	4.22	2.8	3.96	2.77	3.94	3.3	4.16	3.25	4.1

Themes of Pre Training Comments: Increased knowledge, confidence, understanding consistency in RAI-HC, CAPs, and building care plans.
Themes of Post Training Comments: good review, helpful, clarification, plans to change care planning approach, beneficial, suggested to refresh annually, and should be mandatory for all staff.
Training occurred in September, October, November, and December 2016.
100% of the HCPs indicated increased knowledge, comprehension, and confidence in applying their learning in the post training evaluation of learning objectives taught.

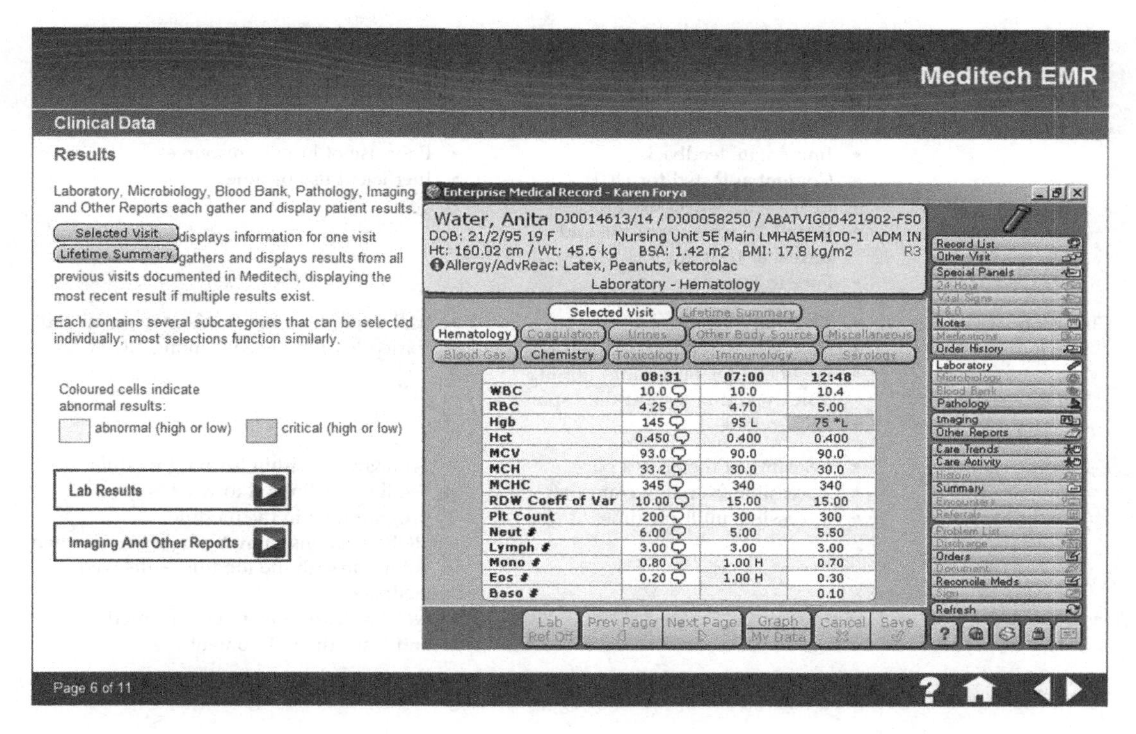

FIGURE 15.4 eLearning Module Sample. *Source: Alberta Health Services. Used with permission.*

We found that as HCPs began to learn a new ICT program, it was beneficial for them to take the time to review the e-learning module (see Fig. 15.4) first to see how it worked and then to review again for the learning content. Although, e-learning modules appeared to be an efficient mode for learning a new ICT, there was still a need for hands-on practice sessions, on-site mentorship, and support to feel confident in applying the ICT in actual clinical work environment. E-learning modules are most meaningful for HCPs when they are built with specific workflow processes in mind.

The education team experienced months of managing challenges such as, time constraints, complex schedules, and limited staff in relation to training practicum nurses. When these challenges presented, the training strategy for practicum nurses had to evolve, as a result e-learning modules were implemented as the primary training delivery. The practicum nurses were then mentored by their preceptors during the practicum instead of attending classroom training.

The needed training included would only require the knowledge related to navigation, documentation, and minimally building care plans. All of these components were available on e-learning modules and could be further reinforced by their preceptors. The practicum nurses found the e-learning modules difficult to follow, and their preceptors communicated that the e-learning modules did not provide the same training experience and foundational knowledge that classroom training provided. The limited retention or application of knowledge delivered via the e-learning modules could be attributed to the lack of embedded

TABLE 15.3 Modes of Training

Modes of Training	Pros	Cons
One to one	• "At the elbow" training • Immediate feedback • Content adjusted for HCP • Learn with workflow process • Beneficial for supportive and refresher training • Face to face	• Cost • Poor use of human resources • Inefficient use of time
Classroom	• Train larger group • Immediate feedback • Beneficial for providing training quickly and in a short time span • Face to face	• Individual learning needs may be missed • Varied learning needs among HCPs
E-learning modules	• Learning at the HCP's pace • Accessible from any computer • Accessible multiple times	• Set answers within learning module • Feedback is limited to what is programmed in the module • HCP questions cannot be addressed when going through the module – delayed feedback • Need to learn how to use the module and learn the ICT content

"authentic learning activities" which have shown to aid HCPs in building their knowledge, skills, and attitudes within their day to day work setting [7]. Even with challenges of using e-learning modules within our organization's experience, they remain a great asset. For any HCPs, having modules available enables them to refresh or renew certain aspects of the ICT thus providing opportunities for continuing competency development [7].

E-learning modules are not the only option that supports independent learning. The development of interactive electronic books (e-books), or multimedia videos or simulation modules also can be effective learning resources. We have found that the media clips need to be short, no longer than two minutes to support quick review. Each of the modes of training delivery has advantages and disadvantages that should be reviewed when developing a training strategy (see Table 15.3). All of these resources should be part of an ICT support library that is accessible at any point of time by the HCPs as well as usable across various device platforms.

SUPPORTING TECHNOLOGY WITHIN WORKFLOWS

The second element contributing to HCPs' learning, competency, and knowledge retention would be workflow considerations for the individual or group being trained to make the content meaningful and applicable in the workplace setting. In ICT education, we originally focused on how to use and navigate the system, but this evolved into presenting material in a different training strategy to make it more meaningful to the HCPs. McLane found that implementation of ICT does have a direct impact on clinician workflows as well as how HCPs interact with each other [6]. Understanding how the ICT interacts with the workflows or clinical practices while the HCPs learns the technology is an important factor in acceptance [9]. HCPs satisfaction with the use of ICT has a direct correlation to the design, which must support the workflows and needs within the clinical setting [6]. To connect with HCPs, the educator needed to consider and present education based on workflows and clinical practices during training. If the clinical workflow was not addressed during the education, there was a high risk of a workaround or improper implementation of the ICT system [17]. The incorrect application of ICT training resulting in workarounds correlates with level 3-Application in Kirkpatrick's model, Bloom's taxonomy, and the competency framework presented in Text Box 15.1, Figs. 15.2, and 15.3. The expected results of training is that the HCP has learned how to use the ICT and to use it safely exactly as it was taught in training. If they are able to do this, a transfer of learning occurs and it would be considered successful learning at level 3.

One opportunity to support the HCPs' ability to understand how the technology interacts with the clinical workflows was through the use of case studies or scenarios. Case studies are a beneficial tool for training, and for evaluating learning as indicated in Kirkpatrick's Model evaluation tools for level 2 in Fig. 15.1 [14]. These case studies described common clinical behaviors such as assessment, care planning, or interventions that the HCPs would use within the practice system to try their new knowledge.

It is the continuing competency in ICT training and learning, that is minimally assessed and monitored. This will often become evident when the HCP is required to transfer knowledge in the workplace, or begin to apply, analyze, synthesize, and evaluate the training to master the ICT system. For some of the HCPs during upgrade and refresher training, this was the first time in five to ten years of their work that knowledge and skills were being assessed and corrected. It is these higher levels of training, and learning that need more consistent evaluation to maintain competency in the application and acceptance of the ICT as illustrated in Level three in Text Box 15.1, Figs. 15.2, and 15.3 [10,13,14]. The longer-term evaluations of learning and transfer of knowledge look at "the degree participants apply what they learned during training on the job" [13], HCPs' ability to "apply knowledge to a new situation," and for demonstrated skills of "using information and methods... in new situations" and "solve problems using skills and knowledge" [10,14].

During one of our ICT upgrades, it was discovered that in an effort to be the most proficient and efficient in their field, HCPs became creative in conducting their work and taking shortcuts or creating workarounds to accomplish this. In addition, nursing specific work was constantly changing and was more chaotic in nature leading to fewer opportunities to be efficient. Often the HCPs did not understand that they had been taught in a specific manner, and the methodology behind the training was meant to ensure the ICT was used correctly [37]. Due to the rise of these shortcuts and workarounds, there was a need to teach the ICT in a manner that was more meaningful to the HCPs. We found one of the best approaches was to understand workflows that the HCPs would encounter in their daily care situations in the workplace, and then for the educators to create presentations that taught HCPs how to use the ICT accurately and appropriately within their scope of practice without taking shortcuts. The shortcuts were a result of HCPs not understanding the rationale as to why specific processes needed to be followed as taught. Once HCPs understood the reasons, there was a decrease in the number of occurrences of shortcuts and more exploration of what could be done differently to achieve better efficiency and effectiveness.

For example, in our ICT system, all interventions added to the care plan must be linked to a problem. The interventions were associated to different patient outcomes such as patient knowledge, patient behavior, and patient status. Depending on the underlying root of the problem, interventions were focused on teaching the patient, offering health care aide assistance in daily tasks, or delivering professional treatments and health monitoring to address the problem. Historically, HCPs were able to select the interventions they felt were applicable to the patient without an associated problem until the ICT upgrade that required those same interventions be added with the root problem. Post training, it was discovered that some HCPs had found another way to add interventions, but they bypassed linking them to a problem. HCPs were instructed not to use this mode, reteaching of the correct method was required to ensure standardized methods to produce an appropriate care plan.

MENTORSHIP ROLES

The third crucial element to contributing to HCPs learning, competency, and knowledge retention would be realizing the impact of utilizing mentors, super users, or champions during and post training for the individual HCPs or learning group. The role of a mentor is key in supporting HCPs through the change management aspects of the introduction of new ICT. The supportive role of a mentor enables the new learners to ask questions around the technology and new processes in a relationship that is safe [25]. Mentors are found at various levels in the healthcare organization and each group performs together to provide an overarching support framework including administrative leadership (e.g., CEO/CIO), clinical leadership within each healthcare discipline (e.g., champions and committees), and bridgers/support staff (translators and educators) [38,39]. Mentors should be those staff with more experience with the ICT and who are able to provide the HCPs with three distinct types of functions: "psychological support..., career-related support..., and role modeling" [40]. Mentors are key to supporting Kirkpatrick's training evaluation model providing opportunities for evaluation of the behavior with direct observations and application of learning [14] (see Fig. 15.1). Other benefits of mentors is with their help to minimize the stressful experiences such as with HCPs learning and acceptance of a new ICT by the creation of peer-to-peer support [41]. We found that both super users and educators had expanded roles beyond delivering ICT support or instruction; additionally, they were important figures in change management. This role required explaining the benefits of the new or upgraded ICT, and helping staff to see the final vision beyond the insecurities of change and the challenges of moving from novice to expert users of the ICT system [19].

The success of HCPs learning not only relies on the delivery mode of the education, but also on post education support. The utilization of super users or champions is a key component in both the classroom and on-site support, immediately and for the long term [39].

Liebe et al. believed having a social network of communication between members was key to supporting acceptance [42]. The development of a mentorship or super user network creates this social environment for new HCP learners of ICT. A review of two of our own ICT implementations of the same system reflects the successes and the challenges of appropriate super user support and the impact on ICT acceptance.

The first ICT implementation project was an example of creating a super user core team and utilizing the team in all phases of supporting HCPs in their learning experience [39,43]. The super users represented each site in the health region and each discipline that would be impacted by the ICT. The super users were trained for several months before implementation began, and when the rest of the HCPs were trained before the system went live, they were involved in the computer lab to assist with training [39,43]. This role solidified the initial training for the super users and benefited their individual sites with competent and confident super users who knew how to troubleshoot through most of the ICT issues, creating strong on-site super user support for the newly trained HCPs. Another positive element of this experience was the regular monthly meetings with the Lead. These meetings kept the super users informed of upcoming changes, provided an avenue for clarifying questions, and a safe place to learn how to correctly use the ICT, contributing to increased knowledge and awareness in healthcare informatics [39].

The second ICT experience was juxtaposition to the first, although it was implemented in the same discipline and health region. There were only two educators to deliver the training for 600 HCPs, the training happened over 3–4 days/week in 2 months, and during the education the super user team did not exist. The new ICT system was already live while HCPs were trained, and it was expected that they begin using the system the next day independently. There was a high level of confusion from HCPs related to how the implementation of the new ICT was occurring, as well as in trying to apply new processes within the ICT while providing patient care. As a result, acceptance levels were low, and resistance was rapidly increasing without the mentorship and positive input from super users in the workplace. In response to the high need for post implementation support at each site, super users were delegated based on availability at sites, versus interest or propensity for informatics and electronic technology. Some were proficient and able to assist their colleagues, while others struggled as much as their peers. Monthly super user meetings were established to begin to build super user confidence and proficiency, but there was varied attendance and some sites still inconsistently attend it. There was poor communication during the ICT implementation, which resulted in mixed messaging, different instructions, and staff frustrated with the super users and educators.

How do we ensure that every experience is going to be like the first ICT implementation? Proficient super users are necessary to provide support in the classroom and continue into the workplace to increase the rate of ICT acceptance [39]. Newly trained HCPs should have a "buddy" program with proficient users, or a super user to observe and then to implement in the "live" ICT system at their workplace. HCPs would benefit from a specific follow up schedule to ensure they are using the system as it was originally taught and to provide an opportunity to discuss their work as it is reviewed before incorrect habits or workarounds have formed. Ongoing support through system upgrades and development of more advanced practices using ICT is essential to successful acceptance and to realizing the promised benefits of these ICTs [5,39].

MONITORING FOR ACCEPTANCE

The fourth significant element to contributing to HCPs learning, competency, and knowledge retention would be identifying the need for monitoring the HCP on elements such as minimum use data elements identified to assess ICT acceptance and implementation. As new ICT is introduced in the clinical workplace, HCPs need to retain the knowledge and skills required to use the technology ensuring safe patient care [44] as part of their daily activities of patient care. Cooper and Zmud's staged model of IT implementation proposes that adoption would be seen early in an organization by accepting the need for the new technology but acceptance and routinization would be the stages when HCPs have committed to using the ICT within their daily work activities (see Fig. 15.5) [45]. Knowledge can be gained via many methods such as those discussed earlier in this chapter. Each method does have challenges for HCPs to retain and apply the training into practice. Monitoring for acceptance or routinization can pose challenges as there are no standards regarding how often evaluation of learning should be occurring in the workplace. When evaluating learning, the tools used at level 3 and 4 of Kirkpatrick's model include on the job application, and job performance as listed in Fig. 15.1 [13]. Bloom's taxonomy and the competency framework identify analysis as level four which expects the learner to demonstrate skills in breaking down information into simpler parts and seeing patterns (see Fig. 15.2 and 15.3) [10,14]. This is when HCPs begin discussing components of a care plan and what would be included for various patient problems versus asking how to navigate through the system, which is more of a novice question. One method to evaluate the retention of knowledge and practice skills is to monitor key elements or activities within the ICT by the HCPs. Understanding what would be defined as key-monitoring elements can vary within ICT. A simple monitoring element could be number

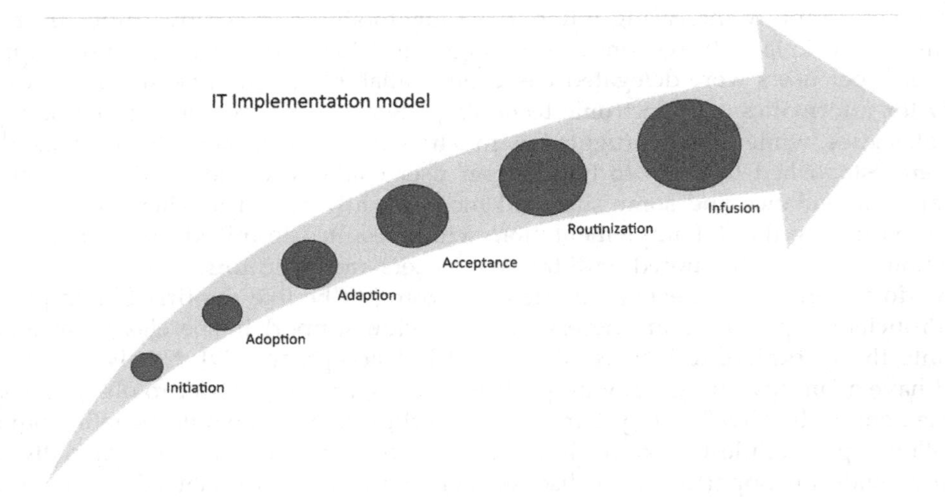

FIGURE 15.5 Staged Model of IT Implementation. Based on Cooper and Zmud Information technology implementation research: a technological diffusion approach. Manag Sci 1990;36:123–39. [45]. *From: Alberta Health Services. Used with permission.*

TABLE 15.4 Minimum Use Elements

Basic Data Elements	Complex Data Elements
Frequency of log in/access	Update patient demographics
Patient look-up	Patient order entry
Patient schedule	Patient results management
Create/print reports	Care plan development
	Documenting patient interventions

of times the HCPs access the system or its components. More advanced monitoring elements within electronic medical records could be minimum use elements derived from actions that ensure documentation of clinical care within the ICT. These could be basic items such as patient demographics to complex items such as patient care plans, clinical orders, and results monitoring that would support consistent and complete electronic clinical chart [46] (Table 15.4). Monitoring elements can contribute to the safe use of the system by triggering various clinical decision support features [47,48]. Monitoring should include elements set by professional regulatory colleges or organization policies. Groups such as College and Association of Registered Nurses of Alberta (CARNA) have set documentation standards that all members are expected to meet such as a plan of care or patient care outcomes [49]. Continuing Care Health Service Standards (CCHSS) [50] exist and audits of patient care plans as part of their standards of care should be performed. The standard of care related to ICT care plans from this group is that a patient's care plan will be accessible to all caregivers involved to be used as a communication tool and point of reference; the patient care needs are evident by how the care plan is built; and each component of the care plan and documentation is up to date. These elements form the basis for the initial ICT training.

Other modes of monitoring that we have used minimally are super user and educator follow-up post training with the individual HCP and the interactions with the ICT or clinical patient records. These methods were beneficial, but due to lack of human resources, we were unable to sustain them, in the future, we would like to try again once we have a stronger super user group and an increase in educator hours. Monitoring the use of the ICT would help to identify knowledge gaps that would indicate the need for refresher training. It would also highlight areas in the training strategy that may be insufficient or unclear to HCPs.

Specific review of various aspects of ICT is another manner to monitor for acceptance, routinization, and change management. During the design and implementation project phase supported through HCPs training, key elements can be defined as measures for clinical acceptance. As monitoring systems are developed to look at these key elements, they should be presented to HCPs as opportunities to celebrate success instead of a means to identify points of failure [6]. The dependence on the individual HCPs to audit their own practice or usage of ICT is inherently challenging. Research has shown that groups such as physicians are "poor at self-assessing their skills and learning needs" [51] and having tools such as reports available are key. Clinical acceptance elements can also be incorporated

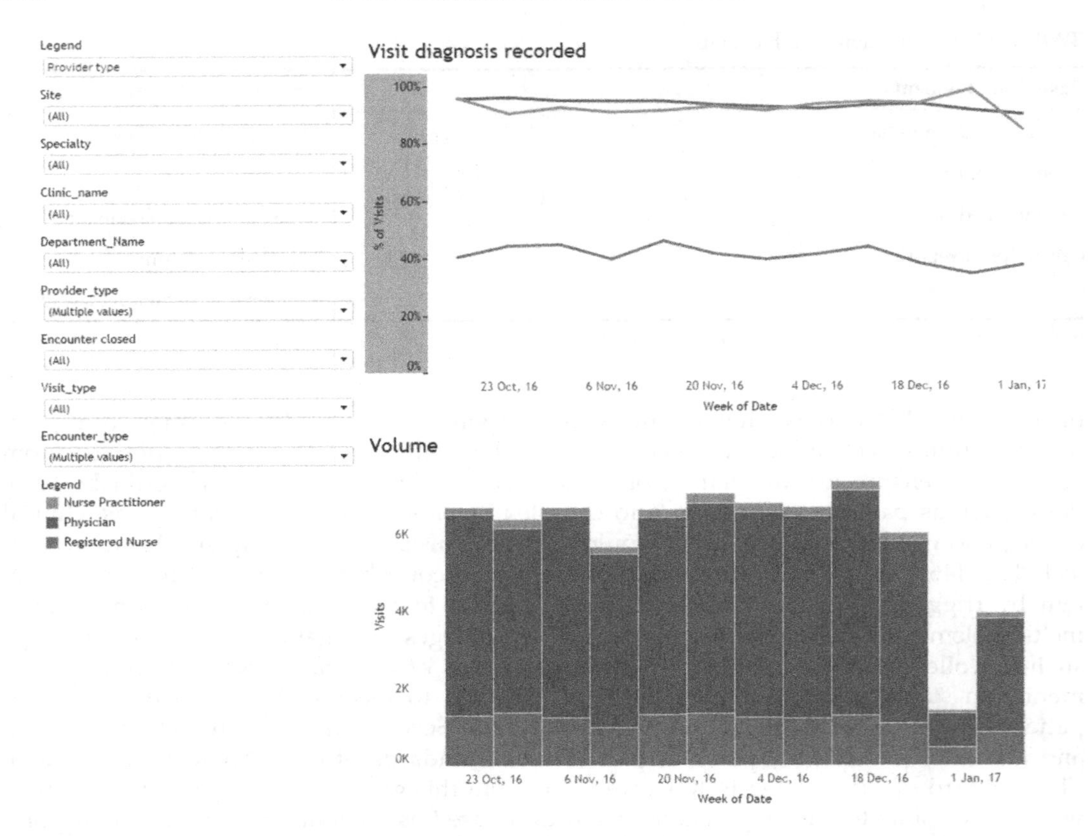

FIGURE 15.6 Adoption Metric Dashboard Sample 1. *From: Alberta Health Services. Used with permission.*

into minimum use guidelines for the various HCPs. As ICT is implemented, having reports or dashboards that monitor these elements enable individual, practice groups, or managers to monitor how acceptance is occurring [52]. A review of individual performance indicators within the ICT is also key in ensuring adequate acceptance, routinization, and competency maintenance. Figs. 15.6 and 15.7 are examples of dynamic dashboards, which use data directly from the ICT clinical patient record and display the actions taken by HCPs where they can compare their practice to others such as within their department or team provide an additional motivation to ensure the minimum use or acceptance elements are being met. For those HCPs who may not be as aware of their weaknesses in the use of the ICT, reports can also illustrate their gaps in knowledge as learning opportunities with managers, mentors, or super users. In other instances, HCPs may not want to ask for help from their manager or department leader when they are aware they have a gap in knowledge or skill for fear of negative [51]. When developing the initial training strategy, it should include evaluations during and immediately after training as well as at designated intervals to assess training, learning, acceptance, and routinization aimed at continued competency.

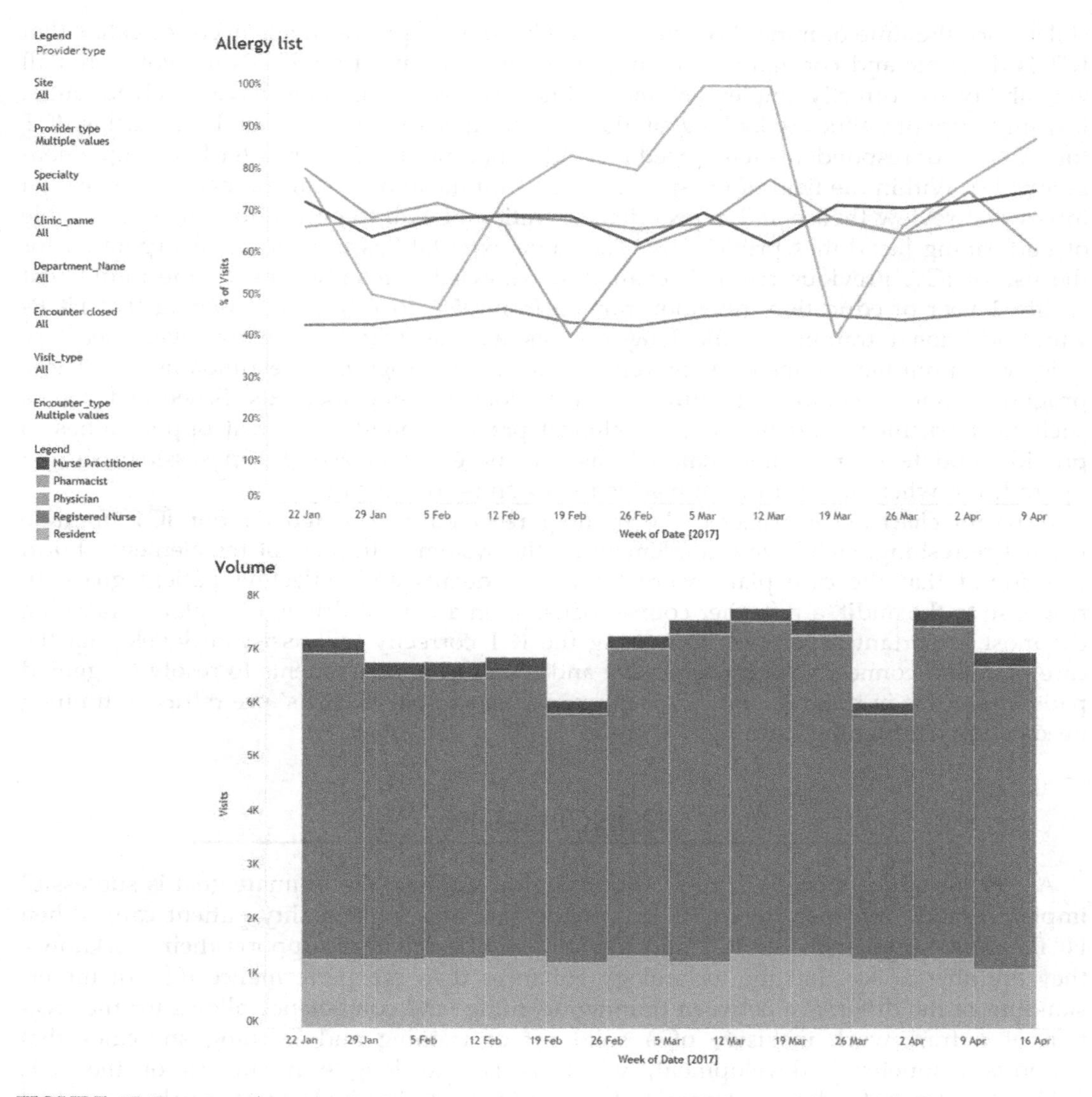

FIGURE 15.7 Adoption Metric Dashboard Sample 2. *From: Alberta Health Services. Used with permission.*

CONTINUING COMPETENCY

The final important element to contributing to HCPs learning, competency, and knowledge retention would be the discovery and identification of continued ICT education to ensure ICT knowledge and skills continue to progress and develop as technology evolves. Continuing competency "seeks to determine whether an individual healthcare professional has continued to provide safe, competent care by maintaining current knowledge and skills since the time of initial licensure" [53]. Health care professionals need to realize that ICT is dynamic and constantly evolving, thus there is a need for HCPs to evolve in skill and ability to correctly implement the technology even after they have received initial learning support. Studies looking at the long-term learning needs of HCPs using ICT found "75% of respondents felt a need for additional training 5 years after EHR implementation" [1]. Within the field of nursing, there are routine and annual competency education topics that review theory and expect demonstration of skill to ensure the nurse is capable of performing her duties proficiently. The same expectations need to be incorporated for the use of ICT. Previous research evaluating knowledge retention over time found that within 1 year of education, retention ranged from 35% to 75% [24]. Knowing that HCPs value additional training in the long-term as well as that knowledge levels decrease quickly, continuing competency or refresh education programs are fundamental. These programs should include a mixture of the various training methods. Scheduled events such as department staff meetings or clinical practice rounds are great opportunities to provide update or refresher materials as are more formalized group sessions during upgrades or when new features are added to existing functionality.

A recent chart audit reflected the training received a year ago for our ICT upgrade needed refreshing. HCPs were implementing the system with most of the elements, but it was found that the care plans were lacking in details and reflecting patient goals. In response to the audit, a refresher course focusing on a patient driven care plan, reiterating the most important aspects of how using the ICT correctly will assist in developing the care plan and connecting comprehensive and monitoring assessments to resolve triggered patient issues is in progress. Fig. 15.5 summarizes a sample of HCPs' pre-refresher training needs and post training learning.

CONCLUSION

As we introduce more ICT into various clinical settings, the ultimate goal is successful implementation and use by HCPS to provide safe and high quality patient care. When HCPs are satisfied with the ICT and the software design that supports their workflows, they are able to see that the technology enhances their job performance [6]. An understanding of the difference between training, learning, and competency allows for the creation of a framework inclusive of a solid set of training and learning strategies that supports competency development, which is key to long-term success of the ICT. Table 15.5 illustrates how a framework can address training strategies such as content development and delivery modes of training, learning strategies with support structures, as well as continued competency. Our experience in the implementations of a variety of ICTs along with literature integration identifies elements that are the top priorities for any training and acceptance strategy.

HCPs will develop a better understanding of the use of ICT when training content development is structured around integrating it into their daily clinical workflows. Inclusion of case studies within the content development training materials for classroom or electronic delivery mode education, and practice time in the training environment enables the HCPs to ask questions or try the new knowledge safely. A variety of training

TABLE 15.5 Summary Training, Learning, and Continuing Competency Table

	Training	Learning	Continuing Competency
Definition	"Organized activity aimed at imparting information and/or instructions to improve the recipient's performance or to help him or her attain a required level of knowledge or skill."	"Measurable and relatively permanent change in behavior through experience, instruction, or study. Whereas individual learning is selective, group learning. . .depend largely on power playing in the group"	"Seeks to determine whether an individual healthcare professional has continued to provide safe, competent care by maintaining current knowledge and skills since the time of initial licensure"
Model/ Framework	• Kirkpatrick training model	• Bloom's Cognitive taxonomy	• Competency framework
Strategies	• Content planning. • Delivery modes. • Use case studies. • Workflow processes—day in the life scenarios. • Organizational expectations.	• Address unique learning needs. • Identify benefits and ease of use. • Identify organizational expectations. • Utilize mentors, super users.	• Content planning. • Delivery modes. • Use case studies. • Workflow processes—day in the life scenarios. • Organizational expectations.
Evaluation	• Post training reaction feedback. • Ability to conduct the steps delivered in training.	• Pre and post training evaluations. • Behavior change with understanding. • Transfer of knowledge to workplace.	• Self and Peer feedback audits. • Standards care record audits. • Licensing body goal review. • Refresher course.
Monitoring	• Assessment delivered during and immediately after training.	• Assessment and audit of care record immediately after training and at designated intervals developed as part of an Evaluation program for immediate post go-live period to 1-year postimplementation.	• Monitor acceptance, application, analysis, synthesis at the workplace. • Assessment and audit of care record at designated intervals as part of an evaluation program

delivery modes such as classroom and e-learning should be available as each offers strengths that support the various learning styles and work schedule unpredictability of HCPs. Grouping HCPs based on similar workflows or practices starts the development of a peer network where HCPs are able to ask questions and share knowledge to improve the group's use of ICT. The size of these groups is key to optimal learning. The educator must be able to assess the individual learning styles and this is a potential challenge in larger class sizes.

A support structure of people and tools needs to be included in the training strategy as learning is not a single event. Mentors and super users continue to provide the peer-to-peer knowledge transfer that is required to increase the level of knowledge retention. Mentors also provide a means to disseminate ICT information and are vital to support updates or changes of the technology. Mentors can also provide the first level of monitoring for acceptance and use of the ICT while at the front line.

Monitoring for acceptance is needed to ensure HCPs develop competency with ICT. Developing a set of elements that are strategic to ensure safe, quality, and practice-supported use of ICT should be incorporated into the competency development plan. Minimum use elements must incorporate HCPs practice expectations as well as other organizational policy requirements such as assessment or care planning documentation. These elements should also be the basis for reporting systems such as practice dashboards or manager reports. Evaluating the results of the monitoring reports provides an opportunity to see where knowledge gaps exist in the use of the ICT. We know that learning for the use of ICT cannot be 100% complete within the single initial training event Gap identification from acceptance or routinization monitoring can identify areas for refresher training. Refresher training can be provided in a variety of modes and groups such as monthly staff meeting updates, noon hours "brown bag" luncheon sessions or through short video clip presentations available to the HCPs at regular intervals or on demand. Refresher courses supporting annual continuing competency evaluation are necessary for best practice and safety. Many HCPs are not able to assess their practices with ICT so will not be able to identify their educational needs with managers or mentors. A dynamic and supportive set of training and learning strategies within a competency framework that meets the ongoing needs of new and existing HCPs' will set the stage for safe, competent HCPs and ICT that is accepted and used by new and existing HCPs in a complex healthcare environment [54].

References

[1] Bredfeldt CE, Awad EB, Joseph K, Snyder MH. Training providers: beyond the basics of electronic health records. BMC Health Serv Res 2013;13:503. Available from: http://dx.doi.org/10.1186/1472-6963-13-503.

[2] Antonacopoulou EP. The paradoxical nature of the relationship between training and learning. J Manag Stud 2001;38:327–50.

[3] WebFinance. Online Business Dictionary—BusinessDictionary.com. n.d. Available from: <http://www.businessdictionary.com/> [accessed 6.01.17].

[4] The Office of the National Coordinator for Health Information Technology. Organizational Responsibilities | SAFER Guides., n.d. Available from: <https://www.healthit.gov/safer/guide/sg002> [accessed 15.08.16].

[5] Price Waterhouse Cooper. The Emerging Benefits of EMR Use in Ambulatory Care in Canada – Executive Summary–Canada Health Infoway. n.d. Available from: <https://www.infoway-inforoute.ca/en/component/edocman/resources/reports/benefits-evaluation/3029-the-emerging-benefits-of-emr-use-in-ambulatory-care-in-canada-executive-summary> [accessed 22.08.16]

[6] McLane S. Designing an EMR planning process based on staff attitudes toward and opinions about computers in healthcare. Comput Inform Nurs 2005;23:85–92.

[7] Liaw SY, Wong LF, Chan SW-C, Ho JTY, Mordiffi SZ, Ang SBL, et al. Designing and evaluating an interactive multimedia web-based simulation for developing nurses' competencies in acute nursing care: randomized controlled trial. J Med Internet Res. 2015;17:e5. Available from: http://dx.doi.org/10.2196/jmir.3853.

[8] Bostrom AC, Schafer P, Dontje K, Pohl JM, Nagelkerk J, Cavanagh SJ. Electronic health record: implementation across the Michigan Academic Consortium. Comput Inform Nurs 2006;24:44–52.

[9] Gagnon M-P, Desmartis M, Labrecque M, Car J, Pagliari C, Pluye P, et al. Systematic review of factors influencing the adoption of information and communication technologies by healthcare professionals. J Med Syst 2012;36:241–77. Available from: http://dx.doi.org/10.1007/s10916-010-9473-4.

[10] SABSA Competency Framework 2013 | SABSAcourses.com. n.d. Available from: <http://www.sabsacourses.com/competency-framework-2013> [accessed 09.01.17].

[11] Committee on the Robert Wood Johnson Foundation Initiative on the Future of Nursing, at the Institute of Medicine, Robert Wood Johnson Foundation, Institute of Medicine (U.S.), editors. Transforming Education. Future Nurs. Lead. Change Adv. Health, Washington, D.C: National Academies Press; 2011, p. 163-219.

[12] Centre for the Advancement of Research & Development in Educational Technology. Module 8: Assessment and Evaluation. n.d. Available from: <http://www.slideshare.net/Cardet1/mod8-ppt1-080609tmty-1/10> [accessed 09.01.17].

[13] Smidt A, Balandin S, Sigafoos J, Reed VA. The Kirkpatrick model: a useful tool for evaluating training outcomes. J Intellect Dev Disabil 2009;34:266–74. Available from: http://dx.doi.org/10.1080/13668250903093125.

[14] Adams NE. Bloom's taxonomy of cognitive learning objectives. J Med Libr Assoc 2015;103:152–3. Available from: http://dx.doi.org/10.3163/1536-5050.103.3.010.

[15] Bates R. A critical analysis of evaluation practice: the Kirkpatrick model and the principle of beneficence. Eval Program Plann 2004;27:341–7. Available from: http://dx.doi.org/10.1016/j.evalprogplan.2004.04.011.

[16] Liu W-I, Chu K-C, Chen S-C. The development and preliminary effectiveness of a nursing case management E-learning program. CIN Comput Inform Nurs 2014;32:343–52. Available from: http://dx.doi.org/10.1097/CIN.0000000000000050.

[17] Coles T, Demunnick G, Masesar MA. Transitioning from implementation to integration: an innovative team approach to support integration of technology into clinical practice. Can J Nurs Inform 2008;3:14–24.

[18] Atreja A, Mehta NB, Jain AK, Harris CM, Ishwaran H, Avital M, et al. Satisfaction with web-based training in an integrated healthcare delivery network: do age, education, computer skills and attitudes matter?. BMC Med Educ 2008;8:48. Available from: http://dx.doi.org/10.1186/1472-6920-8-48.

[19] Jones S, Donelle L. Assessment of electronic health record usability with undergraduate nursing students. Int J Nurs Educ Scholarsh 2011;8. Available from: http://dx.doi.org/10.2202/1548-923X.2123.

[20] Kennedy D, Pallikkathayil L, Warren JJ. Using a modified electronic health record to develop nursing process skills. J Nurs Educ 2009;48:96–100.

[21] Englebardt SP. Technology and Distributed Education. Health Care Inform. Interdiscip Approach. St. Louis, MO: Mosby; 2002. p. 267–84.

[22] Galani M, Yu P, Paas F, Chandler P. Battling the challenges of training nurses to use information systems through theory-based training material design. Stud Health Technol Inform 2014;204:32–7.

[23] Koopman RJ, Steege LMB, Moore JL, Clarke MA, Canfield SM, Kim MS, et al. Physician information needs and electronic health records (EHRs): time to reengineer the clinic note. J Am Board Fam Med 2015;28:316–23. Available from: http://dx.doi.org/10.3122/jabfm.2015.03.140244.

[24] Naidr JP, Adla T, Janda A, Feberová J, Kasal P, Hladíková M. Long-term retention of knowledge after a distance course in medical informatics at Charles University Prague. Teach Learn Med 2004;16:255–9. Available from: http://dx.doi.org/10.1207/s15328015tlm1603_6.

[25] Huryk LA. Factors influencing nurses' attitudes towards healthcare information technology: nurses' attitudes towards technology. J Nurs Manag 2010;18:606–12. Available from: http://dx.doi.org/10.1111/j.1365-2834.2010.01084.x.

[26] McKinney HE, DeSantis S. Nursing informatics and nursing education. Nurs Inform Found Knowl. Burlington, MA: Jones & Bartlett Learning; 2012. p. 403–40.

[27] Police RL, Foster T, Wong KS. Adoption and use of health information technology in physician practice organisations: systematic review. Inform Prim Care 2010;18:245–58.

[28] Yan H, Gardner R, Baier R. Beyond the focus group: understanding physicians' barriers to electronic medical records. Jt Comm J Qual Patient Saf Jt Comm Resour 2012;38:184–91.

[29] DiNinno V, Ciubotaru S. Training Strategy for Physicians using ICT 2016.

[30] Dionyssopoulos A, Karalis T, Panitsides EA. Continuing medical education revisited: theoretical assumptions and practical implications: a qualitative study. BMC Med Educ 2014;14. Available from: http://dx.doi.org/10.1186/s12909-014-0278-x.

[31] Benner P. From novice to expert. Collect Read Relat. Competency-based train.. Geelong: Vic.: Deakin University; 1994. p. 127–35.

[32] Powell AL. Computer anxiety: comparison of research from the 1990s and 2000s. Comput Hum Behav 2013;29:2337–81. Available from: http://dx.doi.org/10.1016/j.chb.2013.05.012.

[33] Gore K.S. How Nursing Educators Address the Differing Learning Styles of Students. Walden University, 2015.

[34] Ketikidis P, Dimitrovski T, Lazuras L, Bath PA. Acceptance of health information technology in health professionals: an application of the revised technology acceptance model. Health Inform J 2012;18:124–34. Available from: http://dx.doi.org/10.1177/1460458211435425.

[35] Hurtubise L, Roman B. Competency-based curricular design to encourage significant learning. Curr Probl Pediatr Adolesc Health Care 2014;44:164–9. Available from: http://dx.doi.org/10.1016/j.cppeds.2014.01.005.

[36] Mc Nelis J. An exploratory study into virtual learning environments as training platforms in the workplace. Int J Adv Corp Learn 2014;7:8. Available from: http://dx.doi.org/10.3991/ijac.v7i3.3985.

[37] Cornell P, Herrin-Griffith D, Keim C, Petschonek S, Sanders AM, D'Mello S, et al. Transforming nursing workflow, Part 1: The chaotic nature of nurse activities. JONA J Nurs Adm 2010;40:366–73. Available from: http://dx.doi.org/10.1097/NNA.0b013e3181ee4261.

[38] Ash JS, Stavri PZ, Dykstra R, Fournier L. Implementing computerized physician order entry: the importance of special people. Int J Med Inf 2003;69:235–50.

[39] Page D. Turning nurses into health IT superusers. Hosp Health Netw AHA 2011;85:27–8, 2.

[40] Wong P, Myers M. Clinical competence and EBP: an educators perspective. Nurs Manag Springhouse 2015;46:16–18. Available from: http://dx.doi.org/10.1097/01.NUMA.0000469358.02437.67.

[41] Allicock M, Carr C, Johnson L-S, Smith R, Lawrence M, Kaye L, et al. Implementing a one-on-one peer support program for cancer survivors using a motivational interviewing Approach: results and lessons learned. J Cancer Educ 2014;29:91–8. Available from: http://dx.doi.org/10.1007/s13187-013-0552-3.

[42] Liebe J, Hüsers J, Hübner U. Investigating the roots of successful IT adoption processes - an empirical study exploring the shared awareness-knowledge of Directors of Nursing and Chief Information Officers. BMC Med Inform Decis Mak 2015;16. Available from: http://dx.doi.org/10.1186/s12911-016-0244-0.

[43] Terry AL, Thorpe CF, Giles G, Brown JB, Harris SB, Reid GJ, et al. Implementing electronic health records: key factors in primary care. Can Fam Phys Médecin Fam Can 2008;54:730–6.

[44] Borycki E, Kushniruk A, Nohr C, Takeda H, Kuwata S, Carvalho C, et al. Usability methods for ensuring health information technology safety: evidence-based approaches. contribution of the IMIA working group health informatics for patient safety. Yearb Med Inform 2013;8:20–7.

[45] Cooper RB, Zmud RW. Information technology implementation research: a technological diffusion approach. Manag Sci 1990;36:123–39.

[46] Conrad D, Hanson PA, Hasenau SM, Stocker-Schneider J. Identifying the barriers to use of standardized nursing language in the electronic health record by the ambulatory care nurse practitioner. J Am Acad Nurse Pract 2012;24:443–51. Available from: http://dx.doi.org/10.1111/j.1745-7599.2012.00705.x.

[47] Lapane KL, Rosen RK, Dubé C. Perceptions of e-prescribing efficiencies and inefficiencies in ambulatory care. Int J Med Inf 2011;80:39–46. Available from: http://dx.doi.org/10.1016/j.ijmedinf.2010.10.018.

[48] Stewart AL, Lynch KJ. Identifying discrepancies in electronic medical records through pharmacist medication reconciliation. J Am Pharm Assoc 2012;52:59–66. Available from: http://dx.doi.org/10.1331/JAPhA.2012.10123.

[49] College and Association of Registered Nurses of Alberta. Documentation Standards for Regulated Members (January 2013)—DocumentationStandards_Jan2013.pdf 2013. Available from: <http://www.nurses.ab.ca/content/dam/carna/pdfs/DocumentList/Standards/DocumentationStandards_Jan2013.pdf> [accessed 22.08.16].

[50] Alberta Health. Continuing Care Health Services Standards 2016. Available from: <http://www.health.alberta.ca/documents/Continuing-Care-Standards-2016.pdf> [accessed 24.08.16]

[51] Klein D, Staples J, Pittman C, Stepanko C. Using electronic clinical practice audits as needs assessment to produce effective continuing medical education programming. Med Teach 2012;34:151–4. Available from: http://dx.doi.org/10.3109/0142159X.2012.644826.

[52] Graham-Jones P, Jain SH, Friedman CP, Marcotte L, Blumenthal D. The need to incorporate health information technology into physicians' education and professional development. Health Aff (Millwood). 2012;31:481–7. Available from: http://dx.doi.org/10.1377/hlthaff.2011.0423.

[53] Burns B. Continuing competency: what's ahead? J Perinat Neonatal Nurs 2009;23:218–27. Available from: http://dx.doi.org/10.1097/JPN.0b013e3181b0ec9f.

[54] Herbert VM, Connors H. Integrating an academic electronic health record: challenges and success strategies. CIN Comput Inform Nurs 2016;34:345–54. Available from: http://dx.doi.org/10.1097/CIN.0000000000000264.

CHAPTER

16

Integrating Health Informatics Into Australian Higher Education Health Profession Curricula

Elizabeth Cummings[1], *Sue Whetton*[2] *and Carey Mather*[1]

[1]University of Tasmania, Hobart, TAS, Australia [2]Sue Whetton Consulting, Launceston, TAS, Australia

INTRODUCTION

This chapter discusses health informatics as an increasingly integral element of the higher education sector for health professions in Australia. It locates this discussion within political, economic, legal, and cultural factors, as mandated by Government legislation, policies, and strategies, that shape the emerging Australian digital health environment. The discussion begins by briefly reviewing the growth of health informatics and e-health systems to their present pervasive level. It then examines implications of this for the education of health professionals in Australia and internationally. Current challenges related to embedding health informatics in the health professional's curriculum, and strategies seeking to address these challenges, are considered. Evidence from educational interventions, in the form of four practical case studies, is used to demonstrate progress in integrating health informatics, clinical information systems, mobile computing, and social networks into health profession higher education.

The key messages are that e-health, health informatics, and health should be viewed as synonymous in contemporary healthcare, and that there is a need to incorporate e-health and health informatics skills in higher education programs globally. These programs need to incorporate a continuum of skills, knowledge, behavior, and attributes to cater for the diverse needs of health and health informatics professionals.

Health Professionals' Education.
DOI: http://dx.doi.org/10.1016/B978-0-12-805362-1.00016-4

DEFINITIONS

Health informatics is a complex field, encompassing a range of academic disciplines and health professions. This diversity has seen some common terms differentially applied. The following definitions of key terms are used in this chapter.

- **Health informatics**: is the appropriate and innovative application of the concepts and technologies of the information age to improve healthcare and health [1].
- **E-health literacy:** the ability to seek, find, understand, and appraise health information from electronic sources and apply the knowledge gained to addressing or solving a health problem [2].
- **Digital professionalism**: is the competence or values expected of a professional when engaged in social and digital communication [3].
- **Continuing professional development**: is "the means by which members of the profession maintain, improve and broaden their knowledge, expertise and competence, and develop personal and professional qualities required through their professional lives" [4:1].
- **Mobile learning or m-learning**: is defined as learning and teaching interactions that use mobile hand-held devices such as electronic notebooks, tablets, or smartphones [5].

PAST CONTEXT

The underlying structures of twentieth century healthcare in Australia have shaped the environment into which health informatics systems were initially introduced. These structures also influenced health professionals' understandings about, and attitudes toward, health informatics and e-health, and shaped ideas about the knowledge, skills, and behaviors required for health professionals. Twentieth century health services in Australia were based around a knowledge and technology intensive biomedical model, which focused on acute, hospital-based care. As the primary source of information gathering, the face-to-face consultation with the family doctor was the centerpiece of this model. Medical knowledge increased dramatically throughout the twentieth century, resulting in specialization of the health professions. Some specialties guarded their areas of expertise to the extent that professional silos emerged. Health services were structured as large, complex, organizations resulting in complex management hierarchies and segregation of information within individual services, departments, or functional areas. It was into this environment that health information and e-health systems began to be introduced. The limitations of these early information systems interacting with the characteristics of healthcare environment, resulted in the development of individual electronic systems that stored and disseminated information within a department or functional area, but had limited capacity to share information across systems [6,7]. Significant spending in e-health in Australia at the turn of the century was focused on ad hoc projects rather than national level strategies [8]. This was in part a reflection of the federation of states in Australia, whereby opportunities for leading the national agenda were provided to all states and territories. While these project-based activities demonstrated the value of health informatics and e-health, they also limited their visibility and impact on the broader healthcare environment. Although health professionals were aware of emerging electronic systems, they were viewed as peripheral rather than integral to healthcare. Indeed, some early studies showed that many health professionals resisted the introduction of information systems, believing that they could disrupt traditional work routines, flows, and relationships [6,9]. For example, Greatbatch and colleagues [9] found that many health professionals viewed the use of health information systems during consultation as interfering with workflows and clinician-patient relationships. Other studies indicated that there was some basis for concern with these early e-health systems. Greatbatch et al. [9] found that the use of a computer system impacted negatively on doctor–patient interactions while Tanriverdi and Iacono [6] found that the introduction of such systems often required changes to traditional workflows. Lack of attention to how the technology would integrate into existing organizational workflows impacted strongly on the extent to which it was accepted.

In these early years, e-health and health informatics knowledge was viewed as not being primarily within the scope of clinical knowledge. Rather, it was seen as the province of the health informatics professional and other technology experts. Where clinical education programs did incorporate health informatics, they tended to focus on the technical aspects of using the tools rather than on how best to integrate these tools into clinical practice. Even this level of training has been sporadic with the literature indicating that informatics literacy among health professionals was mixed [10–12].

CURRENT CONTEXT

Since the beginning of the twenty-first century, changes in models of care together with significant increases in the capabilities of information management systems, have resulted in health informatics and e-health being increasingly seen as central and essential for health service delivery. There has also been an increasing realization that e-health is about more than technology, and that knowledge and skills, rather than just being confined to the health informatics or computer professional, are required across the health professional spectrum [13,14]. The requirement for health informatics knowledge for all healthcare workers has become a focus for many professional bodies nationally and internationally. The major professional bodies such as International Medical Informatics Association (IMIA), European Federation for Medical Informatics Association (EFMI), and Australasian College of Health Informatics (ACHI) each has educational working groups to focus on understanding the needs and requirements of education in the digital health age.

Models of care have changed, but contemporary health services remain knowledge intensive. While diagnostic technology is increasingly used for gathering and sharing this knowledge, the face-to-face consultation continues to be a significant element in the clinical environment. However, this no longer involves only the family doctor. The growth of specializations means that patients may need to provide the same information a number of times. Meanwhile, current trends, including interdisciplinary care, community-based care, and the involvement of patients in decision-making, are challenging specialist and department silos, creating new linkages and opportunities for communication, collaboration, and participation. In addition, contemporary trends toward community-based, interdisciplinary care are seeing some health organizations adopt flatter management structures. These changes have created a demand for increased and more widespread collection and sharing of information. Increasingly sophisticated information systems, the use of e-health and mobile health (m-health) have provided systems that are beginning to meet these needs. The initial project-based activity of health informatics has evolved so that today there is an emphasis on the development of integrated information systems which link the various applications to provide department, organization, or even nationwide and trans-national information systems. This is exemplified by the trend toward the implementation of nationwide electronic health records in many nations.

Another current trend is to provide citizens with access to all or part of their health records through secure portals. This has been achieved with some degree of success in Australia (My Heath Record), Denmark (sundhed.dk), Estonia (Patient Portal), and France (Dossier Medical Personnel). The establishment of these systems introduces an additional requirement for healthcare professionals in terms of their ability to assist citizens with accessing and understanding their health information [15]. Currently, there is limited education provided to healthcare professionals in this area, but research such as that described in Case Study 4 is essential to developing educational packages to aid citizens in their use and understanding of their health information.

Government Policies, Strategies, and Programs

From before the turn of the century, Federal and State governments have sought to incorporate e-health and health informatics by introducing policies, programs, and strategies that aimed to foster the uptake of electronic systems. The Australian Government, often in partnership with State and Territory governments, has introduced a number of e-heath initiatives. Table 16.1 lists the major nationwide initiatives established by the Australian Government.

Health Online: a health information action plan for Australia. Introduced in 1999, this program sought to provide the basis for "a national strategic approach to using information in the health system and to promote new ways of delivering health services, by

TABLE 16.1 Major Australian Government E-health Policies, Strategies, and Programs

Year	Policy/Strategy/Program
1999	Health Online
2001	Better Medication Management System (BMMS)
2002	Health*Connect*
2008	National E-Health Strategy
2012	Personally Controlled Electronic Health Record (PCEHR) renamed My Health Record (MyHR) in 2015
2013	Australian Health Information Workforce report
2016	National Digital Health Strategy for Australia

harnessing the enormous potential of new technologies" [16:6]. *This program focused on technology with limited attention paid to the relevant education of health professionals.*

Better Medication Management System (BMMS) (later known as MediConnect): sought to provide electronic access to patient medication records through the electronic transfer of medication information between doctors, pharmacists, hospitals, and the Health Insurance Commission [17].

Health*Connect*: this program initially commenced in 2002 with fast tracked trials in two of the smaller states or territories. Following the initial trials in Tasmania and the Northern Territory, the national Health*Connect* program commenced in July 2004. This program sought to improve safety and quality in healthcare by enabling improved access to key health information at the point of care through use of electronic communication. Health*Connect* implementations leveraged existing e-health projects and infrastructure; a blueprint to build a national health information network. This strategy saw the Commonwealth Government, in partnership with state and Territory Governments, fund a range of individual projects that sought to improve safety and quality in healthcare by enabling improved access to key health information at the point of care through use of electronic communication, while continuing to focus on discrete projects. However, unlike earlier programs, Health*Connect* emphasized that health informatics was not only about technology: "Health*Connect* was a change management strategy funded through the Commonwealth Government and facilitated through a partnership with State and Territory" [18].

National E-Health Strategy: Informatics skills were emphasized as a workforce issue in 2008, when the Commonwealth Government released the National E-Health Strategy [19]. This strategy adopted a systematic, coordinated, and all-encompassing approach to e-health. It stressed the need to create a healthcare environment 'where consumers, care providers, and healthcare managers can reliably and securely access and share health information in real time across geographic and health sector boundaries [19:8]. It further emphasized that "the only way this can be achieved is through the implementation of world class E-Health capability" [19:8]. The Strategy acknowledged that successfully achieving this implementation required a highly knowledgeable and skilled workforce comprising health informatics professionals, and also health professionals with relevant e-health skills and knowledge. It specifically recommended that "educational institutions such as Universities, vocational training institutions and professional bodies to embed E-Health into their curricula ... equip the workforce of the future with the skills, experience and knowledge to apply E-Health solutions in everyday practice" [19:58].

PCEHR/My Heath Record: The Personally Controlled Electronic Health Record (PCEHR) was launched in 2012 as the national electronic health record for all Australians. This system was developed using a combination of "bottom up" lead implementation projects and "top down" national initiatives that essentially built upon the learnings from previous ad hoc trials and small-scale projects [20]. The underlying business-to-business gateway provides a link between disparate systems and ensures interoperability [21], with the whole system integrating web-based personal health records with a clinical electronic health record system providing shared access to summary data for both consumers and healthcare providers [22]. Due to low uptake, the Government identified the need to move from an opt-in enrollment to an opt-out system, and renamed it the *My Health Record* in 2015. Development is ongoing at this time.

Subsequent policies and strategies [23,24] released by the Australian government continued to emphasize the need for an e-health literate health workforce:

- **Australian Health Information Workforce report** [23]: released in 2013, focused on the need for health information management and informatics professionals, and also highlighted the need for all health professionals to have e-health and health informatics knowledge and skills.
- **National Digital Health Strategy for Australia** [24]: this 2016 strategy stated that
  - All health workers increasingly need a core set of digital competencies to enable them to work safely, effectively, and efficiently.
  - There is an urgent need to identify targeted ongoing education, training, and development. Recognition of the potential impact on the skills of the workforce could come about as a consequence of digital systems deployment.
  - A core set of competencies including information input and retrieval skills, security, privacy, confidentiality and quality management; and some knowledge of project and benefits lifecycles are also viewed as essential.
  - Extra skill sets should also be developed according to clinical level of use and for specialized system use.
  - There is an even greater need for increased capability in health informatics.
  - Trained health informatics workers are necessary at every stage of any digital health reform to ensure successful implementation.

These government policies and strategies emphasized the need for health professionals to have appropriate informatics and e-health skills. They also highlighted the need for national informatics standards. For example, the National Digital Health Strategy recommended the establishment of "a standardized E-Health competency framework for health workers and health IT practitioners providing an understanding of required E-Health knowledge, skills, and attributes for each professional group" [24:58].

Response of Health Professional Bodies

Despite numerous examinations, reports and recommendations (e.g., [25]) in relation to the current state of informatics in health education curricula in Australia there has been limited response by the health professional bodies. Some work has been forthcoming in relation to health informatics competencies at various levels through organizations such as the Royal Australian College of General Practice [26], the Australian Nursing and Midwifery Federation [27], Australian Health Informatics Education Council [28], and Certified Health Informatician Australasia (CHIA) [29]. However, to date the formal requirement for specific health informatics inclusion in the education of health professionals has only been adopted by the Australian Nursing and Midwifery Accreditation Council [30]. For an undergraduate nursing degree to be accredited ANMAC now requires information and communication technology (ICT) competency to be included as part of the core curriculum [30,31]. The development of competency standards does not axiomatically convert to skill development without these becoming a mandated part of all educational curricula. Case Studies 1, 2, and 4 demonstrate how National standards can effect change within curricula and for addressing the need to prepare preregistration and current healthcare professionals in becoming proficient in the use of informatics.

Response of Education Providers

Education providers include institutions such as universities, vocational training institutions, and professional bodies. In Australia a range of education and training opportunities have been introduced to meet e-health and health informatics workforce needs. Contributions include health informatics academic programs, health informatics streams within other academic programs, and health informatics content in related disciplines such as health information management, information systems, and computing programs. Health informatics units or content have also been offered in academic programs for health professionals [13].

Educational planning and curriculum frameworks in the clinical health professions have received much attention nationally and internationally for over a decade [32] with various recommendations for the promotion of ICT literacy and workforce development strategies. Both the 2013 Australian Workforce Report and the 2016 National Digital Health Strategy for Australia emphasized the need to educate health professionals to be digitally literate. Educational programs should seek to meet this need by developing opportunities and minimizing inhibitors to learning in the workplace. However, education of health professionals continues to be sporadic and relatively uncoordinated. This is indicated in the Australian government funded baseline research, conducted by Gray and colleagues [25], into the status of health informatics and e-health education in clinical health profession degrees (i.e., not including nonclinical degrees, such as business engineering, information technology or life sciences). This research identified several localized and laudable curriculum initiatives, but they were largely unknown beyond their home university [25]. Very few degree programs had a systematic approach to teach, assess, evaluate, or audit this aspect of professional education [25]. Research and scholarship in the area was essentially inactive compared to other areas of health professions education.

CHALLENGES IN HEALTH INFORMATICS EDUCATION FOR CLINICIANS

In her analysis of this sporadic evolution, Jolly [33] identified a number of challenges to be addressed for the successful introduction of informatics education for health professionals. Jolly's challenges are used to discuss the current Australian healthcare professional's educational environment in the following sections.

Overloaded Curricula at Undergraduate Level

Healthcare professionals' education, particularly at undergraduate level, has a very full curriculum. This is due to the multiple requirements of the various professional, registration, and accreditation bodies, which have for many years controlled the content of the curricula. Whilst it is widely acknowledged that there is a requirement to adopt alternative pedagogical approaches in the education of healthcare professionals, change is slow occurring. A number of universities in Australia have demonstrated curriculum initiatives that embrace health informatics but these initiatives are not widely recognized and so the benefits are not yet widespread. That said, there remain very few healthcare professional undergraduate degrees with a systematic approach to including and assessing health informatics [34]. Case Study 2 shows how one university met the ANMAC mandate to include informatics within their undergraduate nursing curriculum.

Assumptions that Health Professionals Know How to Use Technology to Understand Health Informatics

There are assumptions made about the technological skills of new health profession students and graduates based upon their access to technology. Bennett et al. [35] found inherent issues in the assumptions that young people are able to safely and effectively use technologies, while a number of research studies demonstrated that "Digital Natives" have only a superficial understanding of technological developments and applications. This is compounded by poor information seeking, retrieval, and analysis skills due to limited knowledge and understanding of the underlying concepts of data collection, storage, and retrieval. Consequently, the lack of technological skills development in education programs can result in a lack of the proficiency in using the technologies and health informatics concepts required when entering the workforce [35–38]. For example, embedding an e-portfolio into an undergraduate nursing program at this university found student comprehension of the purpose and their capability for learning new software was mixed [39]. The assumption that students could utilize learning and teaching software effectively because they use digital technology was ill founded. Long-lead time, provision of web-based resources, webinars and face-to-face training opportunities were insufficient to adequately support some students in becoming proficient, which due to lack of access, hindered their learning during work integrated learning [39]. Nurse supervisors were provided similar resourcing during the lead-up to implementation of the e-portfolio; however, they found additional barriers relating to organization policies that precluded their use of the technology with students at the workplace [40]. Opportunities to develop sustainable support and guidance to embed digital technology in the workplace continues to be hindered by out-moded standards, guidelines, and codes of conduct [41].

Health Informatics Not Well Understood as a Specialization

Whilst there is a growing awareness of the benefits of using ICT in healthcare there remain many healthcare practitioners and educators with limited understanding of health informatics in general and the specialization of health informatics more specifically. This can, and does, impact upon the way health informatics content is developed and delivered in education courses for healthcare professionals. Many "fall into" health informatics almost by accident when they are trying to implement or understand a new technology being introduced into practice. The need for earlier preparation and a more considered approach to introducing health informatics to healthcare practitioners at all levels is required. Recent activities by the Australasian College of Health Informatics and the Health Informatics Society of Australia, particularly in the development of the CHIA certification, are beginning to increase the profile of health informatics as a specialization. Case Study 1 demonstrates how the University of Tasmania has promoted health informatics as a postgraduate specialization.

Lack of Career Path for Health Professionals With Health Informatics Qualifications

Whilst there are roles for Chief Medical Informatics Officers and Chief Nursing Informatics Officers in most states of Australia, these are recent developments and the pathway to these roles is not well established. There is a poor understanding of the current health informatics workforce in Australia, particularly in clinical informatics, as there is no systematic survey that captures data or reports on the health informatics workforce. The exception to this is health information managers and coders who have a specific occupation group title in the Australian Institute of Health and Welfare's National Health Labour Force surveys and reports [42].

Lack of Understanding of the Current Education Staff About Informatics

There is a poor understanding of health informatics amongst many of the health professional education staff. This is in part due to their background in the health professions. Gray and colleagues [34] found that there was a lack of knowledge and teaching experience in health informatics across the board, including lecturers, tutors, and professional placement supervisors. A similar lack of understanding and ability of professional placement supervisors to facilitate the use of health informatics in the workplace has been identified by Mather and colleagues [41,43–45]. Expertise in health informatics is not formally recognized or required in the recruitment of educators, although there has been change in relation to the standards for accrediting degrees [30].

Currently, there is a paradox hindering the implementation of clinical information systems, mobile computing, and social networks into health curricula and healthcare organizations in Australia [44]. Currently, standards, guidelines, and codes of conduct regarding access and use of digital technology in healthcare environments have been outpaced, leading to a situation where timely, easy, and convenient access to health information at the point of care is unavailable to health professionals. The corollary is academic staff, educators, and health professionals must be educationally prepared to teach students how to become digitally literate and prepare them for their healthcare experiences.

Digital professionalism is an aspect of professional identity formation that is overtly and covertly modeled on and off campus by health professionals. To remediate this situation is a need for mobile computing and social networking training to be included as part of the curriculum [41]. Classroom and simulation-based activities can augment learning by enabling students to become work-ready. Teaching digital professionalism in a safe environment can ensure the next generation of health professionals develop and promote positive professional identity regarding the use of digital technologies and social media. Development of capability to be digitally professional by students needs to be demonstrated prior to undertaking work integrated learning to ensure high quality and safe patient-care remains paramount [41]. Health professionals need to be proficient in a range of health informatics tools within the workplace so they can model appropriate and safe behavior to students. The current paradox needs to be ameliorated to minimize risks and promote opportunities to enhance care and improve patient outcomes. Case Studies 3 and 4 are examples of providing opportunities for health professionals to become educationally prepared to use health informatics tools to promote educational outcomes.

STRATEGIES TO MEET THE CHALLENGES

During the development of the e-health postgraduate courses the following strategies were identified and developed to assist in meeting the challenges indicated by Jolly [33] and discussed above. Whilst these strategies were locally discerned, it is argued that they are relevant to any organization that is developing healthcare professional's education in health informatics.

Develop Integrated Programs Linking Health Informatics Research and Teaching Programs

There is a range of excellent health informatics research being conducted in universities and other organizations in Australia. However, this research is rarely linked to advancing informatics in teaching programs. There are examples where educational programs are extended in collaboration with healthcare organizations (see, e.g., [46]), however, this is once again pushing toward research rather than ensuring an ongoing and cyclic research-teaching nexus. This would achieve the joint objectives of establishing health informatics as an identified discipline, building the research profile of health informatics, and ensuring relevant research and teaching is maintained. Additionally, this process could flow into practice-based research to promote continuity across education and health environments that could further promote health informatics as a discipline.

Embed Health Informatics as an Integrated Common Learning Theme Across Health Professional's Education Courses

Embedding health informatics across courses will ensure that students achieve common graduate outcomes in the area. Building health informatics capacity at the undergraduate level will also provide a pathway into research and study at the postgraduate level, strengthening the research presence of the discipline. Case Studies 1 and 2 demonstrate how capacity building can be achieved.

The challenge of advancing health informatics and e-health education in clinical health profession degrees in Australia cannot be satisfactorily addressed by individuals in separate institutions offering bespoke subjects at a local level. From the perspective of regulating the clinical professions, that approach is outside the governance processes for education in the clinical health professions.

The International Medical Informatics Association (IMIA) [47] has developed a set of e-health competencies they believe are required of all health professionals. The Australian College of Health Informatics (ACHI) is customizing these competencies for the Australian healthcare context. These competencies are considered essential if healthcare professionals are to use information processing and communication technologies efficiently and responsibly in effecting safe patient care.

Develop a Health Informatics Culture Within Education and Research Institutions

Essential to the success of these expectations will be the cultivation of a culture of health informatics. This means encouraging clinicians to view health informatics as a normal part of their daily activities rather than as in some way separate from mainstream healthcare. While this involves the development of skills to enable the use of e-health applications, it also requires clinicians have an understanding of the fundamental changes to the structure of healthcare and relevant legal, political, and cultural issues associated with the increasing presence of health informatics. Case studies 1–4 show how change can be implemented to promote embedding informatics into the workplace.

Postgraduate programs in Australian universities are a mixture of coursework, coursework and research, or research only. Postgraduate programs were the first to embrace health informatics education in various formats. There is a requirement to provide early coursework components in most degree types as many students do not have any background in health informatics. By providing a range of postgraduate qualifications in health informatics, it is possible to develop the research-teaching nexus and establish a specialized health informatics workforce as well as a clinical workforce ready to use and embrace health information technologies in their daily work. Case Study 1 provides an example from the largest postgraduate course in Australia.

Undergraduate programs are the starting point for introducing a culture of health informatics. It is in these programs that health informatics and health can be presented as synonymous by incorporating health informatics technologies as a normal element in lectures, discussions, case studies, and assignment work. It is also in undergraduate programs that students can be encouraged to explore the transformations that are occurring in the way information is used to plan, manage, and deliver healthcare services and to critically discuss the legal, political and cultural issues around the use of health informatics and ensure the they achieve the necessary digital competencies to work in the current and future healthcare systems. Case Study 2 provides the example of how integration of informatics and electronic medical records can expand the learning environment and normalize health informatics in the curriculum.

EXAMPLES FROM AUSTRALIAN HIGHER EDUCATION SECTOR

The following four case studies are examples of a range of developments that have been used by the authors to establish and integrate health informatics into Australian higher education health profession curricula in one Australian state.

Case Study 1: e-Health Postgraduate Course Development

The University of Tasmania Health Informatics graduate program commenced in 2002. This academic program was built on the Health Informatics Education and Training Project, an in-service professional development program that was introduced in 1998 for the (then) Tasmanian Department of Health and Human Services (DHHS). In 1997, the Tasmanian government, through its Department of Health and Human Services, in a forward thinking move, allocated substantial funds to new investment initiatives for the support of preventative and primary healthcare projects. A priority for funding was the trialing of new technology. To facilitate this initiative, DHHS needed to ensure that its health professionals were aware of, and prepared to take up, new technology as it became available. This was the impetus for the Health Informatics Education and Training Project. The (then) university's Department of Rural Health joined with the DHHS to plan and implement the project. The project goal was to provide a range of health informatics education materials to ensure that DHHS staff would be prepared to exploit the new health technologies as they became available.

Several of the challenges identified with regard to health informatics education for clinicians were evident in this project. The DHHS showed enterprise by acknowledging the need for health professionals to be "technology-ready." At the same time, the DHHS assumed health professionals were computer literate. Therefore, computing skills were not included in the initial scope of the project. The project evaluation identified that this initial expectation that staff would be computer literate was not valid and that ICT skills would have been a useful and relevant inclusion in the early materials [46]. The evaluation also noted that, despite early marketing aimed at explaining the term "health informatics," the concept remained unfamiliar to many staff and evoked responses ranging from "what's that?" to "no use to me" [46]. The evaluation of this in-service program provided valuable insights for the development of the graduate program. The University of Tasmania's Graduate Program in E-Health (Health Informatics) consists of Graduate Certificate, Diploma, and Masters level courses. A Bachelor of E-Health (Health Informatics) (Professional Honours) is also offered.

In developing a framework for health informatics education, Garde and Hovenga [48] identified three primary health informatics roles: clinicians who need to understand how the use of ICT can support their practice; ICT professionals who need to understand health information systems and how healthcare is delivered; and software developers and engineers who develop new health informatics applications. Meeting the educational needs of this diverse group presents challenges to course developers. The Graduate Program in E-Health (Health Informatics) is aimed primarily at clinicians who are interested in using ICT to deliver health services. Consequently, the courses emphasize knowledge and skill development for health informatics professionals and for health professionals wishing to enhance their understanding of health informatics. All courses are offered part-time, online with no compulsory face-to-face component. This attracts enrollments from across Australia. The program has been developed for a diverse group of students including nurses, general practitioners, allied health professionals, and health administrators, working in a variety of health professions and healthcare environments from acute and community settings in both the private and public sector.

The University of Tasmania program explores the various interpretations of health informatics, ranging from a technological focus to a socio-technical focus, and considers the implications of adopting one or another of these interpretations. The University of Tasmania emphasizes health informatics as a socio-technical discipline and profession. In doing so, it encourages students to critically explore cultural, economic, political, and legal issues associated with the implementation of health information systems. It aims to produce critical and reflective health informatics practitioners, teachers, and researchers. At a more basic level, and drawing on the learnings of the DHHS project, it is not assumed that students are computer literate or cognizant with the definitions and concepts of health informatics. These elements are integrated across the curriculum.

In 2014, a program evaluation was undertaken. The aims of the evaluation were to "establish the location of health informatics graduates and current students in health services across Australia, map the professional location of the graduates, and to explore the impact of the University of Tasmania program on graduates" approaches to the practice of health informatics [13:159]. The results indicate that the program is having an impact on the health informatics workforce. This is indicated by the spread of graduates across all states and territories where they are increasingly taking up health informatics roles, adopting senior positions and contributing to the professional activities of health informatics organizations. Where graduates are health, rather than health informatics professionals, they reported an increased involvement as clinical representatives in health informatics projects and programs. Respondents also reported that the course had provided them with a richer understanding of the domain of health informatics. Three themes were identified. The first identified that the course provided new and advanced skills and knowledge that participants applied at both the operational and strategic level. A second theme identified that the course clarified, structured, and consolidated existing understanding. This view was primarily expressed by respondents working in health informatics prior to commencing the course. A third theme focused on expectations about the potential of health informatics. Respondents indicated the need to think critically about the potential and limitations of health informatics solutions. All three themes shared the belief that the course had given participants confidence to express their views and knowledge and to apply their skills.

This program demonstrates the potential of postgraduate education to successfully address the challenges around the provision of health informatics education for clinicians. It specifically addressed the issues identified in the DHHS project re digital technology skills and awareness of health informatics as a unique specialization.

Case Study 2: Integrating Informatics Into an Undergraduate Bachelor of Nursing Degree

In response to the new accreditation standards for nursing degrees, introduced by ANMAC in 2012 [30], the University of Tasmania is the first university to integrate informatics into their undergraduate curriculum. The course accreditation process takes approximately nine to twelve months to complete and requires development of significant content and examples for approval of the degree against the prescribed competencies.

The University of Tasmania Bachelor of Nursing degree contains 25 individual units and, through the desire to ensure a solid and progressive development of informatics knowledge and skills, 21 of those units contain some degree of informatics content. Those units where informatics is not included are primarily professional placement units where the students will be engaging with technologies within a wide range of healthcare workplaces. The scaffolding of learning throughout this degree is based upon the premise that although students have a range of experiences with technologies this is related to their personal lives or studies and as such they are not experts in the concepts or use of informatics in the workplace. The focus in the initial semester is upon developing professional attitudes to social media and technology use in education and practice; students also develop their information seeking skills in their first semester of study, as this is a core skill required throughout their lives. Each informatics component has been mapped against the content of the individual units and the whole degree requirements. The introduction of an electronic medical record into the second semester of the degree allows students to become familiar with the concepts around data structures, databases, and different types of data entry. Use of the electronic medical record continues throughout their degree with increasing complexity of tasks as they progress. The core premise when developing this degree was that informatics should become integrated throughout the degree, and not be viewed as additional or separate from the healthcare content or context [49].

The final stage, still ongoing, is to identify the educators' informatics educational needs. This will be undertaken in phases commencing with the identification of the current e-health literacy of the educators, leading to the development of a customized approach to delivering initial training in informatics to the educators. The education will be driven by the nurse educator informatics competencies currently under development [50] and will ensure that all educators have a base level of informatics skills.

Case study 3: Virtual Community of Practice Using Twitter

Continuing professional development is mandatory for maintenance of registration as a registered nurse. Access to educational support and resources for clinical nurse supervisors of undergraduate students is necessary to ensure they are adequately prepared for the role. Availability of consistent information about curriculum, university processes, and clinical updates are required to remain contemporary in their role. A study exploring the understanding of the knowledge, skills, behavior, and attitudes regarding the use of social media by clinical supervisors was undertaken. The aim was to develop a virtual community of practice about clinical supervision for the purpose of supporting continuing professional development (CPD) by promoting digital literacy and use of technology of this group of nurses. This case study outlines the findings from evaluation of the study and describes the strategies used to support the development of a virtual community of practice using the microblogging platform known as Twitter.

As an adjunct to the 2012 mandated inclusion of informatics in undergraduate nursing courses, in 2014 ANMAC released an additional note [31] outlining that health technology and informatics needed to be embedded at technical, contextual, and emancipatory levels. The implication being that all stakeholders needed to integrate health technology and informatics into beginning level nurse preparations to be work-ready. Clinical supervisors are well placed to be positive role models, and demonstrate digital professionalism for students. There was a need to educationally prepare supervisors in informatics for when cohorts of students who were emerging digital health professionals undertook clinical placement in healthcare environments.

An online questionnaire was offered for completion prior to, or at one of seven workshops aimed to enable supervisors to develop skills in mobile technology use and promote and engage with the virtual community of practice (CoP). Respondents were predominately female senior registered nurses, aged over 46 years. They reported mentoring students for more than 5 years and the majority of respondents indicated they had also used computers for more than 5 years. More than half owned a smartphone, purchased "apps" and used Skype. The majority also indicated they used social media. These clinicians indicated they did use YouTube; however, few used it more than twice per week. These senior clinicians were well placed to be change champions in their workplaces, by supporting increased levels of digital literacy through role modeling appropriate and safe mobile learning use; using it for learning and teaching; and also being involved with guiding policy development of technology in their organizations. To further explore the clinical supervisors' perspective about using a digital communication strategy the survey asked them about their beliefs about digital technology as a mobile learning strategy. The majority, of supervisors believed they could learn to use digital technology so they could share information in a CoP. Almost three-quarters of respondents indicated that a virtual CoP would always or usually be useful for communicating effectively about clinical supervision.

Results of the survey indicated that a virtual CoP would be useful for communicating about university information and for sharing information with colleagues in the network and communicate with the university. When asked about using the dedicated blog and microblog for communication within the CoP, respondents were less certain with a small number who responded they would never use the comments section on the blog to share information and others indicating they would not share information using the microblog. One-third of respondents indicated they would like to receive weekly information about clinical facilitation from the university; while the other two-thirds preferred to receive university information only as required. The desire for frequent and topical information demonstrated these senior clinicians were keen to remain updated and contemporary in their role. It increases the likelihood of their use of the CoP about clinical supervision.

Technology drives change and these senior clinicians have demonstrated they were keen to engage in effective learning and teaching communication strategies including joining a virtual CoP. Moreover, evaluation of this communication strategy as a method to build capacity of clinical supervisors that translates into improved quality of learning and teaching of students in the workplace enabled promotion of digital literacy and mobile technology. Healthcare organizations need to be mindful of the benefits of information sharing within and between groups of clinicians that are employed within their facilities. The opportunity to develop sustainable support and guidance of clinical supervisors needs to be encouraged. It will enhance opportunities for building capacity and provide a safe, high-quality clinical experience for students. It will also enable clinical supervisors to model digital professionalism, encourage cultural change, and promote improved outcomes for patients and learners in the workplace [51].

Case Study 4: Mobile Learning Opportunities for Clinical Supervisors

Work integrated learning enables theory to inform practice within formal curriculum and cocurricular activities and enables preparation of students for work-readiness. For development of the nursing profession and meeting the requirements of the legislative changes for learning and teaching of undergraduate nurses and for informal learning and CPD has meant there was a need to support nurses to be trained in digital literacy and competency. However, the lack of standards, guidelines, and codes of conduct to support nurses to access and use mobile technology in the workplace continues to impede progress regarding direction and opportunities for workforce development of digital literacy and health technology in the nursing profession [41,43–45,52]. This process is complex due to the mix of generational cohorts, their preferred learning styles, and attitudes toward digital literacy and health technology. Clinical supervisors are role models for students and their behavior around digital literacy and mobile learning known as digital professionalism, has the potential to profoundly affect the attitudes and behavior of their students [52].

This project was embarked on, in response to a needs assessment undertaken in 2009, where supervisors indicated they wanted improved communication with the university regarding content and process of university information, as well as up to date information about clinical supervision. Underpinning these requirements was the Australian e-Health Strategy [19] that identified it was essential to have a workforce skilled in informatics. The more recent Australian Workforce Development Strategy [53] forecasts there would be a need for the increase in the number of registered nurses and so requirements for promoting mobile learning in the workplace became more imperative.

An online questionnaire investigated the use of digital literacy and use of technology of clinical supervisors who were recruited, prior to workshops to develop digital skills to enable participation in a community of practice for clinical supervisors. The findings revealed clinical supervisors were employed in a variety of healthcare settings. Approximately three-quarters were from rural and regional areas with the remainder of respondents employed at tertiary hospitals. Almost half indicated their workplace was an out-of-hospital environment such as primary care. Approximately half of this cohort identified as nurse educators and more than half had supervised students for more than 5 years.

There were a range of potential inhibitors and opportunities for using mobile learning as a workforce development strategy. The inhibitors included age and gender that reflected the national average workforce data for nursing [23]. Specifically, the lower uptake of the use of smartphones by older clinicians who may not understand the capability of mobile learning in the workplace could impede acceptance of mobile technology. Some of these senior clinicians may not understand the value of accessing learning resources anywhere or anytime in the workplace, when it is safe to do so. They also may not realize the number of apps or resources that are now available in digital versions. Another inhibitor was the lack of the use of digital media that could be accessed to augment learning in the workplace. Those who do not access media clips may not realize the quality available or understand that for some learners visual display of information could be a useful mobile learning strategy.

The findings indicated there are potential opportunities for advancing mobile learning as a legitimate nursing function. This cohort reported that almost half used a smartphone and they were familiar with social media. They also purchased apps and used "Skype" for communication. There are opportunities to harness the skills nurses already know how to use, and build on these to introduce any other tools or resources that would enhance their learning and teaching. This group of clinicians has the option to model digital professionalism to peers, colleagues, and students. They also have the chance to teach their colleagues and also learn from those who have different knowledge about digital platforms and health technology. This sharing and collaborative effort has the potential to promote informal learning and enable CPD to develop the workforce at the workplace. There is an opportunity to increase confidence and develop competence in digital literacy, improve teamwork, and collegiality of clinical supervisors.

The majority of these respondents indicated they were from rural and regional areas. Skype could be used to link these practitioners in synchronous discussions. Part-time or shift workers may prefer to use asynchronous methods such as using Twitter or commenting on a blog. Geographically dispersed practitioners could connect with their peers and use mobile learning resources to improve their knowledge, skills and enhance learning. Learning from "experts" or discussion with like-minded others by asking questions or commenting on information posted that could be of benefit in developing a repository of information at different levels within healthcare. For example, it could be information that relates to a particular healthcare setting, such as a unit or ward, it could be at a rural facility or it could be based around nursing topics such as clinical supervision. Even the "lurkers" or nonparticipants could benefit from the dialogue that occurs between users as it raises awareness and has the potential to increase understanding even if this group do not actively contribute to the dialogue.

The high level of penetration of social media in Australian culture opens possibilities for clinical supervisors to link with other clinicians and leaders in their field. The range and use of platforms available means that sharing among the different groups may impede some networking opportunities, while developing others. Mobile learning as a workforce development strategy is here. Changing skill requirements in healthcare settings have created new opportunities for learning within the health professions. The availability of mobile learning opportunities will ensure clinical supervisors remain leaders in their field. Further research into developing and evaluating workforce development opportunities is required to guide policy direction about safe and appropriate use of mobile learning strategies. Over time, a supportive environment could facilitate cultural change and enable clinical supervisors to model appropriate behavior to promote digital professionalism of the next generation of nurses and facilitate improved health outcomes for patients.

IMPLICATIONS

Since the 1970s, when Anderson, Gremy and Pages [54] first wrote about informatics education for healthcare professionals, it has been evident that the inevitable introduction of computers into healthcare would require changes to the health workforce and the educational needs to healthcare professionals. Rapid growth in the adoption of the wide range of health information technologies, and the proliferation of development platforms has enforced the recognition that educational programs for health professionals of all disciplines and levels are required to ensure safe delivery of care. Whilst there have been some international efforts in this area (see, e.g., [55]) there continues to be a lack of consistency of effort or success across the range of levels and requirements. As health professionals increasingly use information technology, it is crucial for all health professionals to possess at least basic informatics skills. To provide these skills, universities need to develop, deliver, and promote relevant health informatics education in a flexible manner that can be tailored to the individual health professional's need [47]. This chapter has provided an overview of the health informatics space in Australia and discussed some of the challenges faced in relation to the education of healthcare professionals. The case studies provided off some understanding of the types of solutions that can be used to provide health informatics education in relation to a diverse population at different levels of their healthcare practice.

CONCLUSION

Increasingly all healthcare professionals require a core set of digital competencies to enable them to work safely, effectively, and efficiently as digitally professional healthcare providers in a complex, rapidly changing, environment. More broadly, to ensure health services are appropriate, responsive, and flexible to the needs of consumers, there is an even greater need for increased capability in health informatics and understanding of data and its use in healthcare. To build the appropriately skilled and digitally literate workforce for the future, foundational professional education needs to include health informatics and be supplemented by continuing professional development, that is supported by digital technology.

This chapter has provided contemporary real-world case studies that demonstrate practical strategies to promote and embed health informatics education within different groups of health practitioners at one Australian university. The development and cultivation of the health informatics research-teaching nexus is a critical requirement to implement excellence in health informatics in higher education for health professionals.

References

[1] IT-014, What is Health Informatics? 2010. Available from: <http://www.e-health.standards.org.au/ABOUTIT014/WhatisHealthInformatics.aspx> [accessed 08.16].

[2] Oh H, Rizo C, Enkin M, Jadad A. What is eHealth?: a systematic review of published definitions. World Hosp Health Serv. 2005;41(1):32–40 [Medline: 102125492] Available from: <http://dx.doi.org/10.2196/jmir.7.1.e1>

[3] Digital Professionalism 2013. Available from: <http://digitalprofessionalism.com> [accessed 08.16].

[4] NMBA. Nursing and midwifery continuing professional development registration standard; 2016. Available from: <http://www.nursingmidwiferyboard.gov.au/Registration-Standards.aspx> [retrieved 04.08.16]

[5] Traxler J. Defining, discussing and evaluating mobile learning: the moving finger writes and having writ. Int Rev Res Open Distance Lear. 2007;8(2):67–75.

[6] Tanriverdi H, Iacono S. Diffusion of telemedicine: a knowledge barrier perspective. Telemed J 1999;5 (3):223–44.

[7] Whetton S, Walker J. The diffusion of innovation: barriers to the uptake of health informatics [online]. In: Webb, Robert (Editor); Ribbons, Robert M; Dall, Veronica. HIC 2002: Proceedings: Improving Quality by Lowering Barriers. Brunswick East, Vic.: Health Informatics Society of Australia, 2002: [156–161]. Available from: <http://search.informit.com.au/documentSummary;dn = 895915616974411;res = IELHEA> ISBN: 0958537097. [cited 21.01.17].

[8] McGill A. The challenges for health informatics in Australia to 2005 [online]. In: Smith, Joy (Editor); Smith, Len (Editor); James, Paul (Editor). HIC 2001: Proceedings. Brunswick East, Vic.: Health Informatics Society of Australia, 2001: [162-166]. Available from: <http://search.informit.com.au/documentSummary; dn = 912405796537918;res = IELHEA> ISBN: 0958537089. [cited 20.01.17].

[9] Greatbatch D, Murphy E, Dingwall R. Evaluating medical information systems: ethnomethodological and interactionist approaches, vol. 14. London: Health Services Management Research; 2001. p. 181. n3.

[10] Abbott PA, Coenen A. Globalization and advances in information and communication technologies: the impact on nursing and health. Nurs Outlook 2008;56(5):238–246.e2. Available from: <http://dx.doi.org/10.1016/j.outlook.2008.06.009> [Medline: 18922277].

[11] Bembridge E, Levett-Jones T, Jeong SY. The transferability of information and communication technology skills from university to the workplace: a qualitative descriptive study. Nurse Educ Today 2011;31(3):245–52. Available from: <http://dx.doi.org/10.1016/j.nedt.2010.10.020> [Medline: 21093125].

[12] Katz-Sidlow RJ, Ludwig A, Miller S, Sidlow R. Smartphone use during inpatient attending rounds: prevalence, patterns and potential for distraction. J Hosp Med 2012 October;7(8):595–9. Available from: <http://dx.doi.org/10.1002/jhm.1950> [Medline: 22744793].

[13] Whetton S, Hazlitt C. Educating the health informatics professional: the impact of an academic program. Stud Health Technol Inform 2015;214:159–66. ISSN 0926-9630.

[14] Pauleen D, Campbell J, Harmer B, Intezari I. Making sense of mobile technology: The integration of work and private life. SageOpen. 2015. Available from: <http://sgo.sagepub.com/content/5/2/2158244015583859>.

[15] Mold F, de Lusignan S, Sheikh A, Majeed A, Wyatt JC, Quinn T, et al. Patients' online access to their electronic health records and linked online services: a systematic review in primary care. Br J Gen Pract 2015;65: e141–51. Available from: http://dx.doi.org/10.3399/bjgp15X683941.

[16] National Health Information Management Council (NHIMAC). Health online: a health information action plan for Australia. Health Inform Manag 2000;29(4):179–81.

[17] Wrobel JP. Are we ready for the better medication management system? MJA 2003;178:448–50.

[18] Australian Government Department of Health. *HealthConnect*, 2011. Available from: <http://www.health.gov.au/healthconnect>.

[19] AHMAC (Australian Health Ministers Advisory Council), National E-Health Strategy for Australia, Victorian Department of Human Services, on behalf of the Australian Health Ministers' Conference (2008).

[20] Cummings E, Cheek C, van der Ploeg W, Orpin P, Behrens H, Condon S, et al. The Cradle Coast personally controlled electronic health record evaluation research plan. Stud Health Technol Inform 2012;178:14–19 ISBN 9781614990772.

[21] Pearce C, Bainbridge M. A personally controlled electronic health record for Australia. J Am Med Inform Assoc 2014;21(4):707–13. Available from: <http://dx.doi.org/10.1136/amiajnl-2013-002068> Epub 2014 Mar 20.

[22] Almond H, Cummings E, Turner P. Australia's personally controlled electronic health record and primary healthcare: generating a framework for implementation and evaluation. Stud Health Technol Inform 2013;188:1–6. Available from: <http://dx.doi.org/10.3233/978-1-61499-266-0-1> ISSN 1879-8365 (2013).

[23] Health Workforce Australia. *Health Information Workforce Report*, 2013. Available from: <https://www.hwa.gov.au/our-work/health-workforce-planning/health-informaticians-specialist-workforce-study>.

[24] eHealth Working Group (EHWG), A National Digital Health Strategy for Australia, July 2016 – June 2019, DRAFT available for comment March 24, 2016, Version 0.2.

[25] Gray K, Dattakumar A, Maeder A, Butler-Henderson K, Chenery H. Advancing e-health education for the clinical health professions final report. Australia: Department of Education and Training; 2014.

[26] Royal Australian College Of General Practitioners. The RACGP Curriculum for Australian General Practice Health Informatics. 2007. Available from: <http://www.racgp.org.au/scriptcontent/curriculum/pdf/informatics.pdf> [accessed 03.19].

[27] ANMF. National Informatics Standards for Nurses and Midwives. 2015. Retrieved April 6, 2016. Available from: <http://anmf.org.au/documents/National_Informatics_Standards_For_Nurses_And_Midwives>.

[28] Australian Health Informatics Education Council, *Health Informatics – Scope, Careers and Competencies*, 2011. Available from: <http://www.ahiec.org.au/Documents.htm>.

[29] CHIA, Certified Health Informatician Australasia Health Informatics Competencies Framework, 1.0, 2013. Available from: <http://www.healthinformaticscertification.com>.

[30] ANMAC. Australian Nursing and Midwifery Accreditation Council Registered Nurse Accreditation Standards [Electronic Version]. 2012. Retrieved July 23, 2013. Available from: <http://www.anmac.org.au/sites/default/files/documents/ANMAC_RN_Accreditation_Standards_2012.pdf>.

[31] ANMAC Health informatics and health technology – an explanatory note 2014. 2014. Available from: <http://www.anmac.org.au/sites/default/files/documents/20150130_Health_Informatics_Technology_Explanatory_Note.pdf> [retrieved 03.08.16].

[32] Hilberts S, Gray K. Education as ehealth infrastructure: considerations in advancing a national agenda for ehealth. Adv in Health Sci Educ. 2014;19:115. Available from: http://dx.doi.org/10.1007/s10459-013-9442-z.

[33] Jolly R, Australia. Dept. of Parliamentary Services. Parliamentary Library (2011). *The e health revolution: easier said than done*. Parliamentary Library [Canberra]. Available from: <http://www.aph.gov.au/library/pubs/rp/2011-12/12rp03.pdf>.

[34] Gray K, Choo D, Butler-Henderson K, Whetton S, Maeder A. Health informatics and e-health curriculum for clinical health profession degrees. Stud Health Technol Inform 2015;214:68–73. Available from: <http://dx.doi.org/10.3233/978-1-61499-558-6-68> ISSN 0926-9630.

[35] Bennett S, Maton K, Kervin L. The 'digital natives' debate: a critical review of the evidence. Brit J Educ Technol 2008;39(5):775–86. Available from: http://dx.doi.org/10.1111/j.1467-8535.2007.00793.x.

[36] Kennedy G, Dalgarno B, Bennett S, Judd T, Gray K, Chang R. Immigrants and natives: investigating differences between staff and students' use of technology. In: Atkinson R, McBeath C, editors. Proceedings of "Hello! Where are you in the landscape of educational technology?", the Annual Conference of the Australasian Society for Computers in Learning in Tertiary Education (ASCILITE 2008). Melbourne, VIC: Australasian Society for Computers in Learning in Tertiary Education; 2008. p. 484–92.

[37] Brown C, Czerniewicz L. Debunking the 'digital natives': beyond digital apartheid, towards digital democracy. J Comp Assist Learning 2010;26(5):357–69. Available from: http://dx.doi.org/10.1111/j.1365-2729.2010.00369.x.

[38] Li Y, Ranieri M. Are "digital natives" really digitally competent?-A study on Chinese teenagers. Brit J Educ Technol. 2010;41(6):1029–42. Available from: http://dx.doi.org/10.1111/j.1467-8535.2009.01053.x.

[39] Mather C. Embedding an e-portfolio into a work Integrated learning environment: The School of Nursing and Midwifery experience/ *EDULEARN12 Conference Proceedings*, July 2nd-4th, 2012, Barcelona, Spain, pp. 4959-4968. ISBN 978-84-695-3491-5 (2012).

[40] Mather C, Marlow A, Cummings E. Web 2.0 strategies to enhance support of clinical supervisors of undergraduate nursing students: an Australian experience. *Proceedings of the 5th International Conference on Education and New Learning Technologies (EDULEARN13)*, July 1–3, 2013, Barcelona, Spain, pp. 3520-3527. ISSN 2340-1117 (2013).

[41] Mather CA, Cummings EA. Issues for deployment of mobile learning by nurses in Australian healthcare settings. Stud Health Technol Inform 2016;225:277–81. Available from: <http://dx.doi.org/10.3233/978-1-61499-658-3-277> ISSN 0926-9630.

[42] AIHW, various workforce data and reports. Available from: <http://aihw.gov.au/workforce-publications/> [accessed 29.01.17].

[43] Mather CA, Marlow AH, Cummings EA. Digital communication to support clinical supervision: considering the human factors. Stud Health Technol Inform 2013;194:160–5. Available from: <http://dx.doi.org/10.3233/978-1-61499-293-6-160> ISSN 0926-9630.

[44] Mather C, Cummings E. Unveiling the mobile learning paradox. Stud Health Technol Inform. 2015;218:126–31. Available from: http://dx.doi.org/10.3233/978-1-61499-574-6-126> ISSN 0926-9630.

[45] Mather CA, Cummings EA, Nichols LJ. Social media training for professional identity development in undergraduate nurses. Stud Health Technol Inform. 2016;225:344–8 ISSN 0926-9630.

[46] Whetton S, Walker J. The health informatics education and training project: final report and evaluation. University Department of Rural Health; 2001.

[47] Mantas J, Ammenwerth E, Demiris G, Hasman A, Haux R, Hersh W, et al. Recommendations of the International Medical Informatics Association (IMIA) on education in biomedical and health informatics (first revision). Methods Inform Med 2010. Available from: http://dx.doi.org/10.3414/ME11-01-0060.

[48] Garde S, Hovenga E, *Australian Health Informatics Educational Framework*. 2006. Available from: <http://www.achi.org.au/docs/Health_Informatics_Educational_Framework_20060326.pdf>.

[49] Cummings EA, Shin EH, Mather CA, Hovenga E. Embedding nursing informatics education into an Australian undergraduate nursing degree. Stud Health Technol Inform 2016;225:329–33. Available from: <http://dx.doi.org/10.3233/978-1-61499-658-3-329> ISSN 0926-9630.

[50] Kinnunen U-M, Rajalahti E, Cummings E, Borycki EM. Curricula challenges and informatics competencies for nurse educators, *Studies in Health Technology and Informatics*. 2017;232:41–48. Availble from: http://dx.doi.org/10.3233/978-1-61499-738-2-41> ISSN 0926-9630.

[51] Mather C, Cummings E. Usability of a virtual community of practice for workforce development of clinical supervisors. Stud Health Technol Inform 2014;204:104–9. Available from: <http://dx.doi.org/10.3233/978-1-61499-427-5-104> ISSN 0926-9630.

[52] Mather C, Cummings E. Mobile learning: a workforce development strategy for nurse supervisors. Stud Health Technol Inform 2014;204:98–103. Available from: <http://dx.doi.org/10.3233/978-1-61499-427-5-98> ISSN 0926-9630.

[53] Australian Workforce and Productivity Agency. Future Focus: Australia's skills and workforce development Strategy. Canberra; 2013.

[54] Anderson J, Gremy F, Pages JC. Education in Informatics of Health Personnel. IFIP Medical Informatics Monograph Series, Vol. 1. Amsterdam: North Holland Publishing; 1974.

[55] Jaspers MW, Gardner RM, Gatewood LC, Haux R, Schmidt D, Wetter T. The international partnership for health informatics education: lessons learned from six years of experience. Methods Inf Med 2005;44:25–31.

STATE AND NATIONAL LEVEL INITIATIVES

CHAPTER

17

Implementing Informatics Competencies in Undergraduate Medical Education: A National-Level "Train the Trainer" Initiative

Rashaad Bhyat[1], Candace Gibson[2], Robert Hayward[3], Aviv Shachak[4], Elizabeth M. Borycki[5], Amanda Condon[6], Gerard Farrell[7], Kalyani Premkumar[8] and Kendall Ho[9]

[1]Canada Health Infoway, Toronto, ON, Canada
[2]Western University, London, ON, Canada
[3]University of Alberta, Edmonton, AB, Canada [4]University of Toronto, Toronto, ON, Canada
[5]University of Victoria, Victoria, BC, Canada [6]Winnipeg Regional Health Authority, Winnipeg, MB, Canada [7]Memorial University, St. John's, NL, Canada [8]University of Saskatchewan, Saskatoon, SK, Canada [9]University of British Columbia, Vancouver, BC, Canada

INTRODUCTION

As discussed in Part 1 of this book, the practice of health care and health professional education are undergoing many changes, in the face of multiple pressures and demands on the curriculum. This is especially true for Medicine, where these pressures are in part due to the rapid pace at which the volume of knowledge is expanding and rapid changes in technology. As Densen noted "the doubling time of medical knowledge in 1950 was 50 years; in 1980, 7 years; and in 2010, 3.5 years. In 2020 it is projected to be 0.2 years—just 73 days." [1] Consequently, there is much more knowledge for medical students to assimilate during their training now than there has been in the past. Furthermore, the nature of

Health Professionals' Education.
DOI: http://dx.doi.org/10.1016/B978-0-12-805362-1.00017-6

medical information exchange itself has also changed, as the rapid changes in information technology (hardware, software, and the evolution of the Internet) have made it possible to hold a very powerful computer literally in the palm of one's hand, and to access virtually limitless amounts of data.

To effectively address the new challenges that result from the expanding use of information and communication technology in health care, it has been suggested that trainee physicians require formal instruction in the domain of medical informatics [2]. In this chapter, we describe a national-level initiative to bring medical informatics knowledge to undergraduate medical education in Canada. It has been proposed that such training would help bridge the gap between traditional medical practice, and the new norms of the information age such as using an Electronic Medical Record (EMR) to generate a prescription in one's practice.

However, influencing medical education on a national scale is not an easy task, given that much of the responsibility for both health care and education in Canada historically lie within the domain of provincial and not federal authority under the Constitution Act of 1867 (the British North America Act). In 2011, a partnership was forged between two organizations with a national focus—Canada Health Infoway (Infoway) and the Association of Faculties of Medicine of Canada (AFMC). The AFMC "represents Canada's 17 faculties of medicine and is the voice of academic medicine" in Canada [3]. The AFMC provides "guidance on continuing, postgraduate, and undergraduate medical education, and. . .national engagement on a wide range of issues." [3] Infoway is an independent, not-for-profit organization funded by the federal government of Canada, with a goal of improving the health of Canadians by accelerating the development, adoption, and effective use of digital health solutions across the country [4]. The goals of the partnership were to raise the profile of medical informatics (eHealth in the medical context) within medical education and to ensure that Canada's next generation of physicians would be equipped with the skills to thrive in a digitally enabled practice setting.

The two organizations created an initial "Physicians-in-Training" project that resulted in a number of early milestones [5,6], including:

- Completing an environmental scan that identified gaps in e-Health training in the medical curricula nationally [2].
- Development of Canada's first eHealth Competencies for Undergraduate Medical Education [7].
- Influencing the inclusion of eHealth concepts in the CanMEDS 2015 Competency Framework. Created by the Royal College of Physicians and Surgeons of Canada, CanMEDS is an educational framework that "describes the abilities physicians require to effectively meet the health care needs of the people they serve." [8]
- Recommendations from the project's leadership committee (composed of subject matter experts from across the country).

In 2015, the AFMC and Infoway partnered on a second project, entitled "Accelerating Engagement: Patient-focused Digital Health Solutions for Faculty and Residents in Medical Education," which set out to address the environmental scan's recommendation to "train the trainers," by creating teaching materials which would form core eHealth curricular content. The first goal of this project, which was also known as the eHealth

Medical Faculty Development initiative, was to create a series of online webinar workshops and supporting educational resources. Together, the webinars and supplementary resources formed an online toolkit that provides medical educators with tangible tools for teaching eHealth concepts to physician learners (students and residents). The toolkit was made freely available to medical educators (and the public) through Creative Commons licensing.

A second major goal of the project was to continue efforts to advocate for awareness and integration of medical informatics concepts into the medical education curriculum. Senior academic, medical informatics experts were designated as "Peer Leaders," and advocated both within and beyond the boundaries of their institutions. Advocacy took on many forms: meetings with and letters to Undergraduate Deans of Medicine and other key faculty; presentations to colleagues both locally and at conferences; and articles and abstracts written and submitted.

During the course of the second phase of the project, AFMC, Infoway, and participating Peer Leaders noted some changes from 2011, when the environmental scan took place:

First, patients have become increasingly interested in accessing their own health information. Many patients want to interact with their physicians in a technology-enabled manner and these "consumer health" applications have become more accessible [9]. Second, concerns over privacy and security have an impact on every aspect of eHealth, and are essential in any core curriculum. Thus, in this project, these two themes—consumer health and privacy and security—were interwoven as connecting threads across four, core domains of eHealth content: (1) Personal and Shared Information Management; (2) Clinical Decision Management; (3) Clinical Information Management; and (4) Health Communication Management. These four domains were based upon medical informatics curricular content developed at the University of Alberta, by Dr. Robert Hayward.

Each of these four core medical informatic domains became a workshop. The four workshops were presented as Continuing Medical Education (CME) sessions for medical educators, multiple times by webinar from November 2015 to March 2016. As noted above, the content of the four workshops is available in the format of an online eHealth Workshop Toolkit Collection, accessible through Creative Commons licensing on the AFMC website [10]. The CME designation of these workshops was a key factor in their success, and will be discussed further under the heading "Outcomes and Results." The following sections of this chapter will provide an overview of the eHealth workshop themes and content, in the context of the project's broader objectives.

FACULTY DEVELOPMENT WORKSHOPS

Personal and Shared Information Management Workshop

Rationale for the Workshop

Health care is an information intensive, knowledge-based business. On a daily basis clinicians use millions of bits of data from patients, the medical literature, knowledge databases, and their own experience, to diagnose, understand, and treat patients [11]. Increasingly those health data are entered and retrieved through an electronic record.

About 2000 health care transactions take place every minute, most quite complex and all requiring information and document flow [12]. Monitoring an e-record could prevent many adverse drug events, provide reminders for timely screening tests and vaccinations, and facilitate chronic disease management.

The past decade has seen a steady increase in adoption and use of electronic records in all health care settings; about 77% of Canadian physicians reported having and using an EMR in the 2014 National Physician Survey [13]. These are the tools that physicians will be using in practices in the future. In fact, the College of Family Physicians of Canada (CFPC) published a white paper [14] that urges the move to electronic records by 2022—in other words, medical students who enter school today will be using electronic tools exclusively.

In a digital world, information on individual patients can be collected and added to a knowledge database that enables improved treatment within a practice. Current knowledge databases are medical literature repositories that have been built up over time. The ultimate goal is to create a cycle of information flow, whereby data from distributed electronic health records (EHRs) are routinely and effortlessly submitted to registries and research databases. Practitioners at the point of care will receive this new knowledge, using a variety of computer-supported decision support delivery mechanisms. This cycle of new knowledge, driven by experience, and fed back to clinicians, has been called a "learning health care system." [15] To enable this, general principles and processes for good information management practice should underlie the collection, maintenance, protection, and optimum use of data within an electronic record.

The workshop on Personal and Shared Information Management (PSIM) was developed to introduce medical educators to the principles of information management on a personal level and outline the application of these principles to information sharing in an educational, research, or clinical environment. As the first workshop in the series, it also provided an overview of eHealth and eHealth terminology. The PSIM workshop has been prepared and delivered by Dr. Candace Gibson. The next few sections describe the content of the PSIM workshop.

Introducing Terms That Describe eHealth

As noted above, the workshop on Personal and Shared Information Management started by introducing some of the concepts and terms that describe eHealth. Interwoven in this discussion is the important theme of privacy, confidentiality, and security of data. This theme is very important to individual users of data, but of crucial importance as people's personal health information is collected, maintained, and shared. Nowhere is this of greater concern than in the use of social media both on a personal and professional level.

In a broad sense, medical Informatics or eHealth in a medical context is the use of computer and information technology in health care, or what happens at the interface of the terminal and user—be it for e-learning; information searching; accessing a patient record for retrieval, entry or use of health data; videoconferencing; telehealth, etc. Our eHealth Leadership Committee defined eHealth as "the use of information and communication technology and innovation to improve or enable health and health care services" [7]. More specific medical informatics/medical eHealth skills and competencies include:

- Basic Computer Skills and Computer Literacy.
- Information Searching and Retrieval Skills or information literacy—concentrating on evidence-based medical resources and application of evidence-based practices.
- Life Long Learning and E-learning.
- Communication skills—incorporating the use of an electronic medical record into the patient encounter; clinical documentation skills in both the paper-based and electronic records; understanding issues of privacy and confidentiality.
- Coordination of Care and Clinical Information Management—use of an electronic record as a communication tool and central data repository for patient health information and a focus for interprofessional collaboration.
- Research and scholarship—using library resources effectively, critically appraising retrieved material, use of bibliographic software, use of spreadsheets, and basic statistics for data analysis.
- Advanced health informatics skills—web pages, patient portals, blogs, communities of practice, and setting up and conducting a telehealth encounter.
- Assessment of technology (educational technologies as well as high tech diagnostic and therapeutic tools).

These skills were presented at the beginning of the workshop as a background for the more specific topic of personal and shared information management and as an overall orientation to the workshop series.

Introduction to the Principles of Personal Information Management

Personal and shared information management forms the basis of subsequent information and knowledge management in health care. Good record keeping and information management processes begin with personal information management (PIM)—the science of how our personal information is acquired, organized, maintained, retrieved, and used [16]. These same principles apply to the effective management and use of health information. Table 17.1 below presents the main areas of PIM; some of the formats, sources, and uses of PIM; as well as critical questions that must be asked, with respect to health information and practices.

PIM and Health Information Management activities share many of the same characteristics. Both involve the identification, acquisition, organization, storage, protection, access, retrieval, dissemination, analysis, and interpretation of all information, regardless of medium and format [17]. In the context of health information, similar as well as other types of data recur within the electronic record including but not limited to: (1) text files (e.g., history and physical examination, family and social history, and various notes); (2) digital images (X-ray, CT, MRI, and US); (3) medications (supporting e-prescribing); (4) calendars (supporting e-scheduling); (5) reminders (for screening tests, vaccinations, and follow-up); (6) Alerts (adverse drug events; allergies); and (7) communications (emails, faxes, messages, and referrals). Thus, the application of best practices of PIM may also facilitate efficient use of health information.

TABLE 17.1 An Overview of Personal Information Management (PIM) within the Context of Medical Education

Main Areas of PIM in Medical Education	Personal Information Formats, Sources, and Uses	Main Questions About PIM
Records, File and information management; privacy, confidentiality, and security.	Personal file collections (digital and physical); Documents/text, music, photos, videos, and similar	How do you manage your own files and folders on your desktop computer, tablet PC, and smartphone?
The use of online learning platforms and "eClassroom" skills; videostreaming lectures; videoconferencing; simulations and virtual patients.	References (including scientific references, websites of interest)	Are items synched across your devices—do you have the most recent version of a document?
The use of personal data and information management software, for example, e-portfolios; PDAs/mobile phones/mobile app resources.	Personal notes/journal/ePortfolio	How do you manage version control?
Information searching and retrieval skills; information evaluation and critical appraisal.	Legal Documents; Education Records/Transcripts	Do you place work on a shared drive?
Accessing information at the point of care from an electronic record	Address books; Lists (including task lists)	Do you have a distinction between those files on your personal folders versus those files that are on your shared drive?
	Significant calendar dates (appointments, meetings, classes, exams, assignments birthdays, anniversaries); Reminders, Alerts	If items are on a shared drive, are they accessible to everyone? Is access based on role?
	Email, SMS, IM message archives; Fax communications, voicemail; RSS/Atom feeds	Can you find things readily?
		Are files and information secure?

Within our medical schools and in clinical practice we use a number of different electronic tools to manage and support that information and knowledge. These include **communication tools** for sending messages and files (e.g., email, web publishing, file-sharing, wikis, SMS, and IM); **conferencing tools** for video/audio conferencing, chat, forums, etc. (e.g., Webex and Skype); **collaborative management tools** for managing group activities (e.g., content management systems, learning management systems (LMS); **Knowledge management tools** for managing references, concept mapping to enhance learning or understanding and form connections across ideas/disciplines, and note taking; and **mobile applications** including mobile devices in medicine and mobile health interventions.

For most of our students, we now demand and expect a minimum level of computer literacy upon graduation from premedical programs (undergraduate degrees) [18]. Within these computer skills, we expect a basic level of information literacy and the framework for understanding, finding, evaluating, and using information. At most medical schools, these health information literacy skills are taught within the first year in a variety of settings—during an initial boot camp or in early introductory sessions; within an epidemiology module or elsewhere as appropriate for searching and answering both research and clinically based questions. Once found, material should be critically appraised and evaluated before accepting as best evidence [7,19].

Our medical students now learn in an "electronic" space—all notes, texts, clinical guides, e-portfolios, even examinations are available online. Each medical school has its own learning management system (LMS) but, in general, they all provide a platform for the organization of content and a one-point access to all courses, modules, or learning materials needed in the program. Additional features include: the ability to handle many different types and forms of data (text, images, audio, and video); forums for discussion and sharing of ideas; private and secure communication tools; file-sharing capabilities; alerts in the form of announcements, reminders; and quizzes, tests, examinations, and assessments of various sorts. Some of these systems provide very good audit functions for tracking use of the LMS. Access is assigned by role: read only (auditing a course); read and post information (students); read, create, revise, delete information (instructors, course coordinators); posting of grades (grade administrators).

Other components of the learning platforms and models can include multimedia such as videoconferencing or videostreaming of live or taped lectures. As medical schools move toward models of distributed education, these technologies are more important for education and linking faculty and students across distant and disparate campuses. These same systems are often also used for medical rounds, faculty education, peer-to-peer networking, and at a clinical level for the linkage of clinicians and patients (for example see the Ontario Telemedicine Network [20]). Use of these electronic platforms and tools may require adaptation on both the part of faculty and students to adjust to this new modality of teaching and learning.

It is now common practice for medical students to maintain a professional activity portfolio to track skills, competencies, and learning experiences, throughout their first year and into clerkship training in third and fourth year that may be linked or accessed through the LMS. In some schools, the practice is introduced at the pre-medical or undergraduate level as a tool for reflection, career planning and mapping, and for building a skills showcase [21,22]. For example, the development and use of an ePortfolio within residency training to instill the practice of life-long learning is key component of the Competency By Design (CBD) initiative of the Royal College of Physicians and Surgeons (RCPSC). Within this personal learning environment, achievements and milestones can be recorded and collected, reflected upon, shared and/or assessed to ensure that certain skills and competencies have been met and adequately mastered [16]. Certification and CME credits can also be tracked through this electronic system. Examples of such ePortfolio systems include the open source tools provided by Mahara and Sakai (compatible with Moodle LMS) or commercial products such as Chalk & Wire.

Introduction to the Principles of Shared Information Management

Increasingly physicians and other health professionals work together in groups and teams and need to share information and knowledge. In the construction of knowledge, we are dependent on the expertise and knowledge of many (e.g., building wikis, etc.). In practice, we can learn from colleagues and share best practices through Communities of Practice, defined by Etienne Wenger as "groups of people who share a concern, a set of problems, or a passion about a topic, and who deepen their knowledge and expertise in this area by interacting on an ongoing basis." [23,24]

Health information can be shared through linked records with other physicians, members of a health care team (shared care plan), and with patients. An increasing number of patient portals provide access to test results, scheduling, and renewal of prescriptions through a hospital's or a physician's information system. Ultimately, patients themselves are sharing their information and experiences with each other (e.g., through platforms such as PatientsLikeMe). As health information systems achieve greater integration, there is a greater need to have policies and procedures in place to oversee information governance and management.

From a technical point of view, there are a few minimum requirements for seamless data sharing and ease of use, including: a portal or single point of sign on; access to applications, files, folders, shared drives; effective collaboration and collaboration tools at the team or organizational level; access to the medical literature and library resources (PubMed, UptoDate, etc.); content management tools for both structured and unstructured data; and most importantly standards for data vocabulary (coding and classification systems), transfer and exchange, technology, documents, the user interface, and privacy, confidentiality, and security. The Canadian Standards Association (CSA) Model Code for Protection of Personal Information, for example, contains the 10 fair information practices that form the basis for federal and provincial privacy laws and professional codes of ethics [25].

Most people are familiar with the concept of a data vocabulary, which is similar to the thesaurus in word processing. In searching for information, we often use a controlled vocabulary, for instance through PubMed's MeSH (medical subject headings) terms, a predefined set of terms assigned to articles that enable more selective searches. MeSH terms are updated on a regular basis to include new concepts, terms, and diseases (e.g., HIV, AIDS, SARS) as they achieve frequent use in the medical literature.

However, the ideas of a nomenclature (a system of "naming" things) and a classification system (the process of arranging into ordered classes or categories of the same type depending on their intended use) may be less familiar [26]. These coding systems often operate in the background in an electronic record system or are explicitly used in coding records within a healthcare facility. In a nomenclature, codes are assigned to medical concepts, and medical concepts can be combined according to specific rules to form more complex concepts (e.g., Systematized Nomenclature of Medicine—Clinical Terms (SNOMED-CT). This leads to a large number of possible code combinations. In a classification system, all possible codes are predefined (e.g., International Statistical Classification of Diseases (ICD), or the Canadian Classification of Health Interventions (CCI)). These codes allow the aggregation of unstructured data into searchable, structured formats that are used for billing or funding purposes, research, or population health measures. The effectiveness of these coding systems depends on how well data have been entered within the clinical documentation. Standardization of the initial data capture can facilitate the optimal use of the data many times.

Digital Professionalism

Social media platforms such as Facebook, Twitter, and Instagram can not only be tools for communication and knowledge generation, but can also be a hazard and misused. Social media can confer a number of personal and professional benefits for clinicians but the risks of compromising patient confidentiality and one's reputation shouldn't be underestimated. There is a tendency to think that today's students—so-called "digital natives" and "digital dependents"—know how to use these technologies and will readily use them optimally in their practice without guidance or training. Would students be allowed to do the same thing with a stethoscope, a sphygmomanometer, or a scalpel? Students require some instruction on paper records with regard to best practices for clinical documentation and reasoning. Why would we not provide similar education for one of the most powerful tools that we have seen in human history? Social media encompasses technology that is reshaping the way commerce works, the way education itself is delivered, the way medicine will be practiced in the twenty-first century, and the ways that individuals will find health information and support during illness and disease. These issues are discussed in detail in Chapters 4 and 6 of this book and were also presented as part of this workshop, including relevant guidelines published by student groups [e.g., Canadian Federation of Medical Students (CFMS), McGill Social Wellness] and professional associations [e.g., Canadian Medical Association (CMA)]. A podcast within the workshop's toolkit also elaborates on the topic of digital professionalism [10]

Reflections and Lessons Learned

The PSIM workshop was a success, based on the 55 live participants, and the total of 77 participants who had accessed the workshop online by March 30, 2016. Furthermore, 68% of respondents to a PSIM postworkshop survey indicated that they planned to integrate what they learned into their teaching activities. More time for workshop content development would have been helpful, and the Webex format that was used to deliver the webinar created limitations to the impact of the workshop. The presenter's overall impression was that workshop participants learned something new about eHealth, and perhaps more importantly, learned where to go for more information.

The PSIM workshop highlighted the fact that only a few attempts have been made to integrate health informatics into the undergraduate medical curriculum (several faculties have developed one-off courses) [2]. One example includes the School of Medicine at Oregon Health & Science University, whose faculty have set out 13 areas of competency for informatics training, mapped these to specific learning objectives and have begun to implement this within their medical school at the early, preclinical and clinical training periods within the curriculum [26]. Hersh et al. describe this initiative in detail in Chapter 13 of this book. For Canada, the opportunity is here to transform the health care system into a learning health system. The topics outlined above, and in the rest of this chapter, comprise medical informatics knowledge that is needed by medical students to work in the health care system of the present and the future. Using the principles outlined in the PSIM and other core eHealth workshops, doctors can become active participants in building, shaping and using their gathered knowledge for diagnosis and treatment, clinical care guidelines, and personalized medicine [27].

Clinical Decision Management workshop

The Clinical Decision Management (CDM) workshop was prepared and delivered by Dr. Robert Hayward. The purpose of the workshop was to provide a conceptual framework for classifying and understanding clinical decision support systems (CDSS), which may help learners to identify opportunities for introducing clinical decision management into medical school curricula.

Workshop's Description

The workshop addressed three main questions: (1) What is clinical decision support (CDS)?; (2) Why does it matter?; and (3) Where to start? The workshop set the following learning objectives for participants to be able to:

- Recognize different types of decision support systems [28]
- Appreciate possible benefits and harms associated with clinical decision support tools [28], and
- Anticipate the relative strengths and weaknesses of different classes of decision support [28].

To meet these objectives, the workshop presented a definition and the key attributes of CDSS. In particular, it highlighted the role of CDSS in linking observational data with health knowledge. It reviewed the importance of supported decisions, as opposed to unsupported decisions, for patient safety and preventing errors of omission (i.e., forgetting to make a decision, or not realizing that there is a decision to make) and errors of commission (i.e., making a decision that is wrong).

The workshop then presented a clinical decision support (CDS) framework composed of three attributes: type, mode, and proximity.

The "**CDS Type**" identifies the manner in which health observations are linked to health knowledge:

- References summarize evidence that can help the decision-maker understand choices and consequences;
- Alerts review health events to identify acceptable data and or meaningful associations between data;
- Reminders anticipate health events, usually prompting for data or actions required to detect or avoid specific health events;
- Assists facilitate application of knowledge to specific patient circumstances for diagnosis, therapy or prediction; and
- Guides.

The "**CDS Mode**" reflects the level of dependence upon clinician decision-makers for decision initiation and follow-through:

- Passive mode—the decision-maker solicits, obtains and decides to act upon CDS;
- Active mode—a CDS tool makes a recommendation without prompting. The decision-maker then decides whether to act upon the CDSS recommendation; and
- Directive mode—a CDS intervention recommends and implements the decision, although the decision-maker may be able to overrule.

Finally, the "**CDS Proximity**" categorizes how close the CDS system comes, in time, place, and person to the point of decision-making:

- Asynchronous tools intervene before or after a decision is made;
- Synchronous tools are available and active at the point of decision making; and
- A mixed state occurs when CDS is synchronous in time but asynchronous with respect to person or place.

All CDSS can be classified using these three attributes of type, mode, and proximity (Fig. 17.1).

Following an overview of the general framework, the workshop presented each of the CDS attributes and characteristics in detail. It then reviewed the issue of integrating CDS with clinical workflow, making the distinction between routine and complex decisions: **Routine decisions** usually involve data already in the system and can therefore be automated, so that decision-support processes are largely "under-the-surface." In contrast, **complex decisions** occur when clinicians are confronted by unusual problems or outcomes; choices are unique, rarely match the sources of evidence, may involve complex values trade-offs, and are at risk for errors of commission.

Finally, the workshop presented evidence for the impact of CDSS, including positive, uncertain, and negative impacts. This set the stage for a discussion of "where to start." The presenter suggested two somewhat conflicting considerations: high impact scenarios and "quick wins."

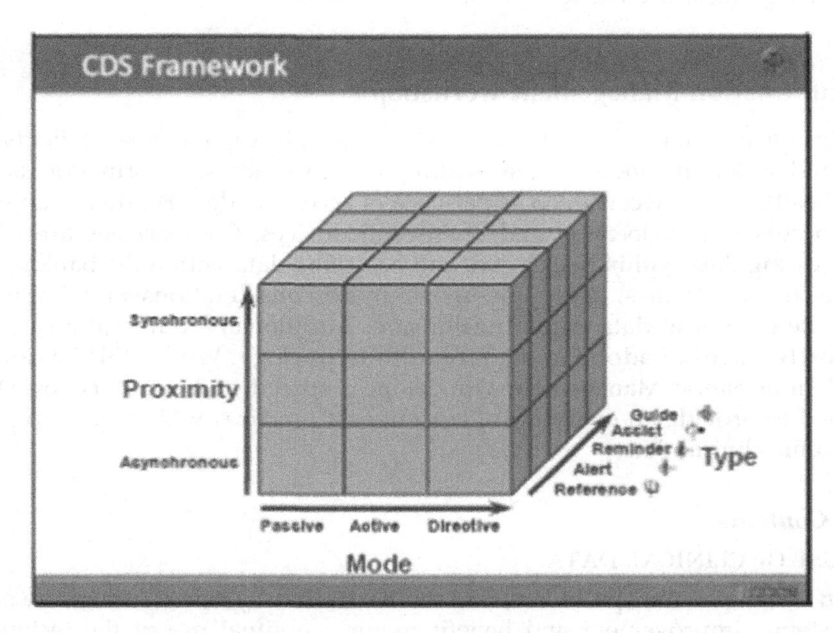

FIGURE 17.1 Clinical decision support framework. *Copyright: Dr. Robert Hayward; used with permission.*

The highest impact of CDSS is found in interventions that: (a) are solidly evidence-based; (b) use assists and guides to manage care of an aging population with chronic, multisystem illnesses; and (c) reduce practice variation through synchronous and directive interventions. However, this type of CDSS is difficult to achieve. Quick wins are more likely to be obtained from simple, credible, and actionable reminders, alerts, and references that: target errors of omission; are provider centric; are close to the point of decision-making; and balance active and passive CDSS.

Reflections and Lessons Learned

Overall, the CDM workshop can be viewed as a success. A total of 61 people participated in the live webinar and the online toolkit materials, including five vodcasts, were accessed 490 times as of March 30, 2016. Survey respondents indicated that they would like and intend to integrate CDM into their teaching, that they enjoyed the presentation, and that they found the online toolkit materials to be very useful to them. In reflecting upon the workshop and survey responses, the presenter concluded that the workshop enabled participants to take home valuable lessons, and had a positive impact on plans to introduce CDM into their teaching, incorporating critical thinking about CDS. He suggested that to be successful, future initiatives need to focus on the needs of educators trying to introduce a new subject area into a crowded curriculum and to offer practical strategies for supplementing and complementing existing curricular content. Finally, the presenter acknowledged the limitations of the online webinar format and suggested that more could be accomplished with face-to-face faculty development and mentorship sessions, although these are harder to arrange and resource.

Clinical Information Management workshop

As the health care environment becomes increasingly digital, how patients, families, clinicians, and other members of the health care team access information is evolving. Outside of health care, the consumer experience of accessing data is often seamless, immediate, and occurs in any location and on several devices. Comparisons are often made between accessing data within health care and accessing data within the banking industry. While there are similarities, there any many more considerations and implications of access and use of clinical data within health care. Additionally, our health care system is still in the early stages of adopting digital health technology. Within this broader context, the Clinical Information Management workshop, prepared and delivered by Dr. Gerard Farrell, aimed to provide an overview of current best practices with respect to physicians' use of electronic clinical data.

Workshop Contents

OPTIMAL USE OF CLINICAL DATA

A state in which patients, providers, and the health care system can measure actual clinical and systemic improvement and benefit requires optimal use of the technology and streamlined access to and use of clinical data.

First, consider the case of a 45-year-old male with a history of hypertension. He presents to his family doctor to follow up on his blood pressure management. His hypertension has been treated with two medications—an angiotensin converting enzyme inhibitor (Ramipril 10 mg daily) and a thiazide diuretic (Hydrochlorothiazide 12.5 mg daily). His blood pressure during the office visit is 160/90. He has a blood pressure machine at home and has been diligent about checking his blood pressure. His home machine saves the readings and connects to a mobile application on his smartphone that stores the readings, generates averages, and prompts him to check his blood pressure at different times. The application also has the ability to generate a report that can be sent via email, to his physician to be incorporated into his medical record. The patient has his phone with him at this appointment and wants to share the information with his physician, especially because the home readings consistently demonstrate blood pressure readings of 120–125/70–75. This information is important for the physician. It may guide treatment decisions in the management of the patient's blood pressure as well as provide a reference point for how the blood pressure measurements change over time. Further, the patient reports having made significant lifestyle changes including dietary changes and increased physical activity. He wants to try reducing the dose of his Ramipril to 5 mg daily and follow his blood pressure to see how it responds.

This common clinical scenario, with an added technological layer, raises many questions. How does this information get into the patient's record? How does the physician record it within the patient's record? How does the physician share the information with others within the health care team? What if the patient has other medical issues and is being followed by specialists; how would they see the home blood pressure readings? How does the community pharmacy know about the change in dose of the medication? What if the patient presents to a walk-in clinic or an emergency room—how will those providers obtain information about recent medication dose changes and baseline blood pressure?

Traditionally the patient would need to present most, if not all, of this information repeatedly, in each of the various clinical settings. Digital health technology should allow information sharing and transfer to be more streamlined and accessed across multiple sites and by providers who need the information to provide care for the patient.

DIGITAL HEALTH AND ELECTRONIC RECORD SYSTEMS

As defined by Canada Health Infoway, "Digital health refers to the use of information technology/electronic communication tools, services and processes to deliver health care services or to facilitate better health." [29] This definition includes all digital health technology, including electronic medical records, electronic health records, personal health records, laboratory information systems, diagnostic imaging systems, electronic prescribing, patient portals, and consumer-directed products such as mobile applications. All of these systems, which are described below, are important and relate to clinical data management, but the following section focuses on three topics: Electronic medical records (EMRs), electronic health records (EHRs), and personal health records (PHRs).

An **electronic medical record (EMR)** is "an electronic record of health-related information on an individual that can be created, gathered, managed, and consulted by authorized clinicians and staff within one health care organization" [30]. In Canada, the term usually refers to the electronic record systems used in physician offices. Appointment scheduling, documentation of clinical visits, a record of laboratory results, diagnostic imaging results, and consultation reports are all stored and accessed electronically. Billing, and communication amongst team members within a given office are also often part of the EMR. An additional key feature of most EMRs is the patient's medical history, including a problem list, a medication list, and allergies. Accuracy of the information contained within any section of the EMR relies on the quality of the information entered into the record by members of the health care team. More advanced features of an EMR include medication interaction verification prior to prescribing a new medication, and reminders for recommended screening tests. Access to the content of the EMR is generally limited to the clinicians and staff within a clinic, providing care to a patient.

An **electronic health record (EHR)**, defined by Canada Health Infoway "is a secure, integrated collection of a person's encounters with the health care system; it provides a comprehensive digital view of a patient's health history" [31]. An EHR generally contains data that is collected from several sources, and that is accessible across sites and health care organizations by health care team members providing patient care. For example, an emergency room (ER) physician might use a provincial EHR to access a patient's past laboratory or diagnostic imaging results, ordered at another hospital, when they present to the ER for assessment. Later, the patient's family physician might use the EHR to access any investigations completed in the ER. Interoperability, the ability of different systems to share information seamlessly with each other, is a key component of an EHR [32]. Ease of access to an EHR is imperative to ensure its optimal use—this could involve minimizing the requirement of multiple logins and reducing the need to access the EHR from another system, outside of the EMR.

Finally, a **personal health record** (PHR) is the information and documentation that a patient keeps, relating to their own health. The PHR is a patient-centric record that includes copies of their laboratory and diagnostic imaging results, summary of clinical visits, and any health data they collect themselves, their home blood pressure monitor or blood glucose readings. Increasingly, personal health information is collected and shared via mobile applications. Important PHR considerations include privacy and accuracy of this information, as well as how this data integrates with EMRs and/or EHRs. Finally, another key consideration is how this information supports clinical decision-making.

CLINICAL INFORMATION SHARING

As previously described, sharing of patient information is an important aspect of clinical care. Patient care rarely occurs without some degree of information sharing, including clinical data exchange, sharing information between clinicians and patients (e.g., via patient portals), and communication within the health care team. Examples include: sharing results of investigations with patients and families; communicating clinical findings or questions to specialists or other team members; and transferring the care of a patient from one clinical team to another.

Clinical data exchange is the transfer of patient information from one system to another. Interoperability is the ability of systems not only to transfer information, but also to have usable data on the receiving end [32]. Both are important concepts and elements of highly functional digital systems in health care. Local, provincial, national, and international policies related to minimum data standards and technology capabilities are key to facilitating the creation and adoption of interconnected systems.

Consider a patient with diabetes who goes for routine blood work, including testing of the Hemoglobin A1c (A1c for short) to assess their glycemic control. The patient's family physician orders the test by checking a box on a digital form in the EMR. The patient then goes to the laboratory to get blood drawn. Once the result is processed, it is sent back electronically to the ordering physician, to be reviewed within the EMR. This simple clinical scenario triggers an exchange of data, and raises questions. For example: How does the EMR know that an A1c test was ordered? How does the laboratory information system indicate that the test is a hemoglobin A1c? How does the EMR understand the data (indicating A1c) once the result returns electronically? How can another provider, working in another office with a different EMR, providing care for the patient, access the result? What if the patient moves—how would the new caregivers access this information? Additionally, how does the patient get access to the results? An understanding of data standards can help to answer these questions. Data standards relating to clinical terminology, messaging, and document exchange help to ensure the electronic systems can facilitate information sharing in a safe, secure, and accurate manner.

Beyond clinical data exchange, patient portals provide a secure interface through which patients can access their laboratory results, diagnostic imaging reports, and other aspects of their record—are becoming more prevalent. In some instances, patients have additional functions available through the portal, including the abilities to: (1) add information to their record; (2) communicate with their provider; and (3) book appointments. As this technology continues to evolve and become more widespread, patient and provider education on use of these tools and their impact on providing care will become more important.

Finally, communication within a health care team is imperative to ensuring quality patient care. When used appropriately, technology can improve communication amongst team members. They require instruction on the best use of all available tools to improve and maintain quality communication within the health care team and with the patient. Concerns about privacy and security have contributed to the slow adoption of communication technologies within health care, evidenced by the persistent use of fax machines and pagers, and varying jurisdictional limitations on the use of email. This state will continue to evolve over time and will determine how health care team members, including patients and families, interact and share information with each other. How best to incorporate this information into an interoperable medical record is also important.

TEACHING CLINICAL DATA MANAGEMENT

Clinical data management—including the ability to use digital health technology to improve patient care—is an important subject area for the medical education curriculum, and indeed for the curriculum of all health sciences students. Students need to learn how to access clinical information electronically in various settings, and understand how these electronic environments share information.

Furthermore, students should know that digital technology in the health care system is in an early stage of its evolution. They should anticipate change over time and can become advocates for changes in policy that can support optimal use of this technology. Medical students and other learners can play an important role in a health care team, by ensuring optimal quality data input into electronic systems. Students regularly contribute to a patient's medical record.

Thus, the knowledge of appropriate methods for documentation in an electronic chart is essential to maintaining the record's clinical accuracy. Appropriate and consistent documentation prepares students for a future state of enhanced interoperability. When working with electronic records and documentation, students must also learn how to integrate digital information into care delivery. Students should understand, for example, how clinical decision support tools can be included in patient care.

Reflections and Lessons Learned

The CIM workshop had 38 live participants and 28 participants who had accessed the workshop online by March 30, 2016, exceeding the project target. A total of 69% of respondents to a postworkshop survey indicated that they planned to integrate information from the workshop into their teaching activities. The CIM workshop presenter reflected that the tight project timelines presented challenges and frustrations to the faculty involved in the workshops. The presenter also reflected that true success for the workshops would be in achieving the integration of medical informatics subject matter into the medical curricula.

Access to the right information, in the right place, by the right provider, at the right time, is essential to providing safe, timely, and quality care. Enabled by optimal data exchange and interoperable systems, digital health technology can facilitate this type of care. The Institute for Healthcare Improvement, describes the "triple aim" in the pursuit of quality improvement in the health care system and in clinical care as improved population health and improved experience of care at lower cost. Incorporating optimal use of digital health technology is key to achieving these goals [33].

As digital health technology improves, it will feature a greater degree of integration and accessibility of systems across care settings. In time, these changes, along with enhanced clinical decision support tools and improved electronic communication, will make it easier to achieve and measure improvements in clinical outcomes. A solid understanding of the concepts outlined in the CIM workshop can prepare physicians to use digital tools in the most effective manner for optimal patient care.

Health Communication Management workshop

The Health Communication Management (HCM) workshop was the last of the series to be developed. In October 2016, the AFMC invited interested Canadian and international medical educators with experience in eHealth to contact them about developing the workshop. A group of five scholars (Drs. Elizabeth Borycki, Sharon Domb, Andre Kushniruk, Shmuel Reis, and Aviv Shachak) responded and were tasked with developing the workshop and supplementary materials. Members of the group included two practicing family physicians (one Canadian and one from the international medical community); and three health informatics professors (Ph.D.)—one person with a clinical informatics, organizational behavior and change management background, one person with a library and information science background, and one person with a computer science and cognitive science background. All of the organizers are involved in either medical or health informatics education and, to a varying extent, in health informatics research. The goal of the HCM workshop was to provide an overview of the various forms of communication facilitated by eHealth technologies in health care settings.

The unique circumstances and characteristics of the HCM workshop created both challenges and opportunities. The first challenge confronting the group was the tight schedule. Joining late in the process meant that the time to develop the workshop was approximately 3 months before the first live webinar was held. It also meant that the group was not involved in the initial development of the concept and had to adapt the workshop's contents and structure to the requirements specified earlier by other peer leaders. However, the group members found the requirements flexible enough that they could incorporate their own insights, knowledge, and experience and contribute original content to the workshop. Effective division of labor, with two people taking leadership and coordination roles, allowed the group to meet the deadline and deliver the workshop as scheduled.

In addition to time constraints, the involvement of five people across three time zones created some coordination challenges. It was hard to schedule virtual meetings and most communications were asynchronous, using email and the file-sharing capabilities of Dropbox. Each member of the group was responsible for developing and presenting on a different topic. Despite this division of labor, overlap in the topics presented was evident in an initial workshop pilot webinar. Also, the webinar was too long and required editing. These issues were resolved before the first live workshop with the intended audience, and the presentation of the "real" live webinars was seamless. Still, workshop organizers suggested that the group process had a number of benefits. First, as reflected in the workshop's content below, the subject matter covered by the HCM group was diverse and required expertise in a number of areas. Working as a group supported this diversity as each of the organizers contributed different topical content and perspectives. The group structure also facilitated rigorous internal review of materials, and the provision of constructive feedback on the contents and presentation. Finally, the organizers suggested that having a number of different voices on the webinars could generate more interest and be more engaging for participants, rather than listening to the same presenter for two hours.

HCM Workshop Contents

The HCM workshop was organized around three different paths of communication, and their implications for the physician medical practice: (1) patient-clinician communication; (2) communication within the health care team; and (3) communication between patients. The workshops' objectives included: Familiarizing the learners with various available communication management tools; helping learners anticipate the opportunities and challenges brought forward by various types of communication tools; identifying opportunities for complementing existing curricula with case-based activities that illustrate the clinical importance of communication management; and providing a range of teaching tips, examples, resources and content in the eHealth Faculty Development Toolkit that facilitated the curriculum development process to address communication management competencies.

In the area of patient-clinician communication, the workshop addressed the challenges and opportunities associated with the use of computers—particularly electronic medical records (EMRs) —in the clinical encounter. Additionally, the workshop presented the factors that affect patient-clinician communication in computerized settings, as well as some tips and best practices to maximize the benefits and minimize the negative impacts. This part of the HCM workshop was built on a large body of knowledge and curricular approaches that are discussed extensively in Chapters 1, 2, 10, and 11 of this book.

Next, the workshop presented a number of technologies that enable and facilitate communication, both within the health care team and with patients. These included text messaging, videoconferencing, social networks, patient portals, personal health records (PHRs), mobile apps, and more. The presentation highlighted both the opportunities associated with the use of these technologies (such as engaging patients in their own care, patient empowerment, e-consults, and e-referrals) and some of the inherent challenges (such as quality and credibility of information, privacy and security, usability, and more).

Following a general discussion of these technologies, the workshop focused more specifically on aspects of professionalism related to the use of email, text messaging, web portals and social media (see Chapters 4–6 of this book). Presenters highlighted the differences between personal and professional use of these technologies; illustrated the challenges, using some high profile examples from the media; presented on some of the national-level guidelines; and discussed educational implications.

The last section of the workshop discussed patients' use of general social media (Facebook, Twitter, etc.), patient-specific social media, and mobile technology to share information with each other. The workshop highlighted the opportunities created by use of these tools, such as patients receiving informational, social, and emotional support. The presenters also noted the opportunity for greater patient engagement through social media. In addition, a discussion occurred on the challenges with patient social media use, such as those associated with maintaining the credibility of information, the promotion of unhealthy behaviors (e.g., antivaccination or proanorexia [34]), self-diagnosis, and new phenomena such as cyberchondria [35,36] and Munchausen by Internet [37].

Reflections and Lessons Learned

Overall, the HCM workshop was considered a success. Despite the short time for preparation and the aforementioned coordination challenges, the HCM group achieved the following: developed and delivered a webinar three times; created a toolkit comprised of two vodcasts, an annotated bibliography, a list of resources and some educational materials, all accessible online through the Canadian Healthcare Education Commons (CHEC) [10]. A total of 48 people participated in the live webinars, and online materials were accessed 34 times as of March 30, 2016. The postworkshop survey highlighted its positive impact, with many respondents indicating: (a) that they would incorporate a discussion on HCM in their teachings; (b) that they would further investigate the use of patient consent forms for the purposes of doctor-patient communication through e-mail; and (c) that they were more aware of, and will consciously consider, the benefits and dangers associated with using electronic communications with their patients. Finally, organizers felt that the benefits of developing and delivering the workshop as a group (i.e., multiple perspectives, diverse expertise, and different voices on the webinar) outweighed the challenges.

However, in reflecting on the workshop, organizers noted a number of challenges. First, the webinar format did not always facilitate discussion among participants. While in one of the webinars, participants actively engaged in discussion, asked questions, and provided examples and views on the workshop's topics, in other webinars they were silent most of the time. Without the ability to see the participants, the webinar format felt somewhat impersonal and lacking immediate feedback on participants' comprehension of, and reaction to, the content delivered. A second challenge was that the workshop had been prepared with physician educators as the target audience in mind. As it turned out, many of the participants were nonclinicians. This created some mismatch, particularly with some polling questions included in the webinar that were not applicable to this audience. Finally, there was very little feedback on the workshop and toolkit beyond the organizing group's process and the pilot webinar with the larger group of peer leaders. Such feedback would help improve the workshop and toolkit material.

OUTCOMES AND RESULTS

The following section draws upon the findings presented in the AFMC eHealth Project Closure Report, created by Barbie Shore and Kate Proctor. The outcomes of this eHealth Medical Faculty Development initiative, both qualitative and quantitative in nature, fall into three categories: initial (or startup) activities, faculty development live webinar workshops, and advocacy activities.

For initial activities, a leadership team congregated in person at a national eHealth Symposium on April 26, 2015, hosted at the Canadian Conference on Medical Education (CCME). The attendees achieved the goal of recruiting Faculty Peer Leaders to the project, and agreed on an approach to create and disseminate workshop modules of core eHealth content. These modules would be the foundation of the eHealth teaching materials and resources that were envisioned as tools to help to "train the trainers." This eHealth Symposium was also critical to sustaining the national momentum of the first project, which had been waning since 2014.

The primary mode of disseminating the content of the eHealth Faculty Development initiative was the series of four workshops described above. These workshops were delivered multiple times, each as online webinars. The webinars were accredited for Continuing Medical Education (CME) or Continuing Professional Development (CPD) credits, which physicians and other clinicians need to maintain their professional licenses. Accreditation was arranged with both the College of Family Physicians of Canada (CFPC) and the Royal College of Physicians and Surgeons of Canada (RCPSC). Achieving accreditation was an involved process that required significant effort from key Peer Leaders, the Schulich School of Medicine & Dentistry at Western University's Continuing Professional Development Office, the AFMC project team, Infoway, and significantly, both the CFPC and the RCPSC. However, accreditation made the webinar workshops more attractive for potential attendees.

In total, Peer Leaders presented 16 webinar workshops, attended by over 200 live participants and recorded workshop materials (podcast or vodcast) were viewed 629 times by March 30, 2016. These results exceeded the project target and overall expectation of a minimum of 240 participants.

Finally, advocacy was considered a crucial component of enabling change via continuous engagement of key stakeholders (who have the power to influence the medical curricula). The eHealth Faculty Development initiative included 95 advocacy events, impacting more than 1700 stakeholders. Examples of the Peer Leaders' advocacy activities at medical schools included: meetings with Deans of the Faculties of Medicine; letters to Deans and key faculty; published articles and successful conference abstract submissions; meetings with and presentations to key faculty members. As the next section of this chapter emphasizes, advocacy activities will be critical if awareness of eHealth education is to be maintained within Canada's faculties of medicine.

In addition to the aforementioned outcomes, the initiative achieved a number of other successes, as noted by Shore and Proctor:

- A key strategic relationship was established with the Medical Council of Canada (MCC), a stakeholder with the potential to be a key change agent through their influence on medical licensure examinations.
- The webinar workshops had a wide reach across medical faculties, with individuals from 15 of 17 faculties in attendance.
- The pre- and postworkshop surveys indicated a trend toward improved awareness of eHealth.
- The initiative established new connections amongst like-minded individuals, many of whom are continuing this work under umbrella of AFMC's eHealth sub-Group.

Perhaps most importantly, the initiative created an online eHealth Workshop Toolkit Collection, a tangible set of resources that can be used by interested medical educators to teach eHealth concepts in their respective faculties of medicine [10].

NEXT STEPS AND FUTURE DIRECTION

Rapid changes in information and communication technologies are having a marked impact on personal, shared, and clinical information management and decision-making. Thus, it is important to acknowledge that eHealth is immersed in a constantly evolving environment. Future directions of eHealth need to take into consideration a number of factors, such as: the capabilities and expectations of the new generation admitted into medical school; easy access to information by learners, clinicians and patients; extensive use of mobile devices and their potential for monitoring and communication; mutual support of colleagues practicing clinically either asynchronously or in real time; the leverage of these technological advantages to support recruitment and retention of recently graduated physicians to practice in nonurban communities. It is therefore vital for eHealth to be given high priority in the education of health professionals. The progress made in Canada in relation to eHealth education has been described in the previous sections. The next steps that need to be taken are in the realm of advocacy, curriculum development, assessment, and training.

While increasing awareness of progress in eHealth in Canada, there needs to be continued and impactful advocacy for eHealth in medical education with the leaders of the medical education continuum (undergraduate, postgraduate, and continuing professional development) in Canadian Medical Schools. Two activities need to occur simultaneously: (1) endorsement of eHealth at the governance levels, from the Deans and Associate Deans of the respective areas of medical education; and (2) coordinated involvement of curriculum planners in these disciplines to identify paths to embed eHealth content into training.

For curriculum development, there is an urgent need for more resourcing to create a national curriculum package or blueprint for eHealth integration (such as cases and simulated records for training), and to help each school adapt to their needs. Also, it is important to involve curriculum development champions to create and modify future curriculum content collaboratively. Support should be provided for a small number of faculty leads to produce the national curriculum package, and help each school adapt to their needs. Regional and local educational materials relevant to the learners should also be fostered and implemented. For example, while the principles of electronic health records (EHRs) and electronic medical records (EMRs) can be discussed in a similar fashion across the country, the choice of local electronic solutions determines trainees' specific training requirements.

Second, assessment, both within medical schools and through national licensing examinations, is an important driver in medicine that identifies competencies required by physicians. Curriculum evaluation will help to sustain a movement toward eHealth integration into medical education, not only by measuring the degree of eHealth education uptake, but also by assessing the impact of such uptake on medical trainees during and after their education, as they transition into practice.

Similar to other skill sets, medical trainees should be assessed regarding their use of eHealth in the domains of attitude, knowledge, skill, and competence. It is critical for the Medical Council of Canada (MCC) to initiate such assessments. Currently, the MCC has created an examination blueprint for eHealth, and the MCC's Population Health, Ethics, Legal, and Organizational Aspects of Medicine (PHELO) test committee [38] is developing eHealth-focused examination questions in collaboration with representatives from the AFMC-Infoway committee. There is now a need for using other forms of assessments such as key feature questions, Objective Structure Clinical Examinations (OSCEs) [39], and ePortfolios to test competence and sustain reflective continuing usage in eHealth.

In relation to training, there is a need for in-person workshops and/or better use of videoconferencing and other forms of technology to facilitate participation. Workshops and resources must be better aligned to learner levels (i.e., beginner, intermediate, and advanced), incorporating both national and local content. In addition, there is a need for French language workshops and materials. Clinician engagement and change management efforts could be enhanced through Continuing Professional Development (CPD) training and the use of appropriate accrediting systems, to incentivize and reward participation by learners of all stripes (from medical students to practicing physicians).

Given the rapidly evolving technological environment, Canadian medical education could benefit from a continuous environmental scan of eHealth integration in curricula and sharing of best practices nationwide. Also, educational scholarship and knowledge translation in this area is needed. As new technologies are adopted by physicians and other health care professionals, eHealth-related standards of practice must change, and new policies must be introduced to minimize risks relating to patient care. Furthermore, policy makers and senior health administrators can add further momentum toward future eHealth adoption by working with educational institutions to ensure continuous and seamless deployment of eHealth technologies in practice, while also supporting appropriate training to help learners to transition into practice. Finally, involvement of patients and their family members to provide their perspectives would further help medical learners to understand how best to partner with patients through eHealth technologies, to support health and wellness.

SUMMARY

As this chapter has outlined, significant efforts have gone into influencing medical education on a national scale in Canada, with regard to the instruction of eHealth or health informatics concepts. Partner organizations the AFMC and Infoway brought together eHealth experts (also functioning as "Peer Leaders") from across the country, who dedicated their time to creating eHealth educational resources for medical educators.

The eHealth Medical Faculty Development project produced a robust, openly accessible, collection of eHealth educational resources. These resources were disseminated first in the form of accredited, live webinar workshops, and then in the form of an online eHealth Workshop Toolkit Collection. The resources served multiple purposes: (1) to educate medical educator peers, or "train the trainers"; and (2) to provide medical educators with tangible tools that they can use to instruct physician trainees (medical students and resident physicians) about eHealth concepts. Two prominent themes within eHealth were interwoven throughout the workshop content: (1) consumer or patient-focused health; and (2) privacy and security. The former is particularly important as physicians are practicing medicine in time in which the patient is playing a greater role in their own care via patient e-services. Patients should play a role in developing future eHealth educational content.

These senior academic eHealth experts, Peer Leaders, were involved in considerable advocacy efforts around the need to integrate eHealth concepts into the medical education curriculum. Overall, the eHealth Medical Faculty Development initiative and its Peer Network structure achieved multiple successes, as outlined in the chapter.

However, the leadership committee acknowledge that these efforts remain unfinished. An extension to this initiative is currently enhancing the online eHealth Workshop Toolkit Collection, adding clinical cases, and classifying materials that are relevant to various learner levels (beginner, intermediate, and advanced). Eventually, the resources should be "plug and play," tools and components that can be customized for and dropped into a course or curriculum. Most importantly, four areas require attention as we move forward: advocacy, curriculum development, assessment, and training. The ongoing engagement of key stakeholders will be essential going forward (such as the MCC).

Long-term sustainability of these efforts will continue to be a challenge, and will require a consistent source of funding. Interprofessional collaboration with a broader coalition of interested groups—such as the Association of Faculties of Pharmacy of Canada (AFPC), the Canadian Association of Schools of Nursing (CASN), and Allied Health groups—could strengthen future efforts.

Acknowledgments

We would like to thank Barbie Shore, Kate Proctor, Canada Health Infoway's Clinical and Change Leadership Group (Susan Sepa, Lynne Zucker), and all the people who contributed to development and delivery of this project.

References

[1] Densen P. Challenges and opportunities facing medical education. Trans Am Clin Climatol Assoc 2011;122:48–58.

[2] Environmental scan of eHealth in Canadian undergraduate medical curriculum. Available from: <https://chec-cesc.afmc.ca/en/afmc-infoway-physician-training-ehealth-curriculum-elearning/resource/environmental-scan-ehealth-canadian-undergraduate-medical-curriculum-analyse-contextuelle-de-la-cybersante-dans-le-programme-denseignement-medical-predoctoral-canadien> [accessed 28.06.17].

[3] <https://afmc.ca/about-afmc> [accessed 28.06.17].

[4] <https://www.infoway-inforoute.ca/en/about-us> [accessed 28.06.17].

[5] Baker C, Charlebois M, Lopatka H, Moineau G, Zelmer J. Influencing change: preparing the next generation of clinicians to practice in the digital age. Healthcare Quarter. 2016;18:4.

[6] Bhyat R. Physicians-in-training initiative helps future docs become comfortable in a digital health environment. ow.ly/T7k12.

[7] eHealth competencies for undergraduate medical education. Available from: <https://chec-cesc.afmc.ca/en/afmc-infoway-physician-training-ehealth-curriculum-elearning/resource/ehealth-competencies-undergraduate-medical-education-hcm> [accessed 28.06.17].

[8] CanMEDS. Physician competency framework. Available from: <http://canmeds.royalcollege.ca/uploads/en/framework/CanMEDS%202015%20Framework_EN_Reduced.pdf> [accessed 28.06.17].

[9] Zelmer J, Hagens S. Patients online: whither from here? HealthcarePapers 2014;13(4):61–5. Available from: <http://www.longwoods.com/content/23863>.

[10] Association of faculties of medicine of Canada. eHealth workshop toolkit collection. Available from: <https://chec-cesc.afmc.ca/en/collections/afmc-infoway-physician-training-ehealth-curriculum-elearning/ehealth-workshop-toolkit-collection> [accessed 15.12.16].

[11] Smith R. What clinical information do doctors need? BMJ 1996;313(7064):1062–8.

[12] Report of the auditor general of Canada to the house of commons. Chapter 4 – Electronic health records. Available from: <http://www.oag-bvg.gc.ca/internet/docs/parl_oag_200911_04_e.pdf> [accessed 28.06.17].

[13] 2014 National physician survey: backgrounder (December 2014). Available from: <http://nationalphysiciansurvey.ca/wp-content/uploads/2014/12/NPS-backgrounder-2014-EN-r.pdf>.

[14] College of family physicians of Canada. A vision for Canada: family practice – the patient's medical home. Available from: <http://www.cfpc.ca/A_Vision_for_Canada/> [accessed 28.06.17].

[15] Delaney B. Envisioning a learning health care system: the electronic primary Care research network, a case study. Ann Fam Med. 2012;10(1)):54–9. Available from: <http://www.annfammed.org/content/10/1/54.long>.

[16] RCPSC Mainport ePortfolio. Available from: <http://www.royalcollege.ca/rcsite/cpd/moc-program/about-mainport-eportfolio-e> [accessed 28.06.17].

[17] Abrams KJ. Ch 1: Introduction in: fundamentals of health information management, 2nd ed. Eds, K.J. Abrams & C.J. Gibson, Ottawa, ON: Canadian Healthcare Association; p. 1–2.

[18] Information literacy competency standards for higher education. (2000). [Brochure]. Chicago: Association of College & Research Libraries. Association of College and Research Libraries, Information Literacy Standards for Science and Engineering/Technology, by The ALA/ACRL/STS Task Force on Information Literacy for Science and Technology. Available from: <http://www.ala.org/acrl/standards/infolitscitech>.

[19] Seago BL, Schlesinger JB, Hampton CL. Using a decade of data on medical student computer literacy for strategic planning. J Med Libr Assoc 2002;90:202–9.

[20] The Ontario Telemedicine Network. Available from: <https://otn.ca> [accessed 28.06.17].

[21] Developing a Pathway for an Institution Wide ePortfolio Program, International Journal of ePortfolio, 2015;5:75–92.

[22] Tochel C, Haig A, Hesketh A, Cadzow A, Beggs K, Colthart I, et al. The effectiveness of portfolios for post-graduate assessment and education: BEME Guide No 12. Med Teach 2009;31(4):299–318.
[23] Wenger E. Communities of practice: learning, meaning, and identity. Cambridge: Cambridge University; 1998.
[24] Endsley S, Kirkegaard M, Linares A. Working together: communities of practice in family medicine. Fam Pract Manag 2005;12:28–32.
[25] Model Code for the Protection of Personal Information (CAN/CSA-Q830-96). Available from: <http://cmcweb.ca/eic/site/cmc-cmc.nsf/eng/fe00076.html> [accessed 28.06.17].
[26] Hersh WR, Gorman PN, Biagioli FE, Mohan V, Gold JA, Mejicano GC. Beyond information retrieval and electronic health record use: competencies in clinical informatics for medical education. Adv Med Educ Pract 2014;5:205–12.
[27] Graham-Jones P, Jain SH, Friedman CP, Marcotte L, Blumenthal D. The need to incorporate health information technology into physician's education and professional development. Health Affairs 2012;31:481–7.
[28] Hayward R. Clinical decision management: vodcast & study guides – (CDM). Available from: <https://chec-cesc.afmc.ca/en/afmc-infoway-physician-training-ehealth-curriculum-elearning/resource/clinical-decision-management-vodcast-study-guides-cdm> [accessed 28.06.17].
[29] Digital health definition. Available from: <https://www.infoway-inforoute.ca/en/what-we-do/digital-health-and-you/what-is-digital-health> [accessed 28.06.17].
[30] The National Alliance for Health Information Technology Report to the Office of the National Coordinator for Health Information Technology on Defining Key Health Information Technology Terms. April 28, 2008. Available from: <http://www.hitechanswers.net/wp-content/uploads/2013/05/NAHIT-Definitions2008.pdf>.
[31] Electronic Health Record definition. Available from: <https://www.infoway-inforoute.ca/en/solutions/electronic-health-records> [accessed 28.06.17].
[32] Interoperability definition. Available from: <https://www.healthit.gov/buzz-blog/meaningful-use/interoperability-health-information-exchange-setting-record-straight/> [accessed 28.06.17].
[33] Institute for Healthcare Improvement, Triple Aim Initiative. Available from: <http://www.ihi.org/engage/initiatives/tripleaim/pages/default.aspx> [accessed 28.06.17].
[34] Gholami-Kordkheili F, Wild V, Strech D. The impact of social media on medical professionalism: a systematic qualitative review of challenges and opportunities. J Med Internet Res 2013;15(8):e184. Epub 2013/08/30.
[35] Starcevic V, Berle D. Cyberchondria: towards a better understanding of excessive health-related Internet use. Expert Rev Neurother 2013;13(2):205–13. Epub 2013/02/02.
[36] White RW, Horvitz E. Cyberchondria: studies of the escalation of medical concerns in Web search. ACM Trans Inf Syst 2009;27(4):1–37.
[37] Pulman A, Taylor J. Munchausen by internet: current research and future directions. J Med Internet Res 2012;14(4):e115. Epub 2012/08/24.
[38] PHELO definition: Population Health, Ethics, Legal, and Organizational Aspects of Medicine. Medical Council of Canada, PHELO Test Committee. Available from: <http://mcc.ca/about/governance/2016-test-committee-membership/#MCCEE> [accessed 28.06.17].
[39] OSCE definition: objective structured clinical examination.

CHAPTER

18

Development and Evaluation of a Statewide HIV-HCV-STD Online Clinical Education Program for Primary Care Providers

Dongwen Wang

Arizona State University, Scottsdale, AZ, United States

INTRODUCTION

Human immunodeficiency virus (HIV), hepatitis C (HCV), and other sexually transmitted diseases (STDs) have raised serious challenges to public health for decades. In the United States, the annual number of people newly infected with HIV, HCV, and STD are estimated at 44,073, 30,500, and 20 million, respectively [1–3]. The total infections are estimated at 1.2 million, 2.7 million, and 110 million [1–3]. Medical costs associated with diagnosis, treatment, and prevention are estimated at $16 billion a year [3]. With the many ongoing clinical trials on medications, vaccines, and behavioral interventions and the frequent updates of clinical practice guidelines (CPGs) based on the findings from medical research, effective dissemination of the latest clinical evidence to community healthcare providers, who work on the frontline fighting these diseases, has become an essential requirement. Since the 1990s, the New York State Clinical Education Initiative (CEI) has been engaging in dissemination of clinical evidence to primary care clinicians providing care to HIV, HCV, and STD patients through its in-person training program [4]. The CEI program aims to increase access to quality healthcare, to expand the base of clinicians who can effectively manage HIV, HCV, and STD patients, to disseminate the latest CPGs, and to foster partnerships between community-based care providers and HIV, HCV, and STD specialists [4].

Health Professionals' Education.
DOI: http://dx.doi.org/10.1016/B978-0-12-805362-1.00018-8

Comparing the traditional classroom or clinic-based approaches to providing continuing professional education, online training is advocated as an efficient platform for rapid dissemination of knowledge to healthcare providers [5–6]. Built on a history of success for two decades to provide in-person training, CEI launched its online education program in 2008. Since then, CEI has developed 300 + multimedia learning modules, 100 + online continuing medical education (CME) and continuing nursing education (CNE) courses, 14 interactive case simulation tools, and various other online resources [7]. These resources have been disseminated to tens of thousands healthcare providers from 170 + countries through web, mobile apps, email newsletters, and online social networks [8]. In this chapter, we describe the development of CEI online resources, their usage by clinicians, and initial assessments on effectiveness and impact of the CEI online education program. The success of CEI and the lessons learned in its development will provide important guidance to future development of online education programs for more effective and timely dissemination of clinical evidence to better serve clinicians'information needs.

BACKGROUND AND RELATED WORK

Online Resource Indexing, Classification, and Linkage through Ontology

When developing large sets of online resources for patient care, medical research, and clinical education, we typically use ontologies to index, classify, and link these resources. For example, the Systematized Nomenclature of Medicine (SNOMED) [9] and International Classification of Diseases (ICD) [10] are used as standard codes for clinical data and medical billings; Medical Subject Headings (MeSH) [11] is used to index medical literatures; and Gene Ontology (GO) [12] is used as a controlled vocabulary for genes and their annotations. For health professional education, MedBiquitous [13] is a standard that covers a wide range of resources, including activity reporting, competencies, curriculum inventory, educational achievement, educational trajectory, healthcare learning object metadata, medical education metrics, performance framework, professional profile, and virtual patients. In addition to these general-purpose large-scale ontologies, specific communities can develop domain ontologies to serve for particular purposes. Such ontologies need to provide sufficient coverage of the domain at the appropriate granularity level in order to be used as indexes [14]. In development of the CEI online training program, we adapted certain concepts from MedBiquitous to represent the structure of multimedia learning modules, and synthesized a domain ontology to index the HIV, HCV, and STD education resources. We report the technical details in "Multimedia Learning Modules" section.

GuideLine Interchange Format and GuideLine Execution Engine

GuideLine Interchange Format (GLIF) is a computer-based representation model of CPGs that can be customized to individual patient cases for personalized recommendations [15]. GLIF provides support for workflow management, clinical decisions, medical tasks, and patient status identification. An associated tool, GuideLine Execution Engine (GLEE), was developed to interpret CPGs encoded in the GLIF format and to apply to individual patient cases [16]. The GLIF/GLEE framework [17] has been used for a wide range of applications, such as childhood immunization [18], cough management [19], influenza vaccination [20], hyperkalemia screening [21], diabetic foot-care [22], major depression screening [23], tobacco cessation [24], and HIV/AIDS patient management [25]. In development of the CEI online resources, we built the interactive case simulation tools through enhancement of the GLIF/GLEE framework. We report the technical details in "Interactive Case Simulation Tools" section.

Online Resource Usage Tracking

Tracking the usage of online resources is an essential requirement to assess their effective dissemination. For the general purpose of usage tracking, Google Analytics [26] is perhaps the most popular tool that can report metrics such as users, usage sessions, and pageviews to serve for specific business goals. Nevertheless, it does not support direct integration with customized usage data. In order to obtain specific user and usage data for medical research and patient care, others developed specific tracking tools [27,28]. Yet even these tools were insufficient in providing detailed data for usability analyses. To assess the usage of CEI online education resources, we therefore had to develop a customized tracking tool to provide detailed data on user behaviors, to characterize the use contexts, and to integrate with the CEI student portal. We report the technical details in "Usage of CEI Online Education Resources" section.

Evaluation of Online Clinical Education Programs

In a previous review by Macznik et al. on online clinical education programs for health professionals in physiotherapy [29], students perceived that the online platform can provide easy access to educational resources and be effective for knowledge acquisition. Yet there were few data on clinicians' intention to use the learned knowledge, change of their clinical practice, and comparison with in-person training. In another review by Pulsford et al. on education programs in end-of-life care for health and social care staff [30], both in-person and online programs were considered useful for enhancing healthcare professionals' skills and perceived preparedness. Yet there was no direct comparison between the in-person and online educations programs. In a third review by Cohen et al. on online nutrition education program [31], there were several quasi-experimental studies to compare online and in-person training that indicated no differences in nutrition knowledge and performance between the two modes. Yet, the results on student satisfaction, motivation, and perceptions were inconclusive. In our evaluation of the CEI online education program, we included metrics on clinicians' intention to use the learned knowledge and to change clinical practice. We also performed a study to directly compare the in-person and online training. We report the technical details in "Evaluation of CEI Online Education Program" section.

DEVELOPMENT OF CEI ONLINE EDUCATION RESOURCES

Multimedia Learning Modules

An important category of CEI's online education resources is the multimedia learning modules [7], which are recorded presentations by HIV, HCV, and STD clinical experts. To produce multimedia learning modules for the CEI online program, we leveraged the resources in the CEI network. Specifically, the CEI in-person training network consists of multiple centers that assemble faculty members with a wide range of clinical expertise in HIV, HCV, and STD. These domain experts develop curriculum for specific training topics and deliver in-person training through lectures, workshops, and apprenticeships. For multimedia learning module production, we recorded the presentations made by domain experts, transformed them into various media formats for delivery through multiple platforms (e.g., web, iOS app, and Android app), and assembled the background materials associated with these learning modules. On the back-end of the system infrastructure, the multimedia learning modules and associated background materials are stored in a Microsoft SQL Server database. On the front-end, the CEI online learning program supports multiple delivery platforms, including: (1) a main website for desktops and tablets; (2) a mobile website for smartphones and other small-screen hand-held devices; and (3) a variety of Android and iOS apps. Depending on a user's access platform, the CEI media server automatically delivers a multimedia learning module in the appropriate format.

To organize the multimedia learning modules for effective resource management and dissemination, we adapted concepts from the MedBiquitous standard for health professional training [13], and defined a simplified structure to represent the important information of a multimedia learning module. Data elements in this structure include: (1) title of a learning module; (2) learning objectives; (3) presenter information, such as name, degree, affiliation, and short biosketches; (4) sponsor organization; (5) associated credits for continuing professional development, such as CME and CNE; (6) dates of original presentation, online posting, and CME/CNE expiration; and (7) the related CPGs. With a significant accumulation of the multimedia learning modules in the CEI program, their categorization under specific topics is a natural solution to facilitate search, browse, and resource management. Given the rapid development in HIV, HCV, and STD research and our primary focus on clinical education, the existing ontologies or controlled medical terminologies cannot cover the wide range of topics and represent sufficient details of the CEI training resources. We therefore decided to develop our own topic list. For this purpose, we formulated a two-level ontology of HIV, HCV, and STD clinical education topics, with 8 first-level categories and a total of 45 classifications, as shown in Fig. 18.1.

Based on this ontology, we assigned each multimedia learning module under a topic that best matches its content. With these annotations, we developed an online tool for browsing and search of specific categories of multimedia learning modules. In addition, we built a system function to identify the multimedia learning modules that are *related* to a specific module a clinician user is currently viewing. In this initial development, the related multimedia learning modules are defined as all those belonging to the same topic of the module that is currently under review. With this system function, we can proactively disseminate the related multimedia learning modules under a specific topic once a

Sexually Transmitted Diseases (STDs)
Hepatitis
- Hepatitis A
- Hepatitis B
- Hepatitis C
- HIV/Hepatitis Co-infection

HIV - Antiretroviral Therapy
- Treatment Strategies
- Resistance Testing
- Adherence to Treatment

HIV - Prophylaxis
- Post Exposure Prophylaxis (PEP)
- Pre-exposure Prophylaxis (PrEP)

HIV - General Medical Management
- Acute HIV Infection
- Physical and Laboratory Assessment
- Health Maintenance (Immunization/ Vaccination/ Nutrition/ Cancer Screening)
- HIV Testing
- Harm Reduction: Sexual Practices
- Harm Reduction: Substance Use
- Harm Reduction: Partner Notification
- Dermatological Manifestations
- Oral Manifestations
- GI Manifestations
- OB/GYN Care
- Renal Manifestations
- Neurological Manifestations
- Pediatric/Adolescent Management
- HIV and Aging

HIV - Complications
- Opportunistic Infections
- Dyslipidemia
- Diabetes
- Obesity
- Bone Disease
- Neoplastic Complications

HIV - Mental Health
- Behavioral Management
- Neuropsychiatric Aspects of HIV Infection
- Triply Diagnosed: Mental Illness, Substance Use and HIV

HIV - Legal Issues/Policies
- HIV Confidentiality
- Domestic Violence
- HIV Reporting
- Competency/Consent

FIGURE 18.1 Ontology of HIV, HCV, and STD clinical education topics.

clinician has expressed an interest in a particular learning module in that category. Fig. 18.2 shows the screenshots of: (1) a list of multimedia learning modules; (2) the details of a specific learning module page; and (3) the system functions to browse the learning modules under a particular topic and to check the related learning modules.

Interactive Case Simulation Tools

Another category of CEI online resources is the interactive case simulation tools, which are used for decision assistance and case simulation on individual patients [32]. These tools are developed as a vehicle to disseminate clinical evidence. On the back-end, they are driven by a knowledge base, typically translated from sources such as the latest CPGs. On the front-end, they support user interactions such as reviewing the process for patient

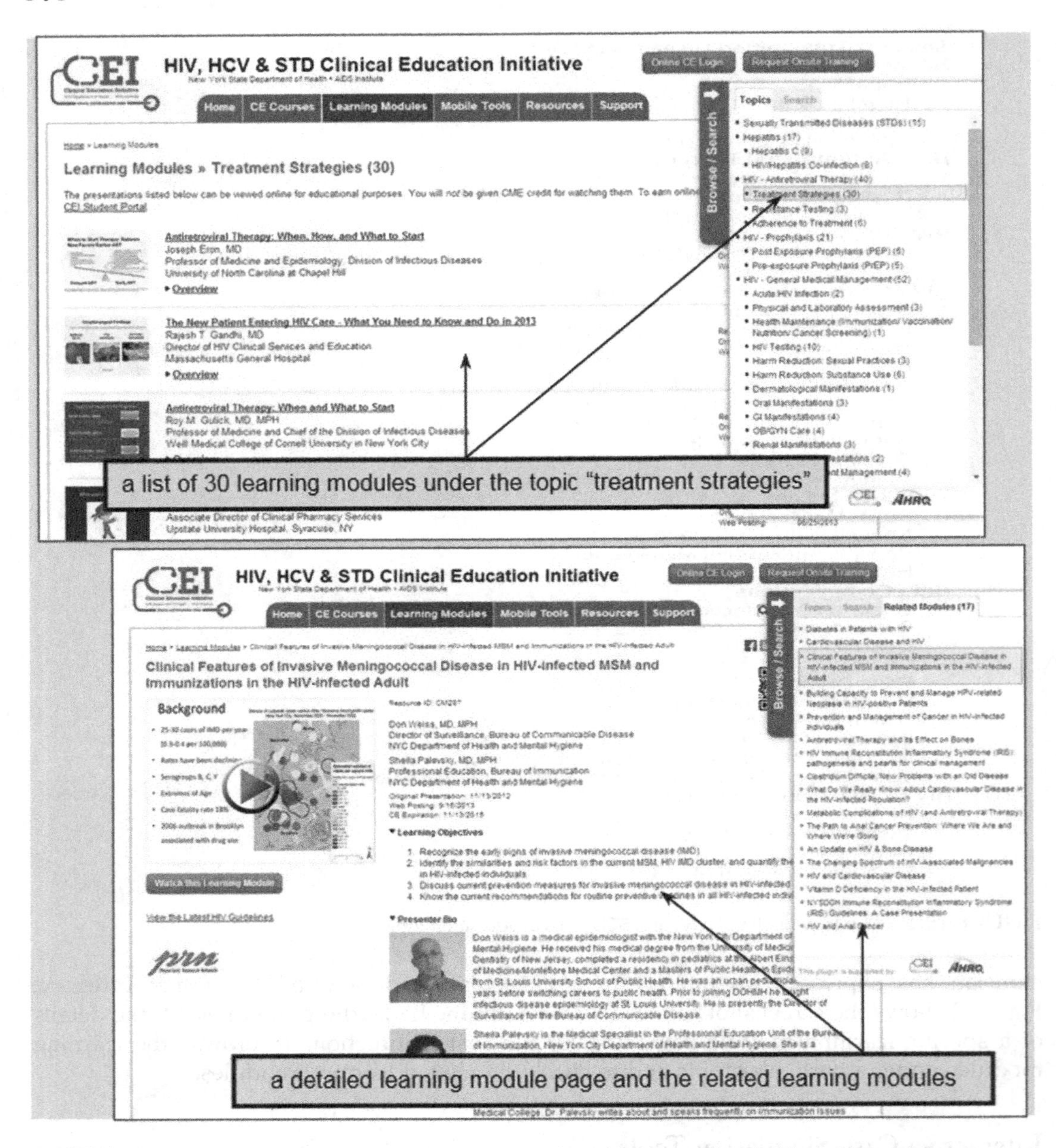

FIGURE 18.2 Screenshots of a list of multimedia learning modules, a detailed learning module page, and system functions to browse learning modules under a particular topic and to check the related learning modules.

management, examining the different options for clinical decisions, and entering case-specific data for individualized recommendations.

The system architecture of interactive case simulation tools is based on enhancements of the GLIF/GLEE framework [15,16]. It consists of five components: (1) a GLEE server

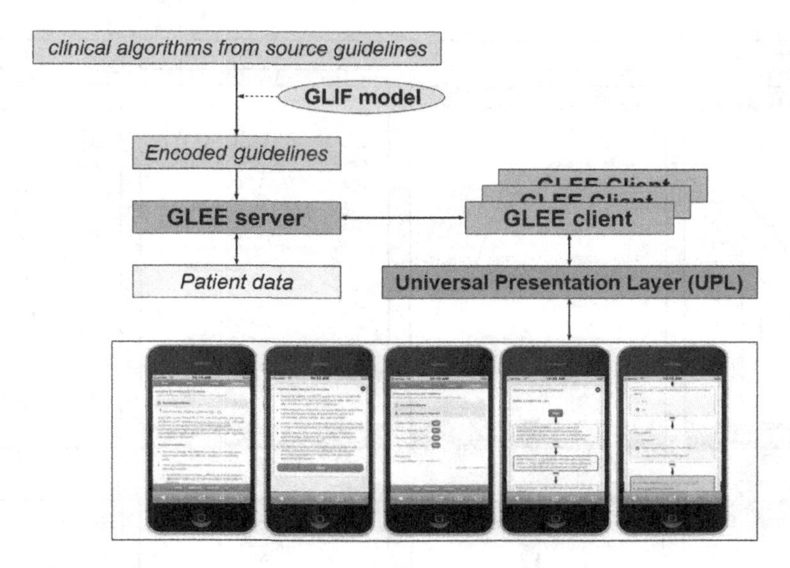

FIGURE 18.3 System architecture of interactive case simulation tools.

hosting computer-interpretable CPGs encoded in the GLIF format; (2) system interfaces to integrate the GLEE server with patient data and execution traces; (3) a set of GLEE clients, each corresponding to an application of a selected CPG to a specific patient case; (4) a Universal Presentation Layer (UPL) to integrate a GLEE client with the user interface of an application; and (5) user interface implementation and integration [25]. The overall system architecture of interactive case simulation tool is shown in Fig. 18.3.

For a specific interactive case simulation tool, we developed two sections of user functions. The first section is *"recommendations."* This section is essentially a set of hyperlinked text pages transformed from the original CPG, highlighting the most important information for patient management. In this section, a clinician user can review the general recommendations, with options to follow the related links to check out the detailed information for specific clinical problems. The second section is *"interactive decision diagram."* In this section, a clinician user can review the process of decision making on predefined sample cases; he/she can also explore his/her own patient cases through interactions with the system. Specifically, a user can check the decision process step by step, review the different decision options, and enter data for a particular patient case to examine the customized recommendations. At any point, a user can go to the CEI website to obtain additional background information. Selected screenshots and user functions from an interactive case simulation tool for insomnia screening and treatment in HIV patients are shown in Fig. 18.4.

Cross-Linkages of CEI Online Resources

The multiple types of online education resources serve for different purposes. Yet there are close underlying connections among these resources. For example, a healthcare provider

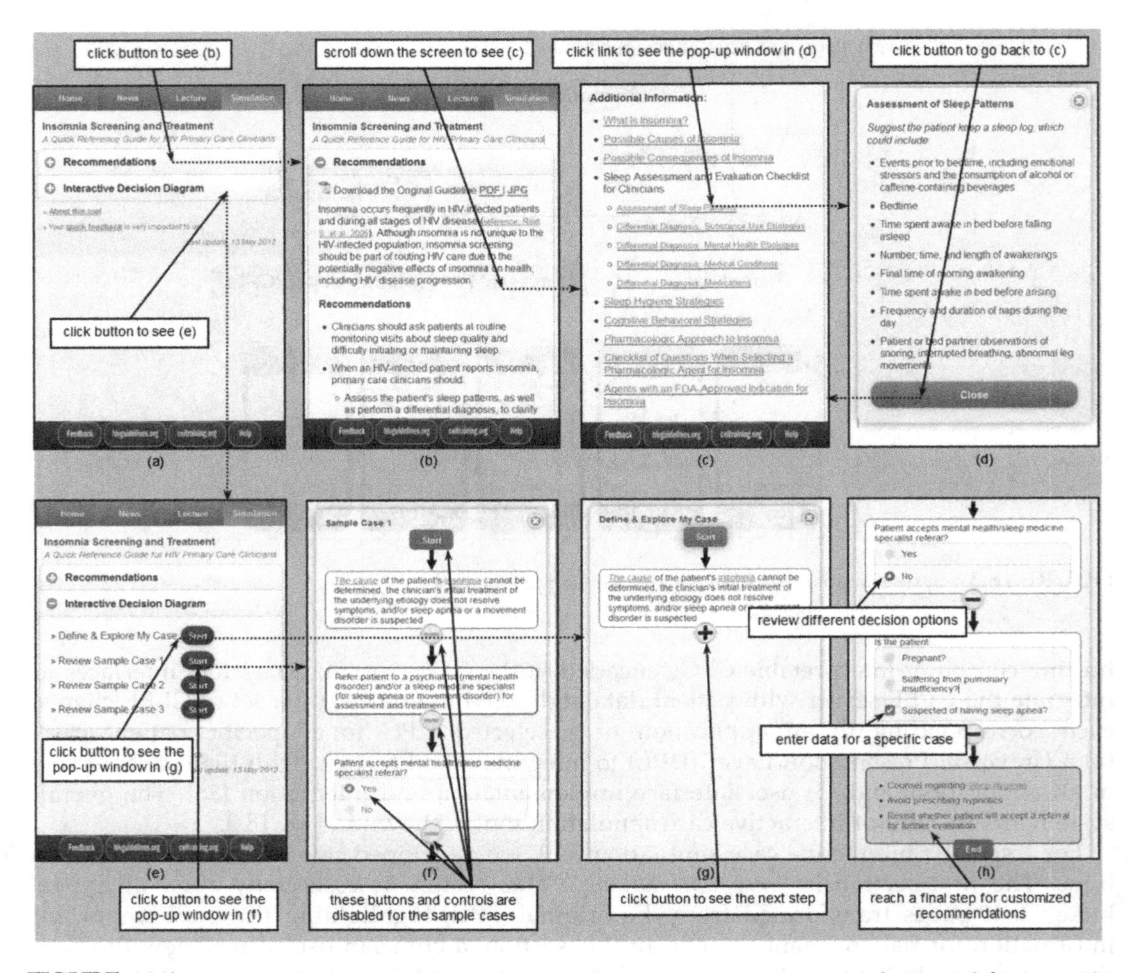

FIGURE 18.4 Selected screenshots and user functions from an interactive case simulation tool for insomnia screening and treatment in HIV patients. *From Le XH, Luque A, Wang D. Assessing the usage of a guideline-driven interactive case simulation tool for insomnia screening and treatment in an HIV clinical education program. Stud Health Technol Inform 2013;192:323–7. © Xuan Hung Le, Amneris E. Luque, Dongwen Wang.*

interested in a specific clinical topic (e.g., "*Hepatitis C*") has found out a specific multimedia learning module on the CEI website (e.g., "*Evaluation and Care of the HCV Patient Prior to or in the Absence of Treatment*"). From the browsing tool shown in Fig. 18.2, he/she may browse the other multimedia learning modules related to the topic of hepatitis C. Meanwhile, this healthcare provider may also be interested in other types of online education resources (e.g., interactive case simulation tools and CPGs) that are related to hepatitis C. These other types of online resources could reside on the same website (e.g., CEI hosts both multimedia learning modules and interactive case simulation tools) or be produced by different programs or sponsors (e.g., HIV Clinical Guideline Program [33], another New York State resource for

HIV care providers). Creating cross-linkages among multiple types of online education resources from different programs has the potential to break down the silos between networks, to provide a comprehensive set of resources in various formats, and therefore to enhance the dissemination of clinical evidence to a wider audience.

As an extension of the concept presented in "Multimedia Learning Modules" section, we initiated a study to create linkages among multimedia learning modules, interactive case simulation tools, and CPGs. For the pilot stage, we leveraged the New York State HIV CEI [4] and Clinical Guideline [33] programs, each of which has its own dedicated audiences. To create the linkages among the different types of resources, we reused the same ontology of HIV, HCV, and STD clinical education topics shown in Fig. 18.1, and assigned each resource to a topic that best matches its content. We developed the cross-linkages among the learning modules, interactive case simulation tools, and the source CPGs through specific topics—all resources under the same topic are *related* to each other. These linkages could be further differentiated as: (1) internal linkages (intra-linkages) among the resources in the same format (e.g., learning module to learning module, CPG to CPG, or interactive case simulation tool to interactive case simulation tool); and (2) cross-linkages (inter-linkages) among the resources in different formats (e.g., learning module to interactive case simulation tool, learning module to CPG, or interactive case simulation tool to CPG). Based on these linkages, we developed a prototype system plug-in for the resource websites of each participating program (e.g., CEI learning module site, CEI interactive case simulation tool site, and Clinical Guideline program site). When a clinician visits a specific resource on one of these websites, the system plug-in is triggered to display: (1) the related resources in the same format (learning modules, CPGs, or interactive case simulation tools) from the current program (through intra-linkages); and (2) the related resources in different formats developed by other participating programs (through inter-linkages).

Based on this concept, we assigned the topics for 180 learning modules, 20 interactive case simulation tools, and 20 selected CPGs. Bridged by the clinical topics, we created 5,560 intra-linkages and 2,664 inter-linkages within and across the three formats of resources, as shown in Table 18.1. A visualization of these linkages is shown in Fig. 18.5. Using these linkages, we developed a prototype system plug-in to display the related resources to enhance resource dissemination, as shown in Fig. 18.6.

TABLE 18.1 Number of Intra-linkages and Inter-linkages Within and Across Different Types of Online Resources

	Learning Modules	Interactive Case Simulation Tools	CPGs
Learning Modules	5,408	–	–
Interactive Case Simulation Tools	1,188	68	–
CPGs	1,288	188	84

From Wang D, Le XH, Luque AE. Identifying effective approaches for dissemination of clinical evidence—correlation analyses on promotional activities and usage of a guideline-driven interactive case simulation tool in a statewide HIV-HCV-STD clinical education program. Stud Health Technol Inform 2015;216:515–9. © Dongwen Wang, Xuan Hung Le, Amneris E. Luque.

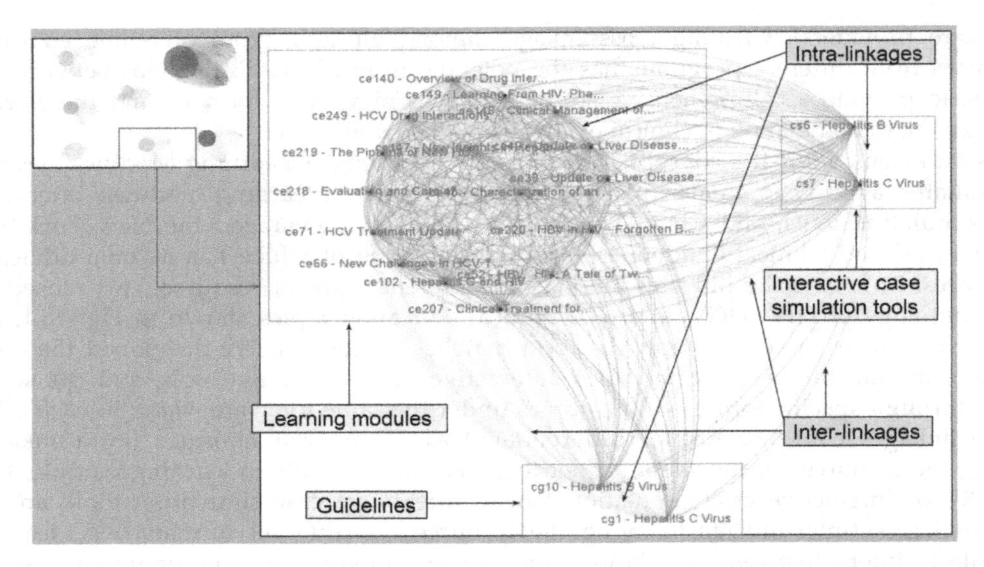

FIGURE 18.5 Visualization of linkages among resources—overview (upper left) and details on a specific part (right).

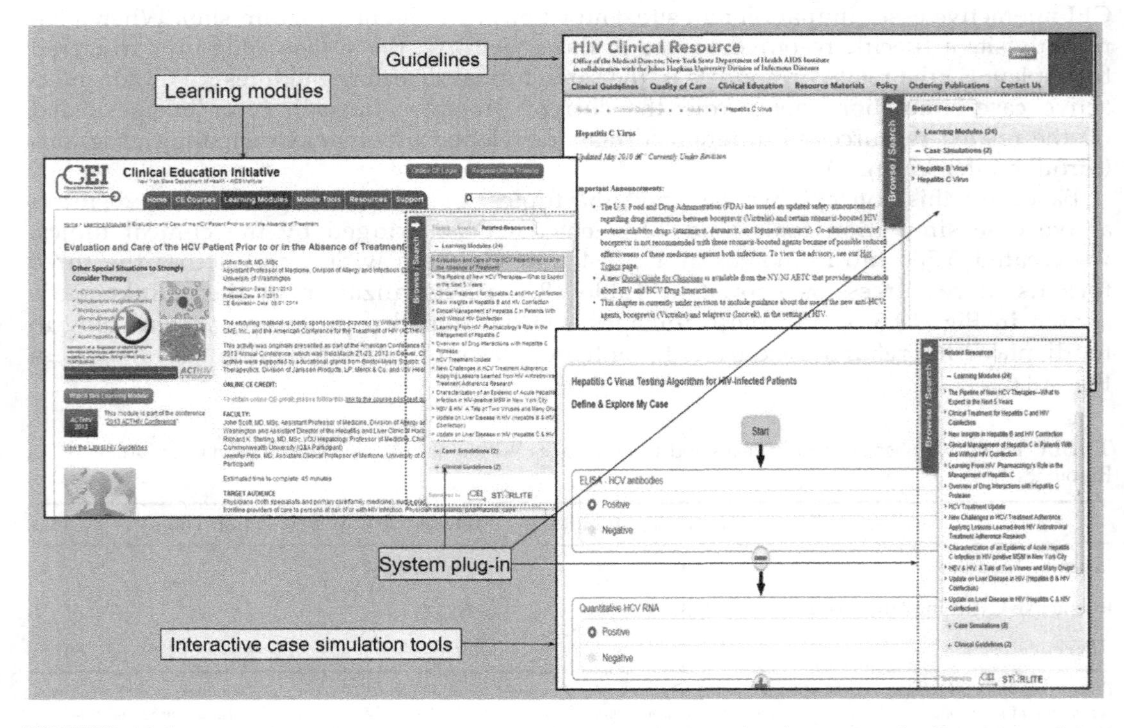

FIGURE 18.6 Screenshots of a prototype system plug-in to display related resources for learning modules, interactive case simulation tools, and CPGs.

USAGE OF CEI ONLINE EDUCATION RESOURCES

Overview of Resource Usage—Web Traffics, Awards of CME/CNE Credits, and App Downloads

For all CEI online education resources, we have been actively monitoring their actual usage since the first day of release. Specifically, we track the visits to CEI website and other web traffic data through Google Analytics [26]. We keep the records of CME/CNE credits awarded to clinicians through CEI online education program [34]. We have also developed customized usage tracking systems to monitor the download of mobile apps and user interactions with interactive case simulation tools [32].

From July 2009 to June 2015, we have recorded 170,123 user sessions and 886,402 page-views for visits to the CEI website from 170 + countries around the world (Fig. 18.7). During the same period, we have awarded a total of 11,300 CME/CNE credits to health-care providers who have successfully completed CEI online courses and passed the exams (Fig. 18.8). Since July 2012, we have registered 8,788 downloads of mobile apps and 102,763 rounds of user interactions with interactive case simulation tools (Fig. 18.9). These data have clearly shown rapid increase of usage over the project years, indicating effective dissemination of the CEI online resources.

Interactive Case Simulation Tool Usage, Use Context, and Usage Patterns

As described earlier, interactive case simulation tools are important CEI resources for dissemination of clinical evidence. Understanding the use contexts and usage patterns of

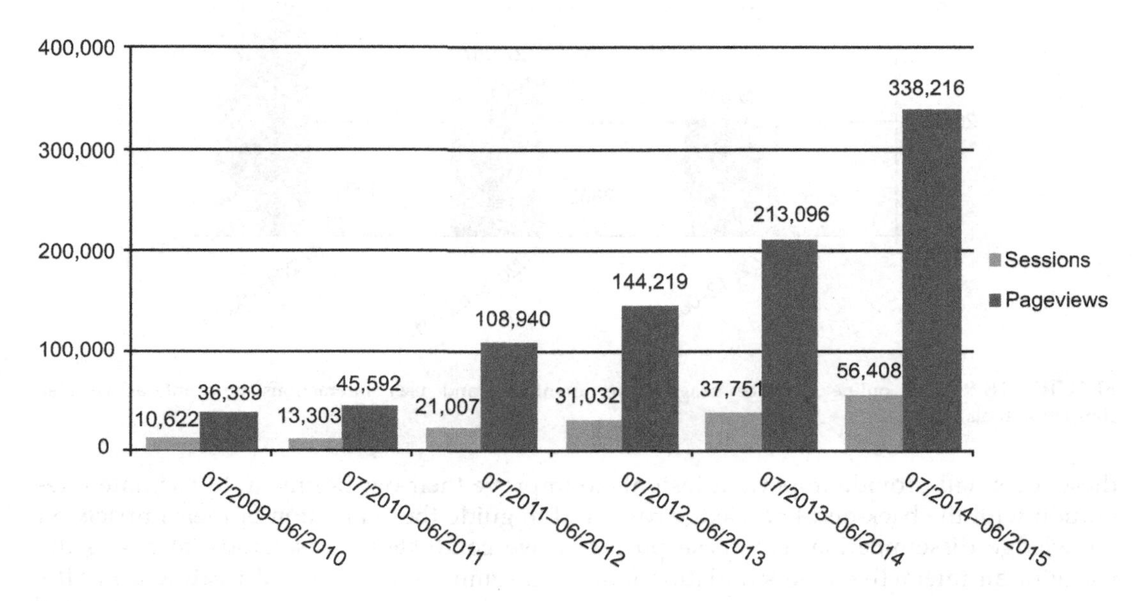

FIGURE 18.7 CEI online resource usage—web traffic increase over project years.

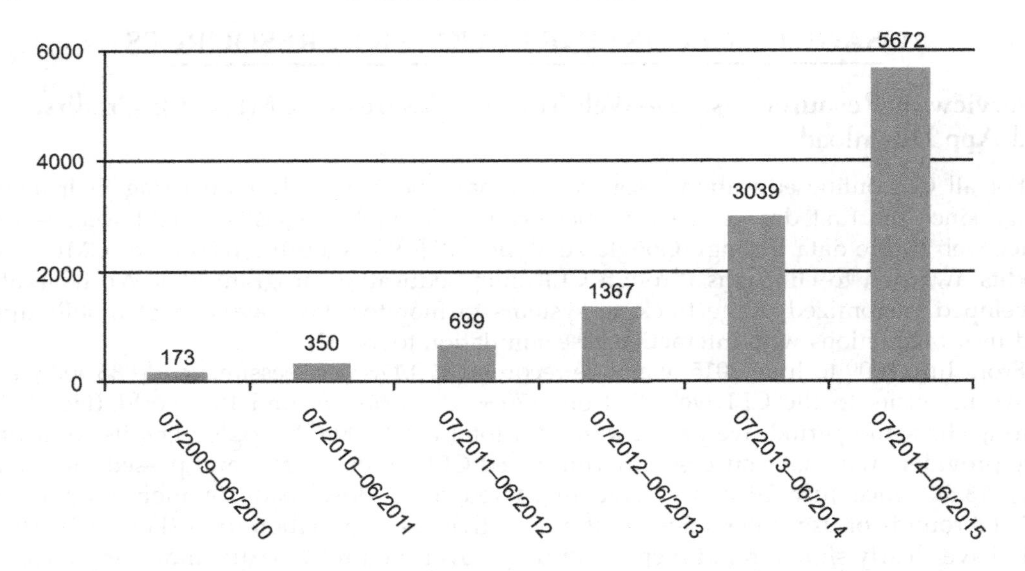

FIGURE 18.8 CEI online resource usage—CME/CNE credits awarded over project years.

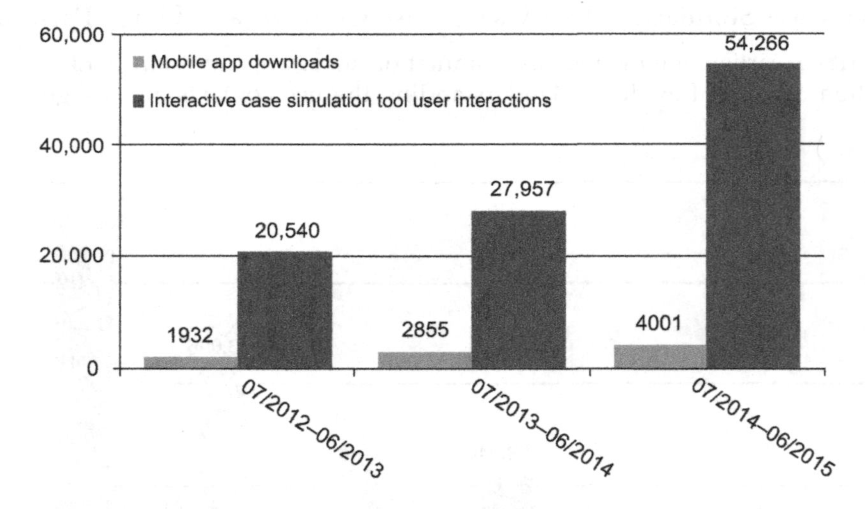

FIGURE 18.9 CEI online resource usage—app downloads and user interactions with interactive case simulation tools.

these tools will provide important insights to improve their development, to facilitate integration with the back-end knowledge base, and to guide the evaluation of their impacts on knowledge dissemination. For these purposes, we conducted a case study to assess the usage of an interactive case simulation tool for insomnia screening and treatment in HIV patients.

To collect system usage data, we built tracking functions to monitor user actions on all screens or functions of this case simulation tool. These user actions include: (1) opening a hyperlink to check the details of a recommendation; (2) pushing a button to go to the next step of the decision process; (3) selecting a specific decision option; and (4) entering the data for a particular patient case. In addition, we captured a user's IP address, usage session ID, and the timestamp associated with each action for detailed analyses of their behaviors. For data analyses, we constructed system usage diagrams as instruments to record and to analyze the usage patterns of the interactive case simulation tool. Usage measures included: (1) frequency of visits to specific screens or functions; and (2) length of stay on specific screens or functions. The technical details for usage pattern analyses can be found in a previous paper [32].

From April 3, 2012 (when the insomnia screening and treatment interactive case simulation tool was released to the public) to October 15, 2012 (when the tool was upgraded to a new version), we recorded a total of 512 visits to this case simulation tool. Among these, 439 (85.74%) were from new users and 73 (14.26%) were from returning users. In terms of access platform, 246 (48.05%) visits came from mobile apps, while the remaining 266 (51.95%) were through web browsers. With regard to the hardware, 274 (53.52%) visits originated from large-screen equipment and 238 (46.48%) were from small-screen devices. From the perspective of the sections of an interactive case simulation tool, 202 (39.45%) visits were in the "*recommendations*" section, 44 (8.59%) were for the sample cases in the "*interactive decision diagram*" section, 223 (43.55%) were for the user-defined cases in the "*interactive decision diagram*" section, and the remaining 43 (8.40%) were cross-over visits spanning over two or more sections.

For average length of stay on a specific screen or function, we recorded: (1) 23.92 seconds for new users and 43.68 seconds for returning users; (2) 13.78 seconds for mobile app users and 38.73 seconds for web browser users; (3) 33.19 seconds for use with large screen equipment and 19.32 seconds for use with small screen devices; and (4) 34.86 seconds in the "*recommendations*" section, 21.91 seconds for the sample cases in the "*interactive decision diagram*" section, 18.73 seconds for the user-defined cases in the "*interactive decision diagram*" section, and 35.09 seconds for the cross-over visits. A summary of the interactive case simulation tool usage data based on frequency of visits and length of stay in specific use contexts is shown in Table 18.2.

Usage pattern analyses on frequency of visits found: (1) returning users were more likely to use large-screen equipment ($P < 0.001$) and to access the interactive case simulation tool through web browsers ($P = 0.011$); (2) visits to sample cases and user-defined cases in the "*interactive decision diagram*" section were more likely from small-screen users ($P < 0.001$) and mobile app users ($P < 0.001$); and (3) small-screen users were more likely to access the interactive case simulation tools through mobile apps ($P < 0.001$). With regard to the average length of stay on a specific screen or function of the interactive case simulation tool, we found: (1) small-screen users had a shorter stay than the large-screen users ($P < 0.001$); (2) mobile app users had a shorter stay than web browser users ($P < 0.001$); and (3) the "*recommendations*" section recorded longer stay than the sample cases and user-defined cases in the "*interactive decision diagram*" section ($P < 0.001$). A summary of interactive case simulation tool usage pattern analyses based on visits and contexts is shown in Table 18.3.

TABLE 18.2 Interactive Case Simulation Tool Usage and Use Contexts

Access Contexts	Frequency of Visits	Length of Stay (Mean ± Standard Deviation, in Seconds)
New users	439 (85.74%)	23.92 ± 61.85
Returning users	73 (14.26%)	43.68 ± 121.45
Mobile apps	246 (48.05%)	13.78 ± 25.93
Web browsers	266 (51.95%)	38.73 ± 97.47
Large screen equipment	274 (53.52%)	33.19 ± 89.46
Small screen devices	238 (46.48%)	19.32 ± 48.30
Recommendations	202 (39.45%)	34.86 ± 63.36
Sample cases	44 (8.59%)	21.91 ± 23.73
User-defined cases	223 (43.55%)	18.73 ± 85.76
Cross-boxes	43 (8.40%)	35.09 ± 79.66

Correlation of Interactive Case Simulation Tool Usage, Use Context, and Resource Dissemination Activities

When analyzing online resource usage, an interesting research question is how the actual usage of these resources relates to promotional activities. To answer this question, we conducted a pilot study to analyze the correlations between the usage of interactive case simulation tools and the resource dissemination activities by the CEI program (e.g., sending out email newsletters, posting to online social networks, and presentation at regional and national conferences). The objectives of this study include: (1) identifying how dissemination activities can increase the usage of online education resources; (2) characterizing the profiles of use contexts that respond to promotions; and (3) directing future development of effective approaches to disseminating clinical evidence.

In this case study, we selected the insomnia screening and treatment interactive case simulation tool as the online resource and the CEI newsletters sent through emails as the promotional activities for analyses. The CEI newsletters are regular updates of the latest CEI clinical and educational resources, such as the newly developed CME/CNE courses, multimedia learning modules, interactive case simulation tools, training events and news, etc. These newsletters contain hyperlinks pointing to the CEI website and other online platforms. When a CEI listserv subscriber receives a CEI newsletter and finds interest in a specific item, he/she can click the hyperlink to check the details of that resource on the CEI website. He/she can also forward the email to colleagues or friends to further disseminate these resources through his/her professional and social networks [35].

To analyze the correlations between interactive case simulation tool usage and promotional activities, we first collected the usage data from April 3, 2012 (when the insomnia screening and treatment ICST was released) to February 4, 2013 (a period of 44 weeks)

TABLE 18.3 Interactive Case Simulation Tool Usage Pattern Analyses Based on Visits and Contexts

	New Users	Returning Users	Mobile Apps	Web Browsers	Large Screen Equipment	Small Screen Devices	Recommendations	Sample Cases	User-Defined Cases	Cross-Boxes	Total
New Users			$P=0.011$		$P<0.001$			$P=0.024$			
Returning Users											
Mobile Apps	221	25			$P<0.001$		$P<0.001$				**246**
Web Browsers	218	48									**266**
Large Screen Equipment	209	65	37	237			$P<0.001$				**274**
Small Screen Devices	230	8	209	29							**238**
Recommendations	178	24	65	137	141	61					**202**
Sample Cases	31	13	36	8	16	28					**44**
User-Defined Cases	193	30	126	97	91	132					**223**
Cross-Boxes	37	6	19	24	26	17					**43**
Total	**439**	**73**	**246**	**266**	**274**	**238**					**512**

*Since the data are symmetric across the diagonal line, we only present the part in the lower-left portion. The *P*-value in each cell of the upper-right portion indicates the statistical significance for its corresponding cell in the lower-left portion.

through tracking of user interactions with the system [32]. We focused on two usage measures, i.e., episodes of use (reflecting the number of audience) and frequency of visits to specific screens or functions of the interactive case simulation tool (reflecting the intensity of use). To characterize the use context, we profiled user interactions by multiple dimensions, including: (1) interactive case simulation tool sections, i.e., user-defined cases, sample cases, recommendations, and cross-boxes; (2) new vs. returning users; (3) large-screen equipment vs. small-screen devices; (4) access through web browsers vs. mobile apps; (5) audience from the United States vs. non-US countries; (6) audience from New York State vs. out of the state; and (7) audience from healthcare organizations, government agencies, and unknown settings. To collect the data of CEI resource dissemination through newsletters, we reviewed all the archived emails, selected those newsletters directly related to interactive case simulation tools, and recorded the date when they were sent.

For data analyses, we segmented the total usage as well as the profiles of use context by the dimensions listed above as bi-weekly time-series. We plotted dissemination activities, i.e., sending CEI newsletters through emails, as a binary variable in a bi-weekly time-series with the same formulation. We then performed correlation analyses between each combination of usage profile/measure and the dissemination activities. We calculated Pearson correlation coefficient (r) to determine the potential correlations. We considered $|r| >= 0.5$ as a strong correlation, $0.5 > |r| >= 0.3$ as a moderate correlation, and $|r| < 0.3$ as a weak or no correlation [36].

A total of 298 episodes and 1,415 rounds of interactions (visits) with the insomnia screening and treatment interactive case simulation tool were recorded during the study period. Meanwhile, five CEI email newsletters were sent during this period to promote the interactive case simulation tools. The detailed distributions of the total interactive case simulation tool usage, its use in specific contexts, and the sending of email newsletters presented as bi-weekly time-series are shown in Fig. 18.10.

Pearson correlation coefficients showed strong correlations between promotional activities and the following usage profiles: (1) usage measured by episodes on total usage; and (2) usage measured by frequency of visits on recommendations, by new users, through web browsers, by audience from the United States, and by audience from healthcare organizations and government agencies. Meanwhile, there were moderate correlations between promotional activities and the following usage profiles: (1) usage measured by episodes on recommendations, by both new and returned users, from large-screen equipment, through web browsers, by audience from the United States, by audience from both inside and outside of New York State, and by audience from all settings; and (2) usage measured by frequency of visits from large-screen equipment, by audiences from both inside and outside of New York State, and on total usage. Weak or no correlations were found between promotional activities and the following usage profiles: (1) usage measured by episodes on sample cases and user-defined cases, from small-screen devices, through mobile apps, and by audience from non-US countries; and (2) usage measured by frequency of visits on sample cases and user-defined cases, from returned users, from small-screen devices, through mobile apps, and by audience from non-US countries. The detailed results of the correlation analyses, including r and P-value for each usage profile, are summarized in Table 18.4.

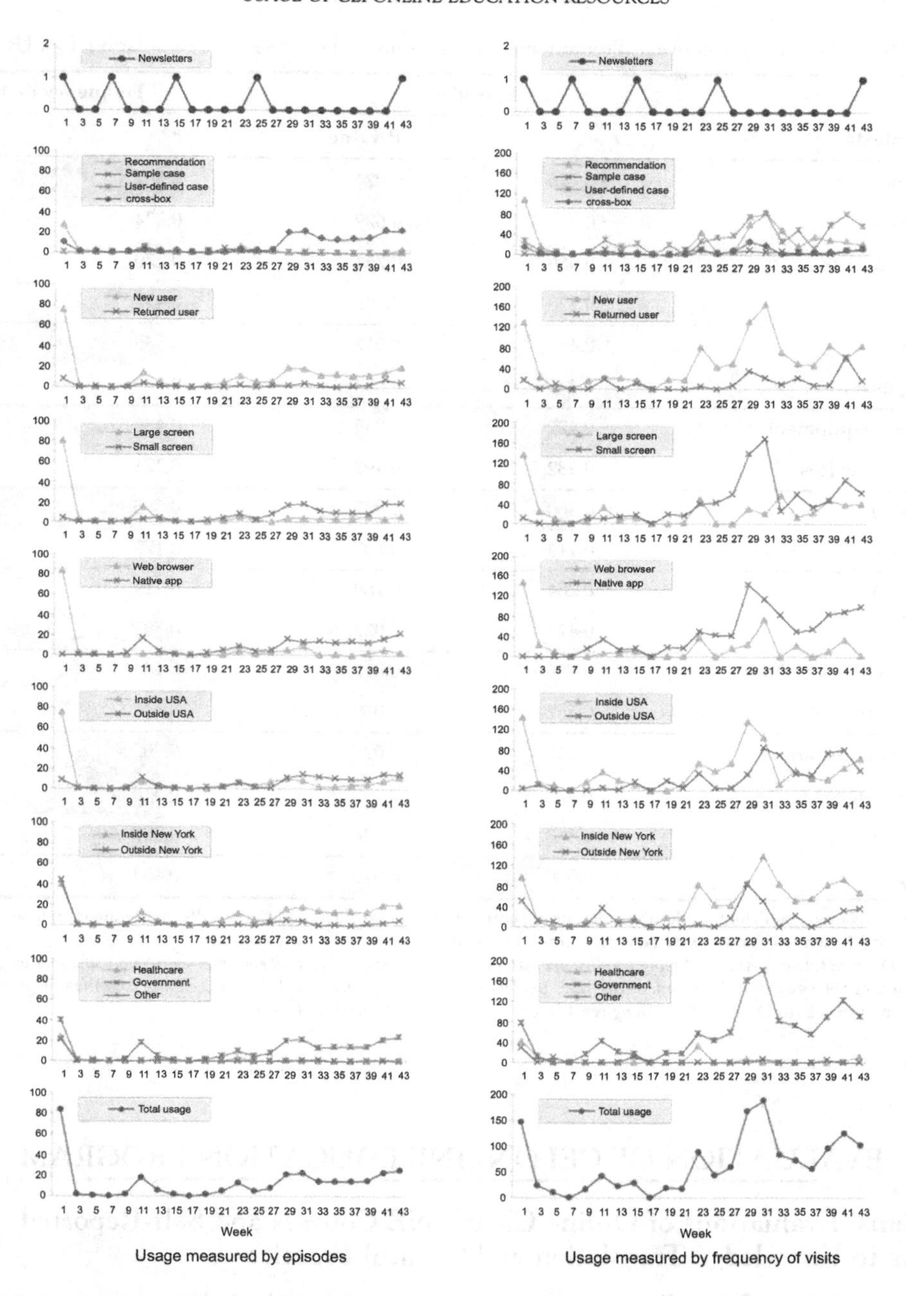

FIGURE 18.10 Sending CEI email newsletters, total interactive case simulation tool usage, and interactive case simulation tool use contexts. *From Wang D, Le XH, Luque AE. Identifying effective approaches for dissemination of clinical evidence—correlation analyses on promotional activities and usage of a guideline-driven interactive case simulation tool in a statewide HIV-HCV-STD clinical education program. Stud Health Technol Inform 2015;216:515–9. © Dongwen Wang, Xuan Hung Le, Amneris E. Luque.*

TABLE 18.4 Correlations Between Promotional Activities and Interactive Case Simulation Tool Usage

	Episodes		Frequency of Visits	
Usage Contexts	*r*	*P*-value	*r*	*P*-value
Recommendations	0.470	0.024	0.551	0.006
Sample cases	− 0.091	0.679	0.224	0.303
User-defined cases	− 0.066	0.764	0.270	0.213
Cross-boxes	0.278	0.198	0.437	0.037
New users	0.492	0.017	0.538	0.008
Returning users	0.446	0.033	0.122	0.579
Large screen equipment	0.422	0.045	0.486	0.019
Small screen devices	0.282	0.192	0.271	0.211
Web browsers	0.460	0.027	0.585	0.003
Mobile apps	0.113	0.607	0.173	0.429
Outside USA	0.284	0.189	0.219	0.315
Inside USA	0.474	0.022	0.502	0.015
Outside New York State	0.486	0.019	0.484	0.019
Inside New York State	0.452	0.030	0.342	0.110
Healthcare organization	0.462	0.027	0.607	0.002
Government agency	0.420	0.046	0.548	0.007
Unknown settings	0.463	0.026	0.325	0.131
Total usage	0.500	0.015	0.484	0.019

*r: Pearson correlation coefficient; dark gray cells: strong correlation ($|r| >= 0.5$); light gray cells: moderate correlction ($0.5 > |r| >= 0.3$); white cells: weak or no correlation ($|r| < 0.3$).

From Wang D, Le XH, Luque AE. Identifying effective approaches for dissemination of clinical evidence—correlation analyses on promotional activities and usage of a guideline-driven interactive case simulation tool in a statewide HIV-HCV-STD clinical education program. Stud Health Technol Inform 2015;216:515–9. © Dongwen Wang, Xuan Hung Le, Amneris E. Luque.

EVALUATION OF CEI ONLINE EDUCATION PROGRAM

Clinicians' Evaluations of Online CME/CNE Courses and Self-Reported Impacts to Knowledge Translation and Clinical Practice

To evaluate the CEI online education program, we conducted a study to characterize the clinician students' background, to assess their feedback on taking the online CME/CNE courses, and to evaluate the potential impact to their knowledge increase and clinical practice [34].

Course Evaluation

L-a. The information and/or skills provided in this event were useful and relevant
[1] Strongly Agree
[2] Agree
[3] Neutral
[4] Disagree
[5] Strongly Disagree
[6] Not Applicable

L-b. The information and/or skills taught in this event were easy to comprehend
[1] Strongly Agree
[2] Agree
[3] Neutral
[4] Disagree
[5] Strongly Disagree
[6] Not Applicable

L-c. The trainer was knowledgeable about the topic
[1] Strongly Agree
[2] Agree
[3] Neutral
[4] Disagree
[5] Strongly Disagree
[6] Not Applicable

L-d. As a result of this event I intend to use the knowledge/skills I have learned today in my clinical practice
[1] Strongly Agree
[2] Agree
[3] Neutral
[4] Disagree
[5] Strongly Disagree
[6] Not Applicable

M. How might the format of this activity be changed in order to be most appropriate for the content presented?
[1] Format was appropriate; no ch
[2] I would like the following format
[3] Include more case-based pr
[4] Add breakouts for subtopics

Gender:*
Male
Female
Transgender: Male to Female
Transgender: Female to Male

Racial/ethinic background:*
Race
American Indian or Alaska Native
Asian
Black or African American
Native Hawaiian or Pacific Islander
White
Other

Highest level of education:*
Doctoral Degree
Master Degree
Bachelor Degree
College Coursework
High School Diploma
Other

Primary professional discipline/occupation:*
Physician
Physician Assistant
Nurse Practitioner
Nurse
Dentist
Dental Hygienist
Pharmacist
Psychiatrist
Psychologist
Nutritionist/Dietician
Counselor
Case/Care Manager
Social Worker

FIGURE 18.11 Partial screenshots of the evaluation questionnaire and clinicians' background from the CEI student portal. *From Wang D, Luque LE. Evaluation of a statewide HIV-HCV-STD online clinical education program by healthcare providers—a comparison of nursing and other disciplines. Stud Health Technol Inform 2016;225:267–71. © Dongwen Wang, Xuan Hung Le.*

We selected a study period from November 3, 2014 (when the new CEI student portal, a centralized function for student and course management, was released) to December 31, 2014 (when the quarterly reporting period ended). As a part of the process for course completion, each clinician was required to provide evaluations on the training. The evaluation measures included: (1) usefulness/relevance of information; (2) easy comprehension; (3) trainer's knowledge; (4) appropriateness of format; (5) knowledge level before and after training; (6) intention to use the learned knowledge; and (7) intention to change clinical practice. Within the student portal, a clinician's personal information (e.g., contact, demographics, and education level) and professional background (e.g., clinical discipline, employment setting, practice years, and patient case load) were collected and stored in the student profile. Partial screenshots of the evaluation questionnaire and clinicians' background from the CEI student portal are shown in Fig. 18.11.

A total of 335 clinician students completed 12 available online CME/CNE courses during the study period. Among these clinicians, 108 (32.24%) were physicians, 53 (15.82%) nurses or nurse practitioners, 23 (6.87%) health program administrators, and 20 (5.97%) case managers. In terms of practice years, 25 (7.46%) of them had more than 30 years' clinical experiences, and 162 (48.36%) had less than 10 years' in practice. With regard to employment setting, 78 (23.28%) were in hospitals, 49 (14.63%) were employees of state or local health departments, 48 (14.33%) had their private or group practices, and 37 (11.04%) worked in community health centers. The detailed report of the clinicians' professional and personal background is shown in Table 18.5.

TABLE 18.5 Clinicians' Professional and Personal Background

Categories	Characteristics	Number and Percentage (n = 335)
Gender	Male	131 (39.10%)
	Female	203 (60.60%)
	Transgender	1 (0.30%)
Race	American Indian/Alaska native	7 (2.09%)
	Asian	36 (10.75%)
	Black/African American	65 (19.40%)
	Native Hawaiian/Pacific Islander	4 (1.19%)
	White	188 (56.12%)
	Other	35 (10.45%)
Ethnic background	Hispanic/Latino	57 (17.01%)
	Not Hispanic/Latino	278 (82.99%)
Highest level of education	Doctoral degree	106 (31.64%)
	Master degree	63 (18.81%)
	Bachelor degree	81 (24.18%)
	College coursework	44 (13.13%)
	High school diploma	24 (7.16%)
	Other	17 (5.07%)
Primary professional discipline or occupation	Physician	108 (32.24%)
	Physician assistant	15 (4.48%)
	Nurse practitioner	16 (4.78%)
	Nurse	37 (11.04%)
	Pharmacist	1 (0.30%)
	Psychiatrist	1 (0.30%)
	Psychologist	1 (0.30%)

(Continued)

TABLE 18.5 (Continued)

Categories	Characteristics	Number and Percentage (n = 335)
	Counselor	8 (2.39%)
	Case/care manager	20 (5.97%)
	Social worker	10 (2.99%)
	Medical/dental assistant	5 (1.49%)
	Lab manager/technician	1 (0.30%)
	Public health professionals	10 (2.99%)
	Health program administrator	23 (6.87%)
	Health educator	10 (2.99%)
	Health profession student	6 (1.79%)
	Other	63 (18.81%)
Years in primary profession or occupation	>30 years	25 (7.46%)
	21–30 years	66 (19.70%)
	11-20 years	82 (24.48%)
	< = 10 years	162 (48.36%)
Employment setting	Community health center	37 (11.04%)
	Hospital/hospital clinic	78 (23.28%)
	Emergency department	7 (2.09%)
	Private practice/group practice	48 (14.33%)
	Managed care organization	9 (2.69%)
	Professional organization	7 (2.09%)
	Substance use center/program	7 (2.09%)
	Mental health center/program	1 (0.30%)
	Nursing/chronic care facility	12 (3.58%)
	State/local health department	49 (14.63%)
	Correctional facility	3 (0.90%)
	School/college-based clinic	11 (3.28%)
	Military/VA	3 (0.90%)
	Not employed	12 (3.58%)
	Other	51 (15.22%)

(Continued)

TABLE 18.5 (Continued)

Categories	Characteristics	Number and Percentage (n = 335)
Location of employment setting	Rural	52 (15.52%)
	Suburban	82 (24.48%)
	Urban	201 (60.00%)
Number of patients per month—HIV/AIDS	0	149 (44.48%)
	1–10	85 (25.37%)
	11–20	22 (6.57%)
	21–40	17 (5.07%)
	41–60	20 (5.97%)
	61–100	15 (4.48%)
	100 +	27 (8.06%)
Number of patients per month—HCV	0	185 (55.22%)
	1–10	80 (23.88%)
	11–20	20 (5.97%)
	21–40	21 (6.27%)
	41–60	10 (2.99%)
	61–100	7 (2.09%)
	100 +	12 (3.58%)
Number of patients per month—other STDs	0	179 (53.43%)
	1–10	68 (20.30%)
	11–20	15 (4.48%)
	21–40	19 (5.67%)
	41–60	23 (6.87%)
	61–100	10 (2.99%)
	100 +	21 (6.27%)

The clinicians' evaluations of the online courses were very positive (useful/relevant: 88.66%; easy comprehension: 87.76%; knowledgeable trainer: 91.04%; appropriate format: 84.18%). With regard to the impact of training, 83.58% indicated intention to use the learned knowledge, 46.27% had at least 1-level of knowledge increase, and 45.78% of those providing direct patient services expressed plan to change their practice because of the training. A summary of the evaluation is shown in Table 18.6.

TABLE 18.6 Clinicians' Evaluation of CEI Online Courses

Measures	Positive Responses
Useful and relevant	297/335 (88.66%)
Easy to comprehend	294/335 (87.76%)
Knowledgeable trainer	305/335 (91.04%)
Format appropriate	282/335 (84.18%)
Intend to use knowledge	280/335 (83.58%)
Increase knowledge	155/335 (46.27%)
Will change practice	76/335 (22.69%)
Will change practice (excluding those not providing patient services)	76/172 (44.19%)

Comparison of Online and In-Person Training

As shown in the previous section, the clinicians had very positive evaluations on the CEI online education program. An interesting research question is to compare the online training with the traditional in-person training. Leveraging the training data in both modes from the CEI student portal, we conducted a study to compare these two formats of training [37].

For this purpose, we selected the CEI trainings delivered from April 1, 2015 to June 30, 2015 in both in-person and online formats. We collected clinicians' self-reported evaluation data from the CEI student portal, with the same set of study measures as those reported previously, i.e., usefulness/relevance, easy comprehension, trainer's knowledge, appropriateness of format, knowledge increase, intention to use knowledge, and intention to change practice.

Four training sessions were delivered in dual modes during the study period: (1) *"STD-HIV inter-relationship"*; (2) *"Treatment for hepatitis C: new tests, new drugs & new recommendations"*; (3) *"Vaginitis"*; and (4) *"Syphilis."* These training were completed by 368 clinicians (in-person 231, online 137). Clinicians' evaluations continued to be very positive for both training formats (useful/relevant: in-person 96.26%, online 91.24%; easy comprehension: in-person 91.59%, online 91.97%; knowledgeable trainer: in-person 96.26%, online 89.78%; appropriate format: in-person 75.48%, online 87.59%). The data also showed significant impacts to knowledge increase and clinical practice (intention to use knowledge: in-person 89.20%, online 83.94%; >=1 level of knowledge increase: in-person 61.32%, online 42.34%; plan to change practice: in-person 50.96%, online 33.71%). Comparing the two formats, in-person training was preferred in terms of usefulness/relevance ($P = 0.048$), knowledgeable trainer ($P = 0.015$), increase of knowledge ($P = 0.001$), and change of practice ($P = 0.009$); online training was preferred with regard to appropriateness of format ($P = 0.006$). No statistically significant differences were found when measuring easy comprehension ($P = 0.899$) and use of knowledge ($P = 0.151$). A summary of the study results is shown in Table 18.7.

TABLE 18.7 A Comparison of In-Person and Online Training

Measures	In-Person Training ($n=231$)	Online Training ($n=137$)	*P*-Values
Useful and relevant	√ 206/214 (96.26%)	125/137 (91.24%)	**$P=0.048$**
Easy to comprehend	196/214 (91.59%)	126/137 (91.97%)	$P=0.899$
Knowledgeable trainer	√ 206/214 (96.26%)	123/137 (89.78%)	**$P=0.015$**
Format appropriate	157/208 (75.48%)	√ 120/137 (87.59%)	**$P=0.006$**
Intend to use knowledge	190/213 (89.20%)	115/137 (83.94%)	$P=0.151$
Increase knowledge	√ 130/212 (61.32%)	58/137 (42.34%)	**$P=0.001$**
Will change practice	√ 80/157 (50.96%)	30/ 89 (33.71%)	**$P=0.009$**

*√: Training format with better performance.

DISCUSSION

Previous research has shown that multimedia resources and case simulation tools can enhance clinical education [38–41]. However, few studies have reported on development of large digital repositories of multimedia learning modules and interactive case simulation tools, and none on their integrations. By leveraging the CEI network and the existing tools such as GLIF [15] and GLEE [16], we were able to scale up the development of multimedia learning modules and interactive case simulation tools to build the CEI digital training repositories. Consequentially, we recorded rapid increase in usage of the CEI online education resources.

When assessing the usage of interactive case simulation tools, our study results have shown that both the "*recommendations*" and "*interactive decision diagram*" sections were frequently used. The latter recorded a higher visit frequency but a shorter length of stay (see Table 18.2). We have also identified specific use contexts and usage patterns of interactive case simulation tools (see Tables 18.2 and 18.3). These findings provided quantitative measures on the actual usage of interactive case simulation tools by clinicians. More important, our correlation analyses have demonstrated connections between the usage of interactive case simulation tools in specific contexts and the promotional activities. These results indicated: (1) when planning for effective resources dissemination, those use contexts with strong correlations should be the primary focus; and (2) those use contexts with weak or no correlations do not need promotions.

In our evaluation of the CEI online education program, we have recorded high levels of satisfaction from the clinicians participating in the program (see Tables 18.6 and 18.7). In addition, the data have shown that a significant proportion of the clinicians intended to use the knowledge learned from the program and to change their practice (see Tables 18.6 and 18.7). More important, our comparative study have indicated that: (1) both online and in-person training can be used effectively to disseminate HIV, HCV, and STD clinical evidence to clinicians; and (2) although online learning is still not equal to the traditional in-person training in certain aspects, it has become a preferred channel for clinicians to update their medical knowledge (see Table 18.7).

There were a few limitations in the studies reported in this chapter. First, in the pilot study on interactive case simulation tool use contexts and usage patterns, we selected a single interactive case simulation tool of insomnia for usage analyses. Generalizability of the findings, therefore, needs to be verified. In fact, we have just completed another pilot study and reconfirmed many findings in usage of another interactive case simulation tool for mental health screening in HIV patient [42]. Additional applications of interactive case simulation tools should be tested to further validate the results from our studies. Second, in analyses of the connections between interactive case simulation tools usage and promotional activities, we focused only on correlations, which could be the first step to determine causality. For the use contexts with strong correlations, it may worth further investigating on the detailed pathways that contribute to the increasing use of clinical and educational resources. Last, in the initial evaluation of the CEI online education program, we only presented the clinicians' self-reported measures on their perception of the program and impact, but did not include analyses on covariates such as the specific training courses or topics, clinician students' personal background (e.g., demographics, education level, etc.), and their clinical practice (e.g., discipline, employment setting, practice years, case load, etc.). In a recent pilot study, we compared the course evaluations by clinicians from a variety of clinical disciplines, and indeed found significant differences [43]. Additional covariate analyses should be performed to identify the important factors that contribute to a successful online clinical education program.

CONCLUSION

We have developed a successful online HIV, HCV, and STD clinical education program through: (1) development, organization, and linkage of a large repository of online education resources; (2) active dissemination of the online resources to healthcare providers and close monitoring of their usage; and (3) effective evaluation of the program through assessment of clinician users, their feedback on specific training courses or topics, and potential impact to knowledge increase and clinical practice. Study data have shown rapid increase in usage of online resources and very positive program evaluations by clinicians. Our future works include: (1) verifying generalizability of the findings on resource usage patterns; and (2) full-scale studies to examine the effectiveness of the CEI online education program and to identify the co-variates that contribute to its success.

Acknowledgments

The studies reported in this chapter were sponsored by the Agency for Healthcare Research and Quality (AHRQ) through grant R24 HS022057, and by New York State Department of Health AIDS Institute through contracts C023557, C024882, and C029086. The content is solely the responsibility of the author and does not necessarily represent the official views of the sponsors. I would like to thank: (1) CEI staff Amneris Luque, Xuan Hung Le, Terry Doll, Matthew Bernhardt, and Monica Barbosu for their contributions to the studies; and (2) AHRQ and CEI program officers Marian James, Beatrice Aladin, Cheryl Smith, Howard Lavigne, Lyn Stevens, and Bruce Agins for their support.

References

[1] Centers for Disease Control and Prevention, HIV/AIDS Basic Statistics. Available from: <http://www.cdc.gov/hiv/statistics/basics.html> [accessed 03.08.16].

[2] Centers for Disease Control and Prevention, Viral Hepatitis—Hepatitis C Information. Available from: <http://www.cdc.gov/hepatitis/hcv/hcvfaq.htm> [accessed 03.08.16].

[3] Centers for Disease Control and Prevention, CDC Fact Sheet—Incidence, Prevalence, and Cost of Sexually Transmitted Infections in the United States. Available from: <https://www.cdc.gov/std/stats/STI-Estimates-Fact-Sheet-Feb-2013.pdf> [accessed 03.08.16].

[4] New York State HIV-HCV-STD Clinical Education Initiative. Available from: <http://ceitraining.org> [accessed 03.08.16].

[5] Fitzgerald C, Kantrowitz-Gordon I, Katz J, Hirsch A. Advanced practice nursing education: challenges and strategies. Nurs Res Pract 2012;854918.

[6] Belda TE, Gajic O, Rabatin JT, Harrison BA. Practice variability in management of acute respiratory distress syndrome: bringing evidence and clinician education to the bedside using a web-based teaching tool. Respir Care. 2004;49(9):1015–21.

[7] Wang D, Le XH, Luque A. Development of digital repositories of multimedia learning modules and interactive case simulation tools for a statewide clinical education program. Proceedings of the 6th international workshop on knowledge representation for health-care. 2014. p. 145–51.

[8] Wang D, Le XH, Luque AE. Identifying effective approaches for dissemination of clinical evidence—correlation analyses on promotional activities and usage of a guideline-driven interactive case simulation tool in a statewide HIV-HCV-STD clinical education program. Stud Health Technol Inform 2015;216:515–19.

[9] SNOMED CT. Available from: <http://www.ihtsdo.org/snomed-ct> [accessed 03.08.16].

[10] WHO–International Classification of Diseases. Available from: <http://www.who.int/classifications/icd/en/> [accessed 03.08.16].

[11] Medical Subject Headings. Available from: <http://www.nlm.nih.gov/mesh/> [accessed 03.08.16].

[12] Gene Ontology Consortium. Available from: <http://geneontology.org/> [accessed 03.08.16].

[13] MedBiquitous. Available from: <http://www.medbiq.org> [accessed 03.08.16].

[14] Cimino JJ. Desiderata for controlled medical vocabularies in the twenty-first century. Meth Inform Med 1998;37(4–5):394–403.

[15] Boxwala AA, Peleg M, Tu S, et al. GLIF3: a representation format for sharable computer-interpretable clinical practice guidelines. J Biomed Inform 2004;37(3):147–61.

[16] Wang D, Peleg M, Tu S, et al. Design and implementation of the GLIF3 guideline execution engine. J Biomed Inform 2004;37(5):305–18.

[17] Wang D, Peleg M. Using GLIF and GLEE to facilitate knowledge management in development of clinical decision support systems. Stud Health Technol Inform 2007;129:2268–70.

[18] Irigoyen M, Findley S, Wang D, Chen S, Chimkin F, Pena O, et al. Challenges of immunization registry reminders at inner city practices. Ambul Pediatr 2006;6(2):100–4.

[19] Wang D, Shortliffe EH. GLEE–A model-driven execution system for computer-based implementation of clinical practice guidelines. Proc AMIA Symp 2002;855–9.

[20] Peleg M, Boxwala AA, Tu S, Zeng Q, Ogunyemi O, Wang D, et al. The InterMed approach to sharable computer-interpretable guidelines: a review. J Am Med Inform Assoc 2004;11(1):1–10.

[21] Wang D. Translating Arden MLMs into GLIF guidelines–a case study of hyperkalemia patient screening. Stud Health Technol Inform 2004;101:177–81.

[22] Peleg M, Wang D, Fodor A, Keren S, Karnieli E. Lessons learned from adapting a generic narrative diabetic-foot guideline to an institutional clinical decision-support system. Stud Health Technol Inform 2008;139:243–52.

[23] Choi J, Currie LM, Wang D, Bakken S. Encoding a clinical practice guideline using GuideLine Interchange Format: a case study of a depression screening and management guideline. Int J Med Inform 2007;76(suppl):S302–7.

[24] Roberts WD, Patel VL, Stone PW, Bakken S. Knowledge content of advance practice nurse and physician experts: a cognitive evaluation of clinical practice guideline comprehension. Stud Health Technol Inform 2006;122:476–80.

[25] Le XH, Luque A, Wang D. Development of guideline driven mobile applications for clinical education and decision support with customization to individual patient cases. Proc AMIA Symp 2012;1828.

[26] Google Analytics. Available from: <http://www.google.com/analytics> [accessed 03.08.16].

[27] Liu N, Marenco L, Miller PL. ResourceLog: an embedded tool for dynamically monitoring the usage of web--based bioscience resources. J Am Med Inform Assoc 2006;13(4):432–7.

[28] Donaldson RI, Ostermayer DG, Banuelos R, Singh M. Development and usage of wiki-based software for point-of-care emergency medical information. J Am Med Inform Assoc 2016;23(6):1174–9.

[29] Macznik AK, Ribeiro DC, Baxter GD. Online technology use in physiotherapy teaching and learning: a systemic review of effectiveness and users' perceptions. BMC Med Educ 2015;15:160.

[30] Pulsford D, Jackson G, O'Brien T, Yates S, Duxbury J. Classroom-based and distance learning education and training courses in end-of-life care for health and social care staff: a systemic review. Palliat Med 2013;27 (3):221–35.

[31] Cohen NL, Carbone ET, Beffa-Negrini PA. The design, implementation, and evaluation of online credit nutrition courses: a systematic review. J Nutr Educ Behav 2011;43(2):76–86.

[32] Le XH, Luque A, Wang D. Assessing the usage of a guideline-driven interactive case simulation tool for insomnia screening and treatment in an HIV clinical education program. Stud Health Technol Inform 2013;192:323–7.

[33] HIV Clinical Resource. Available from: <http://www.hivguidelines.org> [accessed 03.08.16].

[34] Wang D, Luque A, Doll T, Barbosu M, Bernhardt M. Evaluation of a statewide online HIV-HCV-STD clinical education program–characterization of healthcare providers' professional background, self-reported knowledge increase, and intention to change clinical practice. AMIA Annu Symp Proc 2015;1726.

[35] Le XH, Luque A, Wang D. Leveraging information technologies and multiple online platforms to disseminate HIV/AIDS clinical evidences to community healthcare providers. In: Proceedings of the 7th annual conference on the science of dissemination and implementation: transforming health systems to optimize individual and population health; 2014.

[36] Cohen J. Statistical power analysis for the behavioral sciences. 2nd ed. New Jersey: Lawrence Erlbaum Associates; 1988.

[37] Wang D., Luque A., Doll T., Barbosu M., Bernhardt M. Assessment of a New York State program to disseminate HIV-HCV-STD clinical evidence to community healthcare providers—a comparison of in-person and online training. Proceedings of the 8th Annual Conference on the Science of Dissemination and Implementation in Health. 2015.

[38] Tan A, Ross SP, Duerksen K. Death is not always a failure: outcomes from implementing an online virtual patient clinical case in palliative care for family medicine clerkship. Med Educ Online 2013;18:22711.

[39] Sperling JD, Clark S, Kang Y. Teaching medical students a clinical approach to altered mental status: simulation enhanced traditional curriculum. Med Educ Online 2013;18:1–8.

[40] Kobak KA, Craske MG, Rose RD, Wolitsky-Taylor K. Web-based therapist training on cognitive behavior therapy for anxiety disorders: a pilot study. Psychotherapy 2013;50(2):235–47.

[41] Sekiguchi H, Suzuki J, Gharocholou SM, Fine NM, Mankad SV, Daniels CE, et al. A novel multimedia workshop on portable cardiac critical care ultrasonography: a practical option for the busy intensivist. Anaesth Intensive Care 2012;40(5):838–43.

[42] Wang D. Healthcare providers' usage of an interactive case simulation tool for HIV patient mental health screening in a statewide clinical education program. AMIA Annu Symp Proc 2016;1625.

[43] Wang D, Luque AE. Evaluation of a statewide HIV-HCV-STD online clinical education program by healthcare providers–a comparison of nursing and other disciplines. Stud Health Technol Inform 2016;225:267–71.

SECTION VIII

INFORMATION AND COMMUNICATION TECHNOLOGY AS AN EDUCATIONAL TOOL

CHAPTER

19

IS4Learning—A Multiplatform Simulation Technology to Teach and Evaluate Auscultation Skills

Daniel Pereira[1,2,3], Pedro Gomes[1,3], Carla Sá[3], Ovídio Costa[3], Zilma Silveira Nogueira Reis[4], Sandra Mattos[5], Ricardo Cruz-Correia[2,3] and Miguel Tavares Coimbra[1,3]

[1]Instituto de Telecomunicações (IT), Lisboa, Portugal [2]Center for Health Technology and Services Research (CINTESIS), Porto, Portugal [3]University of Porto, Porto, Portugal [4]Federal University of Minas Gerais, Minas Gerais, Brazil [5]Heart Institute of Pernambuco, Pernambuco, Brazil

THE (ALMOST) LOST ART OF AUSCULTATION

The stethoscope is the oldest cardiovascular diagnostic instrument in clinical use (Fig. 19.1). Invented in 1816 by Laënnec, what first was a more 'modest' way to listen to heart sounds than direct application of the ear in a patient's chest, quickly revealed itself as a powerful way to enhance heart sounds and it is still considered the most cost-effective way to perform the first layer of screening of cardiopulmonary disease [1]. By auscultating the heart, we have an understanding of cardiac rate and rhythm, the sound of the closing and, sometimes, the opening of valves, and anatomical abnormalities such as congenital or acquired defects. Heart sounds are caused by turbulent blood flow, while laminar flow is silent. When used properly the stethoscope often enables physicians to make a rapid and an accurate diagnosis without any additional studies. However, cardiac auscultation is in decline and the lack of ability to either hear or interpret a cardiac abnormality starts with medical students and continues through to physicians of different ages [2]. Understanding these sounds is a difficult skill to master. Relevant pathological activity is often soft, short-lived and occurs in proximity to loud, normal activity: a typical murmur is 1000 times softer than normal heart sounds and can last for as little as thirty milliseconds [3].

Health Professionals' Education.
DOI: http://dx.doi.org/10.1016/B978-0-12-805362-1.00019-X

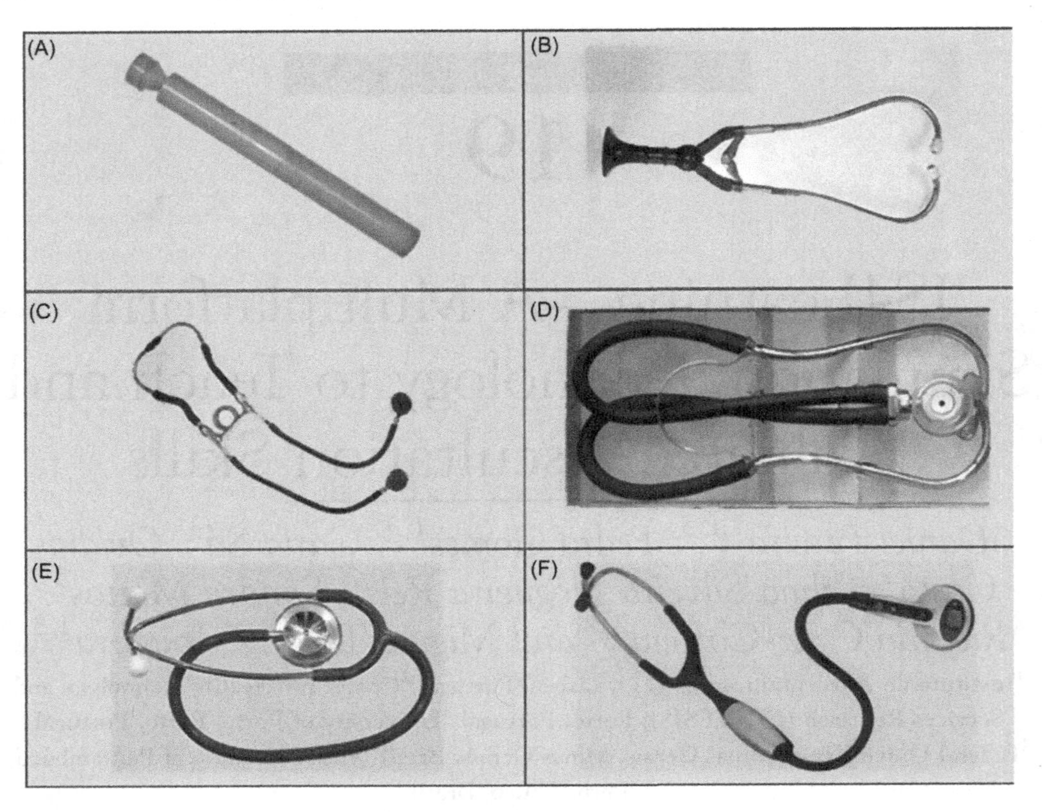

FIGURE 19.1 The evolution of the stethoscope: (A) Original Laennec stethoscope (1816); (B) Cammann binaural stethoscope (1852); (C) Kerr Symballophone (1940); (D) Rappaport-Sprague stethoscope (1960); (E) Original Littmann stethoscope (1961); and (F) Littmann electronic stethoscope model 4000 (2000). *Adapted from Medical Antiques Online [Medical Antiques Online]; used with permission.*

Medical simulation is a rapidly evolving field, bridging the gap between medicine and engineering, and is perceived by many as the future of medical education to prevent the continuously rising costs of training future doctors. Predictions consider that this medical simulation market will reach numbers as high as 2.27 billion USD by 2021 [4]. Motivated by all of this, our main objective in this chapter is to present a perspective on how virtual patient technologies, namely, one called IS4Learning, can address these difficulties in the teaching of auscultation.

Is auscultation becoming a lost art? Some believe the stethoscope has become the forgotten instrument in cardiology [5] and auscultation has lost its importance. Can we really afford to lose this technique and its respective know-how?

Many factors have conspired to limit adequate teaching and maintenance of cardiac auscultation skills. Indeed, the requirements and expectations of junior doctors, with regard to auscultation, are much lower now than in previous generations [6]. A major reason for poor auscultation skills is that teaching methods have changed little in the last 50 years, and the number of medical students today is significantly larger. If we associate this with technological advancements such as ultrasonic imaging and Doppler techniques, cardiac auscultation is receiving less emphasis in teaching and practice [6]. Some technologies have tried and failed to address this limitation, such as audio CDs, websites, and mannequins, since none of these are able to address the real needs of today: the mass teaching and certification of auscultation skills, allowing 500 repetitions per type of cardiac murmur, as defended by Barret et al. [7]. The consequence of all of this is that cardiac auscultation has no structured teaching in three fourths of American internal medicine programs and two-thirds of cardiology programs [8]. This will inevitably lead to poor practice and teaching of this technique at all levels of training. Although not as well documented, the same process of attrition is probably affecting the other cognitive skills of bedside examination [9–11].

THE ESSENTIALS OF AUSCULTATION

Heart and lung auscultation are part of the basic patient physical examination [12]. It is dependent on the knowledge of the human anatomy, and the understanding of the transmission of sounds to the body surface. The observed signs should always be interpreted anticipating anatomical variants, and evaluated within clinical context (e.g., innocent murmurs in children). Auscultation allows the extraction of information from the cardio hemic and respiratory systems, and different steps are required for a comprehensive assessment of the patient status. The physical examination starts when the patient enters the room. Observation of skin coloration, fingers, lips, and other clinical signs (e.g., breathing difficulty) may already convey important information on the patient's state. This, together with the information transmitted by the patient, is the prelude for a guided auscultation [13].

Auscultatory heart sounds include the S1, corresponding to the closure of the mitral and tricuspid valves, and the S2, corresponding to the closure of the aortic and pulmonary valves; extra sounds may be heard including extra systolic sounds (ejection sounds or clicks), extra diastolic sounds (S3, S4, or opening snap), and systolic and diastolic murmurs. Their location of origin and propagation is dependent on the heart morphology, and is the basis for the need of different auscultation areas. To detect and evaluate relative amplitudes in the different auscultation areas has important clinical value for the diagnosis of heart disease. As an example, the timing, shape, location of maximum intensity, radiation, and frequency content of heart murmurs, are useful characteristics in their description to discern between diseases. Heart auscultation is initiated with the patient seated or in the supine position. Inspection of the thorax allows the definition of some important anatomical references that are useful for the adequate placement of the stethoscope: the sternum, the clavicle, and the axillary lines. Auscultatory areas include: the left sternal border of the second intercostal space (ICS), also referred to as aortic area; the right sternal border of the second ICS, also referred to as pulmonic area; the third, fourth and fifth left para-sternal ICS (fourth and fifth also known as tricuspid area); and the fourth and fifth left ICS in the mid-clavicular line, also referred to as the mitral area. Heart sounds and murmurs originated in the four valves may be heard in different areas of the thorax (irradiation) as represented in Fig. 19.2 [13,14].

Some maneuvers may aid in the detection of specific signs such as asking the patient in supine position to partially rollover to the left side (approximate left ventricle to the chest wall), this accentuates the left-side S3, S4, and mitral murmurs from the mitral stenosis. Heart sounds vary within the respiratory cycle, and respiration sounds may also interfere in their assessment (Valsalva maneuver). It is also important to understand the different features of a

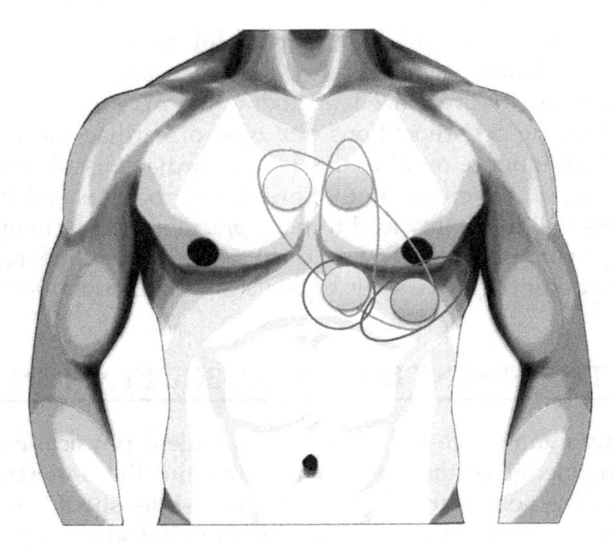

FIGURE 19.2 Main auscultation areas, where the heart sounds and murmurs originating in the different heart valves are more audible: *green*, aortic; *orange*, pulmonic; *blue*, tricuspid; *purple*, mitral.

stethoscope: the diaphragm is better for high frequency sounds such as the S1, S2 and murmurs, while the bell is more sensitive to low frequency signals such as the S3 and S4, applied with low pressure on the patient's' chest. Auscultation of the carotids may also be useful [14].

Concerning the examination of the respiratory system and lung auscultation, it initiates in the same manner, with the patient in the seating position, the thorax is inspected for abnormalities and symmetry during the respiratory cycle, which should be evaluated in terms of rhythm and amplitude, palpating and evaluating the thrill. It is important to identify several anatomical sites such as the suprasternal notch and the trachea (possible deviations), the scapula, the spinous processes, ribs, among others. After inspection, the different auscultation spots (intercostal spaces) are identified, and scanned during a full respiration cycle in the search for altered or absent sounds by comparison with their analogous (top-bottom symmetric search, initiated in the apex). Identification of normal sounds (normal bilateral vesicular sounds, bronchovesicular, and bronchial) or adventitious sounds (crackles. wheezes, rhonchi, pleural rubs, and stridor), might be a challenge. All the sounds should be evaluated within the respiratory cycle (frequency content, amplitude, and timing) to distinguish different pathophysiological processes (e.g., high or low blockade, consolidation). It may also be informative to assess transmitted voice sounds (bronchophony, egophony, and pectoriloquy), asking the patient to repeat different words and assessing their transmission characteristics that vary according to the medium of transmission (airless lung). Lung auscultation should be executed anteriorly, laterally and posteriorly, with the patient breathing with an open mouth. The arms crossed in the front, and the hands over the shoulders increase scapulas spacing [13].

It is clear from this brief description of the auscultation process that different abilities are necessary to perform an auscultation. Proficiency is dependent on the acquisition of three different skills:

- **Positioning**: identification of the auscultation area, based on anatomical references and adjusted to different biotypes

- **Gesture**: the mechanical gesture to adequately place and hold a stethoscope over the auscultation areas
- **Listening**: identifying, and understanding of the sounds by integrating the auscultation signs with the pathophysiology

Acquiring competences in these three axis motivates the development of new technologies for the guided study of healthcare professionals and students. Available tools were proven in the past to be insufficient, with poor detection of pathological signs during an auscultation [15].

HOW DO WE TEACH AUSCULTATION TODAY?

Traditionally, cardiac auscultation has been taught best at the bedside during clinical undergraduate training and in preparation for postgraduate membership examinations. It is an essential component of the clinical examination, but like most clinical skills requires repetition [7] and clinical experience to make an accurate diagnosis. However, as many as three-quarters of USA residents and two-thirds of cardiology trainees no longer receive formal teaching in cardiac auscultation [8]. Several studies have reported an apparent lack of ability of residents to correctly diagnose a cardiac murmur [16] and this decline in competency in cardiac auscultation spreads across the board, from medical students to practicing physicians [17].

If this decline is related to the lack of clinical exposure and deficiencies in current teaching methods, maybe this can be rectified by proposing new methodologies, approaches, and solutions. Recent studies encourage this possibility, showing that the use of computer-based simulation increases the ability and confidence in detecting cardiac murmurs and added heart sounds [18], auscultation skills [19], knowledge of cardiac auscultation, [20] and general improvement in auscultation [21,22].

Simulation-Based Auscultation Training

The first simulator dedicated to auscultation training was Harvey Cardiology Patient Simulator (Fig. 19.3), launched in 1968 which presents several cardiovascular conditions and supports a comprehensive curriculum.

This sophisticated mannequin is able to display a number of cardiovascular indices and underwent rigorous testing as an educational model with pilot studies first reporting promising results as early as 1980 [23,24]. It has since undergone several modifications and remains an important learning resource for healthcare professionals and trainees. Several other simulators have followed the advent of Harvey like Auscultation Trainer with SmartScope [25], Cardiology Patient Simulator "K" ver.2 [26], SimPad Sounds Trainer [27]. An important limiting factor for the widespread use of this technology is its high acquisition and maintenance costs, which prevent it from achieving its promised educational impact. It is common that students during their medical training have a single 2 hours' group session with this technology (Fig. 19.3), usually with the presence of an experienced teacher.

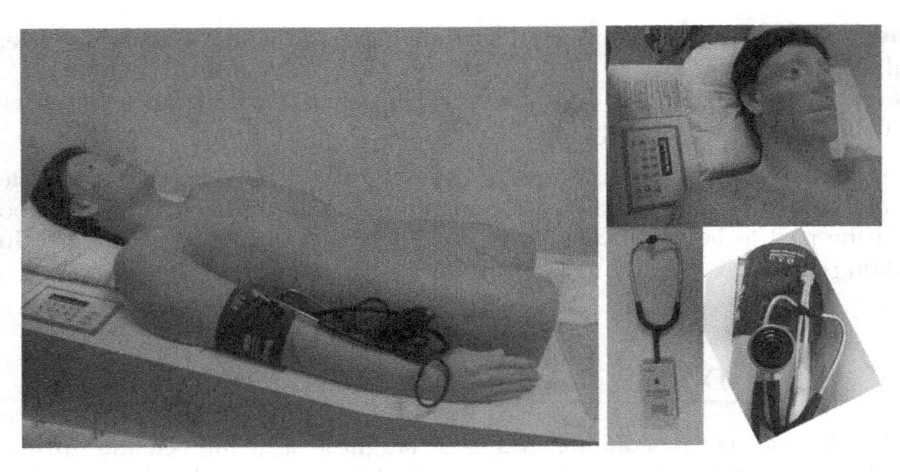

FIGURE 19.3 A modern version of the "Harvey Simulator," which first appeared in 1968.

A number of revision papers have exalted the virtues of high-fidelity patient simulation for training clinical skills. This immersive simulation environment can be used for training of basic technical skills, such as the recognition of heart sounds, but it is most appropriate and cost-effective in the training of the nontechnical skills (or soft-skills) and its application in crisis resource management [4,28,29]. The use of such complex and highly technological environments has advantages over real patient experiences: it encourages the acquisition of skills through experience, in realistic, interactive, and controlled environments; it allows individual or team training in a standardized and graded experience, adapted to the learning needs; it provides the opportunity to deal with rare and/or life threatening situations with no risk to real patients [30]. As opposed to the clinical setting, where errors must be prevented or repaired immediately to protect the patient, in a simulated environment error may be allowed to progress, so as to demonstrate their implications to the trainee, or to enable a quick reaction to rectify them [28].

Can We Rethink the Teaching of Auscultation Using Simulation? Yes We Can!

Without easy access to real patients and to forbiddingly expensive simulators, a medical student can only resort to passively listening to libraries of sounds that are usually synthesized or collected in artificial environments. Would it be possible to explore novel interaction technologies to create virtual patient that can reach a compromise between these two approaches? The IS4Learning is a virtual patient simulator designed for tablet devices that allows students to auscultate a virtual torso with a real electronic stethoscope, listening to real auscultations gathered from real patients in a variety of real environments [31]. Heart sounds libraries can provide students with a rich and realistic set of sounds, but still fail to mimic the required mechanical gestures and stethoscope positioning of a real auscultation exam. The IS4Learning takes advantage of digital stethoscopes and the rich touch interaction available in most portable devices to simulate the touch of the stethoscope with the torso of the patient. When the student touches the screen with the stethoscope tip, a sound is sent the stethoscope corresponding to the one that he would be listening in a real body, training muscular memory and visual positioning. The combination of this virtual environment with a library rich in real auscultations provides the students with personal and portable simulators that will increase their number of training hours as well as the number of different signs a medical student will hear during his training. In order to fully understand how this simulation technology can affect the teaching of auscultation, we need to discuss two distinct topics: understand the details of the various developed technologies, and report pilot experiences in three archetypal learning use-cases such as classroom teaching, e-learning, and self-study.

IS4HEALTH SIMULATION TECHNOLOGIES

The first part of the teaching equation is the simulation technology itself. What are the functionalities that these technologies can provide us? Why are they coherent with the specific needs of training auscultation skills? Do they have relevant limitations that we must account for? A total of three simulation technologies have been researched, developed, deployed, and tested, all of which have a high potential for impact regarding the teaching of auscultation. There is the core technology (IS4Learning) that has been quickly summarized in the last section, a simplified web version (IS4Learning Web) that sacrifices some functionalities for a more appealing price and user pool reach, and a more powerful multi-user simulation technology for enhanced presential learning (IS4Sharing). In this section, we will describe each technology, including its development and results.

The IS4Learning Technology

As summarized in "Can We Rethink the Teaching of Auscultation using Simulation? Yes We Can!" section, the IS4Learning Technology is a virtual patient simulator designed for tablet devices that allows students to auscultate a virtual torso with a real electronic stethoscope. As discussed in "The Essentials of Auscultation" section, motivation for this technology comes from the understanding that there are three fundamental skills for performing auscultation: the ability to identify all the correct auscultation spots in a patient's body, the ability to place and hold the stethoscope correctly in all the identified spots, and the ability to listen and understand the heart and lung sounds produced by a stethoscope. According to Barrett et al. [7], and for the specific case of cardiac auscultation, this is only possible to learn correctly by intense repetition of the gesture and the listening, estimating that a minimum of 500 repetitions are needed per type of murmur for adequate learning. We have discussed earlier that these three skills can be repeatedly trained with either real patients or patient simulators but the costs and teacher needs of these solutions make them unrealistic in practice. At the other end of the spectrum, audio CDs and their online or mobile application variations, besides only training one of the required skills (listening), they decontextualize the learning process and thus limiting effective knowledge translation to reality and clinical practice. Aware of these limitations and inspired by recent

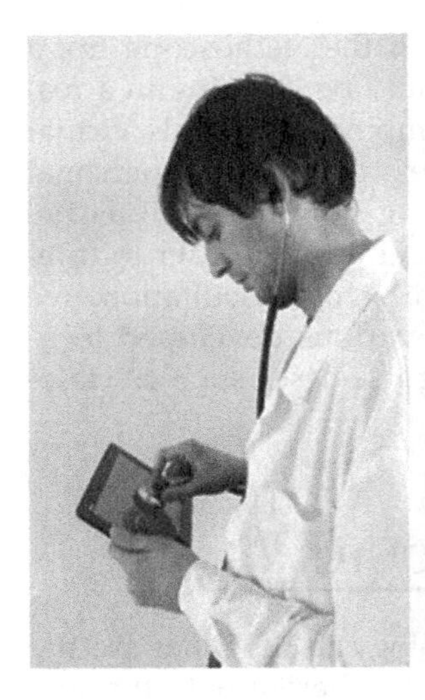

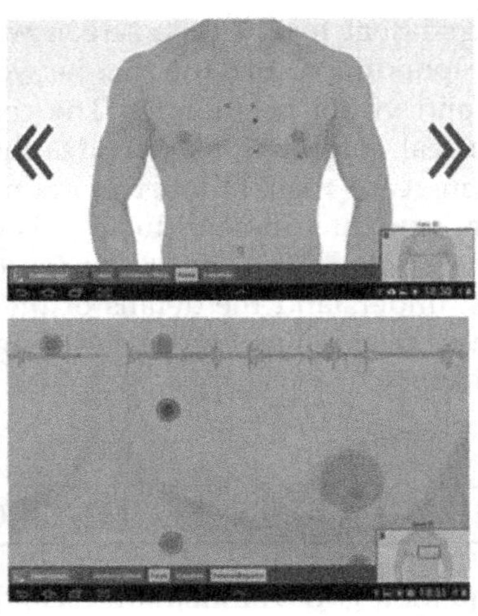

FIGURE 19.4 (left) Picture of a medical student using the virtual simulator "IS4Learning"; (top-right) 2D torso with four sounds ready for auscultation; (bottom-right) Zoomed 2D torso and a depiction of the phonocardiogram of the sound that is being sent and played by the digital stethoscope.

technology innovations in wireless communications and touch-interaction devices, the idea for IS4Learning was born: what if a tablet could function like a virtual torso of a patient, and we could perform auscultation on it with a real stethoscope? (Fig. 19.4).

Engineering this vision into a usable medical simulator required an intense multidisciplinary effort in the areas of human-computer interaction, software engineering, and medical education. The first stage involved research into the interactive process itself and how it should be implemented in a tablet. The simplest decision was to use a 10″ tablet, which was large enough to cover an area in the chest that includes the four main cardiac auscultation spots in a 1:1 scale. Designing the interaction process itself, in order to create an intuitive and instantly usable technology that anyone could pick up and use was not so simple. The following research questions were identified:

- How should we visually show the auscultation spots on the virtual torso?
- What is the clearest way to display a phonocardiogram on the simulator?
- Would a visual depiction of the skeleton be a good option to help a student train palpation, or should we resort to other alternatives such as tablet vibration when touching over a bone?
- What is the simplest, clearest way for a student to 'navigate' the full virtual torso? A zoom function? A swipe slide?

Low-fidelity prototyping was used for this research stage (Balsamiq—balsamiq.com), associated with heuristic evaluation, a form of discount evaluation method that can be used quickly and effectively for this type of prototypes (Nielsen 1994). Some details of this prototype can be seen Fig. 19.5, and evaluation results motivated small minor corrections but were in general very encouraging [31].

Motivated by this success, a second phase of software engineering ensued, whose objective was to implement this software in an Android application format, which could communicate as unobtrusively as possible with an electronic stethoscope. The choice of the latter technology was simple given that a wireless solution was deemed as imperative in

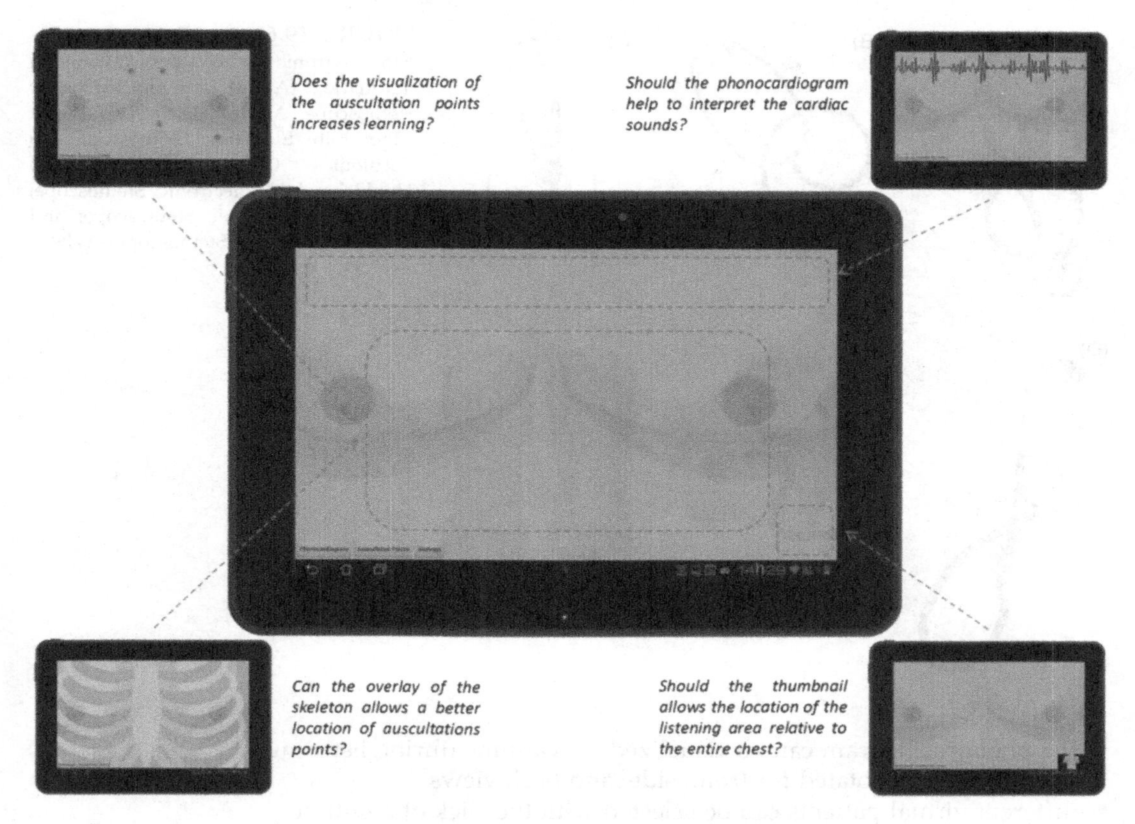

FIGURE 19.5 Preview of the low-fidelity prototype used to test research questions associated with user interaction.

the studies of the first phase, and given the choices available in the market for electronic stethoscopes (Fig. 19.6). Only two options have in-built wireless transmissions capabilities and the 3 M Littmann 3200 not only exhibited much better sound quality during our tests, but has a typical stethoscope ergonomy which enables the adequate training of one of the required skills for auscultation (gesture).

Development of both the application and its integration with the stethoscope was complex, but the final result proved robust, very easy to connect with the stethoscope, and all planned functionalities were successfully implemented (Fig. 19.4):

- touching an auscultation spot in the tablet with the stethoscope plays this sound in the stethoscope
- visual feedback is given as a circle around the stethoscope tip if the user is over a spot or not
- the tablet vibrates when the stethoscope tip is placed over a bone
- palpation is possible by touching the tablet with two fingers and revealing the section of the skeleton underneath

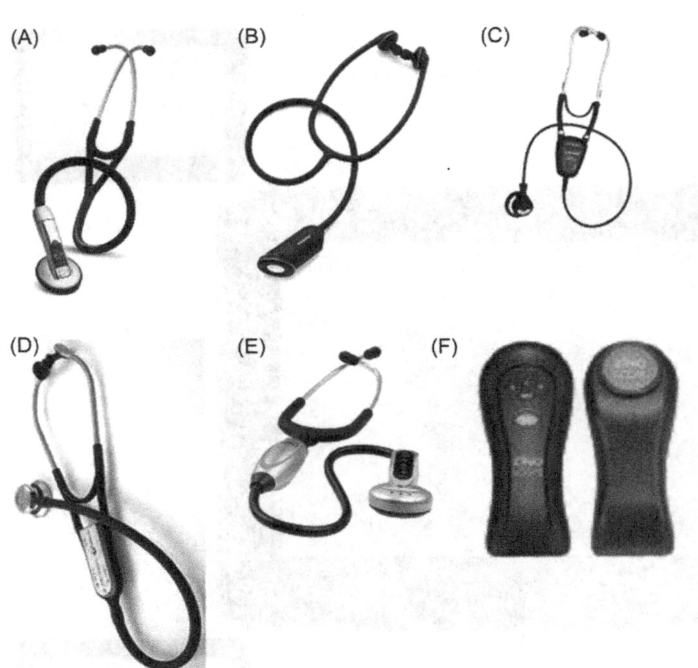

FIGURE 19.6 Six of the most relevant electronic stethoscopes available on the market: (A) 3M Littmann Electronic Stethoscope Model 3200; (B) Welch Allyn Elite Electronic Stethoscope; (C) Cardionics Clinical E-Scope; (D) Thinklabs ds32 Electronic Stethoscope; (E) JABES Electronic Stethoscope; and (F) Wireless Digital Stethoscope—WISE.

- the phonocardiogram can be visualized in real-time during listening
- the torso can be rotated for front, side, and back views
- different virtual patients can be selected with the click of a button

The IS4Learning Web Technology

Although the IS4Learning described in "The IS4Learning Technology" section obtained a very positive reaction from early testers and adopters, medical simulation reality motivated us to create a simpler version of this technology, more adequate to reach a much broader audience. Going back to Barrett et al., we need an average of 500 repetitions per cardiac murmur to achieve satisfactory competence in their detection during clinical practice [7]. Multiply this with the number of cardiac murmurs that must be learned (Barrett performed his studies with only four basic murmurs but if we include more complex ones this number can easily reach 10) and with the high number of medical students that Universities have (the Faculty of Medicine of the University of Porto alone has around 1200 students spread over 6 years of studies), and even if we create auscultation training rooms in Universities, fully equipped with tablets and electronic stethoscopes, this is still a hard target to reach. What if we sacrifice one of the three skills (gesture), while keeping the other two (positioning, listening), but are able to deliver the same interactive learning concept using a web browser and headphones? With this in mind, the IS4Learning Web technology was created (Fig. 19.7).

FIGURE 19.7 A classroom running the IS4Learning Web software in each computer.

From a design perspective, this technology should be a software component that can be easily integrated into either a webpage, or a popular e-learning platform such as Moodle (moodle.org). Simplifications should include the ability to listen to sounds using headphones instead of an electronic stethoscope, and the ability to click on the virtual torso without the need of a touch-interaction device. These choices dramatically reduce the hardware needs of this simulation technology, enabling it to be used in any computer anywhere in the world, be it in a classroom, e-learning, b-learning, or in a self-study environment. Login and user management became mandatory, which actually proved to also be a simple and powerful way to track the learning progress of each individual student. The obvious sacrifice is that there is no easy way to train the student in the way he holds or places the stethoscope, which must be compensated by conventional training. However, the ability to train the two other auscultation skills (positioning, listening) maintains its full effectiveness, with a simpler tool to fit more conventional learning approaches. As we will describe in "Archetypal Auscultation Learning Use-Cases" section, this simulation technology has been tested in a variety of pilots with very promising results, not only hinting at strong pedagogical impact but also provoking intense enthusiasm in both teachers and students.

The IS4Sharing Technology

Bedside auscultation with real patients is one of the most effective ways to learn the three fundamental skills needed, namely, gesture, positioning, and listening. Sadly, this is a complex teaching methodology to implement in reality given its high operational and financial costs. How do we find real patients available for so many hours of training? How do we have enough teachers to be present during these studies? A compromise is usually found in most medical schools in which students have at least one session during their course with a teacher and a real patient. Such a session typically starts with the teacher explaining the auscultation fundamentals and then performing one himself, verbally describing what he is

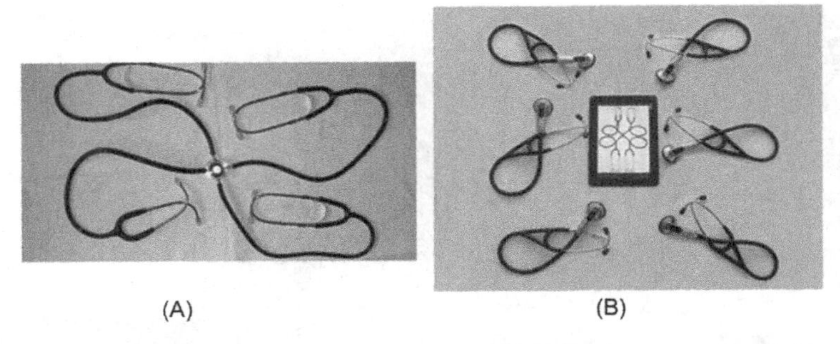

(A) (B)

FIGURE 19.8 Simultaneous auscultation using multiuser stethoscopes. (A) MDF Teaching Stethoscope, (B) Ds Sharing—Wireless digital "octopus."

FIGURE 19.9 Using the IS4Sharing technology to simultaneously listen to the sounds collected by one of a total of seven electronic stethoscopes.

listening and relevant clinical signs such as the presence of murmurs. Students then take turns to repeat this process and discuss what they are listening with the teacher and among themselves. But even this rich, interactive, stimulating learning environment has an important limitation: the listening of these sounds is not simultaneous. In practice, this means that the teacher tries to describe what he is listening, before the students actually listen themselves. And afterward, he must memorize exactly what he heard, so that when the student is listening he can give him pointers to explain, for example, that a cardiac murmur is systolic. And while for cardiac auscultation we have a nicely tuned pump called the heart that produces repeatedly the same sound, this is not true for pulmonary auscultation, in which a patient that tires quickly will soon start producing weaker inspiration and expiration sounds, limiting both explanations and the total number of times students can listen. A solution for this are the multiuser stethoscopes, sometimes called the "octopus" stethoscopes, and described by Davies in 1861 [32], as depicted in Fig. 19.8.

Although this seems like a clever idea, the fact is that sales of this type of "octopus" technology was never very relevant. A possible explanation is that the crowded spaces that bedside learning requires, associated with the inability to make very long tubes due to acoustic energy loss, leads to very poor sound fidelity. What if we could explore the power of wireless transmission of electronic stethoscopes to create a new digital "octopus" technology? Guided by this vision, the IS4Sharing technology was designed, developed, and successfully tested with enthusiastic user reactions (Fig. 19.9).

As functionalities the IS4Sharing technology can transmit an auscultation in real-time to a total of six other electronic stethoscopes, using an Android smartphone or tablet as its wireless communications hub. In order to boost the dynamic nature of group learning, the 'source' stethoscope can be switched with a simple button click, enabling the auscultation by students and teacher, without having to switch stethoscopes among themselves. The limit of six stethoscopes can be expanded by "chaining" more smartphones as support transmission hubs. Such a simple technology opens immense possibilities to the fields of education and medical simulation since there is no reason that this interaction is limited to a bedside setting. We can auscultate virtual patients in a group environment, we can remotely auscultate patients by transmitting these sounds via internet, or we can save the auscultations performed by the doctors and students for posterior analysis. The educational potential of all these functionalities and technologies is clear but does any of it actually enhance the learning process? Some of the more obvious use-cases will be discussed in the following section.

ARCHETYPAL AUSCULTATION LEARNING USE-CASES

The second part of the teaching equation is the learning scenario itself. Which scenarios should we consider for the learning of auscultation skills? What are their limitations? Which of these can benefit from the described technologies? Besides the traditional classroom scenario, in which a teacher shares the same physical space with a number of students and capacitates them using a lecture format, we can consider alternative learning use-cases such as the increasingly popular e-learning or b-learning scenarios, or even a self-learning use-case in which the student has tools to learn by himself and is then subjected to a standardized test for knowledge and competence certification. In this section we will discuss how these three archetypal auscultation learning use-cases can benefit from the novel simulation technologies presented in the last section, presenting whenever possible very promising results from the pilot studies done so far.

Classroom Training

The classroom training of auscultation consists of a theoretical introduction to the auscultation technique, followed by a description of the main findings and their clinical meaning. Usually, digitally recorded cases are provided to allow the assessment of the acquired competences during the class, allowing a more interactive transfer of knowledge.

Use-Case: A teacher is teaching a physiology class for second year medical students, in which he will explain the theory and practice of auscultation. After explaining how the heart produces sounds, what the various heart sounds are, what murmurs are and what causes them, he proceeds to discuss practical examples of normal and pathological heart sounds. Each student then picks up his tablet, and puts on his electronic stethoscope. In order to show normal heart sounds, the teacher calls up a volunteer student, which he then proceeds to auscultate using his own electronic stethoscope. Students see the positioning and observe the gesture, and listen in real-time to the performed auscultation as well as the teacher's comments and remarks. Afterward, the teacher tells them to activate virtual patient 1 for this class and each student tries to auscultate his tablet directly, training positioning, gesture and listening. Finally, the teacher tells them to activate virtual patient 2 and chooses one student to perform auscultation on the tablet. All others listen to it simultaneously. A discussion follows moderated by the teacher about the characteristics of the sounds listened.

Required Simulation Technology:

- IS4Sharing—Enabling simultaneous listening of real and virtual patient auscultations
- IS4Learning—Enabling the auscultation of virtual patients

Pilot Studies: Pilot studies were developed with the proposed simulation technologies, during the Med. Win 2015 conference at the Faculty of Medicine of the University of Porto (Portugal). Three cardiac auscultation workshops (1 hour each) were provided to a total of 4 medical students (Fig. 19.10). Data were collected in a real clinical environment, using an electronic stethoscope Littmann 3200 and the DigiScope Collector [33]. The patient's age ranged from 3 to 76 years. A Cardiologist determined the cardiac findings that would be assessed during the tests, and two resident of Internal Medicine determined the cases that would be assessed.

The workshop was designed as follows: the auscultatory findings of ten real patients, via a virtual torso integrated in an eLearning platform (Moodle) and headphones, were presented. The virtual torso (Fig. 19.11) allowed to listen five auscultatory areas (aortic

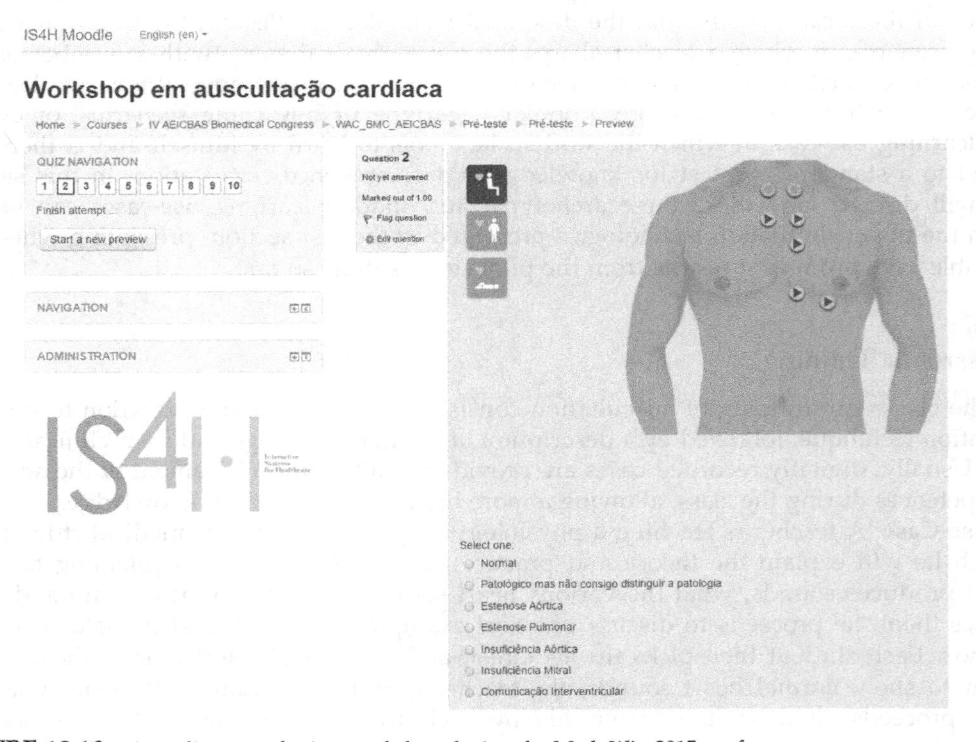

FIGURE 19.10 A cardiac auscultation workshop during the Med. Win 2015 conference.

FIGURE 19.11 IS4Learning Web screenshot.

valve area; pulmonic valve area; Erb's point; tricuspid valve area; mitral valve area) through a Pioneer SE-M521 headphone (frequency response 7–40.000 Hz; impedance 32Ω; sensitivity 97 dB/mW). Each individual user was able to hear the same heart sounds of the patients presented in the workshop, at the same time, but at his own rhythm, as if he or she were listening directly to the patient through his or her own stethoscope.

The workshop began with a pretest to evaluate the student's' skills to diagnose a wide variety of heart sounds and murmurs and characterize the cardiac sounds, followed by a two-parts course: 15 minutes of a theoretical lecture, and 45 minutes during which the auscultatory findings of data from ten real patients were presented and discussed. For each case, the instructor began with a presentation and quick discussion of the patient's clinical history. Following the participants auscultated the provided virtual torso in the web platform, and discussed the final diagnosis according to the auscultation and clinical history provided. The workshop ended with a posttest.

Evaluation of diagnostic proficiency consisted of five cardiac auscultatory events directly recorded from real patients and reproduced through a virtual patient as described above. Test takers had the opportunity to listen to each auscultation area and were asked to characterize the sound and provide a diagnosis. Table 19.1 presents the global results of the workshops. It may be observed that the global grade increased in the posttest as well as the accuracy in the adequate classification of normal sounds. It is also interesting to notice the student's very poor ability when faced with pathological cases, which did not increase with just two hours of training (a slight decrease was observed, possibly due to the added confidence but unimproved skill), highlighting the importance of the 500 repetitions needed, as claimed by Barrett et al. [7]. In accordance to what was observed in other studies, we found a high detection error in the pretest that improved with this simple intervention. We believe this serves as motivation to the introduction of new teaching paradigms for the learning of the auscultation.

TABLE 19.1 Global Results of the Three Cardiac Auscultation Workshops During the Med. Win 2015 Conference

Results	Pretest	Posttest	Variation
Grade	38.10%	43.98%	5.88%
Accuracy in normal cases	41.75%	57.66%	15.91%
Accuracy in pathological cases	21.90%	17.87%	−4.03%

Use Case: e-Learning

E-learning can be defined as the delivery of a learning, training, or education program by electronic means. It involves the use of a computer or electronic device (e.g., a mobile phone) in some way to provide training, educational, or learning material [34]. An e-learning environment can be divided into two categories: synchronous and asynchronous. Synchronous e-learning involves online studies through chat and videoconferencing. This kind of learning tool is real-time. It is like a virtual classroom, which allows students to ask, and teachers to answer questions instantly, through instant messaging. On the other hand, asynchronous learning can be carried out even while the student is offline. Asynchronous e-learning involves coursework delivered via web, email and message boards that are then posted on online forums.

Use-Case: In order to prepare for a mass cardiac pathology screening operation, a regional hospital needs to refresh the knowledge of 100 nurses in a set of bedside skills including auscultation. It also needs a certification mechanism that can assess that these professionals are good enough to accomplish the planned task. Given that these nurses practice in different health centers in different regions, they decide to enroll in an auscultation b-learning course of a remote University that offers this. This course consists of a series of 10 one-hour synchronous lectures in which the students listen to the teacher via Skype for 45 minutes and then selected questions are answered during 15 minutes. During classes, the teacher exemplifies knowledge by asking students to listen to selected auscultation sounds by clicking with the mouse over IS4Learning Web virtual patients in their course Moodle for that class. Students have their own headphones at home for listening, and select spots using their mouse. When the class ends, the teacher tells them to practice selected virtual patient cases in their Moodle before the next class. When the module ends, the teacher can inspect individual student progression by tracking their activity on IS4Learning Web, and certify their auscultation knowledge and skills via a final theoretical and practical exam using virtual patients.

Required Simulation Technology:

- IS4Learning Web—Enabling the auscultation of virtual patients at a student's home and without specialized hardware, and the individual tracking of a student's answers, including the final exam's

Pilot Studies: The Faculty of Medicine of the University of Porto provides a sports medicine specialization coordinated by Professor Ovídio Costa (Cardiologist specialized in Sports Medicine). One of the modules is about Cardiology, including the teaching of auscultation skills. In the past 2 years, a virtual torso embedded in the Moodle platform,

FIGURE 19.12 A classroom equipped with computers transformed in cardiac examination room.

allowed 148 doctors to virtually listen to cases during class, and also enabled students to access them from home, providing a continued learning process. The feedback was extremely positive and most of the students asked for the possibility to continue to access the virtual cases after the end of the course.

Other advantage of this technology is to offer a simultaneous evaluation of the student. A classroom equipped with computers (Fig. 19.12) can be transformed into an examination room for cardiac auscultation, avoiding an individual evaluation through a simulator, or the use of real patients, reducing dramatically the time required to evaluate all students.

Use Case: Self-study

The study of auscultation is highly dependent on knowing the gesture integrated with the recognition of the typical signs of pathology, as early described. One of the most challenging, and also more rich environments, is the bedside learning. Although this is the most interesting modality for the student, it is highly unreliable since there is no certainty on the adequate identification of the auscultation signs. It is desirable that the student trains theses skills extensively before addressing the patient, which is not compatible with the learning paradigm in the classroom training. This motivates the development of self-study tools that may provide a realistic experience, and allow the exhaustive training of different pathologies' identification.

Use-Case: A third-year medical student is preparing for his medical semiology exam in which he will have to auscultate a real patient and identify relevant cardiac and pulmonary auscultation signs, before reaching a diagnostic. He only had around one hour of training in his university's Harvey Simulator and feels unprepared for a real patient evaluation. He books a table in the auscultation room of his university's simulation center, where he can find an electronic stethoscope and a tablet with the IS4Learning software. In here, he trains for a few hours the positioning, gesture and listening of cardiac and pulmonary sounds using virtual patients. Still unsatisfied, he uses his same login to practice at home in his IS4Learning Web account, performing training exams until he can consistently obtain very good marks for most types of signs. He quickly checks the online dashboard of his results, which confirms that his learning progress has been consistently positive and has reached a very high level across all auscultation fields. He is now ready and confident for his medical semiology exam.

Required Simulation Technology:

- IS4Learning—Enabling the auscultation of virtual patients within a simulation center
- IS4Learning Web—Enabling the auscultation of virtual patients at his home.

The learning of auscultation by self-study has been addressed in the past, and there are a few platforms available for the virtual training of auscultation, such as the 3 M Littmann Auscultation Lessons [35], the Blaufuss, and the Easy Auscultation [36]. Some CD-ROM courses are also available. The shortcomings of some of these platforms, is the limited number of cases, no clinical history to train the diagnostic skills, the lack of correlation between anatomical positioning and the clinical findings, and also the use of synthetic data.

Usually, the medical student that is preparing for the semiology exam, has access to only a few hours of simulation tools like the Harvey, and practical experience with only a few patients, which does not allow the training of detection skills facing the existent multiplicity of clinical findings. The preparation is done basically by studying the theory of the gesture, and of the clinical findings within a clinical scenario, which has been proven in the past to be insufficient [8]. This demonstrates the need for more realistic tools that allow the extensive home-based training. Being able to train different pathological cases, from different real patients, training the gesture of the auscultation, and also allowing a self-evaluation of the acquired competences, may improve the auscultation skills. IS4Learning and IS4Learning Web arise in this context, as response to the needs expressed by medical students and other healthcare professionals, that wish to improve their proficiency in auscultation, a simple but yet important first line screening tool for heart and lung disease detection.

AUSCULTATION 2.0

In this book chapter, we have motivated that it is imperative that we rethink the way we teach auscultation to prevent the decay of this art. Interactive systems using electronic stethoscopes provide an interesting compromise between the unavailability of real patients and the high cost of physical patient simulators. A good example is our novel IS4Learning technology, which has been successfully deployed and tested with real students. Can simulation play a pivotal role in the recovery of this (almost) lost art? Yes it can!

Acknowledgments

This work was partially funded by the Fundação para a Ciência e Tecnologia (FCT, Portuguese Foundation for Science and Technology) under the references Heart Safe PTDC/EEI-PRO/2857/2012 and SFRH/BD/80650/2011; and Project I-CITY—ICT for Future Health/Faculdade de Engenharia da Universidade do Porto, NORTE-07-0124-FEDER-000068, funded by the Fundo Europeu de Desenvolvimento Regional (FEDER) through the Programa Operacional do Norte (ON2) and by national funds through FCT/MEC (PIDDAC).

References

[1] Shaver J. Cardiac auscultation: a cost-effective diagnostic skill. Curr Probl Cardiol 1995;20(7):441–530.
[2] Pelech AN. The physiology of cardiac auscultation. Pediatr Clin North Am 2004;51(6):1515–35, vii–viii.
[3] Luisada AA. The functional murmur: the laying to rest of a ghost. Dis Chest 1955;27(5):579–81.
[4] Markets Ma. Healthcare/Medical Simulation Market by Product & Services (Patient Simulator, Task Trainers, Surgical Simulator, Web-based Simulation, Software, Dental Simulator, Eye Simulator), End User (Academics, Hospital, Military)—Global Forecast to 2021; 2016.
[5] O Rourke R, Braunwald E. Physical examination of the cardiovascular system. Harrisons Principles of Internal Medicine, vol. 1. New York: McGraw-Hill; 2001. p. 1255–61.
[6] Tavel ME. Cardiac auscultation: a glorious past—and it does have a future!. Circulation 2006;113(9):1255–9, Epub 2006/03/08.
[7] Barrett MJ, Lacey CS, Sekara AE, Linden EA, Gracely EJ. Mastering cardiac murmurs*: The power of repetition. Chest 2004;126(2):470–5.
[8] Mangione S, Nieman LZ, Gracely E, Kaye D. The teaching and practice of cardiac auscultation during internal medicine and cardiology training. A nationwide survey. Ann Intern Med 1993;119(1):47–54.
[9] Brown RS. House staff attitudes toward teaching. J Med Educ 1970;45(3):156–9, Epub 1970/03/01.
[10] Goetzl EJ, Cohen P, Downing E, Erat K, Jessiman AG. Quality of diagnostic examinations in a university hospital outpatient clinic. Ann Intern Med 1973;78(4):481–9, Epub 1973/04/01.
[11] Wray NP, Friedland JA. Detection and correction of house staff error in physical diagnosis. JAMA 1983;249 (8):1035–7.
[12] Force AT. Recommendations for clinical skills curricula for undergraduate medical education. Association of American Medical Colleges; 2008.
[13] Bickley L, Szilagyi PG. Bates' guide to physical examination and history-taking. Philadelphia: Lippincott Williams & Wilkins; 2012.
[14] Perloff JK. Physical examination of the heart and circulation. PMPH-USA; 2009.
[15] Mangione S. Cardiac auscultatory skills of physicians-in-training: a comparison of three English-speaking countries. Am J Med 2001;110(3):210–16, Epub 2001/02/22.
[16] Alam U, Asghar O, Khan SQ, Hayat S, Malik RA. Cardiac auscultation: an essential clinical skill in decline. Br J Cardiol 2010;17(1):8–10.
[17] Vukanovic-Criley JM, Criley S, Warde CM, Boker JR, Guevara-Matheus L, Churchill WH, et al. Competency in cardiac examination skills in medical students, trainees, physicians, and faculty: a multicenter study. Arch Intern Med 2006;166(6):610–16, Epub 2006/03/29.
[18] Ostfeld RJ, Goldberg YH, Janis G, Bobra S, Polotsky H, Silbiger S. Cardiac auscultatory training among third year medical students during their medicine clerkship. Int J Cardiol 2010;144(1):147–9.
[19] Stern DT, Mangrulkar RS, Gruppen LD, Lang AL, Grum CM, Judge RD. Using a multimedia tool to improve cardiac auscultation knowledge and skills. J Gen Int Med 2001;16(11):763–9.
[20] Vukanovic-Criley JM, Boker JR, Criley SR, Rajagopalan S, Criley JM. Using virtual patients to improve cardiac examination competency in medical students. Clin Cardiol 2008;31(7):334–9.
[21] Criley JM, Keiner J, Boker JR, Criley SR, Warde CM. Innovative web-based multimedia curriculum improves cardiac examination competency of residents. J Hosp Med 2008;3(2):124–33, Epub 2008/04/29.
[22] Tuchinda C, Thompson WR, editors. Cardiac auscultatory recording database: delivering heart sounds through the Internet. Proceedings of the AMIA Symposium. American Medical Informatics Association; 2001.

[23] Ewy GA, Felner JM, Juul D, Mayer JW, Sajid AW, Waugh RA. Test of a cardiology patient simulator with students in fourth-year electives. J Med Educ 1987;62(9):738–43, Epub 1987/09/01.

[24] Gordon MS, Ewy GA, Felner JM, Forker AD, Gessner I, Mcguire C, et al. Teaching bedside cardiologic examination skills using Harvey, the cardiology patient simulator. Med Clin North Am 1980;64(2):305–13.

[25] Simulaids. Life/form® Auscultation Trainer & SmartScope™, Available from: <https://www.simulaids.com/product/170-LF01142>; 2016 [cited 09/03/2017].

[26] Kagaku K. Cardiology Patient Simulator "K"ver.2, Available from: <https://www.kyotokagaku.com/products/detail01/mw10.html>; 2012 [cited 09/03/2017].

[27] Laerdal. SimPad Sounds Trainer, Available from: <http://www.laerdal.com/us/item/200-20150>; 2015 [cited 09/03/2017].

[28] Bradley P. The history of simulation in medical education and possible future directions. Med Educ 2006;40(3):254–62, Epub 2006/02/18.

[29] Issenberg SB, McGaghie WC, Petrusa ER, Lee Gordon D, Scalese RJ. Features and uses of high-fidelity medical simulations that lead to effective learning: a BEME systematic review. Med Teach 2005;27(1):10–28.

[30] Schuwirth LWT, van der Vleuten CPM. The use of clinical simulations in assessment. Med Educ 2003;37(s1):65–71.

[31] Pereira D, Gomes P, Mota É, Costa E, Cruz-Correia R, Coimbra M, editors. Combining a tablet and an electronic stethoscope to create a new interaction paradigm for teaching cardiac auscultation. In: International Conference on Human-Computer Interaction; 2013: Springer.

[32] Abelson D. Stereophonic stethoscope with teaching attachment. Am J Cardiol 2000;85(5):669–71, A11. Epub 2000/11/15.

[33] Pereira D, Hedayioglu F, Correia R, Silva T, Dutra I, Almeida F, et al., DigiScope—Unobtrusive collection and annotating of auscultations in real hospital environments. Engineering in Medicine and Biology Society, EMBC, 2011 Annual International Conference of the IEEE; 2011: IEEE.

[34] Stockley D. E-learning definition and explanation (Elearning, Online training, Online learning). Retrieved March. 2003; 15:2010.

[35] 3M Littmann Stethoscopes. Auscultation lessons. Available from: <http://www.littmann.ca/wps/portal/3M/en_CA/3M-Littmann-CA/stethoscope/littmann-learning-institute/auscultation-education/> [cited 23.03.17].

[36] Easy auscultation. Lessons, quizzes & guides. Available from: <https://www.easyauscultation.com/> [cited 23.03.17].

CHAPTER

20

The Use of Mobile Technologies in Nursing Education and Practice

Hanan Asiri[1] *and Mowafa Househ*[2]

[1]Armed Forces Hospitals Southern Region, Khamis Mushait, Saudi Arabia
[2]King Saud Bin Abdulaziz University for Health Sciences, Riyadh, Saudi Arabia

INTRODUCTION

In recent years, there has been notable growth in the use of mobile technology around the world. The uptake of such devices has penetrated and accelerated human socializing, communication, and entertainment on a global level. Yet, when it comes to the realms of learning and education, a number of scholars have voiced their concerns about the utilization of mobile technology in this domain of human interaction. For example, Herrington et al. (2009) indicated that such conventional tools have been minimally used in learning contexts. Additionally, they pointed out that learning environments, which use these tools, have little theoretical foundation in their use, even though they are so-called "early adopters" willing to utilize this new technology for pedagogical purposes. It is still unclear if using mobile technology in learning has a sound theoretical reasons behind it [1]. In practice, however, many scholars study the way mobile technology has affected many areas of life such as banking, commerce, health, etc. Mobile health technology, which is our concern in this chapter, is known as mHealth. One broad definition for mobile health is the definition of the World Health Organization (WHO) which defines mHealth as the "medical and public health practice supported by mobile devices such as mobile phones, patient monitoring devices, personal digital assistants (PDAs), and other wireless devices" [2].

Nevertheless, tracking back to the beginning of what we call today mobile technology, Gerard Goggin discusses an evolutionary journey that started in the second half of the nineteenth century with the invention of the telephone. In his book "Cell Phone Culture: Mobile Technology in Everyday Life" [3], Goggin describes how such an invention became a well-established way to communicate in various ways with others like family and friends, engaging in social organizations and activities, and even to conduct business by the middle of the twentieth century—at least as it applies to industrially advanced

Health Professionals' Education.
DOI: http://dx.doi.org/10.1016/B978-0-12-805362-1.00020-6

countries. With the introduction of mobile phones, the fixed telephone was replaced and mobile phones largely took its place by the beginning of the twenty-first century. Goggin describes this revolutionary voyage toward mobile technology with statistics about the penetration of mobile technology in different parts of the world and explains further, exploring in more depth how this form of the technological revolution affects our life today. Recent numbers from the 2015 report of the International Telecommunications Unit (ITU) [4] estimated that as of the end of 2015, there were more than 7 billion users in the world with a mobile line. According to the same report, we can see that between 2010 and 2015, there was a steady decrease in the number of fixed telephone subscriptions, in both developed and developing countries. Additionally, in ten years—between 2005 and 2015, in the same countries, mobile-cellular telephone subscriptions have risen dramatically from 992 million and 1.2 billion in 2005 to almost 1.5 and 5.6 billion in 2015 in the developed and developing worlds, respectively. This means an increase in subscriptions by more than 500 million in the developed world and 4 billion in the developing world.

This remarkable penetration of mobile technology in every aspect of our life today includes the sectors of education, medicine, and communication. Focusing on the importance this technology is a necessity to study its effect on the profession of Nursing. Therefore, this chapter is organized to introduce the reader to the rise of mobile technology used in nursing education and practice. Subsequently, we conclude with a brief review of future trends in the use of mobile technologies in nursing education and practice.

THE USE OF TECHNOLOGIES IN NURSING EDUCATION AND PRACTICE

The introduction of technology in its various forms (e.g., fax, computer, and mobile computing) into the domain of nursing has created many opportunities and challenges for both nurse educators and practitioners. The introduction of technology into nursing practice was accompanied by challenges for the nursing profession such as: (1) not feeling comfortable in using new forms of technologies that impact their traditional way of practice; and (2) a fear of jeopardizing human interaction in the nurse–client relationship.

The first reported introduction of computer applications into the professional and scholarly literature of the nursing profession dates back to the early 1970s with the first "computer applications in nursing" reports which began to appear at that time [5]. These systems, that is, Nursing Care Planning systems, were aimed to relieve the documentation burden on the staff and improving the accuracy as well as quality of the plan [6,7].

In her paper titled "An Overview of Nursing Informatics (NI) as a Profession: How We Evolved Over the Years," Asiri (2016) tracks back the beginning of the field of nursing informatics (NI) and provides an international glimpse into how technology has diffused within nursing practice. With the exception of the United States and other western countries, there is little documentation on the historical evolution and progression of technology use in nursing education and practice [8]. Nevertheless, examining how the term "nursing informatics" evolved throughout the years and its impact on nursing practice can serve as a historical record for understanding how technology has historically impacted nursing education and practice.

In 1976, Scholes and Barber first proposed the term "nursing informatics" and the term continued to be used by many scholars in the 1980s including Ball and Hannah (1984) [9], Hannah (1985) [10], and Grobe (1988) [11], and is still used today by many other scholars as well [12,8]. Furthermore, nursing informatics was approved by the American Nurses Association (ANA) in 1992 as a new nursing specialty [13]. The year of 1982 witnessed the international recognition of nursing informatics by the International Medical Informatics Association (IMIA) by establishing Working Group 8 (WG8) intended to represent the nursing interest in the field of informatics. This was followed by an extra step of recognition in 1994 with the transformation of the working group to become a Nursing Informatics Special Interest Group known as the International Medical Informatics Association–Nursing Informatics (IMIA-NI) Work Group [14,8].

As nursing informatics started to gain international recognition, more nurses began working in the field. Murphy (2010) [15] indicates that most NI pioneer nurses, who varied in their age, titles, roles, experience, and responsibility, got into the field of nursing informatics accidently because they wanted to be better clinicians, were involved as project members or educators in hospital IT implementation projects, or were technically curious and interested in trying new technologies.

Other reasons that led nurses to use technology in their work, especially in the United States, was the number of acts and legislation requirements that mandated the demonstration of "meaningful use" in exchange for offering incentives or applying penalties in case of failing to do so. In 2009, the legislation of Health Information Technology for Economic and Clinical Health Act (HITECH Act), which pushed for the adoption of the Electronic Health Record (EHR) was enacted. Accordingly, this necessitated that the healthcare workforce must be prepared to use such technology efficiently and effectively in their work. Tellez (2012) [16] pointed out that because of the HITECH Act, many nursing, health services, and national policy organizations recommended that the Bachelor of Science in Nursing (BSN) Programs in the United States be reviewed to increase the training on different areas related to information, computer competency, and informatics such as information management, information literacy, computer competency, and informatics training. Such revisions to the nursing education program were deemed necessary as more computers were being introduced into the workplace.

Earlier, the 1960s and 1970s witnessed the first appearances of hospital information systems, which spurred other related developments such as the creation of standardized languages for capturing hospital data such as the North American (now International) Nursing Diagnosis Association (NANDA) and the Systematized Nomenclature of Medicine—Clinical Terms (SNOMED-CT), which facilitated the description of the provided care by both nurses and physicians. Also, the formation of different healthcare informatics organizations such as the International Medical Informatics Association (IMIA), the American Medical Informatics Association (AMIA), and the AMIA Nursing Informatics Working Group helped with the introduction of technology into nursing practice and an increasingly complex environment of healthcare [17].

Unlike a paper-based patient file, the benefits that technology offers are greater accessibility to updated and timely data, which can be viewed by more than one clinician at the same time from different locations. Such promise holds the potential for improved communication, which can improve the quality of care provided to the patient. The value of effective communication i.e. accurate, timely, complete, unambiguous, and understood by the recipient is not to be underestimated because it is one of the main patient safety goals for the Joint Commission International (JCI) accreditation body. Accordingly, the most prominent technologies can help improve communication: the EHR along with its different related supporting technologies such as Telenursing, eLearning, and Mobile Technology, which have multiple applications across the various fields of nursing practice. However, our discussion in this chapter will focus only on the development and handling of mobile technology within the nursing education and profession practice.

THE USE OF MOBILE TECHNOLOGIES IN NURSING EDUCATION AND PRACTICE

Before exploring the status of mobile technology in nursing education and practice and how they are being used, a clear definition is necessary. Furthermore, what are the dimensions of mobile Health (mHealth) to be addressed in this chapter? Mobile technology is defined as the "wireless devices and sensors (including mobile phones) that are intended to be worn, carried, or accessed by the person during normal daily activities. mHealth is the application of these technologies either by consumers or health care professionals" [18]. Samples, Ni, and Shaw (2014) define mHealth as the "use of personal wireless communication devices, including mobile phones and smartphones, smartwatches, wireless sensors worn or carried by an individual, tablet computers, and point-of-care devices, to support continuous health monitoring, feedback, and behavior modification of individuals and populations" [19]. Samples et al. also suggest that mHealth provides an opportunity for nurses to empower their patients to become proactive managers of their own care by increasing and improving the awareness and usability of the technologies of mHealth in preventive care such as the self-management of patients with chronic diseases [19].

When it comes to how nurses utilize mHealth in education, a number of researchers have explored this from different angles. For instance, O'Connor & Andrews (2015) [20] reviewed the mobile technology and its use in the clinical nursing education literature. They concluded that the unique features that mobile technology offers—such as creating pervasive and continuous access to data—have provided support for nursing students during their clinical training in the form of instant access to evidence-based materials that improve those students' knowledge as well as their skills. This process can enhance their nursing practice at the point of care. They also recommend that both nurse educators and students should consider thinking about adopting handheld devices in order to augment clinical nursing education and practice. However, this may require dealing with many sociotechnical aspects, which can be barriers to implementing mobile platforms in a clinical setting.

Many nursing bodies and societies around the world such as the International Council of Nurses (ICN), and many National Nurses Associations (NNAs), have called for information technology integration into the nursing curriculum to prepare nursing students for the current modern practice environment of providing evidence-based patient care. Accordingly, many nurse educators and programs have implemented and utilized mobile technology in the classroom [21–27], simulation laboratory [28,26], and clinical settings. [21–25,29,30–35,27] However, issues like lack of Information Technology (IT) support; cost; lack of acceptance and role-modeling among nursing faculty; lack of structured activities and/or assignments designed to encourage the implementation of mobile devices; and last but not least, constraints on the use of mobile technology in clinical settings are a number of obstacles facing educators in nursing informatics [36].

Furthermore, the absence of a clear definition of mobile technology and its boundaries and where they lie in clinical nursing education, is a gap in the current body of knowledge. This may be due to mobile technology's nature in an area that is constantly evolving with the introduction of new technologies. As suggested by Olsson and Gullberg (1991) [37], a standardized definition of mobile technology that specifically relates to the field of nursing education would be helpful to clarify such boundaries and roles. Additionally, this definition could remove any confusion or ambiguity as formalized definitions have done in other areas of nursing education.

For nursing practice, using mobile technology is considered by the profession—especially with the continuously increasing number of mHealth Applications available. Some of the issues that have encouraged nurses to try using mobile technologies include, but are not limited to, an international nursing shortage [38–40] as well as changing health care roles, which result in role blurring and conflict [41,42]. Nursing scholars have indicated several uses for mHealth in nursing practice. These uses are seen, for example, in supporting individual-centered care in the form of self-monitoring of peoples' health-related behaviors and receiving a feedback of such behaviors via mHealth technologies [43,44]. mHealth can be utilized as well in using devices such as wearable fitness sensors that allow consumers to track many data such as their nutrition, calories burned, steps taken, and sleep habits. By doing so, consumers are allowed to monitor their day to day activity in a comfortable and easy manner while seeking at the same time social networking support when they link themselves with common social media platforms like Twitter and Facebook [19]. The nurses' role is pivotal here, as they can help move mHealth into the mainstream health care arena when they understand the potential mHealth has for their patients by incorporating mHealth into their patients' daily preventive care strategies [45]. Also, as noted earlier, nurses can encourage their patients, especially those with chronic diseases to focus on self-management that can result in reduced health care costs and improved health outcomes with lifestyle adjustment and health promotion. This movement toward empowering patients, especially those with chronic illnesses, is important and critical given that two-thirds of the world's population dies from chronic diseases that include respiratory and cardiovascular diseases, diabetes, and cancer [46]. However, for the use of mHealth in nursing practice to succeed, nursing professionals need to clearly understand both the potential as well as the limitations of mHealth technologies, especially, with regards to how mHealth impacts patient care.

CURRENT CHALLENGES AND OPPORTUNITIES OF MHEALTH IN NURSING

Despite the benefits nurses can gain from using mHealth, some issues represent a challenge for nurses using mHealth technology. These challenges can be technical such as poor infrastructure and lack of resources; human resource insufficiency such as lack of qualified health care personnel using such technology; or even behavioral, with related challenges such as users' unfavorable adoption of these apps and varied levels of user satisfaction. In addition, financial and regulatory constraints can also be an issue. Resistance to accepting mHealth technologies, and access to evidence-based materials, can be expected from some health care providers or patients such as elderly long-term patients. Technically, the quality and accuracy of the data collected by mobile technology devices can be compromised by integration barriers. For instance, when a user uses a wristband, he or she can simply move the arm back and forth to mimic steps so that the generated data suggests 15,000 steps were taken for example, while the truth is, the user only manipulated the device while they were inactive [19].

In many countries, authorities impose regulation on healthcare providers to maintain patient safety in using mobile devices for health purposes. For example, the US Food and Drug Administration (FDA) imposes different regulations for medical devices and tracking or self-monitoring tools. According to a 2013 white paper released by the US Department of Health and Human Services FDA, there is a distinction in terms of regulations between "mobile applications," which refers to technologies that allow consumers to log things such as life events, retrieve content that is medical in nature, or communicate with a healthcare provider, and are not currently FDA regulated, as opposed to "mobile medical applications," that is, those technologies that act as either medical devices or as accessories to medical devices, which are subject to FDA regulation [47]. Nurses should be aware of such regulations to be able to make informed recommendations to their patients and their families about which care and type of data are expected to be received when using mHealth tools. Also, such awareness is needed by nursing educators to convey knowledge related to these apps, their pros and cons, for both providers and nursing students correctly and accurately.

When it comes to mHealth opportunities, text messaging, used for example to remind patients to take their prescription medication or attend their appointments, has been considered by Folaranmi (2014) [48]. Also, mHealth has the capability to collect, transfer, and store real-time data, which can be used later for different health care purposes either on an individual level or in an aggregated from. Again, this will largely depend on how important concepts of patient privacy and confidentiality are, and whether the person using these apps has provided informed consent to utilize that information. Yet, overall, mHealth can improve access to basic health care especially in vast and disparate areas such as Sub-Saharan Africa and rural regions. Nevertheless, the severity of these challenges and the degree of the possible opportunities differ from country to country depending on their availability—along with other factors such as demographics and socioeconomics.

Furthermore, nurses' chances to get to know and become experts in using mHealth might be affected by factors such as the resources of the users themselves and the healthcare organization where they work in terms of time, cost, and human resources.

The complexity of the healthcare institution also represents an important factor in determining the opportunities for nurses to learn more and gain practical experience in mHealth. For instance, a nurse that works in a technologically advanced complex hospital that uses mHealth applications for different purposes such as monitoring their chronic diabetic patients' blood glucose levels, will likely be more experienced in using the mHealth technology than a nurse working in an underfunded small healthcare institution. Geographical location also has an impact on chances for nurses as well as other healthcare providers to get a chance to be involved in mHealth activities per a 2011 report by the World Health Organization (WHO) aimed at determining the status of mHealth in its member states [2]. As indicated by the WHO report, higher-income countries such as those located in the European region, which are currently considered the most active in mHealth compared to the least active African counterparts, show more mHealth-related activity than countries with lower-income. Accordingly, a nursing student or practitioner's chances to practice mHealth are affected by a number of interrelated internal and external factors, which should be carefully considered when addressing this issue.

Additionally, when it comes to types of mHealth applications used for different purposes, it is challenging to develop a comprehensive analysis and cover all of them with the fast pace of mHealth development. However, we can classify the main mHealth applications used by the nursing workforce as well as by the patient/layperson into two principal categories:

1. mHealth applications that are being used by nurses to carry on their duties and help them take care of their patients. We can name such applications as Nurse-Centered mHealth applications.
2. mHealth applications that are being used by nurses as well as by their patients and lay people to manage and monitor their health and physiological activities. We can name such applications as Patient/Layperson-Centered mHealth applications.

The following table shows examples of mHealth applications from these two categories. It should be noted that these examples are intended to show the varied categories of current mHealth applications and are not intended as an exclusive reference to this broad and rapidly developing market (Table 20.1).

When it comes to the regulations that control the use of mHealth applications, several countries and regions have tackled this matter and issued regulatory guidelines such as the FDA and the European Union (EU) regulatory frameworks for medical devices and mHealth. These regulations emerged as an attempt to control the side effects of such penetration of mHealth in many countries that may impact patients' safety, privacy, and confidentiality to name a few. With such varied regulations, along with the rapid development in the mHealth market, no nurse can be 100% sure that he or she is fully aware of all the types of mHealth applications found on the market, nor the international regulations controlling them. For example, does application X fall under FDA regulations, which control mobile medical applications, or under mobile applications, which are not regulated? With every day that passes, mobile health technology surprises us with new advances.

Finally, nurses' chances to master the different types of mHealth applications are narrowed as a result of proprietary rights for these applications, which also have affected

TABLE 20.1 Nurse and Person Centered mHealth Applications

Nurse-Centered mHealth Applications	Patient/Layperson-Centered mHealth Applications
Drug references/databases applications to provide information about the drug, dose, interactions, etc.	Diet, weight, and fitness applications such as those tracking the pulse, calories burned, diet trackers, suggested workouts, etc.
Medical calculator applications to calculate for example body mass index (BMI), body surface area, normal lab results, pediatric dosage calculations, etc.	Herbal remedy and recipe apps that provide information about different herbs used for medical purposes.
Educational applications such as Nursing fundamentals, NCLEX exam preparation, Nursing skills like physical assessment and patient history, human anatomy and physiology, Nursing Care Plans-NANDA, etc.	Medical encyclopedias that provide a valuable source of information about varied health conditions in terms of the signs and symptoms, diagnosis, treatment, prognosis, and prevention as applicable.
Nursing specialty applications such as surgical nursing, pediatric nursing, medical nursing, nursing research, etc.	Women and men's health applications; including pregnancy and period tracker apps.
Medical/nursing-topic quizzes, manuals, and games for different educational purposes.	Disease-specific management applications such as hypertension and diabetes management apps.
Nursing references, libraries, and guides.	Medical-topic games for different ages and purposes.

nursing students as well as nursing practitioners in terms of educational opportunities and practice on using other applications such as the EHRs and student portals. For instance, considering the different properties of various mHealth applications, a nursing school cannot afford to effectively teach students about all different types of mHealth applications found on the market, nor will it always allow vendors to create a student portal for learning major applications such as EHR, since every major vendor has a different structure from its competitors.

FUTURE TRENDS IN THE USE OF MOBILE TECHNOLOGIES IN NURSING EDUCATION AND PRACTICE

In 2014, C. Lee Ventola mentioned several uses of mobile health technology by health care professionals in a paper entitled "Mobile Devices and Apps for Health Care Professionals: Uses and Benefits" [49]. These uses include, but are not limited to, managing information such as writing or dictating notes; managing time such as scheduling meetings and appointments; accessing and maintaining patients' health records; communicating and providing consultation for their patients such as social networking, video conferencing, email, texting, voice, and video calling; using mobile devices for information gathering and referencing with medical textbooks, literature, and journals; using mobile devices for clinical decision-making such as clinical treatment guidelines, laboratory test ordering, and medical exams; using mobile devices for patient monitoring such as collecting clinical data, and monitor patient health and location; using mobile devices for medical education and training such as continuing medical education, board exam preparation, case studies, and finally, using it for E-learning and teaching.

In nurse education for instance, there are several mobile apps that provide guidance and exams for several areas of practice such NCLEX-RN (i.e., National Council Licensure Examination-Registered Nurse) exam preparation apps, as well as physical examination and assessment guidance apps. In practice, nurses can use mobile apps for example to calculate drugs before their administration or review normal lab tests to make sure their patients values are within the normal range. These are only examples of some of the functionalities of mobile technology in nursing education and practice which have been a result of the rapid growth in the state of information technology that we witnessed throughout the years. However, more advances are on the way, as nursing as well as other areas of healthcare continue to develop.

When it comes to the future trends expected to transpire within the next five years, the results of a 2015 survey done by research2guidance, can be helpful. The survey, "mHealth App Developer Economics 2015" [50], is considered as the largest global study exploring the current status of mHealth app publishing and with an outlook on the trends in the market for the next 5 years. More than 5000 mHealth practitioners, that is, mHealth decision makers, app developers, and others, participated in the survey. According to the which, health and fitness comprise the majority of health apps comprising of 56% of total applications, followed by medical category apps (44%). It is suggested that the upcoming applications will be more than fitness apps and will provide tracking functionalities such as those designed for diabetic or asthma patients.

Additionally, for the next 5 years, the survey indicated that 85% of mHealth practitioners preferred smartphone devices and rated them as a primary target device, that is, the device that will be the principal target for mHealth applications in a matter of 5 years. Also, in the preferred platform for mHealth apps, the survey revealed that iOS and Android are currently the dominant platforms for mHealth practitioners and are expected to remain in the same position in the near future. Other platforms, such as Windows phone, showed no evidence to change that order. The survey also indicated that automation of data input by mHealth application users will be one of those main trends expected to be seen in the next few years. Accordingly, apps should be connected to sensors (and many apps are already connected this way) in order to feed these apps with activity and biometric data. When it comes to the type of sensors that can be connected to an app, there are seven main types. For example, built-in smartphone sensors (e.g., accelerometer, heart rate measured with the phone's camera) is the category with the highest market potential according to 70% of mHealth practitioners, followed by wearable devices by 53%. The second sensor type that is expected to be important in the next 5 years will be those implemented in a wearable device (e.g., patches or chest belt, wristband) and the ones that transmit data either wirelessly via Bluetooth for instance, or using a side loading method such as cables. Furthermore, the research2guidance survey revealed that in the category of the therapy field preference, the main target group for mHealth application publishers is patients with chronic conditions. Since 2010, diabetes has been considered the therapy field that offers the highest mHealth potential. With 70% of mHealth practitioners rating it, diabetes has the highest standing in terms of business potential for mHealth in the next 5 years. Other conditions such as obesity, which once came close to diabetes in mHealth business potential ratings in 2012, dropped to 38% in 2015. Hypertension, depression, and chronic heart diseases are consistently ranked between the third to fifth positions in the highest business potential for mHealth applications. In the preferred app category, diagnostic applications are now seen by the respondents as the app category that offers the highest mHealth business potential in the next 5 years, surpassing remote monitoring apps and overtaking their place after they dropped by 21 percentage points in mHealth practitioners' rating of business potential since last year. However, the biggest increase in the mHealth business potential rating in the same category was observed in medical condition management applications that effectively support the patient and at the same time go beyond simple functions of tracking and monitoring. Moreover, within the same category, EHR and nutrition applications have recorded a significant rise in their business potential rating for the next 5 years [50].

CONCLUSION

Mobile technology and mHealth applications have affected the way nursing education and practice is carried out. Nurses are moving toward empowering their patients using mHealth applications in order to improve patients' health and assist them to make healthy lifestyle adjustments. By increasing their patients' awareness and usability, these apps may support proactive and evidence-based healthcare. In nursing education, nursing educators and students are utilizing mobile technology in different ways resulting in reported enhancement in the nursing students' academic performance—both in terms of their acquired skills and knowledge. However, integrating mobile technology into nursing education and practice is still facing various challenges and obstacles hindering its optimal infusion into the field of nursing. With continued researcher and patient focus on this topic, many opportunities lie ahead which can be used in improving the conditions of mobile technology and mHealth applications for both patients and their families as well as nursing students, educators, and practitioners.

References

[1] Herrington J, Herrington A, Mantei J, Olney I, Ferry B. Using mobile technologies to develop new ways of teaching and learning. In: Herrington J, Mantei J, Olney I, Ferry B, Herrington A, editors. New technologies, new pedagogies: Mobile learning in higher education [Internet]. Wollongong: University of Wollongong; 2009 Chapter 1. Available from: <http://ro.uow.edu.au/cgi/viewcontent.cgi?article = 1077&context = edupapers>.

[2] World Health Organization Global Observatory for e-Health: mHealth: New horizons for health through mobile technologies: second global survey on eHealth. 2011. Available from: <http://www.who.int/goe/publications/goe_mhealth_web.pdf>.

[3] Goggin G. Cell phone culture: mobile technology in everyday life. New York, NY: Routledge; 2006, [cited 21.04.16]. p. 254.

[4] Information and Communication Technologies (ICTs) Report. Geneva (SZ): International Telecommunication Union (ITU); 2015. p. 8. Available from: <http://www.itu.int/net/pressoffice/press_releases/2015/17.aspx#.V7HKh_I96Uk>.

[5] Ozbolt J, Saba V. A brief history of nursing informatics in the United States of America. Nursing Outlook [Internet] 2008;56(5):199–205. Available from: <http://www.ncbi.nlm.nih.gov/pubmed/18922268> [cited 21.04.16].
[6] Cornell SA, Bush F. Systems approach to nursing care plans. Am J Nurs 1971;71:1376–8. Available from: <http://www.ncbi.nlm.nih.gov/pubmed/5207102>.
[7] Wesseling E. Automating the nursing history and care plan. J Nurs Admin 1972;2(3):34–8. Available from: <http://www.ncbi.nlm.nih.gov/pubmed/4482238>.
[8] Asiri H. An Overview of Nursing Informatics (NI) as a Profession: How we Evolved Over the Years. In: Gilbert J, Azhari H, Ali H, Quintão C, Sliwa J, Ruiz C, Fred A, Gamboa H. BIOSTEC 2016: Proceedings of the 9th International Joint Conference on Biomedical Engineering Systems and Technologies: Volume 5: HEALTHINF; February 21–23, 2016; Rome, Italy. Portugal: SCITEPRESS—Science and Technology Publications, Lda; 2016. p. 200–12.
[9] Ball M, Hannah K. Using computers in nursing. Reston: VA: Reston Publishers; 1984.
[10] Hannah K. Current Trends in nursing informatics: implications for curriculum planning. In Hannah K. L., Guillemin E. J., Conklin, D. N. Nursing uses of computers and information science: proceedings of the IFIP-IMIA international symposium on nursing uses of computers and information science; May 1–3, 1985; Calgary, Alberta, Canada. Amsterdam, Netherlands: Elsevier; 1985. p. 181–7.
[11] Grobe SJ. Nursing informatics competencies for nurse educators and researchers. NLN Publ [Internet] 1988;14:25–40. Available from: <http://www.ncbi.nlm.nih.gov/pubmed/3368303> [cited 17.06.16].
[12] Scholes M, Barber B. The role of computers in nursing. Nurs Mirror Midwives J. [Internet] 1976;143(13):46–8. Available from: <http://www.ncbi.nlm.nih.gov/pubmed/1049093> [cited 19.06.16].
[13] American Nurses Association. The standards of practice for nursing informatics. Washington, DC: American Nurses Publishing; 1995 (Pub. no. NP-100).
[14] Scholes M, Tallberg M, Pluyter-Wenting ESP. International nursing informatics: a history of the first forty years, 1960-2000. 2nd ed. Swindon, England: British Computer Society; 2000. p. 112.
[15] Murphy J. Nursing informatics: the intersection of nursing, computer and information sciences. Nurs Econ [Internet] 2010;28(3):204–7. Available from: <http://www.ncbi.nlm.nih.gov/pubmed/20672545> [cited 25.06.17].
[16] Tellez M. Nursing informatics education past, present, and future. CIN [Internet] 2012;30(5):229–33. Available from: <http://journals.lww.com/cinjournal/Citation/2012/05000/Nursing_Informatics_Education_Past,_Present,_and.1.aspx> [cited 18.06.16].
[17] Englebardt S, Nelson R. Health care informatics: an interdisciplinary approach [e-book]. St Louis, MO: Mosby; 2002. Available from: <https://www.amazon.com/Health-Care-Informatics-Interdisciplinary-Approach/dp/0323014232> [cited 18.06.16].
[18] Doswell W, Braxter B, DeVito Dabbs A, Nilsen W, Klem ML. mHealth: technology for nursing practice, education, and research. J Nurs Educ Pract [Internet] 2013;3(10):99–109. Available from: <http://www.sciedu.ca/journal/index.php/jnep/article/view/2180> [cited 18.06.16].
[19] Samples C, Ni Z, Shaw R. Nursing and mHealth. Int J Nurs Sci [Internet] 2014;330–3. Available from: <http://www.readcube.com/articles/10.1016/j.ijnss.2014.08.002> [cited 18.06.16].
[20] O'Connor S, Andrews T. Mobile technology and its use in clinical nursing education: a literature review. J Nurs Educ [Internet]. 2015;54(3):137–44. Available from: <http://www.ncbi.nlm.nih.gov/pubmed/25693246> [cited 18.06.16].
[21] Beard KV, Greenfield S, Morote E, Walter R. Mobile technology lessons learned along the way. Nurse Educ [Internet] 2011;36(3):103–6. Available from: <http://www.ncbi.nlm.nih.gov/pubmed/21502842> [cited 18.06.16].
[22] Brubaker CL, Ruthman J, Walloch J. The usefulness of Personal Digital Assistants (PDAs) to nursing students in the clinical setting: a pilot study. Nurs Educ Perspect [Internet] 2009;30(6):390–1. Available from: <http://www.ncbi.nlm.nih.gov/pubmed/19999943> [cited 18.06.16].
[23] Chioh MS, Yan CC, Tang KL, Mustaffa SM, Koh MG, Sim TW, et al. The use of the personal digital assistant by nursing students in the classroom and clinical practice: a questionnaire survey. Singap Nurs J [Internet] 2013;40(20):38–44. Available from: <http://connection.ebscohost.com/c/articles/87003099/use-personal-digital-assistant-by-nursing-students-classroom-clinical-practice-questionnaire-survey> [cited 18.06.16].
[24] Cibulka NJ, Crane-Wider L. Introducing Personal Digital Assistants to enhance nursing education in undergraduate and graduate nursing programs. J Nurs Educ [Internet] 2011;50(2):115–18. Available from: <http://www.ncbi.nlm.nih.gov/pubmed/21210606> [cited 18.06.16].

[25] George LE, Davidson LJ, Serapiglia CP, Barla S, Thotakura A. Technology in nursing education: a study of PDA use by students. J Prof Nurs [Internet] 2010;26(6):371–6. Available from: <http://www.ncbi.nlm.nih.gov/pubmed/21078507>.

[26] Swan BA, Smith KA, Frisby A, Shaffer K, Hanson-Zalot M, Becker J. Evaluating tablet technology in an undergraduate nursing program. Nurs Educ Perspect [Internet] 2013;34(3):192–3. Available from: <http://www.ncbi.nlm.nih.gov/pubmed/23914464> [cited 18.06.16].

[27] Wyatt TH, Krauskopf PB, Gaylord NM, Ward A, Huffstutler-Hawkings S, Goodwin L. Cooperative m-learning with nurse practitioner students. Nurs Educ Perspect [Internet] 2010;31(2):109–13. Available from: <http://www.ncbi.nlm.nih.gov/pubmed/20455369> [cited 18.06.16].

[28] Schlairet MC. PDA-assisted simulated clinical experiences in undergraduate nursing education: a pilot study. Nurs Educ Perspect [Internet] 2012;33(6):391–8. Available from: <http://www.ncbi.nlm.nih.gov/pubmed/23346788> [cited 18.06.16].

[29] Hudson K, Buell V. Empowering a safer practice: PDAs are integral tools for nursing and health care. J Nurs Manag [Internet] 2011;19(3):400–6. Available from: <http://www.ncbi.nlm.nih.gov/pubmed/21507112> [cited 18.06.16].

[30] Johansson PE, Petersson GI, Nilsson GC. Nursing students' experience of using a personal digital assistant (PDA) in clinical practice—an intervention study. Nurse Educ Today [Internet] 2013;33(10):1246–51. Available from: <http://www.ncbi.nlm.nih.gov/pubmed/22999410> [cited 18.06.16].

[31] Kuiper R. Metacognitive factors that impact student nurse use of point of care technology in clinical settings. Int J Nurs Educ Scholarsh [Internet] 2010;7(1):1–6. Available from: <http://www.ncbi.nlm.nih.gov/pubmed/20196764> [cited 18.06.16].

[32] Secco ML, Dorion-Maillet N, Amirault D, Furlong K. Evaluation of nursing central as an information tool, part I: student learning. Nurs Educ Perspect [Internet] 2013;34(6):416–18. Available from: <http://www.ncbi.nlm.nih.gov/pubmed/24475605> [cited 18.06.16].

[33] Williams MG, Dittmer A. Textbooks on tap: using electronic books housed in handheld devices in nursing clinical courses. Nurs Educ Perspect [Internet] 2009;30(4):220–5. Available from: <http://www.ncbi.nlm.nih.gov/pubmed/19753854> [cited 18.06.16].

[34] Wittmann-Price RA, Kennedy LD, Godwin C. Use of personal phones by senior nursing students to access health care information during clinical education: staff nurses' and students' perceptions. J Nurs Educ [Internet] 2012;51(11):642–6. Available from: <http://www.ncbi.nlm.nih.gov/pubmed/22978275> [cited 18.06.16].

[35] Wu C, Lai C. Wireless handhelds to support clinical nursing practicum. Educ Technol & Soc [Internet] 2009;12(2):190–204. Available from: <http://www.ifets.info/journals/12_2/14.pdf> [cited 18.06.16]

[36] Ramen J. Mobile technology in nursing education: where do we go from here? A review of the literature. Nurs Educ Today [Internet] 2015;35(5):663–72. Available from: <http://www.ncbi.nlm.nih.gov/pubmed/25665926> [cited 18.06.16].

[37] Olsson HM, Gullberg MT. Nursing education and definition of the professional nurse role. Expectations and knowledge of the nurse role. Nur Educ Today [Internet] 1991;11(1):30–6. Available from: <http://www.ncbi.nlm.nih.gov/pubmed/1994228> [cited 18.06.16].

[38] Buerhaus P, Donelan K, Ulrich B, Norman L, Williams M, Dittus R. Hospital RNs' and CNOs' perceptions of the impact of the nursing shortage on the quality of care. Nurs Econ [Internet] 2005;23(5):214–21. Available from: <http://www.ncbi.nlm.nih.gov/pubmed/16315651> [cited 18.06.16].

[39] Cho S, Ketefian S, Barkauskas V, Smith D. The effects of nurse staffing on adverse events, morbidity, mortality and medical costs. Nurs Res [Internet] 2003;52(2):71–9. Available from: <http://www.ncbi.nlm.nih.gov/pubmed/12657982> [cited 18.06.16].

[40] Needleman J, Buerhaus P, Mattke S, Stewart M, Zelevinsky K. Nurse-staffing levels and the quality of care in hospitals. N Engl J Med [Internet] 2002;346(22):1715–22. Available from: <http://www.ncbi.nlm.nih.gov/pubmed/12037152> [cited 18.06.16].

[41] Kessler I, Heron P, Dopson S, Magee H, Swain D, Askham J. The nature and consequences of support workers in a hospital setting. UK: National Institute for Health Research; 2010, 229.

[42] McCabe S, Burman M. The tale of two APNs: addressing blurred practice boundaries in APN practice. Persp Psych Care [Internet] 2006;42(1):3–12. Available from: <http://www.ncbi.nlm.nih.gov/pubmed/16480412> [cited 18.06.16].

[43] Guo F, Li Y, Kankanhalli MS, Brown MS. An evaluation of wearable activity monitoring devices. Proceedings of the 1st ACM International workshop on personal data meets distributed multimedia, 13. Barcelona, Catalunya, Spain. New York, NY, USA: PDM; 2013. p. 31–4.
[44] Zheng J, Shen Y, Zhang Z, Wu T, Zhang G, Lu H. Emerging wearable medical devices towards personalized healthcare. Proceedings of the 8th International Conference on body area networks. Brussels, Belgium: ICST; 2013. p. 427–31.
[45] Klasnja P, Pratt W. Managing health with mobile technology. Inter [Internet] 2014;21(1):66–9. Available from: <http://dl.acm.org/citation.cfm?id = 2540992&dl = ACM&coll = DL&CFID = 658206969&CFTOKEN = 57096546> [cited 18.06.16]
[46] Sarasohn-Kahn J. A role for patients: the argument for selfcare. Am J Prev Med [Internet] 2013;44(Suppl 1): S16–18. Available from: <http://www.ncbi.nlm.nih.gov/pubmed/23195159> [cited 18.06.16].
[47] U.S. Department of Health and Human Services Food and Drug Administration (FDA). Mobile medical applications: guidance for industry and food and drug administration staff. U.S.: Department of Health and Human Services Food and Drug Administration (FDA); September 25, 2013. p. 44. Available from: <http://www.fda.gov/downloads/MedicalDevices/.../UCM263366.pdf>.
[48] Folaranmi T. mHealth in Africa: challenges and opportunities. Persp Pub Heal [Internet] 2014;134(1):14–15. Available from: <http://www.ncbi.nlm.nih.gov/pubmed/24395839> [cited 18.06.16].
[49] Lee Ventola C. Mobile devices and apps for health care professionals: uses and benefits. PT [Internet] 2014;39 (5):356–64. Available from: <http://www.ncbi.nlm.nih.gov/pmc/articles/PMC4029126/> [cited 18.06.16].
[50] Research2guidance. mHealth App Developer economics 2015 Survey. Berlin, Germany: Research2guidance; November 2015, p. 37. Available from: <http://research2guidance.com/r2g/r2g-mHealth-App-Developer-Economics-2015.pdf>.

CHAPTER

21

Leveraging Social Media for Clinician Training and Practice

Victoria Wangia-Anderson[1] *and Prerna Dua*[2]

[1]University of Cincinnati, Cincinnati, OH, United States [2]Louisiana Tech University, Ruston, LA, United States

AN OVERVIEW OF SOCIAL MEDIA

Definition

The definition of social media has been constantly evolving and remains integral to how we communicate in our digitally advanced society. In the broader sense, it refers to the online tools being used which help people communicate with each other, exchange ideas, respond to posts, write reviews, etc. [1]. It is viewed as "the freely available Web-based platforms that facilitate information sharing of user-generated content, such as social networking sites, media-sharing sites, blogs, microblogs, and wikis" [2]. In the context of healthcare, social media should not only be viewed as any tool that enhances social networking, but also as an electronic tool for provider and consumer empowerment and engagement. Social media facilitates social networking through the electronic exchange of information between at least two parties.

Use

Social media usage continues to grow and surpass other uses of the Internet. Nielsen, a global information and measurement firm has provided some revealing data on social media usage, showing that in 2010, social media sites were accessed at a rate of one out of every 4½ minutes that Internet users were online [3]. The average end-user spent about 6 hours on social media-related sites [3]. Given the large percentage of people involved in social media participation, the healthcare sector has observed considerable changes. Of the people who use social network sites, 23% have followed their friends' personal health experience or updates on the site, 17% of social network site users have used a social

Health Professionals' Education.
DOI: http://dx.doi.org/10.1016/B978-0-12-805362-1.00021-8

networking site to remember or memorialize other people who suffered from a certain health condition and, 15% of social network site users have gotten any health information on the sites [4]. The above statistics prove that a there is interest in using social media for health related issues [4].

The use of social media has seen a considerable increase over the last decade. Social media finds a greater advantage in that it has no geographical boundaries. It is very much efficient in that it can connect individuals who are located at two far ends of the globe at any time [5]. All such advantages could, in the healthcare scenario, translate to a revolutionary way of reaching patients with health education and promotion messages, engaging patients and students from various demographic groups, and changing communication patterns overall [5,6]. Overtime, use of social media has evolved and become more prevalent in the context of teaching and learning, and more educators and students are exploring use of the tools and information. As social media tools continue to evolve, expansive amount of data continues to be produced and shared thus providing opportunities for stakeholders to leverage its capabilities.

Audience

A large percentage of U.S. adults use social networking sites. The Pew Research center estimated that in 2013, 73% of adults online used social networking sites [7]. Studies estimate about one-third using the Internet to diagnose a medical condition [7,8]. Social media use varies across healthcare organizations, but large urban private nonprofit and teaching hospitals have been found to show higher utilization [9].

Preference for social media use has been more pronounced amongst a younger generation, "millennials", but the acceptance by other generational groups continues to grow as well. Today's students are eager to engage in social media at a personal level, and show interest in using social media tools for learning too. It is being widely used by medical students and across the healthcare sector [10,11]. Healthcare professionals are using social media to communicate with other professionals, and in some cases, with patients, as well as are using it for marketing purposes [1]. Patients are also users of social media and use it for a variety of reasons including exchanging opinions, sharing news, and for social support [1]. There are definite differences in the acceptance of social media in clinical practice and education. Clinicians, faculty or students who were deemed as "digital immigrants" ("born before the digital technology boom") are less enthusiastic about using social media than "digital natives" ("born after the digital technology boom") [12].

Categories

Social media has evolved to allow interaction and collaboration, as end-users are now able to see themselves as content creators. Social media falls into the second generation of the World Wide Web (Web 2.0). Web 2.0 technologies have the following key characteristics: they allow for "online reflection, social spaces, online collaboration, social bookmarking, and repository" [13]. Web 2.0 differs from the prior generation by enhancing the ability for end users to collaborate, author and contribute by adding to others' messages. End-users are not just consumers of information but empowered users who can be the editors of the information and can comment and respond. Responses can be expressed as sentiments through symbols that allow end-users to deduce the emotions of the responder.

Social media has evolved to the extent that categories have emerged. Some categories include social networking services, wikis, blogs, microblogs, and media-sharing websites. Rutherford [14] classified them into three categories, namely, social media resources for content sharing and organizing, social media resources for creating and editing websites and social networking sites. As new technologies are emerging in this digital age, more categories will be added to this framework and its scope will be broadened [15].

Examples

Some popular social media sites transcend disciplines and are being used across the healthcare sector by patients, healthcare providers and for medical education. Social media examples discussed below include Facebook, Twitter, Blogs, Wikis, LinkedIn, and YouTube. These six examples were selected because they have proven to be social media platforms that have endured for the past 5–10 years in a world of ever changing technology, and have the potential to empower patients, transform patient engagement, patient-provider interaction, healthcare professional education, and clinical practice.

SOCIAL MEDIA RESOURCES FOR CREATING AND EDITING WEBSITES

This category of social media allows users to create their written content or share videos. The information can be either at length using blogs or tiny blurbs that are used in microblogging.

Blogs are online public logs or journals. Though text is a large component of a blog, other media can be integrated into the blog to enrich the content. Blogs allow end-users to comment publicly. They have been used in the field of medicine since 2004 and are beneficial in the way that they can reach a wide audience [5]. Some physicians use blogs to communicate with the public. Of all the social media tools used thus far in medical education, blogs have been studied the most [12]. *CureTogether* is an example of a blog that helps patients discuss over 500 medical conditions, enabling them to get a better understanding of topics such as symptoms and care treatment. Mayo Clinic provides a blog, "Sharing Mayo Clinic" where patients share their experiences as they visit the clinic for diagnosis and treatment options. Mayo Clinic also keeps its patients up to date by providing information on the innovative technology being adopted at their "Center for Innovation" blog.

Wikis are collaborative writing applications that have consistently gained popularity in many industries including healthcare. Collaborative writing applications are defined as "a category of social media that enables the joint and simultaneous editing of a webpage or an online document by many users" [16]. Wikis allow users to engage with content as original authors and contributors. They can either be in the public domain allowing editing by anyone or be restricted to a target audience. Wikis have been used for education, by patients, and by health professionals; however, little is known about what audience benefits the most [16]. Wikipedia is an online encyclopedia that includes content written by end-users in collaboration. Among many other topics, it includes content on medical and health topics. Users of wikis contribute knowledge, find information, and update content. Wikis are also being used for training clinicians [17].

Twitter is a popular social networking tool that falls in the category of microblogs. Twitter allows users to create messages, share others' messages by "retweeting," follow other end-users to keep abreast of what they are posting and receive small chunks of messages of about 140 characters, referred to as "tweets." To use Twitter, an account has to be established. Twitter is essentially a microblogging tool that is different from any other because it allows for real-time immediate distribution of the short messages. Twitter has contributed to creating virtual learners and in the context of healthcare education, it is heavily utilized at professional meetings or for "live-tweeting" [12]. Twitter chats are organized around a particular topic using hashtags, "a keyword or phrase following the pound sign," thus, eliminating the need for the individuals to gather at particular geographic locations for discussions.

SOCIAL NETWORKING SITES

Social networking sites are defined as "web-based services that allow individuals to (1) construct a public or semipublic profile within a bounded system, (2) articulate a list of other users with whom they share a connection, and (3) view and traverse their list of connections and those made by others within the system" [18].

Facebook is a social networking site that allows users to create profiles of themselves, locate others to connect with in order to extend one's social network. The information posted on Facebook pages varies across users but is often personal in nature. Friends are able to express their views about posts by commenting on posts, liking posts, or using symbols. Healthcare organizations, patient advocacy groups, hospital systems, nonprofit organizations, and educational institutions have established Facebook pages to reach and educate a larger audience on new treatment options, to increase the web traffic to their pages and to increase exposure. Another social networking platform, somewhat similar to Facebook but with a narrower target audience, is Doximity. Doximity was designed specifically for clinicians and has about half a million verified users.

LinkedIn is the world's largest professional network with millions of members in over 200 countries and territories. It was launched in 2003 and is available in 24 different languages. Healthcare professionals and educators can create a profile, view other registered users' profiles, communicate through messaging, search for jobs, join groups, and request other users to join their network. It is utilized by many to find connections to jobs, mentors, business leads, job candidates, industry experts, and business partners. Online participation allows for connection with professional peers, building reputation, self-branding, professional, and personal development, employment recruiting, data and information sharing, tracking industry trends, and following competitors. It is the third most popular social media site after Facebook and Twitter and is free to join.

SOCIAL MEDIA RESOURCES FOR CONTENT SHARING AND ORGANIZING

This category of social media provides platforms to share a wide variety of videos and pictures.

YouTube is a video-sharing and streaming site that allows users to access videos others have created, upload videos, comment on videos posted, rate posted video and, share videos that are posted [19]. YouTube ranks amongst the most visited websites on the Internet [19]. Videos posted can be used in the context of clinical education and clinical practice. For example, educators are using the videos as teaching aides and posting their content on YouTube for others to access for free. Students are using YouTube to supplement their course materials and are able to pose questions to the author who may not necessarily be their instructor. Patients are learning about their conditions and other patients' experiences by watching the videos posted on YouTube. Practitioners in healthcare are viewing videos on YouTube for professional development and in some cases, to support their clinical decision-making, and to disseminate patient education [19]. YouTube is also being used by healthcare organizations and educational institutions to reach out to prospective target audiences, and as a channel for marketing and advertising [19].

Pros of Using Social Media for Education and Practice

Only few studies have investigated the use of social media by students in clinical disciplines. However, some exist and show support for use of social media for learning purposes. A recent study investigated use of social networking sites by medical students and allied health science students through meta-analysis. The study showed a significant number of respondents using social networking sites but a much smaller percentage using the sites for academic purposes [20]. The consensus amongst researchers seems to be that clinical students are using social media sites, but the use of those sites for educational purposes appears to be limited [21–23]. In instances where social media has been used and studied, there is evidence showing that students have found it to improve their learning and in some cases, their academic performance [24].

The improved communication capabilities provided by social media networks has had a considerable impact on training and education. Clinical students who use social media have been found to have a better understanding of the ethics, communication, and professionalism, when compared to their counterparts who do not use social media [25]. Students taking courses in classrooms can now interact outside the classroom through social media. Social media has not only allowed for interaction with colleagues, but has also allowed students to interact with medical experts.

Though only few studies show the impact of social media on learning outcomes, a few have shown correlation between participating in social media technology use and higher grades [26]. Social media has also been found to enhance learning by improving learning engagement, promoting interaction with faculty, and enhancing collaboration [12].

Social media has broken the geographic barrier of communication in that it enables a patient, if need be, to consult a doctor even when they are separated by miles [27].

The technological advances in healthcare have expanded the possibilities with social media. Hospital patient portals have now made it easier for patients to make appointments, check their lab reports, and in some cases, communicate with their providers [27]. Patients are also more engaged with communities across the globe and able to share experiences, support one another, and learn from one another.

Cons of Using Social Media for Education and Practice

The major concern for online sharing of health information is privacy. According to the United States' Health Insurance Portability and Accountability Act (HIPAA), and similar legislation in other countries, the information of the patient has to be protected. Though it does not restrict the dissemination of medical knowledge that is unidentified, communicating with or about patients on social media could be regarded as a breach of privacy policy, which could attract penalties [6]. Also, a patient might lose trust in the organization if he or she learns that their privacy has been compromised due to social media posts [5]. Chretien et al. [28] has discussed the ethical implications of social media on clinicians with the American Medical Association (AMA) principles.

Privacy concerns are a major barrier for clinician educators and practitioners. Studies reveal that social media users are not comfortable identifying as patients online and worry about the privacy of their information [29]. Also, users doubt the credibility of the information found on the Internet about diagnosis or medication of their health issues [2,27,30,31]. Healthcare professionals are showing preference for face-face communication between a patient and a professional rather than virtual meetings [30]. They also feel that online discussions regarding the diagnosis and treatment of illness can lead to more confusing results rather than getting the patient relieved. Many healthcare professionals also lack the time and money to invest on social media networking.

Reliability of the information being provided by social media can be questionable. Most of the health information posted on social media sites is by unknown people or individuals who may not be easily identifiable. Although evidence-based medicine deemphasizes anecdotal reports, social media on the other hand tends to emphasize them, relying on individual patient stories for collective medical knowledge [6].

Healthcare professionals have to be very cautious about the information posted online. Usually younger doctors and medical students are less cautious in matters related to the vulnerability of the content being posted online. For example, health care professionals who prescribe abstinence from alcohol and then post pictures on social media of one's own drunken exploits are very likely to lose a patients' trust [25].

Social media helps in connecting people with similar medical issues. This is beneficial at times, but in most cases the benefits can be undermined by the drawbacks. For example, social media groups help patients in sharing their own medical diagnosis stories, but this at times can be quite misleading. Though more than one person can be diagnosed with similar health problems, the probability of them having the same medical condition and medical history is highly a matter of chance. Further, the treatment and medication for the patients will also vary. Instances where patients have had an unpleasant experience with either the treatment or a certain medication can make their peers very skeptical in accepting any new medical treatment by their health care professionals. Patients who have less than desirable experience during an encounter can convey their displeasure through social media [5]. Such situations might lead to biased opinions by other patients, which can end up harming the reputation of organizations and health care professionals [5]. Social groups can also be addictive in a way that they demand a lot of one's personal time, which might also prove to be unproductive.

Social media sites can also act as a stress to the boundary of one's personal life. Even with strict privacy settings, health care professionals might still receive "friend requests" from clients and other boundary pushing behaviors, which pose challenges to privacy and also to the psychotherapeutic relationship [27,32,33].

Many a times, online doctor consultations can be considered as a social stigma. The relationship developed between a doctor and a patient when they meet in person is certainly compromised by the intervention of social media. The problem/ disease when discussed with the health professional face-to face is perceived to be understood better, which also translates to a better diagnosis. This is one main reason why many doctors and other healthcare professionals abstain from the use of social media for professional reasons [34].

The ability for social media information to be widely distributed either immediately or later creates risk that ought to be discussed by both educators and practitioners. Messages created in poor judgment or inflammatory in nature can surface and result in damaging consequences to individuals and organizations. The messages sent through social media are usually timely with little opportunity for vetting through careful review. Messages can be retracted by deleting the posted message; however, the viral nature of dissemination guarantees that the messages will reach some of the audiences.

SOCIAL MEDIA IN EDUCATION

With the advent of social media during the last two decades, teaching practices have advanced and are centered toward student-oriented pedagogy. The traditional classroom lectures with assignments, essays, and reading material are now supplemented with online videos, podcasts, wikis, and blogs. This section will focus on how social media is being used in education and discuss the opportunities it provides.

Opportunities

Educators must prepare for the expectations and preferences of future students and adjust to ensure that teaching remains effective and student satisfaction is high. Strong evidence exists suggesting preference for student engagement and learning that is enhanced by electronic tools such as social media. A survey conducted in 2011 showed strong support for social media use by students and residents; 94% of medical students and 79.4% of residents in the United States were found to be active online social media users [35]. It is also becoming evident that incoming students are not intimidated by technology even without prior exposure to a specific tool [36]. There is evidence specific to healthcare professional students, showing that these students prefer to receive educational materials online [37]. Though patients, students, and the general public have quickly adopted social media technology for various purposes, healthcare professionals and instructors have not adopted as quickly for professional purposes [35]. We cannot over emphasize the need to change mentality and embrace the social media that digital natives have grown accustomed to, and will expect to continue to use as students and practitioners.

Students, instructors, and practitioners have access to a plethora of information including information generated through social media technologies. They may not always have the ability to evaluate the credibility and relevance of information presented to them and could be misinformed. Organizations, practitioners, and instructors can provide criteria to aid in this process. They can also serve as subject matter experts and assist with the validation of information presented. Students and patients should always proactively consult with instructors and practitioners, respectively. Students are certainly comfortable using social media for personal use, but they may need to receive orientation on how to use them in the context of learning [13]. There is opportunity for faculty to embrace these technologies, adopt them, and train students on how to meaningfully interact and collaborate to achieve the expected learning outcomes. Social media that increases interaction among students, and between students and instructors has positive learning outcomes. Students can be encouraged to distribute content through social media to enable widespread dissemination and allow for a broader audience to redistribute the content and interact in real time. Likewise, instructors can use these preferred channels to distribute content to student groups who would be followers. For example, content can be managed by archiving and retrieving it through hashtags.

There is opportunity for instructors to provide guidance on how students might perceive, address, and respond to others in the context of social media use. Though students today are comfortable with the use of social media technology and not hesitant to apply it beyond their own private uses, coaching on how to effectively communicate messages that enhance learning and also demonstrate ethical considerations is crucial. Students should be able to communicate effectively and persuasively through social media and be responsible users of the technologies. Assignments designed for courses and training should be developed to enhance learning and not solely for socializing.

Instructors interested in using social media in medical and healthcare education might consider offering the following tips to students. They are summarized well as follows: "(1) identify and then reflect on your digital identity and goals for using social media; (2) select a tool based upon goals and the strength of platforms available to support educational activities; and (3) observe and establish comfort first. Think, then contribute (4) make some initial connections and tap into the power of a community; (5) know and apply existing social media guidelines for the responsible use of social media; (6) develop individual guiding principles with which you are comfortable; (7) keep all patient information private; (8) handling "friend" requests from trainees know your options and their consequences; (9) share credible information: disseminate evidence-based health information, enhancing public health; (10) engage, learn, reflect, and teach; (11) research: advance your academic productivity by expanding your professional network; and (12) Mentor and be mentored: demonstrate responsible social media use." [12]

Examples

Examples of social media usage in the realm of medical education are fewer than those for practice. Social media is being adopted for medical education at a slower rate than other disciplines. There are some examples of effective use of social media for education. The online community called QunatiaMed allows for collaboration amongst physicians and is a great resource for students and educators seeking to learn from practitioners and experts. Presentations on various topics are delivered and questions can be posed to the expert.

Instructors have also been using social media to enhance teaching experiences. In one such example, an instructor in a geriatric pharmacotherapy course at the University of Rhode Island used Facebook to encourage class discussions and to connect students with senior citizens who had volunteered to participate in the course. Students at the University of Cincinnati used Twitter to stay abreast of current health care and informatics news and to document their findings in a blog post, which was accessible to other students and the instructor. Tools have continued to be integrated into teaching to create community and provide an avenue for access to current news, for example.

Blogs have been used in the classrooms to increase student and medical trainee interest in clinical disciplines and to promote professionalism and humanism [38]. They have also been used to improve communication and interpersonal skills by promoting reflection, creativity, and collaboration [39]. A study concluded that clinical excellence improved with the use of social media tools. Clinical excellence was defined by domains such as communication and interpersonal skills, professionalism and humanism, knowledge, diagnostic acumen, passion for patient care, scholarly approach, modeling clinical excellence, interfacing with researchers, and skillful negotiation of healthcare [40]. Social media tools have been found to augment student learning, allow for real-time communication, and helped medical students connect with medical experts. The tools fostered collaboration and creativity [38].

Clinical training programs can use social media to facilitate teaching. One study showed that over 90% of emergency medicine residents at over 10 programs used social media for education on a daily basis [41]. Hospital programs such as Mayo Clinic are offering their curriculum electronically through audio recordings and iTunes [41]. Twitter has been found to increase readership of medical journals such as the Journal of the American College of Radiology [41]. A Fellows-in-Training blog was developed to increase interaction with leaders in cardiology at national conferences. The interactions were uploaded to YouTube. Twitter has also been popular and used by fellows to disseminate messages about conferences [41].

As social media use in education increases, we have more evidence of suitable theories that educators might consider as frameworks for developing sound pedagogy. Some learning theories are directly related to social media use in education. Connectivism, constructive approaches, and social cognitive theory are three learning theories an educator seeking to use social media use for teaching and learning might consider [42–44].

Connectivism is grounded in the premise that clinical students will generate knowledge as they understand their experiences. Diverse opinions are valued and learning entails connecting specialized nodes (anything that is connected to another node such as data or organizations) or information sources. Humans are not the sole holders of learning and learning can reside in non-humans. Capacity to be more knowledgeable than one already is critical. Connections have to be sustained for continued learning, currency is the purpose for the learning activities, and decision-making is a learning process and must be flexible to accommodate changing what to learn and the changing information. The connections students make ensure that they are able to remain current and learning begins with the individual student but continues to grow as the learner interacts with others in their networks including instructor and peers as well as organizations [42].

Constructive approaches focus on learning occurring in a social context but more learner-centric. The educator or practitioner is in the role of a facilitator. Communication, exploring, and reflection lead to the learning and learners have their own learning goals. In this case, knowledge is subjectively generated and social media can be viewed as creating this flexible environment where learners can develop this knowledge by enabling the social interaction [43]. A learner practicing connectivism is likely to be participating in constructivism. The difference is brought about by the relationships and networks in connectivism that are not ancillary as they are with the constructive approach, but core to connectivism.

Bandura's social cognitive theory focuses on learning being social and promoted by observing others [44]. Social learning is a cognitive process that can be achieved through direct experiences and observing and modeling or imitation [44]. The theory also posits that humans create social-structural networks and are part of those networks [44]. As individuals use social media to create profiles and connect with friends and colleagues, they are creating these networks. Social interaction is viewed as critical in the cognitive development process. The observation triggers individuals to interpret behavior and adjust their own. Individuals using social media are now not confined to the reality of those in their environment but are influenced by others external to their own sphere.

The interaction opportunity that social media affords to students and educators in healthcare disciplines can allow for enriched discussion, adaptation, knowledge generation, community, and knowledge transfer. These theoretical frameworks can be the basis for promoting use of social media for education and learning. Clearly, students and educators can both be benefactors of these tools and interactions.

SOCIAL MEDIA IN PRACTICE

As the number of e-patients who are actively engaged in managing their health is on the rise, the healthcare practitioners who can assist them with their needs are trying to match up to their expectations as well. This section discusses the dynamic role that healthcare practitioners have established in meeting their patients' needs using social media.

Opportunities

Studies show about one-third of the population in countries such as the United States use social networking to find fellow patients and discuss their conditions, and that about 36% of social network users evaluate the knowledge of fellow users before making a health care decision. Given these statistics, health scientists have realized that the impact of social media on health related issues can be utilized for further development of research opportunities, better patient education and engagement and, clinical training for students [1,25,27,29,45]. Social media tools like Twitter have been shown to be useful for real-time monitoring for various purposes such as substance abuses, foodborne illnesses, and viral diseases [46–49].

Healthcare practitioners can promote use of social media that improves a patient's experiences and wellbeing by for example, promoting tools that patients can use to share stories, and increase communication with their providers and social networks. Clinicians can also use social media for their own training, to share information for better treatment options, to share information for patient empowerment, and to share information in times of crisis.

Examples

Clinicians have been able to use social media to increase their involvement in continuing medical education and those involved in research are identifying patients for clinical trials using social media. About 14% of the hospitals have been known to have a social media and social networking presence to enhance their ability to market and communicate to stakeholders [45]. Hospitals use social media for promotional purposes and to gain feedback regarding patient experiences. There are a number of ways through which healthcare professionals can leverage social media to enhance patient experience by providing quality care and increasing professional networking.

As social networking has evolved so have physician portals where physicians can post a query and get responses from other physicians to seek better treatment options. These online portals are private and provide memberships after checking the credentials. *Sermo is* an online portal exclusively for doctors. It is equivalent to a doctor's lounge where doctors can share experience and insights about medicine. Approximately one out of every six U.S. physicians are members of Sermo [45]. As of August 2016, there were 600,000 credentialed physicians on Sermo from all over the globe. Since Sermo provides an opportunity for the physicians to remain anonymous, physicians nonhesitantly ask questions and respond. Doximity is yet another physician based social networking site, which caters to more than half a million physicians where most are U.S. based. Doximity is HIPAA compliant and lets the physicians share texts and images in a secure environment. Similarly, other popular physician-based social networking sites include but are not limited to QuantiaMD (www.quantiamd.com), The Medical Directors forum (www.medicaldirectorsforum.skipta.com), and Doc2Doc (doc2doc.bmj.com).

There are social networking sites for nurses as well which provide nurses a common platform for networking and to share their experiences. The American Nurse Association, the largest association of nurses in United States provides their platform via ANAnursespace (www.ananursespace.org). Allnurses (allnurses.com) has over 975,000 nurse members to share their experiences on an active forum. Mightynurse (mightynurse.com) is another platform to empower the nurses with ample advice from fellow nurses. Nursinglink (nursinglink.monster.com) is a platform that allows nurses to post their resumes and look for job openings. Professional networking sites are popular amongst pharmacists too. The American Society of health-system pharmacists provide their networking platform via ASHP connect (connect.ashp.org). PhamQD (www.pharmqd.com) and pharmacistsociety (pharmacistsociety.com) are other very common networking platforms amongst pharmacists.

During crisis, healthcare facilities are better able to connect with consumers by providing minute-by-minute update on their capacity, treatment, and emergency room status. This communication is specifically helpful during disease outbreak or a natural disaster by providing live updates from Center for Diseases Control (CDC) or Red Cross. On the flip side, consumers can assist healthcare facilities as well by providing the live status in case of emergency. For example, the bombing during the Boston marathon was on twitter and the trauma services prepared themselves better by postponing elective surgeries and making them available for the injured marathoners. Government agencies such as the Food and Drug Administration (FDA) and the Centers for Disease Control and Prevention (CDC) are using social media to engage the public during product recalls and for H1N1 flu pandemic preparations. Physicians utilize media-sharing sites such as YouTube to educate patients about certain procedures. Patient empowerment assists patients in making informed decisions about their health. Before the advent of Health 2.0, patients were dependent on physicians to provide any necessary information to them regarding their treatment options. There are now social networking websites that are used to create a community of individuals with similar experiences who interact to share experiences and offer support. Examples include: PatientsLikeMe (patientslkeme.com), which helps the patients to track a disease, access information and learn about the real-world experience of patients suffering with the same problem [27], MedHelp (medhelp.com), which is a social network that also has a number of tracking tools for pain, weight, etc.

CrowdClinical.com is a free platform that collects, aggregates, and summarizes health related information in real time. It uses the method mentioned above, combined with supervised machine learning, natural language processing and manual curation to systematically identify and rate (e.g., determine sentiment) Twitter messages that discuss patient experiences receiving health care. CrowdClinical began monitoring tweets recently in early 2015 and continues to track patient experience discussions in real time.

Anonymized data collected from social networks is being used to advance biomedical research, precision medicine and in clinical trials by developing predictive models. The results generated from these models help identify the decisions made by the patients, the trends observed and why those decisions were made. With the data-driven approach, the healthcare industry is at a perfect edge to enhance the quality of care for its patient population by increasing consumer satisfaction and giving way to a healthier population.

DATA ANALYTICS: A NEW OPPORTUNITY FOR EDUCATORS AND PRACTITIONERS

An area of great promise for both healthcare practitioners and educators is data analytics. Data analytics is the science of examining raw data and extracting patterns from it in order to obtain information. Social media analytics refers to the use of analytics to mine for patterns from social media information. It is most commonly used to track sentiments in order to predict the attitude, behavior or needs. Some examples of social media analytic tools include: buffer, hootsuite, Google analytics, brand watch, and 33Across. Volumes of data is being generated by online user groups about clinical, public health, and wellness topics, that can be mined to gain knowledge about various topics for use by healthcare and public health practioners and also by educators. Social media analytics and the results from data mining can be useful to healthcare practitioners seeking to gain insight from the data and make predictions. The big data can be analyzed to detect real-world issues and gain insight about communication patterns and exchanges. The real-world issues can also be enlightening for educators, clinicians, and healthcare organizations as data is user-generated and patterns can be recognized. The results from data analytics can help end-users identify communities in the social network. End users of social media data must be prepared for the nuances of this data, realizing that it is different from traditional data.

Analytics tools have been used to mine social media data pertinent to health and well-being. For example, the Symplur analytics tool was used to analyze Twitter activity associated with indigenous health to gain knowledge about social and emotional wellbeing and mental health of Aboriginal and the Torres Strait Islander population [50]. As we know, people talk about their thoughts and experiences on social media. The "sentiments" of people can be captured and put into effective use by analyzing them and putting them into effective use. Sentiment analysis is a process of understanding user's point of view whether it is positive, negative or neutral by examining the text. The Computational Epidemiology Group in Boston Children's Hospital's Computational Health Informatics Program (CHIP) came up with an idea to perform sentiment analysis of health related tweets, and use this method to help people compare and contrast different hospitals. For example, if there is an extended wait time in the emergency department of one hospital then people may choose another one when they need emergency care. A few years ago, the Boston Children's Hospital team analyzed the reviews of Yelp, and found out some potential relationships between food poisoning and public health officials. Now they are establishing sentiment analysis on Twitter, so that hospital management can better track patient experience metrics. They deployed a system that mines the comments on Twitter, and use sentiments as a proxy to generate hospital satisfaction metric to help patients make decisions. They accomplish this by using artificial intelligence and machine learning classifier to filter out the tweets that have nothing to do with patient experience. They then proceed to use natural language processing to analyze the remaining tweets to distinguish different expressions, and to classify them into different emotions such as angry, happy, and neutral. The results are validated with HCAHPS (the formal U.S. nationwide patient experience survey) and the Hospital Compare 30-day readmission rate, which shows that the hypothesis is mostly correct [51].

CONCLUSION

Social media have been initially popular for socializing. However, over time, social media resources have demonstrated added value by offering more benefits to end-users such as patients, students, and healthcare practitioners. Social media tools allow for active contribution, connectivity, knowledge generation and transfer, community, and collaboration. The acceptance of social media tools by consumers has led to the explosion of use and produced big data available for mining and visualization, creating opportunity for teaching, learning, and practice. Students and the general public are immersed in the use of social media and well versed in multiple technologies creating the challenge for educators to adopt these tools in education. Likewise, practitioners in healthcare are now challenged to explore the use of these tools to engage with their patients and benefit their organizations.

Social media has changed the healthcare delivery system, which followed a top down approach with healthcare practitioners providing the information to the patients. The e-patient population partners in their healthcare needs, voluntarily seeking health information, and thus leading to patient empowerment. The patients who took a passive role in their healthcare needs are now replaced with the digitized e-patients who are active and engaged in their healthcare processes. As social media provides greater transparency and more communication between the patient and provider, there are also ethical and professional implications on the physicians that draw a line for them for ethical considerations.

Practitioners, patients, educators, and students have the moral obligation to consider these ethical implications of their use of social media and should carefully assess messages or content before they are shared or submitted. There is also opportunity for faculty to advocate for ethical use of social media by their students. Guidelines should be developed to govern ethical use of social media ensuring that no harm occurs as a result of using the tools. The guidelines might include specifics about protecting privacy, Internet etiquette (netiquette), and compliance with legal statutes. Guidelines for medical or other clinical students, should include, for example, information about protecting patient health information as students interact and collaborate using social media.

In promoting use of social media for education and practice, we must reiterate the need for practitioners, educators, students, patients, and others, to use social media responsibly and ethically. Users of social media tend to be receptive to sharing content in their personal spheres, and in some cases, the same openness is observed when social media is used in other contexts, with little regard to information security safeguards and confidentiality. Privacy and confidentiality should be prioritized by educators and practitioners. Patients and students should be guided to protect their privacy and confidentiality. Content appropriateness and quality should be assessed before it is shared or relied upon. Content appropriate and quality for personal contacts may not be appropriate or suitable for professional contacts. Social media users have the responsibility to ensure that they have the appropriate authorization to share specific content. Policies should be developed to guide sharing of content and overall social media interaction.

References

[1] Antheunis ML, Tates K, Nieboer TE. Patients' and health professionals' use of social media in health care: motives, barriers and expectations. Patient Educ Couns 2013;92(3):426–31.

[2] Chretien K, Kind T. Social media and clinical care. Circulation [Internet] 2013;127(13):1413–21. Available from: <http://circ.ahajournals.org/content/127/13/1413>.

[3] Nielsen. Social Network Blogs Now Account for One in Every Four and Half Minutes Online [Internet]. 2010. Available from: <http://www.nielsen.com/us/en/insights/news/2010/social-media-accounts-for-22-percent-of-time-online.html> [updated 15.06.10].

[4] Pew Research Center. The social life of health information [Internet]. 2011. Available from: <http://www.pewinternet.org/2011/05/12/the-social-life-of-health-information-2011/> [updated 15.05.11].

[5] Kamel Boulos MN, Wheeler S. The emerging Web 2.0 social software: an enabling suite of sociable technologies in health and health care education. Health Info Lib J [Internet] 2007;24(1):2–23. Available from: <https://www.ncbi.nlm.nih.gov/pubmed/17331140>.

[6] Chou WS, Hunt YM, Beckjord EB, Moser RP, Hesse BW. Social media use in the United States: implications for health communication. J Med Internet Res [internet] 2009;11(4):e48. Available from: <https://www.jmir.org/2009/4/e48/>.

[7] Duggan M, Smith A. Social media update 2013 [Internet]. Washington, DC: Pew Internet & American Life Project; 2013. Available from: <http://www.pewinternet.org/2013/12/30/social-media-update-2013/> [updated 30.12.13].

[8] Fox S, Duggan M. Health Online 2013 [Internet]. Washington, DC: Pew Internet & American Life Project; 2013. Available from: <http://www.pewinternet.org/Reports/2013/Health-online.aspx> [updated 15.01.13]

[9] Griffis HM, Kilaru AS, Werner RM, et al. Use of social media across U.S. hospitals: descriptive analysis of adoption and utilization. J Med Internet Res [Internet] 2014;16(11):e264. Available from: <http://dx.doi.org/10.2196/jmir.3758>.

[10] White J, Kirwan P, Lai K, Walton J, Ross S. 'Have you seen what is on Facebook?' The use of social networking software by healthcare professions students. BMJ Open [Internet] 2013;3(7):e003013.

[11] Usher K, et al. Australian health professions student use of social media. Collegian 2014;21(2):95–101.

[12] Madanick R. Education becomes social: the intersection of social media and medical education. Gastroenterology 2015;149(4):844–7.

[13] Sandars J, Schroter S. Web 2.0 technologies for undergraduate and postgraduate medical education: an online survey. Postgrad Med J 2007;83(986):759–62.

[14] Rutherford C. Using online social media to support preservice student engagement. MERLOT J Online Learn Teach 2010;6(4):703–11.

[15] Mâţă L. Social media tools in initial teacher education. In: Information Resources Management Association, Social media and networking: concepts, methodologies, tools and applications. IGI Global; 2015. https://www.igi-global.com/book/social-media-networking/125529#table-of-contents.

[16] Archambault PM, van de Belt TH, Grajales III FJ, et al. Wikis and collaborative writing applications in health care: a scoping review. J Med Internet Res 2013;15(10):e210. Available from: http://dx.doi.org/10.2196/jmir.2787.

[17] Chu LF, Young C, Zamora A, Kurup V, Macario A. Anesthesia 2.0: internet-based information resources and Web 2.0 applications in anesthesia education. Curr Opin Anaesthesiol 2010;23(2):218–27. Available from: http://dx.doi.org/10.1097/ACO.0b013e328337339c.

[18] Boyd D, Ellison N. Social network sites: definition, history, and scholarship. J Comput Mediat Commun 2008;13:210–30.

[19] Stellefson M, Chaney B, Ochipa K, et al. YouTube as a source of COPD patient education: a social media content analysis. Chronic Respir Dis 2014;11(2):61–71. Available from: http://dx.doi.org/10.1177/1479972314525058.

[20] Guraya S. The usage of social networking sites by medical students for educational purposes: a meta-analysis and systematic review.. N Am J Med Sci 2016;8(7):268–78.

[21] Gray K, Annabell L, Kennedy G. Medical students' use of Facebook to support learning: Insights from four case studies. Med Teach 2010;32:971–6.

[22] Roblyer M, McDaniel M, Webb M, Herman J, Witty JV. Findings on Facebook in higher education: a comparison of college faculty and student uses and perceptions of social networking sites. Inter High Edu 2010;13:134–40.

[23] Erfanian M, Javadinia SA, Abedini M, Bijari B. Iranian students and social networking sites: prevalence and pattern of usage. Pro-Soc Behav Sci 2013;83:44–6.

[24] Bickerdike, A., O'Deasmhunaigh, C., O'Flynn, S., & O'Tuathaigh, C. Learning strategies, study habits and social networking activity of undergraduate medical students. Int J Med Educ 7, 230–236. Available from: <http://doi.org/10.5116/ijme.576f.d074>.

[25] Kivunja C. Innovative methodologies for 21st century learning, teaching and assessment: a convenience sampling investigation into the use of social media technologies in higher education. Int J High Educ 2015;4(2):1–26.

[26] Cheston CC, Flickinger TE, Chisolm MS. Social media use in medical education: a systematic review. Acad Med 2013;88:893–901.

[27] Moorhead SA, Hazlett DE, Harrison L, Carroll JK, Irwin A, Hoving C. A new dimension of health care: systematic review of the uses, benefits, and limitations of social media for health communication. J Med Internet Res [Internet] 2013;15(4):e85. Available from: <https://www.jmir.org/2013/4/e85/>.

[28] Chretien Katherine C, T. Kind. Social media and clinical care. Circulation. 2013;127:1413–21 originally published April 1, 2013. Available from: <https://doi.org/10.1161/circulationaha.112.128017>.

[29] O'Keeffe GS, Clarke-Pearson K. The impact of social media on children, adolescents, and families. Pediatrics 2011;127(4):800–4.

[30] Hawn C. Take two aspirin and tweet me in the morning: how Twitter, Facebook, and other social media are reshaping health care. Health Aff 2009;28(2):361–8.

[31] Batool R, Khan WA, Hussain M, Maqbool J, Afzal M, Lee S. Towards personalized health profiling in social network. 6th Int Conf New Trends Inf Sci Serv Sci Data Min 2012;53:760–5.

[32] Greysen SR, Kind T, Chretien KC. Online professionalism and the mirror of social media. J Gen Intern Med 2010;25(11):1227–9.

[33] Ventola CL. Social media and health care professionals: benefits, risks, and best practices. Pharm Ther 2014;39(7):491–520.

[34] George D, Rovniak L, Kraschnewski J. Dangers and opportunities for social media in medicine. Clin Obstetr Gynaecol 2013;56(11):453–62.

[35] Bosslet GT, Torke AM, Hickman SE, et al. The patient-doctor relationship and online social networks: results of a national survey. J Gen Intern Med 2011;26:1168–74.

[36] Sandars J, Morrison C. What is the net generation? The challenge for future medical education. Med Teach 2007;29:85–8.

[37] Giordano C. Health professions students' use of social media. J Allied Health 2011;40:78–81.

[38] George DR, Dellasega C. Use of social media in graduate-level medical humanities education: two pilot studies from Penn State College of Medicine. Med Teach 2011;33:e429–34.

[39] Varga-Atkins T, Dangerfield P, Brigden D. Developing professionalism through the use of wikis: a study with first-year undergraduate medical students. Med Teach 2010;32:824–9.

[40] Batt-Rawden S, Flickinger T, Weiner J, Cheston C, Chisolm M. The role of social media in clinical excellence. Clin Teach 2014;11:264–9. Available from: http://dx.doi.org/10.1111/tct.12129.

[41] Snipelisky D. Social media in medicine: a podium without boundaries. J Am Coll Cardiol 2015;55 (22):2459–61.

[42] Vygotsky LS. Mind in society: the development of higher psychological processes. Cambridge, MA: Harvard University Press; 1980.

[43] Pritchard A. Ways of learning: learning theories and learning styles in the classroom. Great Britain: David Fulton Publishers; 2005.

[44] Bandura A. Social cognitive theory of mass communication. Media Psychol 2001;3(3):265–99.

[45] Keckley P.H., Hoffman M. Social networks in health care, communication, collaboration and insights. Deloitte Center for Health Solutions, 2010 [Internet]. Available from: <http://www.healthcarevisions.snapmonkey.net/f/2010_Deliotte_Social_Networks.pdf>.

[46] Signorini A, Segre AM, Polgreen PM. The use of Twitter to track levels of disease activity and public concern in the U.S. during the influenza A H1N1 pandemic. PLoS One 2011;6:e19467.

[47] Jashinsky J, Burton SH, Hanson CL, et al. Tracking suicide risk factors through Twitter in the US. Crisis 2014;35:51–9.

[48] Hanson CL, Burton SH, Giraud-Carrier C, et al. Tweaking and tweeting: exploring Twitter for nonmedical use of a psychostimulant drug (Adderall) among college students. J Med Internet Res 2013;15:e62.

[49] Chary M, Genes N, McKenzie A, et al. Leveraging social networks for toxicovigilance. J Med Toxicol 2013;9:184–91.

[50] Sweet M, Geia L, Dudgeon P, McCallum K. #IHMayDay: tweeting for Empowerment and social and emotional wellbeing. Australas Psychiatr 2015;23(6):636–40.

[51] Hawkins JB, Brownstein JS, Tuli G, Runels T, Broecker K, et al. Measuring patient-perceived quality of care in US hospitals using Twitter. BMJ Qual Saf 2015;. Available from: http://dx.doi.org/10.1136/bmjqs-2015-004309.

PART III

EVALUATING STUDENTS AND PROGRAMS

CHAPTER

22

Using Activity Data and Analytics to Address Medical Education's Social Contract

David Topps, Rachel H. Ellaway and Maureen Topps
University of Calgary, Calgary, AB, Canada

INTRODUCTION

The profession of medicine is afforded significant status and autonomy; in return for which, it is expected to ensure that its members are competent and disposed to meeting society's healthcare needs. This relationship is often called the "social contract" [1]. Medical schools are expected to be a major contributor to meeting the social contract for medicine through ensuring the quality of the individuals they are training and entering into the profession. Both medicine and medical education are dynamic systems: they are in a constant state of change and flux, driven by changing practices, technical capabilities, and societal expectations. Increased attention to patient safety, quality improvement, and growing sophistication in the educational sciences have demonstrated a pervasive problem in medical education: a systematic failure to fail individuals who have failed to meet the expected standards for a licensed physician [2–5]. As a result, medical education has to some extent fallen short of its social contract.

This continues to be a problem despite evidence that it is happening and many attempts to mitigate this trend [5]. It is not that the system is unwilling, but rather that it seems incapable to address its own shortcomings. We therefore need alternative approaches that can address the failure to fail problem consistently, reliably, and efficiently. In an age of ambient digital technology, we now have multiple points of contact with learners' activities that generate data that we can sample and use to inform our appraisal of learners' capabilities. We can use this to triage those who need more help to meet the standards of professional medical practice and to identify those who should, in fact, be excluded from the profession.

Health Professionals' Education.
DOI: http://dx.doi.org/10.1016/B978-0-12-805362-1.00022-X

In this chapter, we outline the problem, we consider some of its causes, some of the barriers educators face in trying to address this problem, and we present a vision of how the use of technology, ambient surveillance, and intelligent use of the increasingly large data that is now being collected in and around educational systems [6] can be integrated as an approach to addressing the failure to fail problem [5]. We argue for a reconfiguring of medical education training programs to shift from a complete dependence on human raters to one that makes use of machine-generated data combined with human judgment to sustain the quality of the health professional workforce.

BACKGROUND

Given the high stakes associated with training health professionals, assessment and quality assurance have long been critical components of contemporary medical education. Until recently, progression through medical training was structured around episodic high-stakes exams. More recently, the shift to competency-based medical education has involved a shift to ongoing observation-based assessment completed in the workplace where preceptors make regular written observations of their learners' progress [7]. This growing dependence on increased observation and assessment, by the same individuals who are responsible for mentoring and supporting their learners, is proving to be problematic.

There has been a growing trend in North American medical schools that once a student has passed the hurdles of getting into medical school, it is expected that they will have the required capabilities to complete the training and will therefore make it all the way through to practice. Only egregious acts and significantly substandard performance over an extended period of time are likely to lead to exclusion. This has been exacerbated by the growing effort directed toward the remediation of failing learners [8], a generous range of learning accommodations, and general assumptions that our selection processes have been robust. It has been observed that there is a small, but significant fraction of learners who experience problems and continue to do so throughout their careers [9], who are nevertheless not excluded from professional practice [5]. For some, these tendencies extend backward to earlier education including high school [10]. Although this is not a uniquely North American problem, it is notable that many countries throughout the world admit large numbers of students to medical education, and sustain a significant attrition rate, whittling down the numbers as they craft their candidates [11].

We have observed that this may be partly due to the increasingly litigious nature of our students. Much of our teaching has become community based, which leads to an enriched experience, but we are then reliant on occasional observers to be judges, which they dislike intensely. Many programs have lost excellent teachers because of a poor learner experience—a very sad loss on both sides. Teachers have learned to be wary of failing a learner, fearing complaints about their skills, the learning environment or ability to judge, or simply wishing to avoid the unpleasantness involved [5]. We explored this phenomenon using a standardized underperforming learner encounter (see Box 22.1) and found significant preceptor reluctance to make negative judgments, even about this fictional and unambiguously problematic individual [12].

BOX 22.1

LACKADAISICAL LARRY

We wanted to explore how preceptors responded to behaviors typical of failing or problematic learners. To do this, we constructed a virtual patient case, "Lackadaisical Larry" [12], using an open-source platform called OpenLabyrinth [13]. Larry is a medical student, who exhibits an increasingly pathological series of behaviors, but who generally gets away with things by being nice. You play the case as one of his preceptors and have to reign in his increasing misbehaviors, seeking an optimum strategy. If you are too lax with Larry, he gets worse; if you are too rigid and blatant in your approach, he launches various complaints about you. The case is complex and Larry gets up to all sorts of tricks, not all of which are obvious, but this is an amalgam of real behaviors seen in our own learners. At the end, you are asked to assess whether Larry should pass this rotation, in standard evaluation format. We asked 22 experienced community and academic faculty to play this case, wondering how they would handle it, as a baseline to its wider deployment as a faculty development tool for new community-based preceptors. We were amazed that only three failed Larry. While many assessed him as 'Borderline Unsatisfactory', under our assessment policies, which are clear to preceptors, this is still a pass. Given how deliberately poor a learner Larry was made to be, these responses are disappointing, particularly when you realize that the common reasons given for rationalizing such inappropriate judgments (avoiding conflict, not wanting to raise a complaint process) did not apply because Larry was virtual: he was unable to send a complaint to the Dean about how he had been treated.

This is not to say that preceptors should become more harsh and judgmental; being hawkish tends to inhibit the creation of constructive educational relationships with learners. What we need are more robust processes that allow us to dynamically triage our learners into one of the following three categories:

- Those learners who are performing as well as or better than expected and who will, as far as we can tell, go on to make good doctors. We continue to check that their performance meets or exceeds expectations, but we need to take no additional steps to shape their training experiences.
- Those learners who are struggling to perform as required, and yet demonstrate the aptitudes that will allow them to function as physicians, as long as they receive sufficient support in addressing their shortcomings. These individuals are remediated to the point that they either rise to the category of successful learners, or they fail to thrive or show the right kinds of aptitudes and need to be excluded.

- Those learners whose performance or professional standards fall below what is acceptable and for whom remediation cannot address their shortcomings. These individuals need to be excluded from the profession.

Not only do we need to more accurately and consistently triage learners into the appropriate category, we also need to address issues of fairness for learners in ensuring that any aspects of suboptimal performance are due to learners' shortcomings rather than artifacts that result from the biases of their teachers or the learning environment as a whole. How we identify which category learners are assigned to and how they move between categories is a critical component of medical education, but one that is proving challenging to sustain. So what can we do differently?

It is not simply a matter of better training of preceptors regarding their ability to provide robust assessments of their learners' performances. It is about how we can make the system less dependent on the ratings of individual preceptors, who experience a conflict between acting as both mentor and judge. One answer lies in making better use of the data we capture on and around our learners through their use of the multitude of electronic systems found in contemporary health workplaces and training environments.

There has been much interest in improving observation mechanisms such as procedure logs and 'in-training evaluation of residents' [14,15]. While we have been successful at improving the logistics and processes of observational assessment [16,17], there is, as yet, little evidence of its impact on actual teaching and practice and it seems that very few programs have been able to close this loop. When a deficiency of experience is identified (short of triggering an episode of remediation), it is often left up to the learner to find additional experiences to mitigate these gaps. The quality of data gained with workplace assessments is also questionable. For instance, it has been demonstrated that tick box scores are convenient, but are of limited value in judging progression [18]. Qualitative assessment of free text commentaries can provide greater value, but its utility is limited given the time it takes to process this information. It is simply too much work for most programs to examine these with the level of detail required to extract meaningful conclusions from text-based data [17].

This type of assessment depends on teachers' abilities as observers and witnesses, but, as noted above, it is questionable whether preceptors can maintain objectivity given their other commitments. Studies have questioned the reliability of eye witness testimony and our ability to recall what we think we saw, something that the legal profession is also struggling with [19,20]. The approximate equivalent probability estimate for "beyond a reasonable doubt" is around 90% probability [21]. Indeed, this low level has been such a potential embarrassment that legal scholars in this area have tended to obfuscate such discussions with defenses around things being too complex to be reduced to a single number [22].

In clinical work, while we know that the patient's history rather than examination informs many of our diagnostic conclusions, we often provide our radiological colleagues with minimal history and then regard their observations and opinions with much greater diagnostic weight than perhaps is appropriate in the absence of context [23,24]. The overall clinical presentation and context, combined with the data, matters in our diagnostic assessment of the patient. Perhaps it is time we applied similar standards to our educational diagnoses.

AMBIENT SURVEILLANCE

We increasingly live in a world of ambient surveillance. Even if we are not under the watchful eye of a security camera, everything we do online and many things we do in real life are now tracked in extraordinary detail. Although we like to think we still have some degree of privacy, our use of banking, loyalty cards, and the Internet is fast eroding this [25]. The business models for Google, Facebook, and Amazon, among others, are based on tracking our purchasing and searching habits, what we are interested in and who we communicate with. This is very valuable data. For example, the annual value-add for Google in tracking such data is estimated at around $1200 per person compared to the cost of tracking this data of approximately $2 per person [26]. Cases such as Wikileaks and the Snowden case [27] illustrate the extent to which telecommunications are being tracked. How we drive is also increasingly being tracked constantly by many vehicles, again mostly for safety reasons. Manufacturers can analyze the black boxes built into most vehicles made since 2010 for many dynamic parameters. Car insurance companies advertise discounts for "enabling" such tracking [28].

Why are we not using similar techniques to assess medical learners? Some groups have raised concerns about privacy issues for learning analytics [29]. Others have suggested that such privacy claims may be quite questionable, given our widespread ceding of privacy in the public arena where we have largely surrendered the right to be tracked in return for online convenience [25]. With the higher standards of accountability that are to be expected of the medical profession, it might be more difficult to contend that the public has no right to know about our backgrounds, education and professional behaviors.

It might also seem strange to use tracking activities that appear at first glance to have only a tenuous relationship to learner behaviors. We would argue that there is educational value in this data. For instance, several studies have used activity data to assess the Conscientiousness Index for learners [30–32]. Conscientiousness is regularly used in assessing candidates for job positions, medical school entry, etc. [33–35]. The Conscientiousness Index relates to how good an individual is at completing routine tasks on time and with an appropriate level of detail. Using attributes such as these has somewhat fallen from favor as they have been shown to have limited utility in the selection process for medical school. There are some indications that this may be because we have been trying to use them to select the best candidates, and yet our smartest candidates are likely to be well able to divine what qualities we are looking for and therefore how to answer the questions [36].

There is growing evidence that we should be looking for a different segment: those who consistently rank poorly on the Conscientiousness Index [30–32]. Those learners that do rank poorly tend to go on to need remediation or other help. Early diagnosis in this matter would be enormously helpful because, at present, such learner issues tend to show up late in the program trajectory. What is particularly interesting in this approach is that the metrics needed to detect this are already tracked in standard administrative logistics. Program administrators already track when they receive licensing documents, proof of insurance, vacation requests, and vaccination certificates. In exploring this for our own programs, we realized that we have another readily available source of very accurate data: every action in the electronic medical record (EMR) is electronically time stamped for an audit trail. Most jurisdictions now require this in an EMR. It would be a simple data extraction exercise therefore to look at which users are frequently tardy with routine tasks such as chart completion, results review, or creation of consult letters. Because this all relates to simple relative timings in regard to the time of the patient visit, such a search would not need to reveal patient identities at all. Basic time differences are sufficient for our purposes.

There are many applications that already capture data about the ways that learners behave. For instance, most contemporary medical learners use Learning Management Systems, portfolio systems, assessment systems, and messaging systems. Their devices move in and out of wireless networks. They use libraries, electronic medical records, and social media. Learners already leave a wide trail of data behind them as they progress through training. Our primary challenge is not to identify who is the best, but to identify individuals who need additional help in their programs, and those few who should not progress to a career in medicine at all. This is not a matter of traditional assessment focusing on levels of success but one of screening by focusing on levels and dimensions of failure or indicators thereof.

SMARTER USES OF DATA

Capturing data is insufficient on its own. We need to be deliberate in what we then actually do with it, for instance in identifying behaviors and patterns that indicate potential failure or undesirable characteristics in an individual. The Observer Effect can be a problem as noted by Berwick et al in their assessment of performance metrics on physicians [37]. As noted with the unsuccessful use of certain kinds of psychometrics, smart candidates figure it out. So, how do we overcome such effects? In exploring this, we have turned to a number of important underlying principles to aid our thinking and learning designs:

1. What we do is what matters, not what we say we do;
2. What an observer says we do is not much better;
3. Not everything we do matters;
4. We don't always know up front which things matter;
5. We all make mistakes—embrace them and learn from them.

The focus is increasingly on triangulating activity data streams from different sources. Kneebone and colleagues have combined actors/standardized patients with simple task trainers, such as suturing models, to create a simulation context, which is much more realistic [38]. The learner does not simply have to deal technically with closing the skin edges, but also has to cope with a scared patient who is in pain. Similarly, we have successfully blended virtual patients with high-fidelity mannequins and virtual camera arrays for multisite simulation training [39]. We also developed an Arduino-based sensor system (total cost $55) combined with activity streams from a virtual patient system (OpenLabyrinth) to study stress levels during a timed virtual scenario [40]. The sensor system measures heart rate and minute changes in electrical conductivity of the surface of a finger due to enhanced (but unnoticeable) sweating, detected well before any overt evidence of stress. The important part about blending such modalities is to think about what activities you want to measure and how to integrate the learning experience.

It was striking how sensitive the combination of metrics was in tracking user stress levels. Even expert clinicians in our reference group, some of whom would be characterized as "unflappable," found these situations to be very challenging. It was even more striking that keen observers of the scenario were able to detect the first signs of stress from these cheap sensors before there were any visible signs of stress. Similar studies have been conducted previously [38,41], but these have largely required dedicated labs and tens of thousands of dollars of sensor equipment. This use of cheap, easily available sensors brings such studies down within reach of the average education research budget. The two key concerns in blending such modalities is to think about what activities need to be measured and how to integrate the learning experience around them.

The range of sensors available to undertake this work are increasingly accessible using everyday technologies. For instance, most smartphones now possess sensors that can detect magnetic fields, proximity, light levels, orientation and acceleration, location, temperature, heart rate, and capacitance. These devices also have a sophisticated computational core and a wide variety of networking channels with which to communicate to other systems. This is reflected in the rapid uptake of personal activity tracking in fitness apps, something that has also opened up many new lines of inquiry for medical and education researchers.

Even regular computer terminals, such as those found in clinics and hospitals, can now be used to track individual performance metrics without modification of the hardware. For example, we all have a keyboarding "fist." This was known to Morse code operators over a hundred years ago. We are uniquely identifiable by our typing style or keystroke dynamics, just as we are with our driving style [42,43]. Proper tracking of this might even eliminate such inane requirements, as rotating passwords, a practice that has been demonstrated repeatedly to reduce, not enhance security [44–46].

BIGGER DATA

As we explored the broad variety of activity metrics that were available to us through virtual scenarios, we were struck by the rapid increase in the volume of data that we had available for analysis. At the time of writing, on a single OpenLabyrinth server, we had over a million data points that were amenable to activity stream analytics. From the Google Analytics outputs on our CliniSnips video channel on YouTube [47], we had detailed metrics on more than 4.7 million views of that channel, down to which 10-second segments of which video had been most popular. This was starting to look rather like "Big Data." In the discussion about what constitutes Big Data, there has been debate about various "V"s: Volume, Velocity, and Variety were the most consistent, with others such as Veracity and Value attracting less interest [48]. Velocity refers to the speed at which this data is generated and analyzed in near-real-time. While there are more data generated by educational systems than ever before, this is still tiny in the context of Big Data science. But the rest of the principles still apply, particularly that of Variety.

Distilling educational and professional activities down to a single metric can be misleading. But examining a broad variety of activities can be much more enlightening. One concern is that, if you track activities toward certain targets, then these observations will change the observed behaviors, sometimes with undesirable side effects. We know this to be equally true of learners. For example, when procedure numbers were tracked in an electronic logging system, the number of procedures recorded increased [49]. However, this says nothing about the quality of performance or patient outcome. Moreover, these activity data streams can be manufactured or misrepresentative [50].

However, as we have tracked an ever-increasing variety and number of activity metrics in our various projects [51], an interesting effect has emerged. It becomes harder for weaker learners to merely simulate doing the right thing. As they attempt to increasingly emulate and reproduce the desired activity metrics so that they appear to pass muster, it requires more and more effort. This has led to the interesting observation that either they then progress to properly conduct the tracked activity, with no further need to simulate and dissimilate their results (a reasonable outcome in its own right); or the faked performance is so close to the desired performance in terms of what is actually done that it no longer matters because the outcome is the same. Either way, the learned behavior can be considered successful.

Now, for researchers who use social media to conduct research on the evaluation of learner activities, some additional complications arise. In our work with activity metrics in virtual scenarios, we have been fortunate to have full access to the raw data. For those who have to extract their data from social media resources such as blogs and Twitter feeds, there are concerns about the quality of the data. This is not raw data, but is produced with some considerable opacity about how it is generated (most social media platforms are quite coy about how they generate their various metrics). There is a lack of metadata uniformity across systems (e.g., "Likes" are generated and counted in different ways, yet tend to be conflated across platforms); and that there is a lack of equivalence between "users" and individuals (e.g., a "user" can represent a bot or a group or a corporation, and individuals can have multiple accounts). To help address this, it is also suggested that having a broad variety of metrics ("wide" data as well as big data) may give researchers more ability to seek clarity [52].

TRACKING ACROSS LEARNING MODALITIES

As we noted above, using a variety of metrics and sources can be useful in many ways. However, we should resist the "Lamp Post" phenomenon, persisting in looking for our research keys because this is where the light is, and instead seek metrics that truly reflect the desired activity and outcome. In an education environment, with all its confounding factors and broad range of learning modalities, this can be tough.

For example, in our CliniSnips series [53], we were able to analyze a very large dataset to a fine degree and had some remarkable insights from this. But we were also frustrated in not knowing much about the context in which our videos were used, or much about the participants. Certain attributes could be approximately inferred, but it would have been more compelling if we had been able to integrate these YouTube resources with other activities and platforms such as OpenLabyrinth, Moodle, and WordPress. At the inception of that project, this was not feasible, but subsequently we have found ways in which this can be achieved. Many systems and platforms now make their data accessible via an Application Programming Interface (API). For example, Google has a whole suite of APIs [54], including a set for YouTube, which allows for the analysis of how content is being used and accessed.

The Experience API (xAPI) is a very simple approach to tracking activity metrics [55] using triplet statements consisting of an Actor, Verb, and Object, as illustrated by:

```
Bob Did This
```

Triplet statements are now used quite widely in other fields such as semantic indexing with their Resource Description Framework triple stores [56], so this structure aligns nicely with other web-oriented analytic approaches. This simple data structure obviously has some greater underlying complexity, but it is very amenable to the flexible approaches required in today's world of big data analytics.

An important factor that will contribute to the utility of xAPI statements is standardization. For example, with the Verb component, there is a common catch-all verb, "experienced," but there needs to be some definition of what this means when we say "Alice experienced a discussion." How these statements are defined is well documented in the xAPI reference sites and we will not repeat them here. What is more important to educators is that there are groups like Medbiquitous [57] whose focus is on creating data standards and interoperable APIs so that collaboration is properly supported. Groups of verb definitions are typically described in xAPI "Profiles" and Medbiquitous is actively involved in defining Profiles across a range of learning experiences and simulation related activities.

Integration of data across systems is not new and there have been many previous protocols and standards that were designed to allow different educational applications to communicate. Foremost amongst these was SCORM [58], which was widely heralded when it was introduced in 2004. SCORM was widely adopted by vendors of large Learning Management Systems (LMS) and was a well-defined and sophisticated structure that was very consistent with data practices of the time. SCORM does a very good job of thoroughly describing learning objects and how they can be and should be used. However, many authors find that the detailed metadata that is required for a learning object is beyond their personal capacity to generate. Most learners these days find learning materials in many ways on the web, and generally do not care whether something is approved or accessible via the LMS. They now have a far wider range of resources to explore and consume. As noted earlier in this chapter, focus has shifted from learning objects to activity streams.

Other means of system integration have attempted to smooth the path between information platforms. For example, the IMS Learning Tools Interoperability (LTI) protocol [59] also helps to streamline integration across systems, looking after single-sign-on authentication and who should have access to what. But LTI is very limited in tracking what the learner actually does in any of these systems, with its simple passing of a single completion parameter back to the calling system, once the activity is completed. Other similar protocols have met with similar limitations and are not well able to provide metrics on what users actually do during the learning activity.

So where are all of these xAPI statements stored? In a simple, specialized type of database called a Learning Record Store (LRS), which is designed to suck up data streams from a very large number of disparate sources, in very large volumes. The underlying data architectures follow Big Data principles and are designed to support easy and rapid analytics on even very large datasets. Access to these datasets is kept highly secure but you can also set up federated access between multiple LRSs so that they can talk to each other securely. While simple in principle, this design means that organizations can collaborate on extremely complex, multisite data analytics, with proper management of data ownership.

It is this flexible data architecture that afforded the simple data analytics in our Arduino project [40], where we collected data from a wide variety of disparate sources. The Arduino device, which captured real-time sensor data on heart rate and galvanic skin response as a measure of users' stress levels, did not have to know anything other than where to send the data and how to format the statements. This is very simple in programming terms. During the project, we were also simultaneously collecting xAPI statements from OpenLabyrinth, WordPress, H5P widgets [60], Moodle [61], and even a smartphone camera via a simple, free, integration service called If This Then That [62]. All of these xAPI statements were collected in our GrassBlade LRS [63]. A portion were then federated with another Watershed LRS [64], which has superior analytic capabilities. Using this combination of data sources, it was remarkably easy to detect the crucial stress points as users navigated our virtual scenarios. This kind of data collection and analysis is well within the capabilities of a small education research group and we encourage others to try similar activity stream blending.

TRACKING ACROSS TIME AND CONTEXT

One aspect of integration across eLearning and simulation systems that has challenged us on occasion has been that of tracking learners across time as well as across multiple systems. An advantage to using xAPI and an LRS has been that all entries are time stamped in a standard ISO 8601 format, making it easy to compare sequences of events and exactly when each activity occurred. Note that there are actually two timestamps: when the activity occurred and when it was recorded in the LRS, something that is not always synchronous. This has also allowed us to go back and extract post-hoc activity metrics from some systems like OpenLabyrinth, because all its activities are also time stamped internally. So, in effect, this has enabled recording in real-time, virtual time or asynchronous time.

By virtual time, we are referring to the fact that in some virtual scenarios, we sometimes want to artificially advance the clock, something you cannot do with high-fidelity mannequins, which expect events to be recorded in real-time. So, for instance, in moving from one node to another, you can tell the user something like: "`you are now seeing the patient two weeks later`"; or "`you just ordered a chest x-ray. 30 minutes later the patient returns from radiology. But this time delay has allowed them to bleed further.`" This ability to manipulate virtual time sounds obvious and trivial. But it is very useful as an educational aid to remind learners that some things take longer than they expect, and while the virtual scenario might deliver a lab result at the click of a button, in real life, they will have to wait. These timestamp differences can also be recorded in the LRS.

Another way that we have used a series of virtual patient cases is in pediatrics where the cases [65] are shown as a set, linked over years, presenting at various stages in the child's life. We also placed longitudinal consequences into the virtual scenarios, such that decisions made early in the care of the virtual child make a difference to their presentation later in the series. So if the learner forgets to recommend echocardiogram monitoring of cardiac status in our Trisomy 21 series, the child presents with increased cardiac complications in later years. This is a much longer view of timelines and activity metrics. Of course, we don't make our learners wait years before they can see subsequent episodes in the child's presentation, but even waiting a week between sessions helps to reinforce the importance of continuity of care. There are few teaching modalities available at the time of writing that can address this essential facet of primary care.

The concept of tracking learner behaviors over extended time periods has another important aspect. The examples we have given deal with small groups and relatively short time periods. Medical education has long sought a better means of tracking progress across the whole educational trajectory from medical school through postgraduate training (residency) to continuing medical education. As noted earlier in this chapter, certain clinician behaviors tend to repeat themselves over the years, sometimes with serious adverse consequences. In the summer of 2014, the American Association of Medical Colleges launched Pivio as such a system to track learners over their careers [66]. Despite much fanfare, at the time of writing the uptake has remained low. This may be due in part to the requirement that learners pay for it but what is the immediate value proposition for them? It is perhaps too early to predict whether this will be successful but the early signs have not been promising.

Whatever the implementation issues, we would argue that the principle is a sound one and longitudinal tracking of activities over time may have some considerable merit. We previously described how an LRS can be designed to accept data from numerous sources and that it has proved easy to federate data across multiple LRSs. It has been suggested that one possible use for xAPI statements is that learners can create their own personal LRS, somewhat like a Personal Health Record, into which they can record their own activities and successes. The ability to then federate this data with other LRSs would facilitate sharing of real data about what the learner has actually experienced or how they have chosen to use their time. As noted above, some might fear the loss of privacy and potential surveillance that could be a consequence of such tracking. This is once again countered by the general observation that we have largely abrogated many of our privacy rights to large corporations already in this regard. At least such tracking has a higher purpose than simply selling you more advertising space. A further argument is that public safety trumps personal privacy.

One aspect of xAPI tracking of activity metrics that has been posing some small, but persistent challenges has been that of contextual identifiers. As noted earlier, context is key in many learning situations and, indeed, this is recognized in the xAPI specification. There are specific areas in the Object portion of an xAPI statement that describe the context of the activity. So in a context where several learners are conducting a team Objective Structured Clinical Encounter, using virtual scenarios and high-fidelity mannequins, we potentially have multiple users, multiple overlapping activity streams and information from multiple sensors and devices. While these may be coincident in timing, it is still important that, as these activities occur, they are somehow linked in the LRS using a common context identifier.

Teachers and learners need not concern themselves with such minutiae but those working with the components of the xAPI statements and developing such learning designs do need to consider how such a context identifier will be managed and propagated across multiple activity streams. At the time of writing, there have been several possible and mildly competing approaches. Indeed, it is the nature of open specifications such as xAPI, which allows many different approaches, that at times such practical aspects are left up to vendors and potentially competing standards. To that end, we suggest that those who are considering the use of xAPI for tracking activities in blended virtual scenarios talk to their technologists about how they intend to address context identifiers across the various systems.

THE ROLE OF ANALYTICS IN MEETING THE SOCIAL CONTRACT

What is the overall value of this data and its processing? Why is this worth studying? This has been raised many times over the years in relation to medical education research. Whitcomb raised this in 2002 [67] but in doing so he cited similar questions from 40 years ago. What value does medical education research bring to the table? What do we know about how various approaches to education work and whether they really have an impact on clinical decision making and practices of future physicians? While the question has been raised repeatedly, answering it has proved challenging.

So many educational interventions simply report on user satisfaction or on whether the intervention has some effect compared to no intervention. We already know that generally, doing something is better than nothing [68,69]. So, going back to the question raised at the beginning of this chapter, would it not be better to track what physicians actually do, rather than what they say they do or what their teachers say they do? And then track it longitudinally and broadly so that we can study practice patterns and trends over time, under a broad range of influences, using many different information streams so as to avoid a single target metric effect?

Knowledge translation has attempted to study the factors, which influence the impact of bench science on clinical practice. While there have been some promising inroads, there are so many confounding factors that this is difficult to tease out. Many physicians now actively resist the drive toward clinical practice guidelines, rationalizing this with "the art of medicine" and need for individualized care for a particular patient. In principle, this is central to patient-centered care and Precision Medicine. The increasing worry is that we are not as good at balancing this individualization with our preexisting frames and cognitive biases, as we think we are [70]. We need Precision Education to study, and hopefully to address, this problem.

However, no one algorithm or data source is likely to be conclusive in terms of making significant decisions about an individual's career or their progression through training into the profession of medicine. Indeed, one of the problems of preceptor observation is that it is unidimensional; each observation is an n of one. We need to think more in terms of a 360-degree model of assessment but not simply in terms of multiple observations but including multiple dimensions of data and multiple contributions to inform the assessment process. This in part embraces the issue of what constitutes sufficient evidence to make these high-stakes decisions and to defend them, potentially in a court of law.

We seem to be at a crossroads. On the one hand, our science is showing us, with greater and greater precision, the shortcomings in our abilities to identify and exclude those who should not enter the profession. However, the current solution, and indeed the great hope of competency-based medical education, of preceptors being the primary mechanism of assessment, is limited and risky and is difficult to defend in the face of growing evidence to the contrary.

By spreading the load, by correlating and triangulating multiple sources and indicators of performance and capability, we have better hopes of meeting our social contract for medicine. However, we also recognize the potential for harm that comes from this approach if it were to be used to exclude individuals not on the basis of competence but that of cultural or social acceptability. Robust validation of these approaches, and the ability to challenge them and test them must be part of such development. Spreading the load of information gathering and provision used to inform assessments, by including machines as well as humans, appears to have greater potential to address issues of reliability and bias. As such, this is a research agenda that must be pursued.

There are ethical challenges regarding surveillance of learners, and of preceptors. We argue the balance needs to be struck between the libertarian position that argues for the freedom of the individual and the social responsibility position that argues for the greater good of society. Given that health professionals are increasingly scrutinized in their practice, not as a way of punishing or controlling them, but as a way of ensuring quality assurance and improvement, then extending this to the phase of preparation for practice is perhaps less of an ethical challenge than it might at first seem.

Errors of ignorance (mistakes we make because we do not know enough) can be reconciled. Errors of ineptitude (mistakes we make because we do not make proper use of what we know) cannot [71].

Instead of treating all patients as being similar to the average 70 kg, 70-year-old man, the modal patient in many large multi-center, randomized controlled trials, we now focus on the individualization that is needed in patient-centered care, applying the results of our findings to a particular patient, with all their quirks, foibles, and social context. The same principles can and should be applied in health professional Precision Education [72,73]. To be clear, this is not the same as the concept of "Precision Teaching," which dates back to the 1960s. Precision Teaching did embrace the idea of tuning your lessons to the individual's learning style, but was a much more primitive example, focusing on behaviorist models of learning. We now recognize a big difference between teaching and learning (you can teach your dog to sing, but how well will he learn?).

CONCLUSION

The social contract between medical education and society is currently at risk due to the shortcomings inherent in the expectation that preceptors are the sole source of judgment as to whether their learners are fit to practice. We have set out arguments for a shift in medical education practice to include much greater use of ambient data and surveillance of learners' performance as a mechanism to address the failure to fail problem. We acknowledge that this is in its early stages and much work is still required to render this in a form that can be safely used on a widespread basis throughout the medical education continuum.

References

[1] Cruess R, Cruess S. Teaching medicine as a profession in the service of healing. Acad Med [Internet] 1997;. Available from: <http://journals.lww.com/academicmedicine/Abstract/1997/11000/Teaching_medicine_as_a_profession_in_the_service.9.aspx> [cited 15.08.16].

[2] Rethans J-J, Norcini JJ, Baron-Maldonado M, Blackmore D, Jolly BC, LaDuca T, et al. The relationship between competence and performance: implications for assessing practice performance. Med Educ [Internet] 2002;36(10):901–9 Blackwell Science. Available from: <http://doi.wiley.com/10.1046/j.1365-2923.2002.01316.x> [cited 20.06.16].

[3] Kassebaum DG, Eaglen RH. Shortcomings in the evaluation of students' clinical skills and behaviors in medical school. Acad Med [Internet] 1999;74(7):842–9. Available from: <http://www.ncbi.nlm.nih.gov/pubmed/10429595>.

[4] Cohen GS, Blumberg P, Ryan NC, Sullivan PL. Do final grades reflect written qualitative evaluations of student performance? Teach Learn Med [Internet] 1993;5(March):10–15. Available from: <http://dx.doi.org/10.1080/10401339309539580>.

[5] Dudek NL, Marks MB, Regehr G. Failure to fail: the perspectives of clinical supervisors. Acad Med 2005;80(10 Supplement):S84–7.

[6] Ellaway RH, Pusic MV, Galbraith RM, Cameron T. Developing the role of big data and analytics in health professional education. Med Teach [Internet] 2014;36(3):216–22 Taylor & Francis. Available from: <https://www.ncbi.nlm.nih.gov/pubmed/?term = 24491226> [cited 05.03.16].

[7] Frank JR, Snell LS, Ten Cate O, Holmboe ES, Carraccio C, Swing SR, et al. Competency-based medical education: theory to practice. Med Teach [Internet] 2010;32(8):638–45. Available from: <http://www.ncbi.nlm.nih.gov/pubmed/20662574>.

[8] Kalet A, Chou C. Remediation in medical education: a mid-course correction. New York: Springer; 2014.

[9] Norcini JJ. Current perspectives in assessment: the assessment of performance at work. Med Educ [Internet] 2005;39(9):880–9 Blackwell Science Available from: <http://doi.wiley.com/10.1111/j.1365-2929.2005.02182.x> [cited 07.08.16].

[10] Baldwin DC, Daugherty SR, Rowley BD, Schwarz MR. Cheating in medical school: a survey of second-year students at 31 schools. Acad Med 1996;71(3):267–73.

[11] O'Neill LD, Wallstedt B, Eika B, Hartvigsen J. Factors associated with dropout in medical education: a literature review. Med Educ 2011;45:440–54.

[12] Topps D, Cowan M, Topps M, Popovic A. Lackadaisical Larry [Internet]. OpenLabyrinth. 2013. Available from: <http://demo.openlabyrinth.ca/renderLabyrinth/index/388> [cited 14.08.16].

[13] Topps D, Ellaway R, Corral J. OpenLabyrinth web site [Internet]. 2015. Available from: <http://openlabyrinth.ca/> [cited 12.11.15].

[14] Short JP. The importance of strong evaluation standards and procedures in training residents. Acad Med 1993;68(7):522–5.

[15] Watling CJ, Kenyon CF, Schulz V, Goldszmidt MA, Zibrowski E, Lingard L. An exploration of faculty perspectives on the in-training evaluation of residents. Acad Med [Internet] 2010;85(7):1157–62. Available from: <https://www.ncbi.nlm.nih.gov/pubmed/?term = 20592512>.

[16] Mysore NM, Grad R, Topps D, Hall D. Implementation of handheld procedure tracking in a family medicine residency. Fam Med United States 2009;41(4):235–6.

[17] Topps D, Hall D. Electronic procedure logs: taking it further. Acad Med. United States 2002;77(7):756.

[18] Ginsburg S, Regehr G, Lingard L, Eva KW. Reading between the lines: faculty interpretations of narrative evaluation comments. Med Educ [Internet] 2015;49(3):296–306. Available from: <http://www.ncbi.nlm.nih.gov/pubmed/25693989> [cited 19.02.16].

[19] Kassin SM, Ellsworth PC, Smith VL. The "general acceptance" of psychological research on eyewitness testimony: a survey of the experts. Am Psychol [Internet] 1989;44(8):1089–98 American Psychological Association Available from: <http://doi.apa.org/getdoi.cfm?doi = 10.1037/0003-066X.44.8.1089> [cited 08.08.16].

[20] Wells GL, Olson EA. Eyewitness Testimony. Annu Rev Psychol [Internet]. Annual Reviews 4139 El Camino Way, P.O. Box 10139, Palo Alto, CA 94303-0139, USA. February 2003;54(1):277-95. Available from: <http://www.annualreviews.org/doi/10.1146/annurev.psych.54.101601.145028> [cited 08.08.16].

[21] Thistlethwaite J, Davies H, Dornan T, Greenhalgh T, Hammick M, Scalese R. What is evidence? Reflections on the AMEE symposium, Vienna, August 2011. Med Teach [Internet] 2012;34(6):454–7 Taylor & Francis Available from: <http://www.tandfonline.com/doi/full/10.3109/0142159X.2012.671561> [cited 15.08.16].

[22] Saunders HD. Quantifying reasonable doubt: a proposed solution to an equal protection problem [Internet]. BE Press. Available from: <http://law.bepress.com/cgi/viewcontent.cgi?article = 4389&context = expresso> [cited 01.11.16].

[23] Doubilet P, Herman P. Interpretation radiographs: history effect of clinical. AJR [Internet] 1981;137:1055–8. Available from: <http://www.ajronline.org/content/137/5/1055.long>.

[24] Song KS, Song HH, Park SH, Ahn KJ, Yang IK, Byun JY, et al. Impact of clinical history on film interpretation. Yonsei Med J 1992;33(2):168–72.

[25] Cukier K. Kenneth Cukier: big data is better data | TED Talk [Internet]. TED.com; 2014. Available from: <https://www.ted.com/talks/kenneth_cukier_big_data_is_better_data?language = en>.

[26] Madrigal AC. How much is your data worth? [Internet]. The Atlantic. 2012. Available from: <http://www.theatlantic.com/technology/archive/2012/03/how-much-is-your-data-worth-mmm-somewhere-between-half-a-cent-and-1-200/254730/> [cited 23.08.16].

[27] Greenwald G, MacAskill E, Poitras L. Edward Snowden: the whistleblower behind the NSA surveillance revelations. The Guardian [Internet] 2013; Manchester. Available from: <https://www.theguardian.com/world/2013/jun/09/edward-snowden-nsa-whistleblower-surveillance>.

[28] Privat L. Allianz Italy debuts self-installed consumer telematics service allie [Internet]. GPS Business News. 2012. Available from: <http://www.gpsbusinessnews.com/Allianz-Italy-Debuts-Self-Installed-Consumer-Telematics-Service-Allie_a3903.html> [cited 01.11.16].

[29] Griffiths D, Drachsler H, Kickmeier-Rust M, Steiner C, Hoel T, Greller W. Is privacy a show-stopper for learning analytics? A review of current issues and solutions. In: Learning Analytics Community Exchange [Internet]. Bolton, UK; 2016, p. 32. Available from: <http://www.laceproject.eu/learning-analytics-review/files/2016/04/LACE-review-6_privacy-show-stopper.pdf> [cited 27.10.16].

[30] Kelly M, O'Flynn S, McLachlan J, Sawdon MA. The clinical conscientiousness index: a valid tool for exploring professionalism in the clinical undergraduate setting. Acad Med [Internet] 2012;87(9):1218–24. Available from: <http://www.ncbi.nlm.nih.gov/pubmed/22836848> [cited 10.07.16].

[31] Finn G, Sawdon M, Clipsham L, McLachlan J. Peer estimation of lack of professionalism correlates with low conscientiousness index scores. Med Educ [Internet] 2009;43(10):960–7 Blackwell Publishing. Available from: <http://doi.wiley.com/10.1111/j.1365-2923.2009.03453.x> [cited 08.08.16].

[32] McLachlan JC, Finn G, Macnaughton J. The conscientiousness index: a novel tool to explore students' professionalism. Acad Med [Internet] 2009;84(5):559–65. Available from: <http://content.wkhealth.com/linkback/openurl?sid = WKPTLP:landingpage&an = 00001888-200905000-00009> [cited 08.08.16].

[33] Morgeson FP, Campion MA, Dipboye RL, Hollenbeck JR, Murphy K, Schmitt N. Are we getting fooled again? Coming to terms with limitations in the use of personality tests for personnel selection. Pers Psychol [Internet] 2007;60(4):1029–49 Blackwell Publishing. Available from: <http://doi.wiley.com/10.1111/j.1744-6570.2007.00100.x> [cited 08.08.16].

[34] Costa PT. Work and personality: use of the NEO-PI-R in industrial/organisational psychology. Appl Psychol [Internet] 1996;45(3):225–41 Blackwell Publishing. Available from: <http://doi.wiley.com/10.1111/j.1464-0597.1996.tb00766.x> [cited 08.08.16].

[35] Saucier G, Goldberg LR. Assessing the big five: applications of 10 psychometric criteria to the development of marker scales. In: Big Five Assessment. 2002, p. 29-58.

[36] Baez B. Personality tests in employment selection: use with caution—Cornell HR review [Internet]. Cornell HR Review. 2013. Available from: <http://www.cornellhrreview.org/personality-tests-in-employment-selection-use-with-caution/> [cited 01.11.16].

[37] Berwick DM. Measuring physicians' quality and performance. JAMA [Internet] 2009;302(22):2485–6. Available from: <http://jama.ama-assn.org/content/302/22/2485.full.pdf>.

[38] Arora S, Sevdalis N, Nestel D, Woloshynowych M, Darzi A, Kneebone R, et al. The impact of stress on surgical performance: a systematic review of the literature. Surgery [Internet] 2010;147(3):318–30 Elsevier. Available from: <http://linkinghub.elsevier.com/retrieve/pii/S0039606009006175> [cited 18.06.16].

[39] Ellaway RH, Topps D, Lachapelle K, Cooperstock J. Integrating simulation devices and systems. Stud Heal Technol Inf [Internet] 2009;142:88–90. Available from: <http://www.ncbi.nlm.nih.gov/pubmed/19377120>.

[40] Meiselman E, Topps D, Albersworth C. Medbiq xAPI workshop. In: Strothers V, Greene P, editors. Medbiq Annual Conference [Internet]. Baltimore, MD; 2016. Available from: <http://www.slideshare.net/topps/medbiq-xapi-workshop2b>.

[41] Hardy L, Mullen R, Jones G. Knowledge and conscious control of motor actions under stress. Br J Psychol [Internet] 1996;87(4):621–36 Blackwell Publishing. Available from: <http://doi.wiley.com/10.1111/j.2044-8295.1996.tb02612.x> [cited 18.06.16].

[42] Monrose F, Rubin A. Authentication via keystroke dynamics. Proceedings of the 4th ACM conference on Computer and communications security – CCS '97 [Internet]. New York, New York, USA: ACM Press; 1997. p. 48–56. Available from: <http://portal.acm.org/citation.cfm?doid = 266420.266434> [cited 01.11.16].

[43] Dobson-Mitchell S. McMaster is watching you. Macleans [Internet]. 2010. Available from: <http://www.macleans.ca/education/uniandcollege/mcmaster-is-watching-you/>.

[44] Schneier B. Frequent password changes is a bad security idea [Internet]. Schneier on Security. 2016. Available from: <https://www.schneier.com/blog/archives/2016/08/frequent_passwo.html> [cited 10.08.16].

[45] Goodwin D. Frequent password changes are the enemy of security, FTC technologist says | Ars Technica [Internet]. Ars Technica. 2016. Available from: <http://arstechnica.com/security/2016/08/frequent-password-changes-are-the-enemy-of-security-ftc-technologist-says/> [cited 10.08.16].

[46] Zhang Y, Monrose F, Reiter MK. The security of modern password expiration: an algorithmic framework and empirical analysis.

[47] Topps D, Helmer J, Ellaway R. YouTube Clinisnips channel [Internet]. YouTube. 2007. Available from: <http://www.youtube.com/user/clinisnips> [cited 31.10.16].

[48] Kobielus J. The four V's of big data [Internet]. IBM Big Data & Analytics Hub. 2013. Available from: <http://www.ibmbigdatahub.com/infographic/four-vs-big-data> [cited 18.06.16].

[49] Topps D. Praxis Pointers: reciprocal publication of practical pearls. Med Educ England 2005;39(5):527.

[50] Ellaway RH, Pusic M, Yavner S, Kalet AL. Context matters: emergent variability in an effectiveness trial of online teaching modules. Med Educ 2014;48(4):386–96.

[51] Topps D, Horne T, Topps M, Popovic A. Rushing Roulette: a case series [Internet]. OpenLabyrinth. 2014. Available from: <http://demo.openlabyrinth.ca/renderLabyrinth/index/578> [cited 14.08.16].

[52] Bruns A. Social media in research: from "big data" to "wide data" | Snurblog [Internet]. Snurblog. 2016. Available from: <http://snurb.info/node/2126> [cited 14.08.16].

[53] Topps D, Helmer J, Ellaway R. YouTube as a platform for publishing clinical skills training videos. *Acad Med*. 2013. http://dx.doi:10.1097/ACM.0b013e31827c5352.

[54] Google. Google APIs [Internet]. Google. 2016. Available from: <https://console.developers.google.com/apis/> [cited 31.10.16].

[55] Smothers V, Topps D, Meiselman E. Learning experience | MedBiquitous Consortium [Internet]. Medbiqitous. 2016. Available from: <http://www.medbiq.org/learning_experience> [cited 01.02.16].

[56] World Wide Web Consortium (W3C). W3C resource description framework (RDF) [Internet]. W3C wiki. 2004. Available from: <https://www.w3.org/RDF/> [cited 31.10.16].

[57] Greene P, Smother V. MedBiquitous consortium | advancing the health professions through technology standards [Internet]. 2001. Available from: <http://www.medbiq.org/> [cited 01.11.16].

[58] Dodds P. SCORM overview [Internet]. Advanced distributed learning. 2004. Available from: <https://www.adlnet.gov/adl-research/scorm/>.

[59] IMS Global: Learning tools interoperability (LTI) [Internet]. IMS Global Learning Consortium. 2012. Available from: <http://www.webcitation.org/6arGC9Thy> [cited 16.06.13].
[60] Griff S-T, Jorgensen P, Petterson F, Mars T. H5P technology [Internet]. Github. 2016. Available from: <https://h5p.org/> [cited 31.10.16].
[61] Topps D, Ellaway R. OpenLabyrinth moodle LMS [Internet]. Moodle cloud. 2016. Available from: <https://openlabyrinth.moodlecloud.com/> [cited 31.10.16].
[62] If This Then that service [Internet]. 2010. Available from: <https://ifttt.com> [cited 31.10.16].
[63] Topps D, Agrawal P, Meiselman E. OpenLabyrinth GrassBlade LRS [Internet]. Next software solutions. 2016. Available from: <http://openlabyrinth.ca/grassblade-lrs> [cited 31.10.16].
[64] Downes A, Rustici M. Watershed LRS [Internet]. Rustici software. 2015. Available from: <https://www.watershedlrs.com> [cited 31.10.16].
[65] Bannister S, Bendiak G, Johnston N, Lad J, Topps D. Canuc-Peds VP cases [Internet]. OpenLabyrinth. 2014. Available from: <http://demo.openlabyrinth.ca/collectionManager/viewAll/12> [cited 14.08.16].
[66] Pelletier S. Technology in academic medicine: Pivio® taps technology to organize students' personal portfolios – October 2014 – AAMC reporter – Newsroom – AAMC [Internet]. AAMC reporter. 2014. Available from: <https://www.aamc.org/newsroom/reporter/october2014/412150/pivio.html> [cited 10.08.16].
[67] Whitcomb ME. Research in medical education: what do we know about the link between what doctors are taught and what they do? Acad Med [Internet] 2002;77(11):1067–8. Available from: <http://www2.tulane.edu/som/ome/upload/What-do-we-know-the-link-between-What-Doctors-are-Taught-and-What-they-do.pdf> [cited 04.07.16].
[68] Norman G. Data dredging, salami-slicing, and other successful strategies to ensure rejection: twelve tips on how to not get your paper published. Adv Heal Sci Educ [Internet] 2014;19(1):1–5 Springer: The Netherlands. Available from: <http://link.springer.com/10.1007/s10459-014-9494-8> [cited 24.08.16].
[69] Cook DA, Hatala R, Brydges R, Zendejas B, Szostek JH, Wang AT, et al. Technology-enhanced simulation for health professions education. JAMA [Internet] 2011;306(9):126–35. Available from: <http://jama.jamanetwork.com/article.aspx?doi = 10.1001/jama.2011.1234> [cited 24.08.16].
[70] Zwaan L, Monteiro S, Sherbino J, Ilgen J, Howey B, Norman G. Is bias in the eye of the beholder? A vignette study to assess recognition of cognitive biases in clinical case workups. BMJ Qual Saf [Internet]. January 29, 2016. Available from: <http://qualitysafety.bmj.com/content/early/2016/05/04/bmjqs-2015-005014.abstract> [cited 30.05.16].
[71] Gawande A. The Checklist Manifesto: how to get things right. New York: Metropolitan Books; 2010. p. 209.
[72] Hart S. Precision education initiative: moving toward personalized education. Mind, Brain Educ 2016;10:209–11.
[73] Topps D. Precision education – does one size fit all? [Internet]. OpenLabyrinth home page. 2016. Available from: <http://openlabyrinth.ca/precision-education-does-one-size-fit-all/> [cited 14.08.16].

CHAPTER

23

Evaluating Educational Interventions for Health Professions in the Digital Age

Aviv Shachak[1], *Shmuel Reis*[2], *With contribution from: Kathleen Gray*[3], *Dawn Choo*[3], *Kerryn Butler-Henderson*[4], *Sue Whetton*[4] *and Anthony Maeder*[5]

[1]University of Toronto, Toronto, ON, Canada [2]The Hebrew University of Jerusalem, Jerusalem, Israel [3]University of Melbourne, Melbourne, VIC, Australia [4]University of Tasmania, Hobart, TAS, Australia [5]Flinders University, Bedford Park, SA, Australia

INTRODUCTION

As we saw in the previous chapters, various educational interventions have been designed and implemented to prepare health care professionals to practice in the information and Information and Communication Technology (ICT)-rich environment of the twenty-first century. However, for interventions to be effective there is a need for more evidence: we need to know what interventions work best, for what, for whom, and under which conditions. Evaluation is critical for providing this evidence. In this chapter, we will discuss the evaluation of educational interventions in general and of interventions related to challenges of the digital era in particular. We start by defining evaluation and discussing the relationship between evaluation and research, Then, we will describe the three intervention types (simple, complicated, and complex), and the different implications of each. Next, the five attributes of evaluation (reliability, validity, feasibility, acceptability, and educational impact) will be introduced, and the various levels of evaluation discussed with an emphasis on the need to consider cost-effectiveness and the process that learners go through. Subsequently, we will highlight the challenges involved in evaluating educational interventions in general and the specific issues pertaining to informatics-related interventions. Finally, we will use case studies from our own and other's research to illustrate these issues.

Health Professionals' Education.
DOI: http://dx.doi.org/10.1016/B978-0-12-805362-1.00023-1

WHAT EVALUATION IS AND HOW IT RELATES TO RESEARCH?

Evaluation has been defined as "the systematic assessment of the worth or merit of some object" or "the systematic acquisition and assessment of information to provide useful feedback about some object" [1]. Although the methods used in evaluation are often the same as general research methods, one distinction is that the focus of evaluation is on providing useful feedback and quality improvement [1,2]. Evaluation may be formative—that is, providing on-going feedback on performance or utility with the purpose of improving curriculum construction, teaching, and learning—or summative, that is, used for determining completion of a unit, course or a program with the "primary goals [of] grading or certifying students, judging the effectiveness of the teacher, and comparing curricula" [3]. Forms of formative evaluation include needs assessment, implementation evaluation, process evaluation, and more. Summative evaluation includes, but is not limited to, outcomes evaluation, impact evaluation, cost-effectiveness analysis, and cost-benefit analysis [1].

EVALUATION OF EDUCATIONAL INTERVENTIONS: SIMPLE, COMPLICATED, AND COMPLEX INTERVENTIONS

Health, healthcare, education in general, and health professions education in particular, and even the Internet are complex domains, that is, they can be neither described nor managed simplistically [4–10]. These domains are usually not amenable to simple interventions where there is a problem, a straightforward solution, and an intervention that implements the change (such as an antibiotic to cure an infectious disease or an application to assess cardiovascular risk). Sometimes, they may present complicated problems (i.e., multifactorial) that require elaborate, albeit predictable interventions (e.g., marketing the above-mentioned antibiotic or disseminating the application). Most of the time, they represent complex challenges affecting systems that are usually self-organizing and may respond to an intervention by unintended consequences (e.g., lifestyle modifications or the introduction of an electronic health record—EHR—to an institution).

> Complexity science, the study of complex adaptive systems, does not consist of a single theory, but rather encompasses a collective of theories and constructs that have conceptual integrity among themselves. .... Perhaps the most important early event was the discovery of "chaos" (Gleick, 1987). Chaos theory presented two propositions that were attractive to social scientists: Small, seemingly inconsequential events, perturbations, or changes can potentially lead to profound, large scale change; and What appears to be random may in fact have an underlying orderliness to it. In addition, the word "chaos" itself was something that in a vernacular sense resonated with current reality [11].

Complex challenges require an acknowledgment and the acquisition of theory and practice of complexity, that is, complex systems have emergent properties; can change when unstable and when the intervention is "at the edge of chaos." A striking example is the state of the deployment of electronic health record (EHR) systems in the United States. While EHRs are finally becoming more prevalent, their implementation is associated with marked resentment and frustration among practitioners to the extent that the New England Journal of Medicine has published a paper entitled: "Transitional Chaos or Enduring Harm? The EHR and the Disruption of Medicine" [12]. As we will see in the following sections, this complexity, amongst other issues, can make the evaluation of educational interventions related to use of ICT in health care a challenging task.

THE FIVE ATTRIBUTES: RELIABILITY, VALIDITY, FEASIBILITY, ACCEPTABILITY, AND EDUCATIONAL IMPACT

The requirements of an educational evaluation include five attributes [13]:

- Reliability: addresses the fidelity to the origin when the intervention is replicated;
- Validity: its fit with the construct evaluated;
- Acceptability: how it is received by those evaluated;
- Feasibility: how problematic the application of the evaluation may be; and
- educational impact: how the evaluation drives learning in the particular knowledge or performance domain that is under evaluation.

McKinley et al. [14] describe how these attributes influence design:

> The design of any assessment process is a compromise between ... five attributes. For example, maximizing reliability, validity, acceptability, and educational impact will increase costs but reduce feasibility, and vice versa. Thus, the particular emphasis given to each attribute is critically dependent on the purpose of the assessment. In formative assessment, for example, validity and educational impact are more important than high reliability, but, in any regulatory assessment to determine fitness to practice, reliability and validity are paramount.

An example of this perspective is provided in case study 3 (Text Box 23.3). However, if we go back to the example in the former section, the deployment of EHRs in the United States is a major top-down countrywide organizational intervention, which includes an educational component. Obviously, providers need to be trained to make optimal use of the software. This training is almost universally technical and ranges from 30 minutes to a few days. We are not aware of a formal evaluation of such educational interventions, but propose—based on the literature and extensive discussions—that the reliability of these interventions is unknown (it needs to be measured), the validity is probably fair (it must be useful for functioning in the new environment), that acceptability is good (as EHR training is often mandated by the employers), and the feasibility reasonable (it is happening in large numbers all over the United States). Finally, the educational impact is most probably driving the providers to look at the use of EHRs as a technical issue. If organizational leaders and vendors would have included training in the additional challenges the computer is bringing to the clinical encounter (e.g., communication skills; see Chapters 1, 2, and 10 in this book) both the validity and the educational impact would probably change positively. This may also have a positive impact on the acceptability of the EHR by clinicians (though this is not part of evaluating the educational intervention but rather an organizational aspect).

LEVELS OF EVALUATION

Crossley et al. [15] describe the principles of good assessment design. They underline the definition of the intended focus as specifically as possible and carefully establishing "the scope of situations the result is intended to generalize." To achieve this goal, educators and researchers need to contemplate what to evaluate. In making this decision, it is useful to consider the level of evaluation. In his model from the field of Professional Development, Kirkpatrick [16,17] suggested breaking the evaluation into four levels or steps. These levels are:

1. Reaction: "a measure of how participants feel about the various aspects of the training program (topic, speaker, schedule, etc.)" [16].
2. Learning: "a measure of the knowledge acquired, skills improved, or attitudes changed due to training" [16].
3. Behavior: "a measure of the extent to which participants change their on-the –job behavior because of training" [16].
4. Results: "a measure of the final results of training (increased sales, higher productivity, bigger profits, reduced costs, improved quality, less employee turnover, etc.)" [16].

Adapting this model to Health Professions Education (HPE), Barr et al. suggest adding a fifth level (Level 4b): "benefits to patients or clients" which "covers any improvements in the health and well being of patients/clients as a direct result of an education programme" [18].

The above levels of evaluation focus mostly on outcomes. However, it has been argued that outcome-focused models "provide little insight into a program's 'theory'—that is, the processes by which it yields desired outputs and outcomes—and they do not capture unintended outcomes" [19]. By conducting a mixed-methods study, researchers were able to uncover the experience of trainees who participated in a clinician-scientist continuing education program and the impact it had beyond the original program logic [19].

In this respect, it may also be useful to consider the social and situated nature of learning. Situated Learning Theory posits that learning often happens naturally and is embedded within the context in which it occurs, in culture, and in activity [20–22]. Brown, Collins, and Dugoid [22] add the idea of cognitive apprenticeship, which "supports learning in a domain by enabling students to acquire, develop, and use cognitive tools in authentic domain activity." This type of learning is embedded in many practices of HPE including problem based learning, clinical rounds, simulation, and clerkship. Furthermore, learning is a social activity that rarely happens in isolation. Communities and networks contribute to learning by creating a commitment and identity within a group of people who share and exchange resources, narratives, concepts and views (communities); and "enable access to a rich web of information sources offering multiple perspectives and dialogues, responses to queries, and help from others" (networks) [23]. From this perspective, informed by their work on Communities of Practice, Wenger et al. [23] suggested a framework for evaluating the value of communities and networks composed of 5 cycles:

- **Cycle 1.** *Immediate value:* the activities and interactions between members have value in and of themselves.

- **Cycle 2.** *Potential value:* the activities and interactions of cycle 1 may not be realized immediately, but rather be saved up as knowledge capital whose value is in its potential to be realized later.
- **Cycle 3.** *Applied value:* knowledge capital may or may not be put into use. Leveraging capital requires adapting and applying it to a specific situation.
- **Cycle 4.** *Realized value:* even applied new practices or tools are not enough. A change in practice does not necessarily lead to improved performance, so it is important to find out what effects the application of knowledge capital is having on the achievement of what matters to stakeholders....
- **Cycle 5.** *Reframing value:* this happens when learning causes a reconsideration of how success is defined. It includes reframing strategies, goals, and values....

For each of these cycles they proposed a set of questions and typical indicators, both quantitative and qualitative. There are notable similarities between Kirkpatrick's levels of evaluation and the cycles proposed by Wenger et al. However, the latter focuses specifically on the value of networks and communities and on both outcomes and the processes that lead to them. In a later modification of the framework, Wenger and Trayner highlight the importance of feedback (learning) loops between cycles [24].

Evaluation that focuses not just on outcomes but also takes processes into consideration may be especially important in interventions aimed at improving clinicians' competence in the digital age. Applying mixed methods, as suggested by parker et al. [19], or using the framework and tools proposed by Wenger et al. [23] may provide valuable insight into what works, what does not work, and why—especially considering the social and situated nature of learning in HPE and the dearth of valid and reliable measure instruments to assess outcomes (see below).

The ultimate goal of educational interventions is to achieve high level goals (e.g., Kirkpatrick's level 4 or Wenger et al.'s Cycles 4 or 5) from transfer of learning to the practice environment to "impact on the structure and process of the organization of care and health outcomes" [25]. While there are numerous studies that evaluate educational interventions delivered via technology (e.g., e-learning modules and mobile technology) there are not many published evaluations of interventions aimed at preparing clinicians for the challenges of the digital era. For example, Flickinger et al. [26] describe the development of a workshop to teach medical students about aspects of professionalism related to social media (for more on this topic see Chapters 4–6 in this book). Formative evaluation of the workshop was conducted by collecting students' feedback in the form of anonymous written comments. Mesko et al. [27] developed and evaluated a workshop to enhance digital literacy for medical students and evaluated its impact using pre- and post-workshop self-reported surveys of skills and attitudes. Responses suggest that students acquired new skills and that the workshop was perceived as useful to them. Finally, Silverman at al. [28] describe the evolution of biomedical informatics curriculum in a medical school in Phoenix, Arizona. They evaluated the curriculum first by comparing it to published competencies, then through standard course evaluations and finally, pre- and post-intervention survey of self-reported skills and attitudes related to informatics. While standard course evaluations had low response rate and were inconclusive, the pre-post surveys showed improvement in skills and attitudes as well as higher scores in half of the items compared to a similar student cohort that did not get informatics training [28].

These few examples from the literature demonstrate projects in which evaluation was integrated into the educational program with valuable lessons learned and some promising results. However, they all remain at relatively low levels of evaluation (Kirkpatrick's levels 1 and 2 and Wenger at al.'s cycles 1 and 2). Evaluating educational interventions at higher levels is more complicated, first because the impact of the intervention may not be immediate. As Morrison [2] noted, "sometimes the full impact of the curriculum may not be known until sometime after the student has graduated." Second, there could be many intervening factors, including organizational culture, norms, workflow, and the position, responsibility and authority of the learner within the health care organization. These intervening factors make it difficult to isolate the impact of the educational intervention. Finally, it is often hard to follow the learners into the workplace after they have completed their course of education to evaluate the final results of training and impact on patient care. However, our position is that as we move forward, educational interventions related to the challenges of using health information technology (HIT) in health care need to be evaluated at all, and not just the lower, levels in order to provide the evidence needed to inform future initiatives.

COST-EFFECTIVENESS EVALUATION

As health care systems around the world face increasing financial constraints, it is becoming highly important to consider cost-effectiveness in evaluating interventions directed at improving healthcare professionals' competence in the digital environment. The same skills may be taught using various methods ranging from didactic lecture-based to high fidelity simulation. For example, we can teach clinicians about the challenges and best practices of using the EHR in the patient-clinician encounter by using didactic lectures, case studies, critical review of videotaped encounters (with real or standardized patients), role plays, simulations with standardized patients [29], or computer-based simulation [30].

In determining which approach to choose, educators need to consider the educational impact of the intervention, the cost of development and implementation—including time and other resources needed, the scalability of the method, sustainability, and any potential cost benefits that may result from the long-term impact of the intervention on quality and safety. In the above example, a lecture-based intervention was as effective as simulation with standardized patients—although several issues may have affected the evaluation (see Text Box 23.1 below) [29]—and a computer-based virtual patient simulation showed promising potential albeit at a preliminary and low level evaluation [30]. Of these interventions, the lecture based is the least expensive to develop. Both simulation methods can be costly but arguably, a computer simulation may be more scalable. Until the educational impact is fully understood and a thorough cost assessment is being conducted, it will remain unclear which of these approaches provides the best outcomes relative to the required investment. Research of educational interventions for HPE in the digital era is still in its infancy. In searching the literature for this chapter, we found some studies that evaluated the impact of using ICT to deliver educational content (mostly web-based e-learning

TEXT BOX 23.1

CASE STUDY: THE ISRAELI PATIENT DOCTOR COMPUTER COMMUNICATION STUDY

In this study, the researchers examined the effect of a simulation-based training intervention aimed at enhancing Family Medicine residents' patient-physician communication skills in a computerized office setting. As already mentioned, in this RCT, the intervention group was not superior to control, which received lecture-based training. In trying to understand this finding, several explanations come to mind. First, it is possible that the result is real and that simulation-based training is not superior to lecture-based training on this topic. A second possible explanation is that, because of the relatively small number of participants ($N = 36$) there was not enough statistical power to detect the difference between control and intervention groups. Based on their power calculation, the authors aimed at recruiting 24 participants for each group (i.e., a total of 48 participants) but were only able to recruit 18 for each group. Finally, the authors propose that the lack of statistical difference between control and intervention may be the result of the way the experiment was designed: both pre- and post-test employed six simulations with standardized patients and thus, the pre-test may have been sufficient to induce learning in the control group too. In subsequent design (in progress), the researchers have resorted to videotaping of encounters in the workplace and hope to eliminate this confounder. This case highlights some of the challenges in evaluating informatics-related educational interventions: difficulty to recruit a sufficient number of participants to supply the required statistical power and, potentially, unexpected effect of the measurement itself on learning which requires rethinking the study design.

Reis, Sagi, Eisenberg, Kuchnir, Azuri, Shalev, & Ziv. The impact of residents' training in Electronic Medical Record (EMR) use on their competence: report of a pragmatic trial. Patient Educ Couns, 2013 93(3), 515-521.

interventions) but very little on interventions intended to improve clinicians' competence in the information and ICT-rich environment of today. As discussed above, most of these evaluations remained at a low level, and none of which included cost-benefits evaluation.

CHALLENGES IN CONDUCTING EVALUATIONS

We began this chapter by discussing the importance of evaluation; however, it is not always easy to perform. In this section, we discuss some of the challenges to evaluating educational interventions in general, and of interventions related to the practicing in information and ICT-rich environments in particular. These challenges include study design, small scale, lack of planning for evaluation from the outset, and lack of specific measure instruments.

Evaluation Study Designs

Both HPE and Health Informatics are strongly influenced by the paradigm of Evidence Based Medicine (EBM), or Evidence Based Health Care (EBHC), in which randomized controlled trials (RCTs) and systematic reviews of RCTs are considered the highest level of evidence for evaluation purposes [31,32]. RCTs certainly have their benefits but they also have some disadvantages. For one, RCTs are best for evaluating the outcomes of an intervention but they are not well suited for understanding how and why educational interventions work (or do not work) and unintended outcomes, which as discussed above are also important to understand.

Another challenge with RCTs is that they often take a long time. This is especially critical in evaluating eHealth applications, and by extension educational interventions related to eHealth: by the time it takes from inception to completion of an RCT, technologies may become obsolete, new technologies emerge, and the knowledge on which the eHealth application is based change substantially. Meanwhile, the intervention is essentially 'locked down' for any technological development [33,34]. Therefore, it has been argued that there is "a need for eHealth research strategies that both increase the pace of research and also produce higher-value interventions that will be more effective than the ascientific application of technology (e.g., by app developers) and therefore, remain appealing and effective despite tumultuous change" [34].

Finally, new knowledge is gained during the intervention, which can be applied for quality improvement. However, locking down the intervention in an RCT does not allow for modifications that could enhance it [34]. In this respect, evaluation may stand in the way of innovation and quality improvement.

Educational interventions aimed at enhancing clinicians' competence in eHealth are subject to the same challenges and thus also require more rapid and efficient evaluation study designs. Below we discuss a number of study designs including factorial designs, quasi-experimental designs, and participatory action research. Table 23.1 summarizes the pros and cons of each.

- Factorial designs: Factorial designs are often used in engineering research. The advantage of factorial designs is that multiple factors can be tested on the same study, thus requiring smaller sample size and less time than what would be needed for a series of separate RCTs [33]. This is very useful for understanding what components of the intervention are most important as well as the interaction effect of various factors, although testing interaction effects may require larger sample size.
- Quasi experiments: Quasi experiments, aka field experiments or non-randomized pre-post intervention studies, are common in both education and health informatics research. This type of studies is particularly useful when randomization is not possible, when a small sample size is expected, and when there is pressure to implement the intervention fast [35]. These conditions often apply to health information technology implementation as well as educational programs. The disadvantage of quasi-experimental designs is that because of the lack of randomization and difficulty to control for confounding variables it may be hard to infer causality [35].

TABLE 23.1 Pros and Cons of Various Evaluation Study Designs

Study Design	Pros	Cons
Randomized controlled trial (RCT)	Strong validity	Time consuming
	Randomization reduces the likelihood of systematic differences between groups	May be costly
	Blinding prevents potential bias	Lock down innovation and application of knowledge for the trial period
		Not so good for evaluating processes
Factorial designs	Allow testing multiple factors in the same study	External validity may be a challenge
	Require smaller sample size than a series of RCTs would	
	Testing of interaction effects	Small sample size may result in type 2 error, especially for testing interaction effects
Quasi experiments	Useful when randomization is not possible	Lack of randomization—and in some designs of control group—compromises validity
	Could be applied when small sample is expected	
	Normally faster than RCTs	
Participatory action research	Focus on action and informing change	Time consuming
	Strong collaboration between researchers and participants	Potential for bias

- Participatory action research (PAR): participatory action research and participatory evaluation research have been applied in various fields including education, nursing, and public health [36,37] but have not yet gained traction in health informatics. PAR differs from other, more traditional, forms of research in several aspects. First, its goal is not simply to describe the world and create new knowledge and understanding of it, but to drive change in practice [36–38]. Second, participation of those being researched in the research process is a key attribute of PAR, to the extent that the boundaries between the researchers and research subjects are often blurred: "The researched cease to be objects and become partners in the whole research process: including selecting the research topic, data collection, and analysis and deciding what action should happen as a result of the research findings" [37]. As Reason [36] suggests, this is not only because it is hard to change practice without involvement of those engaged in it, but also from an ethical standpoint: "People have a right and ability to contribute to decisions that affect them and to knowledge that is about them." Finally, the research process in PAR often is not well defined from the outset but develops and evolves over time through "a reflective cycle, whereby participants collect and analyze data, then determine what action should follow. The resultant action is then further researched and an iterative reflective cycle perpetuates data collection, reflection, and action as in a corkscrew action" [37]. These three characteristics—action orientation, participation, and reflective cycles of data collection, reflection, and action, make PAR particularly appealing for evaluation of educational interventions. The downside, however, is that it can be a time consuming process.

Small-Scale Projects

In many projects and initiatives aimed at improving clinicians' competence in the digital era, the number of participants is fairly small to begin with. From an educational perspective, this may be an advantage because small groups allow for meaningful interactions between instructors and learners, amongst learners, and of the learners with educational materials provided. It also makes the intervention easier to manage from administrative and logistic standpoint. However, a small number of participants may pose a threat for evaluation. First, with a small number of participants, there may not be sufficient power to detect statistically significant differences between control and intervention groups or before and after an intervention (type II error), particularly if response rate is low or if there is high attrition rate during the intervention. Second, findings based on a small N often are not generalizable as the sample may substantially differ from the population it is supposed to represent. Thus, albeit lean and efficient, thereby addressing the time challenge discussed above, and although they may provide valuable insights, small N studies may compromise validity and generalizability. Two of the case studies, the Israeli Patient Doctor Computer Communication and of an interprofessional informatics online course for clinical degrees in Australia, illustrate these challenges (Text Boxes 23.1 and 23.2).

Planning for Evaluation in Advance

The best time to start planning for evaluation is probably at the beginning when deciding on the objectives of the intervention. Then we can think about how we will evaluate the extent to which these objectives have been met. However, to carefully plan for evaluation and conduct it in a systematic, rigorous, manner takes time. Many times, it may seem that the evaluation stands in the way of doing. Particularly in a fast changing and rapidly growing domain such as eHealth there may be a pressure to act fast and implement educational programs as soon as possible. Add to it that funding for eHealth-related educational projects often is approved on the last minute and is provided for a short period only, and the end result is that evaluation is pushed down the list of priorities for the project. If done at all, it often remains at a low level, for examples, Kirpatrick's Level 1 or 2 or Wenger and Trayner's Cycle 1 or 2. Although sometimes it is possible to get back to it and conduct retrospective evaluation, there are elements of the interventions that cannot be

TEXT BOX 23.2

CASE STUDY: AN INTERPROFESSIONAL INFORMATICS ONLINE COURSE FOR CLINICAL DEGREES: EVALUATION OF EDUCATIONAL IMPACT

Authors: Kathleen Gray PhD[1], Dawn Choo MSc[1], Kerryn Butler-Henderson PhD[2], Sue Whetton PhD[2], Anthony Maeder PhD[3]

[1] University of Melbourne, [2] University of Tasmania, [3] Flinders University

Corresponding author: Dr. Kathleen Gray, Health and Biomedical Informatics Centre, Level 1, 202 Berkeley St, University of Melbourne, VIC 3010 Australia +61 3 8344 8936 kgray@unimelb.edu.au

This study evaluated an intervention that was developed as a result of evidence of a shortfall in health informatics and eHealth curricula available to students in most clinical health profession degrees nation-wide in Australia. Course development was done by a national alliance of academics and practitioners; it targeted postgraduate learning; it aligned the curriculum with nationally recognized health informatics competencies; and it implemented the curriculum in a fully online format.

Learning was designed to be somewhat self-directed, guided by a detailed study guide and a choice of weekly activities over the 12 weeks of a semester-long program. Structured group learning was incorporated through asynchronous interactions with the teaching team and other students, based on students lodging their work and reviewing others' work in discussion forums each week. Students received instructor feedback individually each week via the online commenting and grading tool in the learning management system.

Before proceeding to seek credit-bearing status for this program through a university Academic Programs committee, a realistic trial was required. The trial aimed to evaluate students' learning and the student experience in relation to a key feature of the learning design, namely, its shared curriculum for interprofessional learning across different clinical professions and different universities.

The number of trial participants had to be large enough to create a meaningful group learning experience, but small enough so that it did not overload the project team with teaching and assessment work. Recruitment and retention of trial participants was challenging since it targeted students pursuing degree studies, that is, busy people seeking academic credit for their work. A call for expressions of interest to undertake a free trial of a non–credit-bearing course was disseminated nationally through professional organizations, education networks, and social media channels. Applicants had to be either current university postgraduate students enrolled in a clinical health science degree or recent graduates (2010 onward).

Expressions of interest were received from 41 suitable applicants. Eighteen participants from four states in Australia were selected and began the course and seven of them completed it. This attrition was explainable, considering that completion rates in free online education are known to be much lower than in traditional methods of educational delivery; for example, fewer than 10% of students are expected to complete free online learning modules in

TEXT BOX 23.2 (*cont'd*)

Massive Open Online Courses (MOOCs). Of the seven participants who completed the course, three were studying full time, two were working full time, and two were studying and working part time. They were enrolled in a range of health degrees: dementia care ($n = 1$), health administration ($n = 1$), medicine ($n = 1$), nursing ($n = 2$), and psychology ($n = 2$). Of the 11 noncompleting participants, the main reason given was inability to keep up with the course workload in addition to other work or study commitments.

Data collection methods had to be suitable to scale up for a larger more comprehensive trial at a later date, but at the same time they could not intrude on student learning in such a way as to bias the experience or otherwise distract students from a postgraduate level of learning. During their progress through the course, participants completed two activities designed to collect data about their learning: tests and assignments.

A test of informatics knowledge was administered in the first and final weeks of the trial as a hurdle requirement, that is, necessary to complete the course but not graded. The first week test asked students to try to list and briefly describe ten things about clinical informatics and eHealth that were relevant to their professions and careers. The final week test asked them to revise this list. In revising it they had to order the items from greatest to least relevance, and to show how they used learning activities and learning resources to explain and rank each list item. Participants were required to do graded assignments of two kinds, which combined to calculate a final result. Short pieces of weekly writing were marked by the teaching team member facilitating that week's learning. A prerecorded project presentation due in the last week was double marked by teaching team members.

Participants also completed two surveys designed to collect data about additional factors that might have influenced their learning experience. The pretrial survey had three sections: demographic details, a measurement of internet skills, and attitudes toward interprofessional learning. The posttrial survey asked participants about their perceptions of factors, which facilitated their achievement of learning outcomes, levels of motivation and engagement, and their overall satisfaction with the course. Data on student experiences from the posttrial survey showed that six out of seven participants either agreed or strongly agreed that they were satisfied with the course. Participants reported numerous helpful aspects of the course to be: learning as a multidisciplinary cohort; the opportunity to choose learning activities which appealed to the individual's learning style; the provision of relevant links and resources; the breadth of content coverage; the literature review assignment; reading the discussion forums; remote and self-guided learning; and a timed, structured approach which included weekly tasks.

The findings had no statistical power. So it was particularly important to extract thematic insights. We had collected rich qualitative data. For example, Participant J, studying Dementia Care at one university, said: *"If you are confused about all the hype regarding an electronic health record or the meaning of eHealth as it relates to your field, this is the subject* [course] *for you. Although*

TEXT BOX 23.2 (*cont'd*)

my interest is in aged care, I am more interested in how eHealth can support home carers. The subject is flexible in that it allows you to choose which topics are of most interest to you. Additionally, the discussion board with other clinicians provides enlightenment as to the eHealth knowledge of other health professionals. I found the comments by researchers particularly informative. The weekly 200-word assignments were easy enough to do; however, the literature review was more difficult. This subject is a starting point to see if you would like to delve further into the study of this field. If not, at the very least, you will have acquired enough knowledge to have a meaningful conversation with the information technology gurus who are attempting to change the way healthcare is delivered."

One evaluation approach was to do case-by-case analysis on each individual participant's tests and assignments, along with data from her/his pre- and post-surveys. These cases could be presented using rich description and cross-case comparisons were made. Another evaluation approach was to do a linguistic analysis and inductive thematic analysis on aggregated student data about learning and influences on learning. Examples of such findings are as follows.

Statements from the preliminary test of participants' eHealth and health informatics knowledge were categorized in two ways to gain insights into the starting point for learning in the group. Analysis by sentiment type showed that most statements had positive connotations (71%), as one might expect from participants who had volunteered to take this course. However, some statements showed uncertainty (15%) or negativity (14%) about eHealth and health informatics, confirming that this group at the outset had the capacity for critical thinking about this subject area that is expected in postgraduate study.

When categorized according to seven broad health system performance themes, statements showed a wide spread of emphasis. The "Efficiency and Sustainability" theme predominated (e.g.: *"I know eHealth is economically confronting initially, however the long term benefits in terms of sustainable health care and increased productivity suggests the benefits well outweigh the potential financial risks"*), followed by the "Continuity of Care" theme (e.g.: *"I know clinical informatics and e-health can enable patient's higher autonomy in their care – the development of patient portals and individual health records enables patients access to their health information. This provides patients with a sense of greater control and involvement in their care"*).

Statements in the seven participants' posttests of their eHealth and health informatics knowledge showed a greater number of quotations to support opinions, illustrating that participants were able to apply learning resources. In contrast to the preliminary test statements—mainly unsubstantiated opinions, anecdotal observations, and untested assumptions—the posttest statements' use of evidence and examples in the arguments showed that development of a postgraduate level of subject knowledge had occurred.

The seven participants' posttests also used a greater number of words than their pretests (2344 words compared to 1852 words, in total). There was an increase in the number of different words used in their posttest statements (818 unique words, compared 612 in the pretests). Similarly,

TEXT BOX 23.2 *(cont'd)*

there was a slight increase in the mean lexical density from 33.05% in pretest statements to 34.9% in the posttest statements, indicating the ability to utilize a higher volume of content-related concepts by the end of the course. In addition, new salient words emerged in their posttest statements, indicating learning of specialized terminology and concepts related to health informatics. For example, in the post-tests there were five occurrences of the word "interoperability," a word not seen at all in the pretest statements. This analysis of finishing statements gave further evidence that the course effectively supported these students to build detailed eHealth and health informatics knowledge.

Evaluation findings such as those presented here, although not large scale or conclusive, were sufficient to indicate that the course had met its aim of providing a meaningful interprofessional learning experience and that it should go forward as a credit-bearing postgraduate course. In future, larger numbers of participants will enable the trial evaluation methods to be used to generate more substantial research findings.

captured unless planned well in advance. The case study below of a Canadian national level intervention (Text Box 23.3) illustrates this challenge.

The Need for Measure Instruments

The field of health informatics or eHealth is emergent and rapidly evolving. In recent years, there have been a number of efforts to define the set of competencies required for health-care professionals. These initiatives include for example the medical subspecialty board certification in clinical informatics in the United States [39,40] and the Entry-to Practice Nursing Informatics Competencies by the Canadian Association of Schools of Nursing (CASN) [41]. However, there are not yet many measure instruments to assess these competencies, and the success of training interventions to teach them. While existing tools from other domains can be adopted and adapted, there is a great need for new and specific tools. Development of valid and reliable measure instruments is a time consuming process (see the e-SEGUE case study in Text Box 23.4 below), but until sufficient number of these tools are in place our ability to evaluate the outcomes of informatics-related educational interventions will remain limited. To support this goal, we propose creating an online repository where researchers can find and share measure instruments.

In this section, we highlighted some of the challenges of evaluating educational interventions in general, and some of the specific challenges for health informatics related projects. Table 23.2 summarizes the challenges and some of the means to address them.

TEXT BOX 23.3

CASE STUDY: THE ASSOCIATION OF FACULTIES OF MEDICINE OF CANADA (AFMC)—CANADA HEALTH INFOWAY (INFOWAY) EHEALTH FACULTY DEVELOPMENT

In Chapter 17, Bhyat et al. describe the AFMC-Infoway faculty development program, a first of its kind intervention, which included a series of 4 eHealth faculty development workshops (utilizing a "train the trainer" model). Each workshop was delivered three times as a webinar as well as provided a toolkit that was made available online. The evaluation framework included an online participant feedback in the form of pre- and post-survey as well as some measures of participation such as number of workshop participants and access of the available online resources. While a lot of consideration went into defining the required competencies and planning the intervention to address these requirements, the evaluation was not very well integrated into the design. After the administration of the program it became clear that while the number of participants in the live webinars is known, the exact number of participants who accessed the online resources is not, mainly because only the number of times a resource was accessed could be measured but it is not clear how many of these are unique entries or how many of them were by the same people who participated in the live webinars. In addition, due to the limitations of the medium, it was difficult for instructors and evaluators to determine whether participants were in fact present throughout the webinar, for how long, and whether they were actively engaged in watching and listening. The rate of filling the feedback survey ranged from 50% to 92%, depending on the workshop, and it was difficult to drill down in order to better understand perceptions of the intervention and its impact. Thus, in spite of expert design and delivery, the intervention's outcomes remain largely unknown. This will hopefully be amended in subsequent iterations of the program. This is an illustration how evaluation needs to be integral to any intervention's design and perhaps even be the first item to be designed in detail, with the rest of the program flowing in reverse from the evaluation plan.

SUMMARY AND FUTURE DIRECTIONS

We have delineated the complexity of educational interventions' evaluation in general and in health IT-related programs in particular. Many of these interventions to date have not taken full advantage of the available robust theoretical infrastructure that could guide their evaluation. Thus, the main conclusions informing our thoughts for the future are straightforward: evaluation should be integral to intervention design and budget planning; various research designs, albeit not a strong level of evidence as RCTs, may provide

TEXT BOX 23.4

CASE STUDY: E-SEGUE: A NEW TOOL TO ASSESS PATIENT-CLINICIAN-COMPUTER COMMUNICATION

The e-SEGUE is a new measure instrument designed to capture patient-clinician communication in computerized settings. It is based on an existing and extensively used communication instrument—the SEGUE [42], adding the computer related communicative additional skills that were not included when the original tool was constructed. It was constructed and validated through a consensus process and eventually by applying the tool to a simulation database where its feasibility was demonstrated. It is comprised of 22 items (selected out of 27 by the aforementioned consensus process) and was demonstrated to be reliable, valid, and feasible. However, until acceptability and educational impact are evaluated the tool has not passed all the needed steps to make it mature for application.

Assis-Hassid, S., Heart, T., Reychav, I., Pliskin, J., & Reis, S. (2015). Enhancing patient-doctor-computer communication in primary care: towards measurement construction. Israel Journal of Health Policy research, 4 (1), 4–14.

TABLE 23.2 Summary of Challenges and Means to Address Them

Challenge	How It May Be Addressed
Research design	Consider whether an RCT is feasible or the best approach for the evaluation; try to evaluate both outcomes and processes; consider alternative designs to RCTs
Small-scale projects	Adapt the evaluation study design to the expected sample size; conduct power calculation; remember that small-N studies may still yield valuable insight for the purpose of planning and quality improvement
Lack of planning for evaluation	Advocate for evaluation; plan the evaluation from the outset and allocate appropriate resources to it as part of the project plan
Lack of specific measure instruments	Adapt existing tools to the context of educating health professionals about the challenges of the information age; develop and validate new tools as needed; creating a shared repository online of measure instruments

valuable insights and therefore should be considered. For evaluating the outcomes of an intervention, the recommended steps in program evaluation need to be followed meticulously: after proof of concept and pilot, power calculations should be undertaken and followed; the evaluation should be as neutral as possible and carry as little educational impact as possible (see the Israeli Patient-Doctor Computer Communication case study in Text Box 23.1). For understanding the processes by which interventions work, participatory action research with iterative cycles of data collection, reflection, and action or other forms of iterative designs that delineate the full assessment plan first and move to portraying the intervention itself as flowing from its evaluation may be preferred options. To capture both outcomes and processes, mixed methods designs and multifaceted evaluations, where quantitative methods are used to measure impact while qualitative ones aim to explain it may be one way to overcome the difficulties discussed above. This approach is demonstrated for example by the Australian interprofessional online informatics course (Text Box 23.2). Another route may be the application of complexity informed evaluation methods that may fit the complexity of the interventions. Finally, specific new measure instruments and other evaluation protocols need to be developed, validated, and applied with full rigor so that the results are trustworthy, generalizable, and provide strong evidence to inform quality improvement and future interventions.

References

[1] Trochim W.M. The research methods knowledge base 2006. Available from: <http://www.socialresearchmethods.net/kb/> [cited 15.06.16].

[2] Morrison J. ABC of learning and teaching in medicine: evaluation. BMJ. 2003;326(7385):385–7 Epub 2003/02/15.

[3] Bloom BS, Hastings JT, Madaus GF. Handbook on formative and summative evaluation of student learning. New York: McGraw-Hill; 1971.

[4] Bleakley A. Blunting Occam's razor: aligning medical education with studies of complexity. J Eval Clin Pract 2010;16(4):849–55.

[5] Mennin S. Complexity and health professions education: a basic glossary. J Eval Clin Pract 2010;16 (4):838–40.

[6] Mennin S. Complexity and health professions education. J Eval Clin Pract 2010;16(4):835–7.

[7] Mennin S. Self-organisation, integration and curriculum in the complex world of medical education. Med Educ 2010;44(1):20–30.

[8] Hafferty FW, Levinson D. Moving beyond nostalgia and motives: towards a complexity science view of medical professionalism. Perspect Biol Med 2008;51(4):599–615.

[9] Plsek PE, Greenhalgh T. Complexity science: the challenge of complexity in health care. BMJ 2001;323 (7313):625–8.

[10] Plsek PE, Wilson T. Complexity, leadership, and management in healthcare organisations. BMJ 2001;323 (7315):746–9.

[11] Begun J, Zimmerman B, Dooley K. Health care organizations as complex adaptive systems. In: Mick S, Wyttenbach M, editors. Advances in health care organization theory. San Francisco: Jossey-Bass; 2003.

[12] Rosenbaum L. Transitional chaos or enduring harm? The EHR and the disruption of medicine. N Engl J Med 2015;373(17):1585–8.

[13] Epstein RM, Hundert EM. Defining and assessing professional competence. JAMA 2002;287(2):226–35.

[14] McKinley RK, Fraser RC, Baker R. Model for directly assessing and improving clinical competence and performance in revalidation of clinicians. BMJ 2001;322(7288):712–15.

[15] Crossley J, Humphris G, Jolly B. Assessing health professionals. Med Educ 2002;36(9):800–4.

[16] Kirkpatrick DL. Revisiting Kirkpatrick's four-level model. Train Dev 1996;50(1):54–9.

[17] Kirkpatrick DL. Evaluation. In: Craig RL, editor. The ASTD training and development handbook. New York: McGraw-Hill; 1996. p. 294–312.

[18] Barr H, Freeth D, Hammick M, Koppel I, Reeves S. Evaluations of interprofessional education: a United Kingdom review for health and social care. The United Kingdom Centre for the Advancement of Interprofessional Education with The British Educational Research Association, 2000.

[19] Parker K, Burrows G, Nash H, Rosenblum ND. Going beyond Kirkpatrick in evaluating a clinician scientist program: it's not "if it works" but "how it works.". Acad Med 2011;86(11):1389–96.

[20] Lave J. Cognition in practice: mind, mathematics, and culture in everyday life. Cambridge; New York: Cambridge University Press; 1988.

[21] Lave J, Wenger E. Situated learning: legitimate peripheral participation. Cambridge England; New York: Cambridge University Press; 1991;138.

[22] Brown JS, Collins A, Duguid P. Situated cognition and the culture of learning. Educ Res 1989;18(1):32–42.

[23] Wenger E, Trayner B, de Laat M. Promoting and assessing value creation in communities and networks: a conceptual framework. Ruud de Moor Centrum, Open University of the Netherlands., 2011, p. 18.

[24] Wenger E, Trayner B. Strategic evaluation of network activities. 2014. Available from: <http://wenger-trayner.com/resources/publications/strategic-evaluation-of-network-activities/> [cited 17.06.16].

[25] Barr H, Hammick M, Koppel I, Reeves S. Evaluating interprofessional education: two systematic reviews for health and social care. Brit Educ Res J 1999;25(4):533–44.

[26] Flickinger ET, O'Hagan T, Chisolm SM. Developing a curriculum to promote professionalism for medical students using social media: pilot of a workshop and blog-based intervention. JMIR Med Educ 2015;1(2):e17.

[27] Mesko B, Győrffy Z, Kollár J. Digital literacy in the medical curriculum: a course with social media tools and gamification. JMIR Med Educ 2015;1(2):e6.

[28] Silverman H, Cohen T, Fridsma D. The evolution of a novel biomedical informatics curriculum for medical students. Acad Med 2012;87(1):84–90.

[29] Reis S, Sagi D, Eisenberg O, Kuchnir Y, Azuri J, Shalev V, et al. The impact of residents' training in Electronic Medical Record (EMR) use on their competence: report of a pragmatic trial. Patient Educ Couns 2013;93(3):515–21.

[30] Shachak A, Domb S, Borycki E, Fong N, Skyrme A, Kushniruk A, et al. A pilot study of computer-based simulation training for enhancing family medicine residents' competence in computerized settings. Stud Health Technol Inform 2015;216:506–10.

[31] Centre for Evidence Based Medicine. Oxford Centre for Evidence-based Medicine – levels of evidence. 2009. Available from: <http://www.cebm.net/oxford-centre-evidence-based-medicine-levels-evidence-march-2009/> [cited 17.06.16].

[32] Vandenbroucke JP. Observational research, randomised trials, and two views of medical science. PLoS Med. 2008;5(3):e67.

[33] Baker BT, Gustafson HD, Shah D. How can research keep up with eHealth? Ten strategies for increasing the timeliness and usefulness of eHealth research. J Med Internet Res 2014;16(2):e36.

[34] Mohr CD, Schueller MS, Riley TW, Brown HC, Cuijpers P, Duan N, et al. Trials of intervention principles: evaluation methods for evolving behavioral intervention technologies. J Med Internet Res 2015;17(7):e166.

[35] Harris AD, McGregor JC, Perencevich EN, Furuno JP, Zhu J, Peterson DE, et al. The use and interpretation of quasi-experimental studies in medical informatics. J Am Med Inform Assoc 2006;13(1):16–23.

[36] Reason P. Choice and quality in action research practice. J Manag Inquiry 2006;15(2):187–203.

[37] Baum F, MacDougall C, Smith D. Participatory action research. J Epidemiol Community Health 2006;60 (10):854–7.

[38] Holter IM, Schwartz-Barcott D. Action research: what is it? How has it been used and how can it be used in nursing? J Adv Nurs 1993;18(2):298–304.

[39] Hersh WR, Gorman PN, Biagioli FE, Mohan V, Gold JA, Mejicano GC. Beyond information retrieval and electronic health record use: competencies in clinical informatics for medical education. Adv Med Educ Pract 2014;5:205–12.

[40] Detmer DE, Shortliffe EH. Clinical informatics: prospects for a new medical subspecialty. JAMA 2014;311 (20):2067–8.

[41] Canadian Association of Schools of Nursing (CASN). Nursing informatics entry-to-practice competencies for registered nurses. Available from: <http://www.casn.ca/2014/12/casn-entry-practice-nursing-informatics-competencies/> [cited 15.06.16].

[42] Makoul G. The SEGUE Framework for teaching and assessing communication skills. Patient Educ Couns 2001;45(1):23–34.

Glossary

Activity metrics Tracking of what we do in a learning environment, such as moving to a page, watching a segment of a video, responding to a question, or opening a scenario.

Affect-oriented A person (clinician) who is emotion focused. The recognition and validation of emotions is central to caring for a patient.

Ambient surveillance The concept of ubiquitous, permanent monitoring of everything we do and say.

Attentive behavior Attention to verbal and nonverbal signals.

Barcode medication administration A technology used to facilitate the safe administration of medication(s) through scanning various barcodes associated with patient identifiers, medications, and sometimes the health professional.

Blockchain A database of records distributed over a network. The records (called blocks) are added to the database in a linear chronological order and contain time stamps and link to the previous block. Originally developed for the digital currency Bitcoin, blockchain is protected from altering previous data in that it has no single point of failure and cannot be controlled by a single entity.

Clinical decision support system (CDSS) A computer application that supports health care providers in making clinical decisions about diagnosis, treatment or testing. CDSS bring best practices and research knowledge to the point of care and range from relatively simple applications such as reminders and drug interaction alerts, through rule-based and guideline-based systems, to complex artificial intelligence and cognitive computing systems.

Clinical practice guideline Systematically developed statements to assist practitioners and patients in making decisions about appropriate health care for specific clinical circumstances.

Competency The ability to use a knowledge or a skill in order to meet a professional or learning need.

Continuing competency The level of skill and ability to continue to safely and correctly implement a technology after initial training.

Computer-associated silence A specific type of silence that occurs among humans when one or more human participants in an interaction is distracted by or focused on inputting or receiving information from an electronic device.

Cultural competence The understanding of values, beliefs, and health behaviors of people from other regions. Cultural competence is of special importance for providers who care for patients from diverse cultural and geographic backgrounds.

Digital communication Electronic transmission of information that has been encoded digitally.

Digital literacy The capabilities of individuals, drawing on a broad range of interrelated skills, including literacy in technology, information, communication, social and visual arenas to live, learn, and work in a digital society.

Digital professionalism A behavior expected from a professional that adheres to legal and ethical standards when engaged in online communication and social media.

eHealth (or e-Health) Healthcare initiatives and practice supported by electronic or digital media.

eHealth literacy The ability of individuals to obtain, process, evaluate and use health information acquired through electronic resources, especially from the Internet.

Electronic-Clinical Evaluation Exercise (e-CEX) Modeled after the mini-CEX used in clinical medicine (to measure clinical competence), the e-CEX is an evaluation tool created to assess EMR specific communication skills and consists of a 10-item, 90-point tool for use in direct or video-taped observation assessment by trained observers.

Electronic Health Record (EHR) A computer system that enables clinicians across multiple health care organizations to document and view patient information. To effectively enable information exchange across health care settings, EHRs must adhere to common terminology and information exchange standards. In some countries the term refers to the hospital-based electronic record systems (as opposed to community-based primary care settings systems that are referred to as EMR—electronic medical record).

Electronic Medical Record (EMR) A computer system that enables clinicians to document and review patient information. Managed and controlled by a single health care organization. In some countries, the term refers to the electronic record systems in community-based primary care settings (as opposed to hospital-based systems referred to as EHR—electronic health record).

Electronic medication administration record (eMAR) A term used to describe a computer-based record for tracking and documenting patient medications.

Evaluation The systematic process of assessing the value or merit of an object and providing useful feedback about it. In the context of educational interventions evaluation seeks to assess the efficacy and impact of the intervention at various levels: participants' reactions, learning, behavior, and results. Evaluation should be reliable, valid, acceptable, feasible, and facilitate educational impact.

e-SEGUE A measure instrument designed to capture patient-clinician communication in computerized settings. It is based on an existing and extensively used communication instrument—SEGUE (Set the stage, Elicit information, Give information, Understand the patient's perspective, End the encounter).

GuideLine Interchange Format (GLIF) A computer-based representation model of clinical practice guidelines. Guidelines encoded in GLIF can be interpreted and executed by an associated tool—GuideLine Execution Engine (GLEE).

Group Objective Structured Clinical Examination (GOSCE) A simulated clinical interaction in which a learner interacts with a standardized patient (SP), and co-learners are able to both observe the interaction as a group and provide feedback. A trained facilitator typically helps structure the debriefing session, which can consist of an informal feedback from the SP and learners as well as a more formal review of an assessment instrument.

Health literacy The extent to which individuals demonstrate the cognitive capacity and social skills that determine their motivation and ability to locate, interpret, appraise, and use health information and services to promote and maintain good health and make appropriate health decisions.

HUMAN LEVEL A mnemonic created to capture the best practices regarding patient-centered EMR use, such as "Honoring the Golden Minute" and the importance of making the start of the visit technology free, instead beginning with greeting the patient and asking them what their concerns are prior to engaging with technology.

Information literacy An individual's ability to recognize when information is needed and having the ability to locate, understand, evaluate and effectively use the needed information, including information found from online sources.

Interactive case simulation tools Online tools for clinical decision assistance and case simulation on individual patients. These tools are developed as vehicles to disseminate clinical evidence, with a back-end driven by a knowledge base and a front-end supporting user interactions such as review of patient management process, examination of decision options, and entry of case-specific data for individualized recommendations.

Integrated training delivery A mode of delivery where eLearning modules are used to introduce the information and communication technology (ICT) before attending classroom based educational session.

Learning The tangible, measurable permanent change in individuals who have received instruction or studied to improve their knowledge or skills.

Learning analytics The study of data, its capture and reporting, in the context of learners, learning systems and environments.

Learning record store A specialized database, design to record activity statements in an education environment. May or may not be attached to a learning management system.

MedBiquitous A consortium founded by Johns Hopkins Medicine that develops technology and terminology standards for health professions education. It employs XML and Web services in the development of standards for activities, resources, and more.

Mobile Health (mHealth) Use of mobile devices (e.g., mobile phones and tablets) to support health related activities.

Multimedia learning module Educational materials on a specific topic with incorporation of multimedia elements such as images, animations, audios, videos, and texts.

Multitasking interference increase in time to complete and or decreased quality of task completion when two or more tasks are attempted simultaneously.

Nursing informatics competency The ability of nursing students to draw upon various types of knowledge, skills and other internal and external resources to address specific problems in nursing practice.

Participatory medicine An approach to health care that seeks to involve patients, caregivers and healthcare professionals in all aspects of one's care.

Personal Health Information A legal term referring to identifying information about an individual that is related to his or her physical or mental health or the individual's utilization of health services. Many countries have legislation in place to protect personal health information.

Personal Health Record (PHR) An electronic system containing health information of a person that is controlled and managed by that individual. A PHR may exchange information with other systems such as a primary care electronic medical record, hospitals' electronic health records, lab, pharmacy and potentially other people's PHRs, provided that permission for such exchange has been granted by the record's owner.

Personal Information Information that is recorded about an individual such as sex, age, family status, education, blood type, etc. Even if such information is deidentified, it may not ensure the individual's anonymity.

Privacy Breach Unauthorized access, disclosure of or loss of personal information. Faulty operational breakdowns and business procedures may be the cause.

Privacy Impact Assessment A tool that is used to assess potential privacy risks of new or redesigned services and programs. This work may include an assessment of electronic systems in order to design strategies that can be used to mitigate, eliminate, or reduce privacy risks to a level that is acceptable.

Privacy Management Program programs that help organizations nurture a culture of privacy and satisfy regulators of their compliance with privacy legislation. An appropriately designed and implemented program includes policies, privacy awareness training, risk assessments, metrics, and systems for monitoring privacy risk of the organization as well as access controls and system user audits.

Rural health disparities Health concerns specific to individuals living in rural regions versus urban areas, including limited geographic access to healthcare facilities and healthcare providers, a higher prevalence of chronic diseases, and typically lower levels of health literacy.

Simulation A teaching strategy where students interact and refine their learning through a practical and realistic clinical situation, where students are required to interact and/or perform certain skills.

Social media A cluster of internet based applications that allow users to share and create content, acquire feedback in real time, and to adjust content. Social media allows for interaction, participation, sharing, reciprocity, connectedness, information openness, creativity, collaboration, and autonomy.

Super user A person with advanced knowledge of the information and communication technology (ICT) who provides mentorship to a learner thought the change management aspects of the introduction of the new ICT.

Task-oriented A person (clinician) who is problem focused; problem-solving action is central

Telemedicine The implementation of health-related technology to facilitate interactions remotely between providers and patients placed in different geographic regions.

Triadic relationship A term used to describe the interaction between the patient, clinician, and computer in the clinical encounter. It highlights the fact that the computer is an active player in the consultation that both clinicians and patients may interact with.

Triangle of trust positioning the computer screen, patient and provider at three ends of a triangle configuration such that shared viewing of the screen is achieved by all parties. Also referred to as inclusive computerized clinical setting.

Training instruction that is provided "on the job" to help the recipient acquire and enhance knowledge or skills.

Index

Note: Page numbers followed by "*f*" and "*t*" refer to figures and tables, respectively.

D

E

F

P

T

U

V

Printed in the United States
By Bookmasters